Activated Carbon for Water and Wastewater Treatment
Integration of Adsorption and Biological Treatment

Ferhan Çeçen and Özgür Aktaş

Related Titles

Gottschalk, Christiane / Libra, Judy Ann / Saupe, Adrian

Ozonation of Water and Waste Water
A Practical Guide to Understanding Ozone and its Applications

Second, completely revised and updated edition
2009
Hardcover
ISBN: 978-3-527-31962-6

Fränzle, Stefan / Markert, Bernd / Wünschmann, Simone

Introduction to Environmental Engineering

2012
Softcover
ISBN: 978-3-527-32981-6

Duke, Mikel / Zhao, Dongyuan / Semiat, Rafael (eds.)

Functional Nanostructured Materials and Membranes for Water Treatment
Series: New Materials for Sustainable Energy and Development

2012
Hardcover
ISBN: 978-3-527-32987-8

Peinemann, Klaus-Viktor / Pereira Nunes, Suzana (eds.)

Membrane Technology
Volume 4: Membranes for Water Treatment Series: Membrane Technology (Volume 4)

2010
Hardcover
ISBN: 978-3-527-31483-6

Gonzalez, Catherine / Greenwood, Richard / Quevauviller, Philippe P. (eds.)

Rapid Chemical and Biological Techniques for Water Monitoring
Series: Water Quality Measurements

2009
Hardcover
ISBN: 978-0-470-05811-4

Ganoulis, Jacques

Risk Analysis of Water Pollution
Second, revised and expanded edition

2009
Hardcover
ISBN: 978-3-527-32173-5

Ganoulis, Jacques / Aureli, Alice / Fried, Jean (eds.)

Transboundary Water Resources Management
A Multidisciplinary Approach

2011
Hardcover
ISBN: 978-3-527-33014-0

Ferhan Çeçen and Özgür Aktaş

Activated Carbon for Water and Wastewater Treatment

Integration of Adsorption and Biological Treatment

WILEY-VCH Verlag GmbH & Co. KGaA

The Authors

Prof. Ferhan Çeçen
Bogazici University
Inst. of Environmental Sciences
34342 Istanbul
Turkey

Dr. Özgür Aktaş
TUBITAK-MRC
Environment Institute
41470 Gebze, Kocaeli
Turkey

■ All books published by **Wiley-VCH** are carefully produced. Nevertheless, authors, editors, and publisher do not warrant the information contained in these books, including this book, to be free of errors. Readers are advised to keep in mind that statements, data, illustrations, procedural details or other items may inadvertently be inaccurate.

Library of Congress Card No.: applied for

British Library Cataloguing-in-Publication Data
A catalogue record for this book is available from the British Library.

Bibliographic information published by the Deutsche Nationalbibliothek
The Deutsche Nationalbibliothek lists this publication in the Deutsche Nationalbibliografie; detailed bibliographic data are available on the Internet at <http://dnb.d-nb.de>.

© 2012 Wiley-VCH Verlag & Co. KGaA, Boschstr. 12, 69469 Weinheim, Germany

All rights reserved (including those of translation into other languages). No part of this book may be reproduced in any form – by photoprinting, microfilm, or any other means – nor transmitted or translated into a machine language without written permission from the publishers. Registered names, trademarks, etc. used in this book, even when not specifically marked as such, are not to be considered unprotected by law.

Typesetting MPS Limited, a Macmillan Company, Chennai
Printing and Binding Fabulous Printers Pte Ltd
Cover Design Adam Design, Weinheim

Printed in Singapore
Printed on acid-free paper

Print ISBN: 978-3-527-32471-2
ePDF ISBN: 978-3-527-63946-5
ePub ISBN: 978-3-527-63945-8
mobi ISBN: 978-3-527-63947-2
oBook ISBN: 978-3-527-63944-1

Contents

Preface *xvii*
List of Abbreviations *xxi*
Acknowledgement *xxvii*

1 **Water and Wastewater Treatment: Historical Perspective of Activated Carbon Adsorption and its Integration with Biological Processes** *1*
Ferhan Çeçen
1.1 Historical Appraisal of Activated Carbon *1*
1.2 General Use of Activated Carbon *3*
1.3 Application of Activated Carbon in Environmental Pollution *4*
1.3.1 Activated Carbon in Drinking Water Treatment *4*
1.3.2 Activated Carbon in Wastewater Treatment *5*
1.3.2.1 Municipal Wastewater Treatment *6*
1.3.2.2 Industrial Wastewater Treatment *6*
1.3.3 Applications of Activated Carbon in Other Environmental Media *7*
1.3.3.1 Remediation of Contaminated Groundwater and Soil *7*
1.3.3.2 Treatment of Flue Gases *7*
1.3.3.3 Water Preparation for Industrial Purposes *7*
1.3.4 Integration of Activated Carbon Adsorption with Biological Processes in Wastewater and Water Treatment *7*
1.3.4.1 Wastewater Treatment *7*
1.3.4.2 Water Treatment *8*
1.3.5 Improved Control of Pollutants through Integrated Adsorption and Biological Treatment *8*

Activated Carbon for Water and Wastewater Treatment: Integration of Adsorption and Biological Treatment.
First Edition. Ferhan Çeçen and Özgür Aktaş
© 2011 WILEY-VCH Verlag GmbH & Co. KGaA, Weinheim.
Published 2011 by WILEY-VCH Verlag GmbH & Co. KGaA

2	**Fundamentals of Adsorption onto Activated Carbon in Water and Wastewater Treatment** 13
	Özgür Aktaş and Ferhan Çeçen
2.1	Activated Carbon 13
2.1.1	Preparation of Activated Carbons 13
2.1.2	Characteristics of Activated Carbon 14
2.1.3	Activated Carbon Types 15
2.1.3.1	Powdered Activated Carbon (PAC) 15
2.1.3.2	Granular Activated Carbon (GAC) 16
2.2	Adsorption 16
2.2.1	Types of Adsorption 16
2.2.2	Factors Influencing Adsorption 18
2.2.2.1	Surface Area of Adsorbent 18
2.2.2.2	Physical and Chemical Characteristics of the Adsorbate 18
2.2.2.3	pH 19
2.2.2.4	Temperature 20
2.2.2.5	Porosity of the Adsorbent 20
2.2.2.6	Chemical Surface Characteristics 21
2.2.3	Kinetics of Adsorption 22
2.2.3.1	Transport Mechanisms 22
2.2.4	Adsorption Equilibrium and Isotherms 24
2.2.5	Single- and Multisolute Adsorption 27
2.2.5.1	Single Solute Adsorption 27
2.2.5.2	Multisolute Adsorption 28
2.3	Activated Carbon Reactors in Water and Wastewater Treatment 30
2.3.1	PAC Adsorbers 30
2.3.2	GAC Adsorbers 30
2.3.2.1	Purpose of Use 30
2.3.2.2	Types of GAC Adsorbers 30
2.3.2.3	Operation of GAC Adsorbers 31
2.3.2.4	Breakthrough Curves 34
2.4	Activated Carbon Regeneration and Reactivation 37

3	**Integration of Activated Carbon Adsorption and Biological Processes in Wastewater Treatment** 43
	Ferhan Çeçen and Özgür Aktaş
3.1	Secondary and Tertiary Treatment: Progression from Separate Biological Removal and Adsorption to Integrated Systems 43
3.1.1	Activated Carbon in Secondary Treatment 46
3.1.1.1	PAC 46
3.1.1.2	GAC 46
3.1.2	Activated Carbon in Tertiary Treatment 46

3.1.2.1	PAC 46	
3.1.2.2	GAC 47	
3.2	Fundamental Mechanisms in Integrated Adsorption and Biological Removal 47	
3.2.1	Main Removal Mechanisms for Organic Substrates 47	
3.2.1.1	Biodegradation/Biotransformation 48	
3.2.1.2	Sorption onto Sludge 52	
3.2.1.3	Sorption onto Activated Carbon 53	
3.2.1.4	Abiotic Degradation/Removal 53	
3.2.2	Main Interactions between Organic Substrates, Biomass, and Activated Carbon 54	
3.2.2.1	Retention of Slowly Biodegradable and Nonbiodegradable Organics on the Surface of Activated Carbon 56	
3.2.2.2	Retention of Toxic and Inhibitory Substances on the Surface of Activated Carbon 56	
3.2.2.3	Concentration of Substrates on the Surface of Activated Carbon 58	
3.2.2.4	Retention of Volatile Organic Compounds (VOCs) on the Surface of Activated Carbon 58	
3.2.2.5	Attachment and Growth of Microorganisms on the Surface of Activated Carbon 59	
3.2.2.6	Biological Regeneration (Bioregeneration) of Activated Carbon 59	
3.2.3	Behavior and Removal of Substrates in Dependence of their Properties 59	
3.3	Integration of Granular Activated Carbon (GAC) into Biological Wastewater Treatment 59	
3.3.1	Positioning of GAC Reactors in Wastewater Treatment 59	
3.3.2	Recognition of Biological Activity in GAC Reactors 63	
3.3.3	Conversion of GAC into Biological Activated Carbon (BAC) 64	
3.3.3.1	Removal Mechanisms and Biofilm Formation in BAC Operation 65	
3.3.3.2	Advantages of BAC Over GAC Operation 66	
3.3.3.3	Advantages of the BAC Process Compared to Other Attached-Growth Processes 66	
3.3.4	Main Processes in BAC Reactors 66	
3.3.5	Types of GAC Reactors with Biological Activity (BAC Reactors) 67	
3.3.5.1	Fixed-Bed BAC Reactors 67	
3.3.5.2	Expanded- and Fluidized-Bed BAC Reactors 68	
3.4	Integration of Powdered Activated Carbon (PAC) into Biological Wastewater Treatment 69	
3.4.1	Single-Stage Continuous-Flow Aerobic PACT® Process 70	
3.4.1.1	Development of the PACT Process 70	
3.4.1.2	Basic Features of the Activated Sludge Process 70	

3.4.1.3	Characteristics of the PACT Process 71	
3.4.1.4	Process Parameters in PACT Operation 72	
3.4.2	Sequencing Batch PACT Reactors 73	
3.4.3	Anaerobic PACT Process 74	
3.5	Biomembrane Operation Assisted by PAC and GAC 74	
3.5.1	Membrane Bioreactors (MBRs) 74	
3.5.2	The PAC-MBR Process 75	
3.5.3	Membrane-Assisted Biological GAC Filtration – the BioMAC Process 76	
3.6	Observed Benefits of Integrated Systems 76	
3.6.1	Enhancement of Organic Carbon Removal by Activated Carbon 78	
3.6.2	Enhancement of Nitrification by Activated Carbon 78	
3.6.2.1	Inhibition of Nitrification 79	
3.6.2.2	Nitrification in the Presence of PAC 79	
3.6.3	Enhancement of Denitrification by Activated Carbon 80	
3.6.4	Effect of Activated Carbon Addition on Inorganic Species 80	
3.6.5	Enhancing Effects of Activated Carbon in Anaerobic Treatment 81	
3.6.6	Properties of Biological Sludge in the Presence of Activated Carbon 81	
3.6.6.1	Improvement of Sludge Settling and Thickening 81	
3.6.6.2	Improvement in the Dewaterability of Sludge 82	
3.6.7	Effect of PAC on Membrane Bioreactors (MBRs) 82	
3.6.7.1	Importance of Microbial Products 82	
3.6.7.2	Effect of Activated Carbon on Microbial Products 83	
3.6.7.3	Effect of Activated Carbon on Membrane Filtration 85	
3.7	Regeneration of PACT and BAC Sludges 86	
4	**Effect of Activated Carbon on Biological Treatment of Specific Pollutants and Wastewaters: Laboratory- and Pilot-Scale Studies** 95	
	Özgür Aktaş and Ferhan Çeçen	
4.1	Treatment of Industrial Wastewaters 95	
4.1.1	Pharmaceutical Wastewaters 95	
4.1.2	Paper and Pulp Wastewaters 97	
4.1.3	Petroleum Refinery and Petrochemical Wastewaters 98	
4.1.4	Textile Wastewaters 99	
4.1.5	Other Industrial Wastewaters 100	
4.2	Removal of Specific Chemicals 104	
4.2.1	Volatile Organic Compounds (VOCs) 104	
4.2.2	Phenols 106	
4.2.3	Pharmaceuticals and Endocrine Disrupting Compounds (EDCs) 109	
4.2.4	Pesticides and Polychlorinated Biphenyls (PCBs) 109	

4.2.5	Priority Pollutants	109
4.2.6	Dyes	110
4.2.7	Organic Pollutants in Secondary Sewage Effluents	111
4.2.8	Other Chemicals	112
4.3	Landfill Leachate Treatment	113
4.3.1	Leachate Treatment in PAC-added Activated Sludge Systems	114
4.3.2	Leachate Treatment Using the PAC-MBR Process	118
4.3.3	Leachate Treatment in Biological Activated Carbon (BAC) Media	118
5	**Combination of Activated Carbon with Biological Wastewater Treatment at Full Scale**	**127**
	Özgür Aktaş and Ferhan Çeçen	
5.1	Full-Scale PACT Systems	127
5.1.1	Full-Scale PACT for Industrial Effluents	128
5.1.1.1	Organic Chemicals Production Industry	128
5.1.1.2	Synthetic Fiber Manufacturing Industry	130
5.1.1.3	Propylene Oxide/Styrene Monomer (PO/SM) Production Wastewater	130
5.1.1.4	Refinery and Petrochemical Wastewaters	131
5.1.1.5	Treatment of Priority Pollutants	134
5.1.1.6	Treatment of Pharmaceutical Wastewaters	135
5.1.2	PACT for Co-treatment of Domestic and Industrial Wastewaters	136
5.1.3	PACT for Landfill Leachates	136
5.1.4	PACT for Contaminated Groundwaters	138
5.1.5	PACT for Reuse of Domestic Wastewaters	139
5.1.6	PACT for Contaminated Surface Runoff Waters	139
5.2	Biological Activated Carbon (BAC) Filtration at Full Scale	140
5.2.1	BAC Filtration for Reuse Purposes	140
5.2.1.1	Treatment of Sewage for Reuse in Agriculture	140
5.2.1.2	Reclamation of Domestic Wastewater for Drinking Purposes	140
5.2.1.3	Reclamation of Domestic and Industrial Wastewaters for Nonpotable Uses	141
6	**Modeling the Integration of Adsorption with Biological Processes in Wastewater Treatment**	**145**
	Ferhan Çeçen	
6.1	Modeling of GAC Adsorbers with Biological Activity	145
6.1.1	Introduction	145
6.1.2	Fundamental Processes around a Carbon Particle Surrounded by a Biofilm	146

6.1.3	Benefits of Integrated Adsorption and Biological Removal: Link to Fundamental Processes *155*	
6.1.4	Modeling Approaches in GAC/BAC Reactors *158*	
6.1.4.1	Mass Balances in GAC/BAC Reactors *158*	
6.1.4.2	Characterization of BAC Reactors by Dimensionless Numbers *164*	
6.1.5	Prevalent Models in BAC Reactors Involving Adsorption and Biodegradation *173*	
6.1.5.1	Initial Steps in Modeling: Recognizing the Benefits of Integrated Systems *173*	
6.1.5.2	Consideration of Substrate Removal and Biofilm Formation *173*	
6.1.5.3	Integration of Adsorption into Models *174*	
6.1.5.4	Modeling in the Case of High Substrate Concentrations *175*	
6.1.5.5	Modeling the Case of Very Low Substrate (Fasting) Conditions *176*	
6.1.5.6	Modeling the Step Input of Substrate in a Three-Phase Fluidized-Bed Reactor *177*	
6.1.5.7	Consideration of Carbon Properties in Modeling *177*	
6.2	Modeling of the PACT Process *178*	
6.2.1	Mass Balances in the PACT Process *178*	
6.2.2	Mass Balance for PAC in the PACT Process *178*	
6.2.2.1	Determination of Biomass or Carbon Concentration in a PACT Sludge *180*	
6.2.3	Models Describing Substrate Removal in the PACT Process *181*	
7	**Bioregeneration of Activated Carbon in Biological Treatment** *189*	
	Özgür Aktaş and Ferhan Çeçen	
7.1	Mechanisms of Bioregeneration *189*	
7.1.1	Bioregeneration Due to Concentration Gradient *189*	
7.1.2	Bioregeneration Due to Exoenzymatic Reactions *191*	
7.1.3	Bioregeneration Due to Acclimation of Biomass *193*	
7.2	Offline Bioregeneration *194*	
7.3	Concurrent (Simultaneous) Bioregeneration in PACT and BAC Systems *195*	
7.4	Dependence of Bioregeneration on the Reversibility of Adsorption *195*	
7.5	Other Factors Affecting Bioregeneration *198*	
7.5.1	Biodegradability *199*	
7.5.2	Chemical Properties of Substrate *200*	
7.5.3	Carbon Particle Size *200*	
7.5.4	Carbon Porosity *201*	
7.5.5	Carbon Activation Type *202*	
7.5.6	Physical Surface Properties of Carbon *202*	

7.5.7	Desorption Kinetics	*203*
7.5.8	Substrate–Carbon Contact Time	*203*
7.5.9	Concentration Gradient and Carbon Saturation	*204*
7.5.10	Biomass Concentration	*205*
7.5.11	Dissolved Oxygen Concentration	*206*
7.5.12	Microorganism Type	*206*
7.5.13	Substrate and Biomass Associated Products of Biodegradation	*207*
7.5.14	Presence of Multiple Substrates	*208*
7.6	Determination of Bioregeneration	*209*
7.6.1	Investigation of the Extent of Reversible Adsorption	*211*
7.6.2	Use of Adsorption Isotherms	*211*
7.6.3	Direct Measurement by Using Adsorption Capacities	*213*
7.6.4	Direct Measurement by Solvent Extraction	*213*
7.6.5	Quantification of Bioregeneration in Simultaneous Adsorption–Biodegradation	*213*
7.6.6	Measurement of Biodegradation Products	*214*
7.6.7	Use of Respirometry in Aerobic Systems	*215*
7.6.8	Investigation by Scanning Electron Microscopy (SEM)	*216*
7.7	Bioregeneration in Anaerobic/Anoxic Systems	*216*
7.8	Models Involving Bioregeneration of Activated Carbon	*217*
7.8.1	Modeling of Bioregeneration in Concurrent Adsorption and Biodegradation	*218*
7.8.2	Modeling of Bioregeneration in Single Solute Systems	*220*
7.8.3	Modeling of Bioregeneration in Multicomponent Systems	*225*
7.8.4	Modeling of Offline Bioregeneration	*228*
7.8.5	Modeling the Kinetics of Bioregeneration	*228*
8	**Combination of Activated Carbon Adsorption and Biological Processes in Drinking Water Treatment** *237*	
	Ferhan Çeçen	
8.1	Introduction	*237*
8.2	Rationale for Introduction of Biological Processes in Water Treatment	*238*
8.3	Significance of Organic Matter in Water Treatment	*238*
8.3.1	Expression and Fractionation of Organic Matter	*239*
8.3.1.1	Expression of Organic Matter in Terms of Organic Carbon	*239*
8.3.1.2	Measurement of the Biodegradable Fraction in NOM	*239*
8.3.1.3	Nonbiodegradable Dissolved Organic Carbon (non-BDOC or NBDOC)	*243*
8.3.1.4	Fractionation of NOM in Terms of Molecular Size and Chemical Behavior	*243*
8.4	Removal of NOM in Conventional Water Treatment	*244*

8.4.1	Rationale for NOM Removal 244
8.4.2	Extent of NOM Removal 245
8.5	Use of Activated Carbon in Water Treatment 246
8.5.1	Powdered Activated Carbon (PAC) Addition 246
8.5.2	Granular Activated Carbon (GAC) Filtration 247
8.6	Biological Activated Carbon (BAC) Filtration 247
8.6.1	History of BAC Filtration in Water Treatment 247
8.6.2	Combination of Ozonation and BAC Filtration 249
8.6.3	Current Use of BAC Filtration in Water Treatment 249
8.7	Adsorption and Biodegradation Characteristics of Water 250
8.7.1	Raw Water NOM 250
8.7.2	Impact of Ozonation on NOM Characteristics 250
8.7.2.1	Increase in the Biodegradability of NOM 251
8.7.2.2	Change in the Adsorbability of NOM 255
8.7.3	Determination of Adsorption and Biodegradation Characteristics of Water 255

9	**Removal of NOM, Nutrients, and Micropollutants in BAC Filtration** 265
	Ferhan Çeçen
9.1	Removal of Organic Matter 265
9.1.1	Main Mechanisms 265
9.1.2	Breakthrough Curves 265
9.1.2.1	Initial Stage of Operation: Removal by Adsorption 265
9.1.2.2	Intermediate and Later Stages of Operation 266
9.1.3	Bioregeneration of BAC Filters 272
9.2	Factors Affecting the Performance of BAC Filtration 273
9.2.1	Comparison of GAC with Other Media 273
9.2.1.1	Biomass Attachment 273
9.2.1.2	Removal Performance 273
9.2.2	Importance of GAC Grade 274
9.2.2.1	Coal- and Wood-Based GAC 274
9.2.2.2	GAC Activation 274
9.2.3	Empty Bed Contact Time (EBCT) and Hydraulic Loading Rate (HLR) 274
9.2.4	Filter Backwashing 276
9.2.5	Effect of Temperature 276
9.2.6	Effect of Oxidant Residuals 277
9.3	Performance of BAC Filters: Organics Removal 277
9.4	Performance of BAC Filters: Nutrient Removal 285
9.4.1	Nitrification in BAC Filters 285
9.4.1.1	Importance of Ammonia Removal 285

9.4.1.2	Factors Affecting Nitrification in BAC Filters	*285*
9.4.2	Denitrification in BAC Filters	*286*
9.5	Removal of Micropollutants from Drinking Water in BAC Systems	*288*
9.5.1	Occurrence of Organic Micropollutants in Water	*288*
9.5.2	Competition Between Background NOM and Organic Micropollutants	*288*
9.5.3	Adsorption of Organic Micropollutants onto Preloaded GAC	*290*
9.5.4	Effect of GAC Characteristics on Adsorption	*290*
9.5.5	Adsorption and Biological Removal of Organic Micropollutant Groups in BAC Filtration	*291*
9.5.5.1	Pesticides	*291*
9.5.5.2	Pharmaceuticals and Endocrine Disrupting Compounds (EDCs)	*291*
9.5.5.3	Geosmin and MIB	*292*
9.5.5.4	Microcystins (Toxic Algal Metabolites)	*293*
9.5.5.5	Other Organic Micropollutants	*294*
9.5.6	Removal of Precursors and Disinfection By-Products (DBPs) in BAC Filtration	*294*
9.5.6.1	Removal of DBP Precurcors in BAC Filters	*295*
9.5.6.2	Removal of Trihalomethanes (THMs) and Haloacetic Acids (HAAs) in BAC Filters	*297*
9.6	Removal of Ionic Pollutants in BAC Filtration	*298*
9.6.1	Nitrate Removal	*298*
9.6.2	Bromate Removal	*298*
9.6.3	Perchlorate Removal	*301*
9.7	Integration of PAC and GAC into Biological Membrane Operations	*302*
9.7.1	Effect of PAC on Membrane Bioreactors	*302*
9.7.2	BAC Filtration Preceding Membrane Bioreactor Operation	*303*
9.8	Integration of GAC into Groundwater Bioremediation	*303*
9.9	Biomass Characteristics in BAC Filtration	*304*
9.9.1	Microbial Ecology of BAC Filters	*304*
9.9.2	Control of Biofilm Growth in BAC Filters	*305*
9.9.3	Determination of Biomass and Microbial Activity in BAC Filters	*305*
9.9.4	Determination of Microorganisms by Classical and Molecular Microbiology Methods	*306*
9.9.4.1	Heterotrophic Biomass and Activity in BAC Filters	*306*
9.9.4.2	Nitrifying Biomass and Activity in BAC Filters	*308*
9.9.5	Microbiological Safety of Finished Water	*309*

10		**BAC Filtration Examples in Full-Scale Drinking Water Treatment Plants** *319*
		Ferhan Çeçen
10.1		Limits for BDOC and AOC as Indicators of Re-growth Potential in Water Distribution *319*
10.2		BAC Filtration Experiences in Full-Scale Surface Water Treatment *320*
10.2.1		Mülheim Plants, Germany *320*
10.2.2		Leiden Plant, Amsterdam, the Netherlands *320*
10.2.3		Plants in the Suburbs of Paris, France *322*
10.2.4		Ste Rose Plant in Quebec, Canada *323*
10.2.5		Plant in Zürich-Lengg, Switzerland *323*
10.2.6		Weesperkarspel Plant, Amsterdam, the Netherlands *324*
10.2.7		Drinking Water Treatment Plants, Bendigo, Castlemaine, and Kyneton, Victoria, Australia *326*
10.3		New Approaches in the Evaluation of Ozonation and BAC Filtration *327*
10.4		BAC Filtration Experiences in Full-Scale Groundwater Treatment *327*

11		**Review of BAC Filtration Modeling in Drinking Water Treatment** *331*
		Ferhan Çeçen
11.1		Substrate Removal and Biofilm Formation *331*
11.2		Modeling of BAC Filtration *333*
11.2.1		Models Emphasizing Biological Processes in Biofilters *334*
11.2.1.1		Initial Models in Biofiltration *334*
11.2.1.2		The CHABROL Model *335*
11.2.1.3		Uhl's Model *335*
11.2.1.4		The Model of Wang and Summers *337*
11.2.1.5		The Dimensionless Empty-Bed Contact Time Concept *338*
11.2.1.6		The BIOFILT Model *339*
11.2.1.7		Consideration of Multiple Species Inside Drinking Water Biofilters *341*
11.2.2		Models Integrating Adsorption and Biological Processes *342*
11.2.2.1		Initial Models *342*
11.2.2.2		Models Involving Adsorption, Biomass Development, and Biodegradation *343*
11.2.3		Models Describing the Removal of Micropollutants *346*
11.2.3.1		Pesticides/Herbicides Removal *346*
11.2.3.2		Perchlorate Removal *347*

11.2.3.3	Microcystin Removal	348
11.2.3.4	THM Removal	348
12	**Concluding Remarks and Future Outlook** *353*	
	Ferhan Çeçen and Özgür Aktaş	
12.1	Overview of Applications in Wastewater and Water Treatment: PACT and BAC Systems	353
12.2	Further Research on Removal Mechanisms and Micropollutant Elimination	355
12.2.1	Wastewater Treatment	355
12.2.2	Drinking Water Treatment	358
12.3	Further Research on Regeneration of Activated Carbon	360
12.3.1	Importance of Activated Carbon Grade	360
12.3.2	Bioregeneration of Activated Carbon	361
12.3.2.1	Discussion of the Term 'Bioregeneration' and Hypotheses	361
12.3.2.2	Bioregeneration of Activated Carbon in the Case of Micropollutants	362
12.3.2.3	Bioregeneration Conditions	362
12.3.2.4	Bioregeneration of Activated Carbon in Drinking Water Treatment	363
12.3.3	Physicochemical Regeneration of Biological Activated Carbon	363

Index *365*

Preface

Purpose of the Book

The subject of the book is the 'integrated application of activated carbon adsorption with biological processes in water and wastewater treatment.' The enhancement of biological mechanisms by activated carbon adsorption merits serious study since it has been shown to be an effective method for the elimination of various organic and inorganic environmental pollutants.

During my studies over many years on adsorption and biodegradation, alone or together with my students, I realized with some surprise that the existing books on these subjects were either addressing activated carbon adsorption or biological treatment, but not both together in an integrated manner. I also realized that few books contained single chapters that dealt with either PACT or BAC processes. While invaluable work had been published in the form of papers or chapters, all these addressed some specific aspect of integrated treatment, and there was no comprehensive book that gave a detailed account of the different aspects of integrated adsorption and biological treatment. Thus, my idea to write this book originated from my own needs, and the idea was then supported by the book's co-author, Özgür Aktaş. Our goal in writing the book is to provide the reader with a document that attempts to present in a unified way most of the material to date on this subject.

The project necessitated an extensive literature survey spanning a period of approximately 40 years, from the beginning of integrated treatment in the 1970s to the present day. As a result, the book was inevitably expanded to include a wide range of topics. It covers the positioning of various integrated adsorption and biological removal treatment systems within the water and wastewater treatment train, describes how various pollutants can be removed, highlights the mechanisms that underlie the improved performance in small- and full-scale integrated systems, and extensively discusses to what extent pollutants can be eliminated from water or wastewater and what other side advantages are to be expected. However, for a full understanding, only to look at the underlying mechanisms and extent of removal would be inadequate. Therefore, we have also attempted to describe and analyze suspended- or attached-growth reactors involving PAC or GAC in mathematical terms. In this context, models pertaining to integrated water

---

Activated Carbon for Water and Wastewater Treatment: Integration of Adsorption and Biological Treatment.
First Edition. Ferhan Çeçen and Özgür Aktaş
© 2011 WILEY-VCH Verlag GmbH & Co. KGaA, Weinheim.
Published 2011 by WILEY-VCH Verlag GmbH & Co. KGaA

and wastewater treatment systems are also discussed. Results from small- and full-scale water and wastewater treatment are best understood if mechanisms and mathematical analyses of integrated adsorption and biological treatment systems are considered jointly.

In preparing this book, the assumption was made that the reader is equipped with basic knowledge of environmental science and technology, particularly the basics of adsorption and biological treatment. Thus, the book is not an elementary textbook, but is intended for people who are already involved with adsorption and/or biological processes. The principal readership of this book will be in the academic community, to whom the book will hopefully be useful. Some parts of the book may also be used for teaching of graduate level courses in environmental and chemical engineering.

The hope is that the book will appeal to people from both science and engineering disciplines. For scientists, who generally deal with fundamentals, it may be of interest to see the true value of integrated processes in practice. To practicing people, such as engineers operating water or wastewater treatment plants, who are mostly concerned with results, the book should provide a fundamental understanding of the main mechanisms in integrated adsorption and biotreatment. It should also serve as a work of reference for all those engaged in institutions relating to water quality, activated carbon production, and activated carbon adsorption.

Organization of the Book

Chapter 1 provides a brief overview of the history of activated carbon and its use in the water and wastewater treatment sector, and also gives a brief introduction to integrated adsorption and biological treatment.

Chapter 2 is an introductory chapter covering the fundamentals of adsorption and adsorption systems used in water and wastewater treatment. Since the basics of adsorption and adsorber systems are well explained in the existing literature, the chapter focuses on the main aspects of adsorbers that are also integrated into biological systems.

Chapters 3–7 are mainly devoted to the integration of activated carbon adsorption with biological processes in wastewater treatment.

Chapter 3 addresses the integration of activated carbon in biological wastewater treatment. It first highlights the progression from adsorption to concurrent adsorption and biological removal in wastewater treatment. The basic idea in this chapter is to give a clear idea of the main mechanisms underlying the observed positive effects in integrated systems. After this, the improvement of organics removal, removal of volatile pollutants, nitrification, denitrification, and anaerobic digestion in integrated systems are discussed. The chapter also includes the impact of activated carbon on biological sludge. Following the basic mechanisms, two basic processes are discussed that currently integrate the merits of activated carbon adsorption and biological removal in a single unit: the suspended-growth PACT process and the attached-growth biological activated carbon (BAC) process.

Another specific application, the coupling of membrane bioreactors (MBRs) with activated carbon, is also considered.

Chapter 4 concentrates on the effect of activated carbon in biological removal of pollutants. This chapter focuses on the removal of specific compounds as well as the extent of pollutant reduction in various types of wastewater, such as industrial wastewaters and landfill leachates, which contain many inhibitory, toxic, slowly degradable or nonbiodegradable pollutants.

Complementary to **Chapter** 4, which discusses the results and experiences from laboratory- and pilot-scale studies, **Chapter** 5 provides examples of full-scale PACT and BAC applications in wastewater treatment.

Chapter 6 addresses the modeling of combined adsorption and biological wastewater treatment systems. Relevant background information on mass transport, biodegradation, and adsorption processes is discussed in order to throw light on the complex interactions between adsorption and biological removal. The chapter then looks at the basic models that have been developed for attached-growth (BAC) and suspended-growth (PACT) systems.

Chapter 7 deals with bioregeneration of activated carbon, a very important phenomenon in all integrated systems. Bioregeneration is defined as the renewal of the adsorptive capacity of activated carbon by microorganisms in order to provide further adsorption. This chapter provides a comprehensive analysis of various aspects of GAC and PAC bioregeneration and the models describing bioregeneration.

The second part of the book, extending from **Chapter** 8 to 11, focuses on issues related specifically to drinking water treatment and addresses the integration of activated carbon adsorption with biological removal in this field.

Chapter 8 addresses the rationale for the introduction of biological processes into water treatment in general, highlighting the development and role of Biological Activated Carbon (BAC) Filtration. The significance of Natural Organic Matter (NOM) in drinking water treatment is discussed. This chapter also includes detailed information about the importance of ozonation, a treatment step that often precedes BAC filtration. The adsorption and biodegradation potential of raw and ozonated waters are discussed, as these properties have a strong influence on subsequent BAC filtration.

Chapter 9 deals with the removal of NOM, nutrients, and various organic and inorganic micropollutants in BAC filtration. Both removal mechanisms and the extent of removal are discussed. The chapter also includes a section on the characteristics and determination of biomass in BAC filters and on the safety of finished water.

Chapter 10 addresses the full-scale application of BAC filtration of drinking water. The chapter first discusses the limits set for the re-growth potential of water. It covers experiences from different water treatment plants and exemplifies the extent of the reduction of organic and inorganic pollutants that can be achieved.

Chapter 11 addresses the modeling of BAC filtration in drinking water treatment. Since the fundamentals of mass transport, biodegradation, and adsorption are discussed in **Chapter** 6, the only issues covered in this chapter are those pertaining specifically to drinking water treatment. The chapter provides an overview of drinking water biofiltration models that have been developed to describe NOM and micropollutant removal.

Chapter 12 provides an overview of some of the issues discussed in the book, and highlights the need for further research on integrated adsorption and biological removal in water and wastewater treatment.

Suggestions for the Reader

In writing this book, the attempt was made to treat each chapter as a stand-alone topic while at the same time not impairing the cohesiveness of the whole subject. Therefore, in almost all chapters, frequent cross-referencing to other chapters is provided.

Bearing in mind that not all chapters are necessary for every reader, the following suggestions are made:

For a reader who wants to acquire a general idea about adsorption and its historical evolution in combination with biological processes in water and wastewater treatment, it may be sufficient to concentrate on **Chapter** 1.

Chapter 2 presents the fundamentals of adsorption and the use of adsorbers in the water/wastewater field. It can be read independently of other chapters and is perfectly suited to a reader who is interested in adsorption only, and not in biological processes.

Chapter 3 is the key to the comprehension of the chapters that follow, and the reader who wishes to gain a fundamental understanding of the mechanisms and synergism associated with integrated adsorption and biological removal is strongly advised to study it.

The reader who is only interested in wastewater treatment and wants to focus on operational and practical aspects of integrated adsorption and biological removal needs to read **Chapters** 2, 3, 4 and 5 only.

To acquire further insight into the mechanism and mathematical description of integrated adsorption and biological removal, the reader is advised to refer also to **Chapters** 6 and 7, which address modeling of integrated systems and bioregeneration of activated carbon, respectively.

In general, the reader who is interested in the combination of activated carbon adsorption with biological processes in drinking water treatment should refer to **Chapters** 8 to 11. For this reader it would also be helpful to read **Chapter** 3 first. **Chapter** 8 provides general information on BAC filtration in drinking water treatment, and detailed information on this subject is provided in **Chapter** 9. The reader who is interested in practical aspects of BAC filtration and wants to learn about possible full-scale applications of the process should refer to **Chapter** 10, while the reader who is interested in the mathematical formulation of BAC filtration of drinking water is advised to examine the models presented in **Chapter** 11, but, before this, it would be most useful to read **Chapter** 6.

Ferhan Çeçen

List of Abbreviations

16S rDNA	Small Sub-Unit (SSU) rDNA
16S rRNA	Small Sub-Unit (SSU) rRNA
2,4-D	2,4-Dichlorophenoxyacetate
2-FB	2-Fluorobenzoate
5-Fu	5-Fluorouracil
ABS	Acrylonitrile Butadiene Styrene
AC	Activated Carbon
AFLP	Amplified Fragment Length Polymorphism
AMO	Ammonia Monooxygenase
amoA	Ammonia Monooxygenase gene
AOB	Ammonia Oxidizing Bacteria
AOC	Assimilable Organic Carbon
AOCl	Adsorbable Organic Chlorine
AOP	Advanced Oxidation Process
AOX	Adsorbable Organic Xenobiotics (Halogens)
APHA	American Public Health Association
AS	Activated Sludge
ASTM	American Society for Testing and Materials
ATP	Adenosine Triphosphate
B(GAC)	Biological Granular Activated Carbon
BAC	Biological(ly) Activated Carbon (used primarily for granular activated carbon)
BAC-FBR	Fluidized-Bed Reactor containing biological GAC (BAC)
BAF	Biological Aerated Filter
BASM	Biodegradation/Adsorption–Screening Model
BDOC	Biodegradable Dissolved Organic Carbon
BET	Brunauer, Emmett, Teller
BFAC	BioFilm on Activated Carbon (Model)
Bi	Biot Number
BioMAC	Biological Membrane Assisted Carbon Filtration
BKME	Bleached Kraft Pulp Mill Effluent
BOD	Biochemical Oxygen Demand
BOM	Biodegradable Organic Matter

Activated Carbon for Water and Wastewater Treatment: Integration of Adsorption and Biological Treatment.
First Edition. Ferhan Çeçen and Özgür Aktaş
© 2011 WILEY-VCH Verlag GmbH & Co. KGaA, Weinheim.
Published 2011 by WILEY-VCH Verlag GmbH & Co. KGaA

BP	Bromophenol
BPA	Bisphenol A
BSF	Biological Rapid Sand Filtration
BTEX	Benzene, Toluene, Ethylbenzene, Xylene
BTX	Benzene, Toluene, Xylene
BV	Bed Volume
CAA	Chloroacetaldehyde
CB	Chlorobenzene
CBZ	Carbamazepine
CF	Continuous-Flow
CMF	Continuous Microfiltration
CMF-S	Continuous Microfiltration-Submerged
COD	Chemical Oxygen Demand
CP	Chlorophenol
CSTR	Continuous-flow Stirred Tank Reactor
CUR	Carbon Usage Rate
Cytr	Cytarabine
Da	Damköhler Number
DAF	Dissolved Air Flotation
DBP	Disinfection By-Product
DCA	Dichloroethane
DCE	Dichloroethene
DCF	Diclofenac
DCM	Dichloromethane
DCP	Dichlorophenol
D_g	Solute Distribution Parameter
DNP	Dinitrophenol
DO	Dissolved Oxygen
DOC	Dissolved Organic Carbon
DOM	Dissolved Organic Matter
DTPA	Diethylene Triamine Pentaacetic Acid
DZP	Diazepam
E2	17 β-Estradiol
E3	Estriol
EBCT	Empty-Bed Contact Time
EC	Expanded Clay
EDC	Endocrine Disrupting Compound
EDTA	Ethylene Diamine Tetraacetic Acid
EE2	17 α-Ethinylestradiol
EEM	Excitation-Emission Matrix
EOCl	Extractable Organic Chlorine
EPS	Extracellular Polymeric Substances
ESEM	Environmental Scanning Electron Microscopy
F/M	Food to Microorganism (ratio)
FA	Free Ammonia

FBR	Fluidized-Bed Reactor
FNA	Free Nitrous Acid
GAC	Granular Activated Carbon
GAC/BAC	Granular /Biological Activated Carbon (with no clear distinction)
GAC-FBR	Fluidized-Bed Reactor packed with Granular Activated Carbon
GAC-MBR	GAC added Membrane Bioreactor
GAC-SBBR	GAC Reactor operated as a Sequencing Batch Biofilm Reactor
GAC-UASB	Upflow Anaerobic Sludge Blanket packed with Granular Activated Carbon
GAC-UFBR	Upflow Fixed-Bed Reactors packed with GAC
GC-MS	Gas Chromatography-Mass Spectrometry
HAA	Haloacetic Acid
HAA5	Five Haloacetic Acids
HAAFP	HAA Formation Potential
HLR	Hydraulic Loading Rate
HMW	High Molecular Weight
HMX	High Melting eXplosive
HPC	Heterotrophic Plate Count
HRT	Hydraulic Retention Time
HSDM	Homogeneous Surface Diffusion Model
IAS	Ideal Adsorbed Solution
IAST	Ideal Adsorbed Solution Theory
IBP	Ibuprofen
IBPCT	Integrated Biological-Physicochemical Treatment
IC_{50}	Concentration leading to 50% inhibition in biological tests
ISIAS	Improved Simplified Ideal Adsorbed Solution
IUPAC	International Union of Pure and Applied Chemistry
KATOX	A process leading to accelerated oxidation in the presence of activated carbon
LC_{50}	Concentration leading to lethal effect in 50% of biological species
LCA	Life Cycle Assessment
LCC	Life Cycle Cost
LDF	Linear Driving Force
LMW	Low Molecular Weight
MADAM	Michigan Adsorption Design and Applications Model
MAP	Microbially Available Phosphorus
MBR	Membrane BioReactor
MDBA	Multiple-Component Biofilm Diffusion Biodegradation and Adsorption model
MEP	Metabolic End Products
MF	Microfiltration
MIB	2-Methylisoborneol
MLSS	Mixed Liquor Suspended Solids
MLVSS	Mixed Liquor Volatile Suspended Solids
MNP	*m*-Nitrophenol

MTBE	Methyl-Tert-Butylether
MTZ	Mass Transfer Zone
MW	Molecular Weight
NBDOC	Nonbiodegradable Dissolved Organic Carbon
NMR	Nuclear Magnetic Resonance
NOB	Nitrite Oxidizing Bacteria
NOM	Natural Organic Matter
NP	Nitrophenol
NPDES	National Pollutant Discharge Elimination System
NPEs	Nonylphenol Ethoxylates
NPX	Naproxen
OCD	Organic Carbon Detection
OCPSF	Organic Chemicals, Plastics and Synthetic Fiber
OND	Organic Nitrogen Detection
OUR	Oxygen Uptake Rate
PAC	Powdered Activated Carbon
PAC-MBR	PAC added Membrane Bioreactor
PACT	Powdered Activated Carbon Treatment
PACT®	Registered Powdered Activated Carbon Treatment
PAE	Phthalate Ester
PAH	Polycyclic Aromatic Hydrocarbon
PBR	Packed-Bed Reactor
PCE	Perchloroethylene (Tetrachloroethylene)
PCP	Pentachlorophenol
PCPs	Personal Care Products
PCR	Polymerase Chain Reaction
PDM	Pore Diffusion Model
Pe	Peclet number
PFR	Plug Flow Reactor
PNP	p-Nitrophenol
POC	Particulate Organic Carbon
POP	Persistent Organic Pollutant
POTW	Publicly Owned Treatment Works
PPCPs	Pharmaceutical and Personal Care Products
PZG	Plane of Zero Gradient
RBF	River Bank Filtration
RDX	Royal Demolition eXplosive
RFB	Recycle Fluidized-Bed
RWW	Rheinisch-Westfälische Wasserwerksgesellschaft
SAT	Soil Aquifer Treatment
SBBR	Sequencing Batch Biofilm Reactor
SBR	Sequencing Batch Reactor
SBR-PACT	Powdered Activated Carbon Treatment in a Sequencing Batch Reactor
SCFB	Semi-Continuously Fed Batch (Reactor)

SCOD	Soluble Chemical Oxygen Demand
SCR	Specific Cake Resistance
SEC	Size Exclusion Chromatography
SEM	Scanning Electron Microscopy
Sh	Sherwood number
SIAS	Simplified Ideal Adsorbed Solution
SMP	Soluble Microbial Products
SOC	Synthetic Organic Compound
SOUR	Specific Oxygen Uptake Rate
SRF	Specific Resistance to Filtration.
SRT	Sludge Retention Time
SS	Suspended Solids
SSF	Slow Sand Filtration
St	Stanton number
STP	Sewage Treatment Plant
SUVA	Specific Ultraviolet Absorbance
SVI	Sludge Volume Index
SVOC	Semi-Volatile Organic Compound
TCA	Trichloroethane
TCB	Trichlorobenzene
TCE	Trichloroethylene
TCOD	Total Chemical Oxygen Demand
TDS	Total Dissolved Solid
THM	Trihalomethane
THMFP	Trihalomethane Formation Potential
TKN	Total Kjeldahl Nitrogen
TMP	Transmembrane Pressure
TN	Total Nitrogen
TOC	Total Organic Carbon
T-RFLP	Terminal-Restriction Fragment Length Polymorphism
TSMSBM	Transient-State Multiple-Species Biofilm Model
TSS	Total Suspended Solids
TTHM	Total Trihalomethanes
TVH	Total Volatile Hydrocarbons
UASB	Upflow Anaerobic Sludge Blanket
UF	Ultrafiltration
USEPA	United States Environmental Protection Agency
UV	Ultraviolet
VOC	Volatile Organic Compound (Carbon)
WAO	Wet Air Oxidation
WAR	Wet Air Regeneration
WWTP	Wastewater Treatment Plant
XOC	Xenobiotic Organic Compound

Acknowledgement

This work would not have been possible without the Bogaziçi University, Istanbul/Turkey, to which I owe much, first for my education and later for its support during my activities as a faculty member. Most of the information presented in this book was accumulated during the last 20 years of my teaching and research at the University, much of it supported through grants from the Research Fund. Similarly, I am indebted to our national research council TÜBİTAK for its support of several research projects in these years.

I also owe much to the Institute of Environmental Sciences of Bogaziçi University, which offered me an academic career, and to the direct and indirect support of all members of the Institute. I owe special thanks to Prof. Nadim Copty for his careful examination of some chapters, particularly for his critical reading and invaluable suggestions made for the modeling chapters (Chapters 6 and 11).

I am grateful to all of my former thesis students with whom I had the opportunity to work, who helped me realize my ideas and contributed to many of the concepts and results presented in this book. Of these students, two in particular have done great work that has directly contributed to the book:

Özgür Aktaş, who is also the co-author of this book, worked mainly on wastewater treatment and produced his MSc thesis, under my supervision, on 'Powdered activated carbon addition to activated sludge in the treatment of landfill leachate.' After this, his PhD thesis, also under my supervision, 'Bioregeneration of activated carbon in the treatment of phenolic compounds,' contributed much to the literature.

Kozet Yapsakli, a PhD student, wrote her thesis on 'Application of biological activated carbon (BAC) in drinking water treatment' under my supervision. This study further extended our understanding of integrated adsorption and biological removal as applied to drinking water treatment. Some examples, described in Chapters 8 and 9 of this book, are derived from these studies. I also thank her for her careful reading of the chapters in this book, particularly those related to drinking water treatment.

Special thanks go to my research assistant Ayşe Gül Geyik who has worked tirelessly on literature surveys, drawing, and formatting.

Activated Carbon for Water and Wastewater Treatment: Integration of Adsorption and Biological Treatment.
First Edition. Ferhan Çeçen and Özgür Aktaş
© 2011 WILEY-VCH Verlag GmbH & Co. KGaA, Weinheim.
Published 2011 by WILEY-VCH Verlag GmbH & Co. KGaA

Finally, I am indebted to my family members, who supported me throughout my life in all aspects. I dedicate this book to my father, now deceased, who had a great influence on my personality and on my path to becoming an academician.

Ferhan Çeçen

1
Water and Wastewater Treatment: Historical Perspective of Activated Carbon Adsorption and its Integration with Biological Processes
Ferhan Çeçen

1.1
Historical Appraisal of Activated Carbon

Activated carbon is broadly defined to include a wide range of amorphous carbon-based materials prepared in such a way that they exhibit a high degree of porosity and an extended surface area [1]. Moreover, all non-carbon impurities are removed and the surface is oxidized. Although today the term 'activated carbon' is taken for granted, a long time elapsed before it became generally adopted.

The use of activated carbon in its current form has only a short history. On the other hand, according to records, the use of carbon itself dates back to ancient times. The earliest known use of carbon in the form of wood chars (charcoal) by the Egyptians and Sumerians was in 3750 BC [2]. At that time, charcoal was used for various purposes such as reduction of ores in the manufacture of bronze, domestic smokeless fuel, and medicinal applications [3]. In Egyptian papyri dating from 1550 BC we find the first citation of the use of charcoal for the adsorption of odorous vapors – from putrefying wounds and the intestinal tract. The ancient Greeks used charcoal to ease the symptoms of food poisoning [4]. The beneficial effect was due to the adsorption of the toxins emitted by ingested bacteria, thereby reducing their toxic effects.

Hindu documents dating from 450 BC refer to the use of sand and charcoal filters for the purification of drinking water. Recent studies of the wrecks of Phoenician trading ships led to the discovery that drinking water was stored in charred wooden barrels in order to keep the water fresh [4]. In the time of Hippocrates (ca. 460–370 BC) and Pliny the Elder (AD 23–79) wood chars were employed for medicinal purposes [5]. In about 157 BC carbons of vegetable and animal origin were applied in the treatment of many diseases [2]. A Sanskrit text around 200 AD recommends the use of filtration of water through coal after storing it in copper vessels and exposing it to sunlight, providing probably one of the earliest documents describing the removal of compounds from water in order to disinfect it [6].

Activated Carbon for Water and Wastewater Treatment: Integration of Adsorption and Biological Treatment.
First Edition. Ferhan Çeçen and Özgür Aktaş
© 2011 WILEY-VCH Verlag GmbH & Co. KGaA, Weinheim.
Published 2011 by WILEY-VCH Verlag GmbH & Co. KGaA

In the fifteenth century, during the time of Columbus, sailors used to blacken the insides of wooden water barrels with fire, since they observed that the water would stay fresh much longer. It is likely that people at that time proceeded by intuition only, without having any insight into the mechanisms of the effect; these mechanisms were recognized beginning from the eighteenth century.

In the eighteenth century, carbons made from blood, wood, and animals were used for the purification of liquids. The specific adsorptive properties of charcoal (the forerunner of activated carbon) were first observed by Scheele in 1773 in the treatment of gases. Later, in 1786, Lowitz performed experiments on the decolorizing of solutions. He provided the first systematic account of the adsorptive power of charcoal in the liquid phase [7]. In those days, the sugar refining industry was looking for an effective means of decolorizing raw sugar syrups, but the wood charcoals then available were not particularly effective because of their limited porosity [4]. However, a few years later, in 1794, an English sugar refinery successfully used wood charcoal for decolorization. This application remained a secret until 1812 when the first patent appeared in England [2], although from 1805 wood charcoal was used in a large-scale sugar refining facility in France for decolorizing syrups, and by 1808 all sugar refineries in Europe were using charcoal as a decolorizer [4].

In 1811 it was shown that bone char had an even higher decolorizing ability for sugar syrups than wood char. Consequently, a switch took place from wood charcoal to bone char in the sugar industry. In 1817 Joseph de Cavaillon patented a method of regenerating used bone chars, but the method was not entirely successful. In 1822 Bussy demonstrated that the decolorizing abilities of carbons depended on the source material, the thermal processing, and the particle size of the finished product. His work constitutes the first example of producing an activated carbon by a combination of thermal and chemical processes. Later in the nineteenth century, systematic studies were carried out on the manufacture and regeneration of bone chars by Schatten in Germany and the application of charcoal air filters for removing vapors and gases in London sewers by Stenhouse [4].

In 1862, Lipscombe prepared a carbon material to purify potable water. This development paved the way for the commercial applications of activated carbon, first for potable water and then in the wastewater sector. In 1865 Hunter discovered the excellent gas adsorption properties of carbons derived from coconut shells. It is remarkable that the term 'adsorption' was first introduced by Kayser in 1881 to describe the uptake of gases by carbons [4].

Activated carbon was first produced on an industrial scale at the beginning of the twentieth century, and major developments then took place in Europe. However, at the beginning of the twentieth century activated carbon was only available in the form of powdered activated carbon (PAC). The Swedish chemist von Ostreijko obtained two patents, in 1900 and 1901, covering the basic concepts of chemical and thermal (or physical) activation of carbon, with metal chlorides and with carbon dioxide and steam, respectively [7]. In 1909, a plant named 'Chemische Werke' was built to manufacture, for the first time on a commercial scale, the powdered activated carbon Eponit® from wood, adopting von Ostrejko's gasification approach [8]. Other activated carbons known as Norit® and Purit®

were produced in this plant by the activation of peat with steam. The NORIT company, a manufacturer in Holland, first appeared in about 1911 and became widely known in the sugar industry [5]. The powdered activated carbons were used at that time mainly for decolorizing solutions in the chemical and food industries.

On an industrial scale, the process of chemical activation of sawdust with zinc chloride was carried out for the first time in an Austrian plant at Aussing in 1914, and also in the dye plant of Bayer in 1915 [9]. This type of activation involved pyrolytical heating of the carbonaceous material in the presence of dehydrating chemicals such as zinc chloride or phosphoric acid [10].

In parallel to the developments in Europe, in the United States the first activated carbon was produced from black ash, a waste product of soda production, after it was accidentally discovered that the ash was effective in decolorizing liquids [5]. The first commercial production of activated carbon in the United States took place in 1913 [11]. Activated carbon in the form of PAC was used for the first time in 1928 by Chicago meat packers for taste and odor control [12].

The use of poisonous gases in the First World War paved the way for the development and large-scale production of granular activated carbon (GAC). These carbons were used in gas masks for the adsorption of poisonous gases. Subsequently, they were used for water treatment, solvent recovery, and air purification. After the First World War, considerable progress was made in Europe in the manufacture of activated carbons using new raw carbonaceous materials such as coconut and almond shells. The treatment with zinc chloride yielded activated carbons with high mechanical strength and high adsorptive capacities for gases and vapors. Later, in 1935–1940, pelletized carbons were produced from sawdust by zinc chloride activation for the recovery of volatile solvents and the removal of benzene from town gas. Nowadays, the zinc chloride process of chemical activation has been largely superseded by the use of phosphoric acid [4].

1.2
General Use of Activated Carbon

Nowadays, activated carbon finds wide application in many areas, but especially in the environmental field. Aside from environmental pollution control, activated carbon is mainly used in industry in various liquid and gas phase adsorptions [1]. Among liquid phase applications one can list food processing, preparation of alcoholic beverages, decolorization of oils and fats, product purification in sugar refining, purification of chemicals (acids, amines, gylcerin, glycol, etc.), enzyme purification, decaffeination of coffee, gold recovery, refining of liquid fuels, purification in electroplating operations, purification in the clothing, textile, personal care, cosmetics, and pharmaceutical industries, and applications in the chemical and petrochemical industries. Gas phase applications include recovery of organic solvents, removal of sulfur-containing toxic components from exhaust gases and recovery of sulfur, biogas purification, use in gas masks, among others. Activated carbon is also used in medical and veterinary applications, soil improvement, removal of pesticide residues, and nuclear and vacuum technologies.

1.3
Application of Activated Carbon in Environmental Pollution

Although the use of carbon-based materials dates back to ancient times, the use of activated carbon in its current form began in the second half of the twentieth century as a consequence of the rising awareness of environmental pollution. Today, activated carbon is very often utilized in the removal of various organic and inorganic species from surface water, groundwater, and wastewater.

1.3.1
Activated Carbon in Drinking Water Treatment

Adsorption by activated carbon is employed today in drinking water treatment for various purposes. An overview of historical development shows that the first application of activated carbon in the form of GAC was in the year 1910 in Reading, England for the purpose of dechlorination of chlorinated water [12]. In the 1930s and 1940s, in particular in Europe, water works used high chlorine doses for the disinfection of water following the growing pollution of surface waters. Often, GAC filtration was used for dechlorination purposes. However, the dechlorination in these filters cannot be regarded as an adsorptive process since the removal of chlorine depends on a catalytic reaction taking place on the carbon surface. However, the use of GAC for dechlorination purposes was abandoned a long time ago because of the formation of additional haloforms and other chlorine compounds within filters [7].

The use of activated carbon in water treatment for removal of substances responsible for taste and odor dates back to the late 1920s [11]. The undesirable taste and odor in drinking water was mainly attributed to the presence of chlorophenols formed in water as a result of the chlorination of phenols at the disinfection stage [7].

PAC was used for the control of taste and odor in drinking water for the first time in the USA in 1929–1931 [7]. The first GAC filters were installed in Germany in 1929 and in the USA in 1930 for taste and odor removal. By 1932 about 400 water treatment works in the USA were adding PAC to their water to improve taste and odor, and this number increased to 1200 by 1943. The first major GAC filter for public water supply was installed in the USA at the Hopewell, VA, water treatment plant in 1961 [12]. By 1970 the number of waterworks which added PAC to their units or used GAC adsorbers was estimated at 10000 worldwide [7]. In later years, PAC adsorption for water treatment was also integrated with Dissolved Air Flotation (DAF), in which PAC served as an adsorbent for various pollutants and was subsequently floated to the surface by DAF [13].

When activated carbon was used in granular or powdered form in the early 1960s in water treatment, the main aim was the removal of taste and odor. In Europe, where surface waters were heavily polluted, early breakthroughs of odor-causing species were observed in GAC filters, necessitating frequent regenerations. Intensive investigations beginning in the early 1960s revealed that pretreatment of water with ozone was an effective solution to this problem since it extended the GAC bed life. The well-known Mülheim process was developed as a result of these efforts [7]. Details of this process can be found in Chapter 8.

Currently, problems in drinking water treatment extend beyond the scope of taste and odor control. Much attention is being paid to the regulation and control of numerous organic and inorganic compounds in water. Concerns about the presence of Synthetic Organic Compounds (SOC) arose in 1960s. Beginning in the 1970s it was recognized that disinfection of water with chlorine gas or chlorine-containing compounds led to the generation of organic compounds, collectively termed Disinfection By-Products (DBPs), which were suspected of having adverse effects on health [7]. In this regard, Natural Organic Matter (NOM) constitutes the key group of organics acting as precursors for DBP formation. It was also shown that pretreatment of water with ozone led to inorganic hazardous by-products such as bromates. For many decades, adsorption onto activated carbon has appeared to be one of the most reliable methods of NOM and DBP control. This type of treatment is usually conducted in GAC filters. These are usually placed after sand filtration and before disinfection, but, depending on the characteristics of the water and the object of the treatment, GAC filters may also be positioned at other locations within the treatment train.

The presence of synthetic organic contaminants in surface and groundwaters is largely attributed to the discharge of municipal and industrial wastewaters into receiving waters in treated or untreated form. The increased use of fertilizers and pesticides in agriculture is another factor contributing to pollution. Further, discharges into surface waters from non-point sources such as urban runoff also add to pollution.

Raw waters taken from surface and groundwater supplies contain many organic compounds such as phenols, pesticides, herbicides, aliphatic and aromatic hydrocarbons and their chlorinated counterparts, dyes, surfactants, organic sulfur compounds, ethers, amines, nitro compounds, and newly emerging substances such as Endocrine Disrupting Compounds (EDCs). More than 800 specific organic and inorganic chemicals have been identified in various drinking waters, and many more are suspected to be present [1]. Therefore, concerns are frequently expressed about the presence of these compounds, which can be present at levels as low as ng L^{-1} or µg L^{-1}. Because of their proven or suspected health and environmental effects, great efforts are made to control and/or remove them, and one of the major methods of doing this is by adsorption onto activated carbon.

1.3.2
Activated Carbon in Wastewater Treatment

The groups of organics that are generally amenable to adsorption onto activated carbon include pesticides, herbicides, aromatic solvents, polynuclear aromatics, chlorinated aromatics, phenolics, chlorinated solvents, high-molecular-weight (HMW) aliphatic acids and aromatic acids, HMW amines and aromatic amines, fuels, esters, ethers, alcohols, surfactants, and soluble organic dyes. Compounds having low molecular weight (LMW) and high polarity, such as LMW amines, nitrosamines, glycols, and certain ethers, are not amenable to adsorption [11].

Many compounds falling into these categories are encountered in the effluents of various industries and to some extent also in municipal wastewaters and drinking water supplies. Activated carbon has gained importance especially since the mid 1960s as an adsorptive material in the treatment of municipal and industrial wastewaters.

1.3.2.1
Municipal Wastewater Treatment

The first full-scale advanced (tertiary) wastewater treatment plant incorporating GAC was put into operation in 1965 in South Lake Tahoe, California. The use of GAC beds as a unit process became common in the tertiary treatment train [11]. The purpose in employing GAC was to reuse the effluent of municipal wastewater treatment plants for purposes such as industrial cooling water, irrigation of parks, and so on.

Physicochemical treatment options involving PAC adsorption were also tested in lieu of biological treatment. The idea was that primary settling was followed by coagulation and PAC adsorption, settling and perhaps filtration. However, secondary treatment could not be replaced with a merely physicochemical process because of cost [11].

Today, GAC filtration or PAC-assisted membrane operation are mainly conducted as a tertiary treatment step to remove dissolved and refractory organic matter from secondary sewage effluent. The main goal remains to be the reuse of effluent for various purposes.

1.3.2.2
Industrial Wastewater Treatment

Activated carbon adsorption is most commonly applied in industrial wastewater treatment to meet stringent regulations for discharge into receiving waters. In industrial wastewater treatment, activated carbon adsorption can be utilized as a separate unit process. It may be placed after various physicochemical treatment steps such as coagulation/clarification, filtration, and dissolved air flotation. Another alternative is to use activated carbon adsorption prior to biological treatment to remove compounds which might be toxic to a biological system. However, the most widely adopted procedure is to place activated carbon adsorption as a tertiary or advanced treatment step subsequent to biological treatment for removal of refractory organics. To some extent this procedure may also be effective in the removal of inorganics.

Nowadays, activated carbon finds wide application in the treatment of wastewaters generated from industries such as food, textile, chemical, pharmaceutical, pesticides and herbicides production, coke plant, munitions factories, petroleum refineries and storage installations, organic pigments and dyes, mineral processing plants, insecticides, pesticides, resins, detergents, explosives, and dyestuffs. It is also employed in the treatment of sanitary and hazardous landfill leachates.

1.3.3
Applications of Activated Carbon in Other Environmental Media

1.3.3.1
Remediation of Contaminated Groundwater and Soil

Groundwaters are significantly polluted with organic and inorganic substances as a result of industrial spills, accidents, discharges, and so on. Activated carbon adsorption is often employed in remediation of groundwaters for drinking purposes. In groundwater remediation, activated carbon may either directly adsorb contaminants or remove them after their transfer into the gas phase by air sparging or stripping.

Today, activated carbon is also applied in the remediation of contaminated soils. Remediation of soils contaminated with petroleum hydrocarbon and other substances involves the use of thermal desorption methods. The resulting off-gases containing Volatile Organic Compounds (VOCs) are usually treated with PAC or GAC. In contaminated soils, PAC may also be used as a soil additive to immobilize organic contaminants.

1.3.3.2
Treatment of Flue Gases

Activated carbon also finds application in the purification of flue gases such as those emerging from incinerators, and in the removal of gases such as radon, hydrogen sulfide, and other sulfur compounds from gas streams [2].

1.3.3.3
Water Preparation for Industrial Purposes

Activated carbon is utilized in industrial facilities for the production of the water required for various plant items such as steam generators, heat exchangers, cooling towers, and also in the production of ultrapure water.

1.3.4
Integration of Activated Carbon Adsorption with Biological Processes in Wastewater and Water Treatment

Nowadays, adsorption and biological processes for the control of various pollutants generally take place in separate unit, whereas the combination of adsorption and biological processes in the same reactor is relatively less common and more complicated. The main purpose of this book is to provide an insight into this system of integrated application, whose usefulness has been clearly recognized since the 1970s, in both wastewater and water treatment.

1.3.4.1
Wastewater Treatment
1.3.4.1.1 Combined Suspended-Growth Processes
The Powdered Activated Carbon Treatment (PACT) process The PACT process is essentially a modification of the activated sludge process by the addition of PAC. The application of concurrent adsorption and biodegradation in the same

suspended-growth reactor is an effective alternative for the removal of biodegradable and nonbiodegradable compounds. The PACT system has also been adopted for anaerobic treatment.

Integration of PAC adsorption with membrane processes In recent years PAC has been integrated into Membrane Bioreactors (MBR) to bring about a positive effect on contaminant removal and to prevent membrane biofouling.

1.3.4.1.2 Combined Attached-Growth Processes
The Biological Activated Carbon (BAC) Process While PACT is a modification of a suspended-growth process, BAC is essentially a biofilm process that is based on the establishment of biological activity in a GAC adsorber by gradual attachment of microorganisms and development of a biofilm.

Since 1970s, both PAC- and GAC-based biological processes have been applied in the treatment of industrial wastewaters such as organic chemicals, petrochemicals, refineries, textiles/dyes, in joint treatment of municipal and industrial wastewaters for discharge or reuse purposes, and in the treatment of sanitary and hazardous landfill leachates. Detailed discussion of such applications is presented in Chapters 3–7 of this book.

1.3.4.2
Water Treatment

Since the 1970s, a gradual development has taken place in the direction of integrating adsorptive and biological processes in the treatment of surface water or groundwaters. In this regard, BAC filtration is a well-known unit process that combines the merits of adsorption and biological removal in the same reactor. While the majority of GAC adsorption applications target the removal of natural and/or anthropogenic organic compounds, BAC filtration is also suited to some extent for the elimination of inorganics such as ammonia, perchlorate, and bromate. The characteristics of this unit process is extensively addressed throughout Chapters 8–11.

1.3.5
Improved Control of Pollutants through Integrated Adsorption and Biological Treatment

Water and wastewaters are multicomponent mixtures. As such, it is impossible to measure the presence and removal of a large number of compounds present in treatment or remediation systems. Therefore, in the case of organic compounds, monitoring is commonly carried out by the use of sum (collective) parameters such as TOC, DOC and UV absorbance, and the BOD and COD parameters, in water and wastewater treatment, respectively (Table 1.1).

Over the years, more specific parameters have been developed. In the characterization of waters and wastewaters one of the widely used parameters is

1.3 Application of Activated Carbon in Environmental Pollution

Table 1.1 Reduction in parameters or pollutant groups achieved by adsorption, biological removal, or integrated means.

Parameters/pollutant groups	Importance in water (W), groundwater (GW) or wastewater (WW) treatment	Activated Carbon Adsorption	Biological Removal or Transformation	Removal in Integrated Adsorption and Biological Treatment
Organic pollutants				
Oxygen demand parameters				
BOD	WW	F to G	G to E	E
COD	WW	G	L to G	G to E
Organic carbon parameters				
TOC	W, WW, GW	G	L to G	G to E
DOC	W, GW, WW	G	L to G	G to E
VOC[e]	W, GW, WW	L to G	L to G	G to E
BDOC	W, GW	F to G	G to E	E
AOC	W, GW	F to G	E	E
Other organic parameters				
AOX	W, GW, WW	G to E	L to F	G to E
UV_{254}	W, GW, WW	G	L	G to E
Organic Pollutant Groups				
THMs	W, GW	G	P to L	G
HAAs	W, GW	G	G	G to E
Pesticides	W, GW	F to E	L to F	G
Pharmaceuticals	W, WW	F to E	F to E	G to E
Endocrine Disrupting Compounds (EDCs)	W, WW	G	P to G	G to E
Chlorinated hydrocarbons	GW, WW	F to E	L to G	G to E
Inorganic Pollutants				
Bromate	W	L to F[a]	F to G[d]	G[d]
Perchlorate	GW	N	G	G to E
Ammonia	W, WW	N	G	G to E
Nitrate	GW, WW	N	G	G to E
Heavy Metals	W, GW, WW	P to G[b]	F[c]	G[c]

E: excellent, G: good, F: fair, L: low, P: poor, N: none
[a] reduction at surface
[b] dependent on type and conditions
[c] possible transformation and/or biosorption
[d] uncertain
[e] Volatile Organic Carbon: Volatile Organic Compounds are measured based on the surrogate 'carbon.' They can also be measured as individual compounds.

referred to as Adsorbable Organic Xenobiotics (AOX), and represents halogenated organics that have a high affinity towards activated carbon. This parameter is most often used to indicate the chlorinated organic compounds (AOCl).

There are also definitions that are specific to water treatment. In this context, one can list the Disinfection By-Products (DBPs). Specific groups among DBPs are referred to by terms such as Trihalomethanes (THMs) and Haloacetic Acids (HAAs). The respective formation potentials of these groups are abbreviated as THMFP and HAAFP.

In addition to these, in recent decades, a large number of new compounds have been detected in water and wastewater media. Since in raw and finished waters these compounds, referred to as micropollutants, are present at $\mu g\ L^{-1}$ or $ng\ L^{-1}$ levels, the sum parameters mentioned above prove to be useless in their monitoring. Due to this fact, efforts are made to monitor them individually by advanced analytical techniques. Within this context, one can list various pharmaceuticals and EDCs that have received a great deal of attention in the last decades.

Various pollutants found in water and wastewater systems are amenable to either adsorption or biological degradation or transformation. Still, a number of them can be removed by both adsorptive and biological means. Combination of activated carbon adsorption and biological processes in the same unit often offers a synergism, in that a higher removal is achieved than expected from adsorption or biodegradation alone. For many pollutants that are considered to be slowly biodegradable or even nonbiodegradable, integration of adsorption with biological removal may provide the opportunity for biological degradation. This integrated approach can also enable the effective elimination of micropollutants at trace levels.

Various organic and inorganic pollutants are encountered in surface waters, groundwaters, and wastewaters. Table 1.1 provides a brief overview of the relative reduction achieved in parameters or specific groups by means of adsorption, biological activity or integration of both. However, the evaluation presented here is rather general, relative, and qualitative. Comprehensive discussion of the elimination mechanisms, the synergism in integrated adsorption and biological removal, the laboratory-, pilot- and full-scale studies, and the modeling of integrated adsorption and biological removal in wastewater and water treatment is presented throughout Chapters 3–11.

References

1 Bansal, R.P. and Goyal, M. (2005) *Activated Carbon Adsorption*, CRC Press, Taylor & Francis Group, 6000 Broken Sound Parkway NW, Suite 300 Boca Raton, FL, USA 33487-2742.

2 Inglezakis, V.J. and Poulopoulos, S.G. (2006) *Adsorption, Ion Exchange and Catalysis: Design of Operations and Environmental Applications*, Elsevier Science & Technology.

3 Menendez-Diaz, J.A. and Gullon, I.M. (2006) Types of carbon adsorbents and their production, in *Activated Carbon Surfaces in Environmental Remediation*

(eds T.J.Bandosz), Elsevier, Printed in the Netherlands, pp. 1–47.
4 www.caer.uky.edu/carbon/history/carbonhistory.shtml. University Of Kentucky, Center for Applied Energy Research, *History of Carbon* (accessed 29 January 2010).
5 Hassler, J.W. (1963) *Activated Carbon*, Chemical Publishing Co., Inc., New York, N.Y., USA.
6 www.carbonit.com. Carbonit Brochure-Westa Gruppe, CARBONIT Filtertechnik GmbH, Salzwedel, Germany.
7 Sontheimer, H., Crittenden, J., and Summers, R.S. (1988) *Activated Carbon for Water Treatment*, 2nd edn, Forschungstelle Engler – Bunte-Institute, Universität Karlsruhe, Karlsruhe, Germany.
8 Dabrowski, A. (1998) Adsorption – its development and application for practical purposes, in *Adsorption and Its Applications in Industry and Environmental Protection. Studies in Surface Science and Catalysis*, vol. **120** (ed. A. Dabrowski), Elsevier Science B. V., pp. 3–68.
9 Dabrowski, A. (2001) Adsorption-from theory to practice. *Advances in Colloid and Interface Science*, **93**, 135–224.
10 Purcell, P.J. (2006) Milestones in the development of municipal water treatment science and technology in the 19th and early 20th centuries: part I. *Water and Environment Journal*, **19** (3), 230–237.
11 Hendricks, D. (2006) *Water Treatment Unit Processes: Physical and Chemical*, CRC Press. Printed in the USA.
12 Hung, Y.T., Lo, H.H., Wang, L.K., Taricska, J.R., and Li, K.H. (2005) Granular activated carbon adsorption, in *Physicochemical Treatment Processes, Handbook of Environmental Engineering*, vol. **3** (eds L. K. Wang, Y.T. Hung and N. Shammas), Humana Press Inc., Totowa, New Jersey, USA pp. 573–633.
13 Wang, L.K. (2007) Emerging flotation technologies, in *Advanced Physicochemical Treatment Technologies, Handbook of Environmental Engineering*, vol. **5** (eds L.K. Wang, Y. T. Hung and N. Shammas) Humana Press Inc., Totowa, New Jersey, USA pp. 449–489.

2
Fundamentals of Adsorption onto Activated Carbon in Water and Wastewater Treatment

Özgür Aktaş and Ferhan Çeçen

2.1
Activated Carbon

Activated carbons are porous carbonaceous adsorbents. A large variety of organic solutes (Table 2.1) can be removed from water and wastewater by adsorption onto activated carbon, and some inorganic solutes can also be removed by this means. The porous surface of activated carbon adsorbs and retains solutes and gases, the amount of material adsorbed being potentially very large because of the great internal surface of activated carbon. Activated carbon has a high adsorptive surface area (500–1500 m^2g^{-1}), while the pore volume ranges between 0.7 and 1.8 cm^3g^{-1}. It is mainly used in the form of powdered activated carbon (PAC) or granular activated carbon (GAC).

2.1.1
Preparation of Activated Carbons

Commercially available activated carbons are prepared from materials having a high carbon content such as coal, lignite, wood, peat, nut shell, coconut shell, lignin, petroleum coke, and synthetic high polymers. The manufacturing process comprises two phases, carbonization and activation. The carbonization process includes drying and heating to remove undesirable by-products such as tar and other hydrocarbons. The carbonaceous materials are then pyrolyzed and carbonized within a temperature range of 400–600 °C in an oxygen-deficient atmosphere. This removes the volatile low-molecular-weight fraction and causes the material to undergo an activation process. Activation can be achieved thermally by the use of oxidation gases such as steam at above 800 °C or carbon dioxide at higher temperatures. Chemical activation, on the other hand, involves impregnation of the raw material with chemicals such as phosphoric acid, potassium hydroxide, and zinc chloride [1–3]. The term activation refers to the development of the

---

Activated Carbon for Water and Wastewater Treatment: Integration of Adsorption and Biological Treatment.
First Edition. Ferhan Çeçen and Özgür Aktaş
© 2011 WILEY-VCH Verlag GmbH & Co. KGaA, Weinheim.
Published 2011 by WILEY-VCH Verlag GmbH & Co. KGaA

Table 2.1 Classes of organic compounds adsorbed on activated carbon (adapted from [5]).

Organic chemical class	Examples
Aromatics	Benzene, Toluene, Ethylbenzene, Xylene
Polynuclear Aromatics	Naphthalene, Anthracenes, Biphenyls
Chlorinated Aromatics	Chlorobenzene, Polychlorinated biphenyls
Phenolics	Phenol, Cresol, Chlorophenols, Nitrophenols
High-molecular-weight hydrocarbons	Gasoline, Kerosene
Chlorinated aliphatics	Trichloroethylene, Carbon tetrachloride
Aliphatic and aromatic acids	Tar acids, Benzoic acids
Ketones, esters, ethers, and alcohols	Hydroquinone, Polyethylene glycol
Surfactants	Alkyl benzene sulfonates
Soluble organic dyes	Methylene blue, Indigo carmine

Table 2.2 Basic properties of common materials used in the manufacture of activated carbon (adapted from [5]).

Raw material	Carbon (wt%)	Volatiles (wt%)	Ash (wt%)
Wood	40–45	55–60	0.3–1.1
Nut shells	40–45	55–60	-
Lignite	55–70	25–40	5–6
Coal	65–95	5–30	2–15
Petroleum coke	70–85	15–20	0.5–0.7

adsorption properties of carbon. Micropores are formed during the activation process, the yield of micropore formation being usually below 50% [3].

The raw material has a very large influence on the characteristics and performance of activated carbon obtained. Raw materials such as coal and charcoal have some adsorption capacity, but this is greatly enhanced by the activation process [4]. The common materials used in the production of activated carbon and the basic properties of the activated carbons produced are summarized in Table 2.2.

2.1.2
Characteristics of Activated Carbon

Activated carbon is composed of microcrystallites that consist of fused hexagonal rings of carbon atoms, this structure being quite similar to that of graphite. The spaces between the individual microcrystallites are called pores. Micropores of activated carbon, where most of the adsorption takes place, are in the form of two-dimensional spaces between two parallel crystalline planes with an interlayer distance of 3–35Å. The diameter of the microcrystallites is roughly nine times the width of one carbon hexagon. Functional groups that terminate the microcrystallite planes interconnect the microcrystallites. Adsorption occurs on the planar surfaces of the microcrystallites and at the functional groups on the edges of the planes [3, 6, 7]. Adsorption on the microcrystallite planes predominantly results from van der Waals forces. On the other hand, adsorption at the edges of the microcrystallite occurs because of chemical bonding [7].

Activated carbon surfaces generally contain various oxygen complexes. These complexes arise from raw materials as well as from chemisorption of oxygen during the activation process. Oxygen complexes on the activated carbon surface exist mainly in the form of four different acidic surface oxides, namely strong carboxylic, weak carboxylic, phenolic, and carbonyl groups. Also, there are basic groups such as cyclic ethers. Activation at higher temperatures results in a basic surface. The presence of surface oxides adds a polar nature to activated carbons. Thermal treatment of carbons in an inert atmosphere or a vacuum can remove these surface oxide groups [3].

Surface functional groups play an important role in the adsorption of various organic molecules. For example, aromatic compounds can be adsorbed at the carbonyl oxygens on the carbon surface according to a donor–acceptor complexation mechanism. The carbonyl oxygen acts as the electron donor, while the aromatic ring of the solute acts as the electron acceptor. Adsorption also occurs by hydrogen bonding of the phenolic protons with surface functional groups and by complexation with the rings of the microcrystallite planes [7]. Activated carbon may contain large quantities of minerals, mostly calcium, sulfate, and phosphate ions. These groups as well as the acidic or basic organic surface functional groups influence the activated carbon surface properties [8]. The importance of surface chemistry on adsorption and its reversibility in integrated adsorption and biological processes are discussed in Chapter 7.

Activated carbon also contains some ash derived from the raw material, the amount of ash ranging from 1% to 12%. The ash mainly consists of silica, alumina, iron oxides, and alkaline and alkaline earth metals. The ash in the activated carbon increases its hydrophilicity. This is advantageous when PAC is used for water treatment because PAC does not stick on the reactor walls if the ash content is high [3].

2.1.3
Activated Carbon Types

2.1.3.1
Powdered Activated Carbon (PAC)

PAC is made up of crushed or ground carbon particles such that 95–100% of it will pass through a designated sieve of 0.297 mm according to the American Water Works Association Standard, or 0.177 mm according to ASTM D5158 [5]. PAC is generally produced from wood in the form of sawdust, the average particle size of PAC being in the range of 15–25 μm. PAC finds wide application in the treatment of both drinking water and wastewater. In wastewater treatment, it is either added to activated sludge or is contacted with wastewater in a separate unit. PAC may also act as a coagulant for colloidal fractions in the liquid phase. The regeneration of PAC can be rather difficult, because colloidal particles have to be separated from water before regeneration [3].

2.1.3.2
Granular Activated Carbon (GAC)

GAC is usually in the form of crushed granules of coal or shell. GAC may also be prepared by granulation of pulverized powders using binders such as coal tar pitch [3]. Unlike PAC adsorption from the liquid phase, the rate-limiting step in GAC adsorption from the liquid phase often becomes the intraparticle diffusion (Section 2.2.3). GAC particles have sizes ranging from 0.2 to 5 mm. GAC is designated by mesh sizes such as 8/20, 20/40, or 8/30 for liquid phase applications and 4/6, 4/8, or 4/10 for vapor phase applications [5]. Particle sizes in the range of 12/42 mesh are advantageous for liquid phase adsorption [3].

GAC filters, as shown in Section 2.3.2, are widely used in purification processes for drinking water, groundwater and wastewater as an advanced treatment step, particularly for the removal of toxic organic compounds. In some GAC applications in drinking water and wastewater treatment, a microbiological film can form on the particles. Thereby, biological removal of pollutants is combined with GAC adsorption. The resulting material, called biological activated carbon (BAC), is extensively discussed in later chapters.

2.2
Adsorption

Adsorption is considered to be an important phenomenon in most natural physical, biological, and chemical processes, and activated carbon is the most widely used adsorbent material in water and wastewater treatment.

Adsorption is the accumulation or concentration of substances at a surface or interface. The adsorbing phase is termed the adsorbent, and the material being adsorbed the adsorbate. Adsorption can occur between two phases, namely liquid–liquid, gas–liquid, gas–solid, or liquid–solid interfaces. When activated carbon is used, the adsorbing phase is a solid. In activated carbon adsorption, a hypothetical interfacial layer exists between the solid and fluid phases. This layer consists of two regions: that part of the fluid (gas or liquid) residing in the force field of the solid surface and the surface layer of the activated carbon.

Adsorptive surface reactions occur as a result of the active forces within the phase or surface boundaries. These forces result in characteristic boundary energies. The surface tension developed at the surface of the liquid phase results from the attractive forces between the molecules of liquid. The liquid (solvent) molecules have a smaller attractive force for the solute molecules than for each other. Hence, a solute that decreases surface tension is concentrated at the surface. The phenomenon of increased concentration of the soluble material in a boundary or surface is commonly referred to as adsorption [9]. The term 'adsorption' refers to the process in which molecules accumulate in the interfacial layer, whereas desorption denotes the reverse.

2.2.1
Types of Adsorption

In most types of water and wastewater treatment practice, two primary driving forces result in the adsorption of a solute from a solution onto a solid phase.

The first driving force is related to the lyophobic (solvent disliking) character of the solute. The most important factor for the intensity of adsorption is the solubility of a dissolved substance. A hydrophilic substance likes the water system and tends to stay there. Hence, it is less adsorbable on a solid phase. Contrarily, a hydrophobic substance tends to be adsorbed rather than staying in water. Complex organic contaminants, such as humic acids have both hydrophobic and hydrophilic groups, so that the hydrophobic part of the molecule is adsorbed, whereas the hydrophilic part tends to stay in the solution.

The second driving force for adsorption is the affinity of the solute for the solid due to electrical attraction of the solute to the solid. This type of adsorption can occur as a result of van der Waals attraction or chemical interaction with the adsorbent. These forces result due to the high affinity of the solute for the particular solid [9].

The adsorption induced by van der Waals forces is generally termed physical adsorption or physisorption, in which intermolecular attractions take place between favorable energy sites. Physisorption is independent of the electronic properties of the adsorbate and adsorbent molecules. Exchange of electrons does not occur in physisorption. The adsorbate is attached to the surface by relatively weak van der Waals forces. In physisorption, multiple layers may be formed which have similar heats of adsorption. Physical adsorption is predominant at temperatures below 150°C, and is characterized by a relatively low adsorption energy – at most a few kcal mol^{-1}. As a rule, it is a reversible process that occurs at a lower temperature close to the critical temperature of an adsorbed substance. The adsorbate is less strongly attached to a specific site in physisorption compared to chemical adsorption [5]. Therefore, physically adsorbed molecule is free to move within the interface [9].

In chemical adsorption, or chemisorption, the adsorbate undergoes chemical interaction with the adsorbent. Chemisorption involves an exchange of electrons between specific surface sites and solute molecules, a chemical bond being formed. Chemically adsorbed adsorbates are not free to move on the surface or within the interface. Chemical adsorption is characterized by a high adsorption energy (tens of kcal mol^{-1}), since the adsorbate forms strong localized bonds at active centers on the adsorbent. Chemical adsorption is more predominant at high temperatures compared to physisorption, because chemical reactions proceed more rapidly at higher temperatures. Generally, only a single molecular layer can be chemically adsorbed [5, 9]. Table 2.3 gives a comparison of physical and chemical adsorption.

These two forms of adsorption interact together in adsorption. Adsorption of organic molecules exhibit a large range of binding energies. Lower energies are usually associated with physisorption, whereas higher energies are associated with chemisorption. However, it is generally very difficult to distinguish between physisorption and chemisorption [9]. The reversibility of adsorption and bioregeneration of activated carbon in integrated adsorption and biological treatment is discussed in Chapter 7, and the particular importance of the type of adsorption (physical or chemical) is emphasized.

Table 2.3 Comparison of physical and chemical adsorption.

	Physisorption	Chemisorption
Coverage	Mono or multilayer	Monolayer
Nature of adsorption	Nondissociative and reversible	Often dissociative, may be irreversible
Specifity to adsorption sites	Nonspecific	Very specific
Temperature range	Near or below the condensation point of the gas	unlimited
Temperature dependence of uptake (with increasing T)	Decreases	Increases
Adsorption enthalpy	5–40 kJ mol^{-1}	40–800 kJ mol^{-1}
Kinetics of adsorption	Fast	Very variable, often slow
Desorption	Easy by reduced pressure or increased temperature	Difficult – high temperature is required to break bonds
Desorbed species	Adsorbate unchanged	Adsorbate may change

2.2.2
Factors Influencing Adsorption

Activated carbon adsorption is not a uniquely homogeneous process, but is rather dependent on the various factors outlined below. These factors also play a significant role in the integration of adsorption with biological processes (Chapters 3–11).

2.2.2.1
Surface Area of Adsorbent

The extent of adsorption is generally considered to be proportional to the specific surface area. Specific surface area is that proportion of the total surface area which is available for adsorption. The more finely divided and more porous adsorbents would be expected to yield more adsorption per unit weight of adsorbent. The surface can be characterized either as external when it involves bulges or cavities with width greater than depth or internal when it involves pores and cavities that have depth greater than width.

2.2.2.2
Physical and Chemical Characteristics of the Adsorbate

In general, the adsorbability of a compound increases with increasing molecular weight and increasing number of functional groups such as double bonds or halogens [4].

Larger molecules are adsorbed onto activated carbon better than smaller molecules. The degree of solubility of the solute is also of primary concern for adsorption. There is an inverse relationship between the extent of adsorption of a particular solute and its solubility in the solvent from which the adsorption occurs.

High solubility means that the solute–solvent bonds are stronger than the attractive forces between the solute and the adsorbent.

Polarity of the adsorbate is another important factor. A polar solute is preferably adsorbed by a polar adsorbent, whereas a nonpolar solute is more easily adsorbed by a nonpolar adsorbent. Activated carbon adsorbs nonpolar molecules better than polar molecules. An increase in solubility reflects a greater affinity between the solute and the solvent, and acts to oppose the attraction exerted by the carbon. Consequently, any change that increases solubility may be associated with reduced adsorbability. Thus, polar groups (characterized by an affinity for water) usually diminish adsorption from aqueous solutions.

Dissociation constants of weak acids and bases also influence the extent of their adsorption. The reason is that the extent of ionization is determined by the dissociation constant. The adsorption of ionic and molecular forms differ such that the latter are much better adsorbed compared to the former. Greater dissociation constant results in higher amount of the ionic form which is usually not adsorbable on activated carbon.

The presence of substituent groups also affects the adsorbability of organic chemicals (Table 2.4). The introduction of a second or third substituent often extends the influence of the first substituent. The influence of a substituent group depends on the position occupied, for example, ortho, meta, and para [10]. Branched chains are usually more adsorbable than straight chains. On the other hand, an increasing length of chain results in an increase in adsorption capacity [11]. Spatial arrangements of atoms and groups in a molecule influence adsorbability. Aromatic compounds are in general more adsorbable than aliphatic compounds of similar molecular size.

2.2.2.3
pH

Organic molecules form negative ions at high pH values, positive ions at low pH values, and neutral molecules at intermediate pH values. Adsorption of most organic materials is higher at neutral conditions. In general, liquid phase adsorption of organic pollutants by activated carbon is increased with decreasing

Table 2.4 Influence of substituent groups on adsorbability of organic compounds (adapted from [10]).

Substituent group	Influence
Hydroxyl	Generally reduces adsorbability
Amino	Generally reduces adsorbability
Carbonyl	Variable effect depending on host molecule
Double bonds	Variable effect depending on host molecule
Halogens	Variable effect depending on host molecule
Sulfonic	Generally reduces adsorbability
Nitro	Generally increases adsorbability

pH. This results from the neutralization of negative charges at the surface of the carbon at low pH values. Neutralization of negative charges reduces the hindrance to diffusion and leads to more active adsorption sites. The extent of this effect varies with the type and activation technique of activated carbon. The differences in pH values may also arise due to acidic or basic surface functional groups on activated carbon. These groups could be freed by simple contact with distilled water rather than the fixed surface functional groups. An inverse relationship has previously been reported between adsorption capacity and surface acidity [12, 13].

2.2.2.4
Temperature

Adsorption involves specific relations between the properties of activated carbon and the solute. Therefore, the quantitative effects of temperature are not the same with all carbons and solutes. The extent of adsorption should increase with decreasing temperature because the adsorption reactions are exothermic. However, increased temperature also increases the rate of diffusion of the solute through the liquid to the adsorption sites, which eventually leads to an increased adsorption. An important difference in the adsorption of solutes versus gases is found in the role of temperature. An increase in temperature increases the tendency of a gas to escape from the interface and thus diminishes adsorption. However, in adsorption from the liquid, the influence of temperature on solvent affinities is more dominant [10].

2.2.2.5
Porosity of the Adsorbent

The adsorption performance is dependent on the condition of internal surface accessibility. A very important and decisive property of adsorbent materials is the pore structure. The total number of pores and their shape and size determine the adsorption capacity and even the rate of adsorption. The significance of pores in adsorption processes largely depends on their sizes. Most of the solid adsorbents possess a complex structure that consists of pores of different sizes and shapes. Total porosity is usually classified into three groups. According to the IUPAC recommendation, micropores are defined as pores of a width not exceeding 2 nm, mesopores are pores of a width between 2 and 50 nm, and macropores are pores of a width greater than 50 nm. The above classification is widely accepted in adsorption literature. A further classification involves ultra-micropores, which are pores of a width less than 0.7 nm [5].

The mechanism of adsorption on the surface of macropores does not differ from that on flat surfaces. The specific surface area of macroporous adsorbents is very small. Therefore, adsorption on this surface is usually neglected. Capillary adsorbate condensation does not occur in macropores [14]. Macropores do not adsorb small molecules by volume but by surface. For example, phenol adsorption on macropores was reported to be less significant than that on meso- and micropores [15].

In the case of mesopores, the adsorbent surface area has a distinct physical meaning. Mono- and multilayer adsorption takes place successively on the surface of mesopores, and adsorption proceeds according to the mechanism of capillary adsorbate condensation. Specific surface area, pore volume, and pore size distribution are basic parameters characterizing mesopores. Mesopores and macropores play an essential role in the transport of adsorbate molecules through the micropores [14].

The sizes of micropores are usually comparable to those of adsorbate molecules. All atoms or molecules of the adsorbent can interact with the adsorbate species. This is the fundamental difference between adsorption in micropores and larger pores like meso- and macropores. Consequently, adsorption in micropores is essentially a pore-filling process. Therefore, volume is the main controlling parameter for adsorption in micropores [14]. Considering also that most of the total surface area is found in the micropores and the contribution of macropores to the total surface area is very low, the contribution of micropores to adsorption would be expected to be higher. Also, the higher sorption energy in micropores makes them more available adsorption sites for relatively low-molecular-weight organics [15].

2.2.2.6
Chemical Surface Characteristics

The adsorption performance is dependent on chemical surface characteristics of the adsorbent [16]. Heterogeneity of the activated carbon surface significantly contributes to adsorption capacity [13]. Heterogeneity of the carbon surface arises particularly due to surface oxygen groups which, although present in relatively small amounts, affect surface properties such as surface acidity, polarity or hydrophobicity, and surface charge. The presence of heterogeneous oxygen groups on the carbon surface is known to reduce the adsorption capacity due to adsorption of water onto these groups via hydrogen bonding [13]. It has been stated that increasing the number of oxygen-containing surface functional groups increases the polarity of carbon surfaces. Therefore, the selectivity of carbon surface for water increases, and thus adsorbed water clusters may block carbon pores [12]. Such water clusters can prevent pollutant access to hydrophobic regions on the carbon surface, reduce interaction energy between the pollutant and carbon surface, and block pollutant access to micropores [17, 18]. The negative influence of surface oxides has also been attributed to the depletion of the electronic π-band of graphite-like layers resulting in consequent lowering of van der Waals forces of interaction [18] and reduced oxidative polymerization of phenols [19]. On the other hand, surface oxides consisting of carbonyl groups enhance adsorption of aromatic solutes such as phenol and naphthalene [11].

Chemically activated carbons have a pore surface which is less hydrophobic and more negatively charged. Chemical treatment of activated carbon increases the quantity of acidic surface functional groups, whereas thermal treatment results in a decrease in the number of acidic surface functional groups [8]. Adsorption capacities of thermally activated carbons have been found to be relatively higher than those of chemically activated carbons [15, 16]. This is certainly caused by the activation method, which sometimes results in different surface characteristics. Oxygenating the carbon surface decreases its affinity for simple aromatic compounds in the case of chemical activation. Extensive oxidation of carbon surfaces leads to a large decrease in the adsorbed amounts of phenol, nitrobenzene,

benzene, and benzenesulfonate [20]. The formation of functional groups is dependent on activation temperature. For example, phenolic and lactone functional groups appear at activation temperatures up to 400 °C [11].

2.2.3
Kinetics of Adsorption

In an adsorption system, equilibrium is established between the adsorbent and the adsorbate in the bulk phase. The definition of adsorption kinetics is the rate of approach to equilibrium. Adsorption equilibrium does not appear instantaneously because the rate of adsorption is usually limited by the following mass transport mechanisms and depends both on the properties of the adsorbent and the adsorbate [21]. Figure 2.1 shows the main transport mechanisms of an adsorbate to a carbon surface and the adsorption phenomena.

2.2.3.1
Transport Mechanisms
2.2.3.1.1 **Bulk Solution Transport (Advection)** Adsorbates must first be transported from the bulk solution to the boundary layer of water (liquid film)

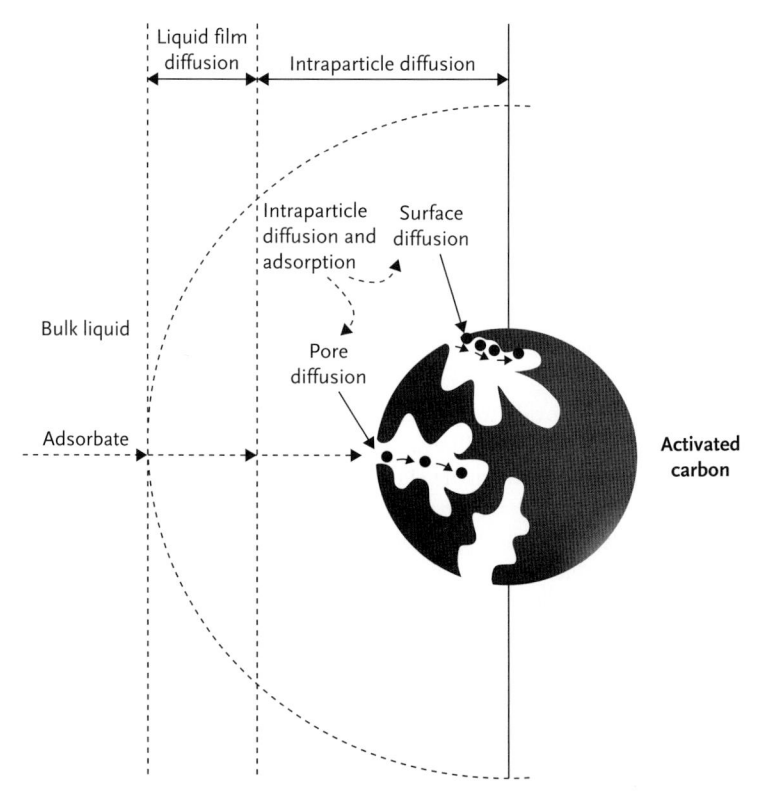

Figure 2.1 External and intraparticle (internal) transport of an adsorbate in an activated carbon particle.

surrounding the activated carbon particle. In quiescent water where the activated carbon is in a suspended state this transport may occur through diffusion. It may also occur through turbulent mixing such as during flow through a packed bed of GAC, or mixing of PAC in a water treatment unit [22].

2.2.3.1.2 External Diffusion In adsorption, there may be an additional resistance to the mass transfer to the carbon surface associated with the transport of adsorbate from the bulk of solution across the stationary layer of water, called the hydrodynamic boundary layer, liquid film or external film, that surrounds the adsorbent particles, as shown in Figure 2.2. Mass transport through this layer occurs by molecular diffusion for which the driving force is the concentration difference. The rate of this diffusion depends on the hydrodynamic properties of the system.

2.2.3.1.3 Intraparticle (Internal) Diffusion Intraparticle diffusion involves the transfer of adsorbate from the surface of a particle such as the activated carbon to sites within the particle. It is independent of hydrodynamic conditions in a system, but depends on the size and pore structure of the particle [23]. Intraparticle diffusion may occur by pore diffusion, which is the molecular diffusion of solutes in fluid-filled pores. Additionally, there is another type of intraparticle diffusion termed the surface or solid diffusion. Surface diffusion is the diffusion of solutes along the adsorbent surface after adsorption takes place [5]. Surface diffusion only occurs if the surface attractive forces are not strong enough to prevent surface mobility of molecules. It is most likely to be significant in porous adsorbents with a high surface area and narrow pores [22]. Generally, 'macropore' and 'micropore' diffusion are termed 'pore diffusion' and 'surface diffusion,' respectively. Pore diffusion and surface diffusion often act in parallel in the interior of the adsorbent particle [24]. As outlined in Chapter 6, in modeling of the adsorption of pollutants from water or wastewater, the surface diffusion is usually assumed to be the dominant intraparticle transport mechanism [23].

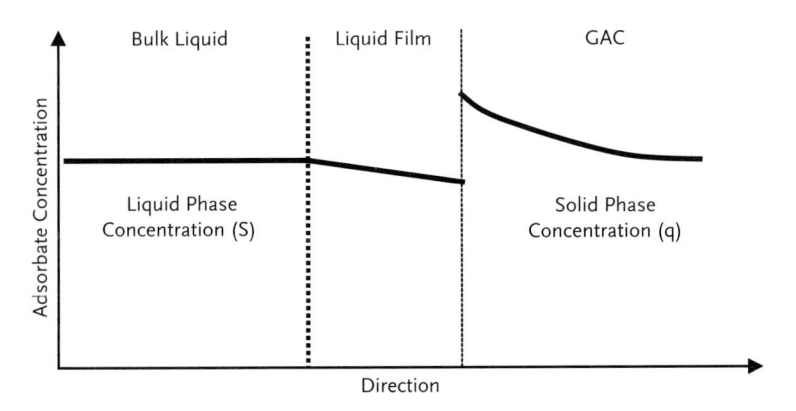

Figure 2.2 Typical concentration profile of an adsorbate in the liquid and solid (carbon) phases in the case of a liquid film over carbon surface.

2.2.3.1.4 **Adsorption** After the transport of the adsorbate to an available site, an adsorption bond is formed. In the case of physical adsorption, the actual physical attachment of adsorbate onto adsorbent is regarded as taking place very rapidly. Therefore, the slowest step among the preceding diffusion steps (advection, external diffusion or internal diffusion), called the rate-limiting step, will control the overall rate at which the adsorbate is removed from solution. However, if adsorption is accompanied by a chemical reaction that changes the nature of the molecule, the chemical reaction may be slower than the diffusion step and may thereby control the rate of removal [20].

2.2.4
Adsorption Equilibrium and Isotherms

Adsorption is usually investigated in stirred batch reactors, where the fluid concentration is uniform and there are no feed or exit streams. In the case of a reversible adsorption, molecules continue to accumulate on the surface until the rate of the forward reaction (adsorption) equals the rate of the reverse reaction (desorption). When these two rates are equal to each other, equilibrium has been reached and no further accumulation will occur [20]. Thus, in batch adsorption, after a sufficiently long time, the adsorbent and the surrounding fluid reach a dynamic equilibrium. Figure 2.2 shows the concentration profile of an adsorbate during adsorption equilibrium.

At the position of equilibrium, there is a defined distribution of adsorbate between the adsorbent and fluid. The partition ratio (w_p) indicates the amount of adsorbate mass adsorbed per amount of initial total mass in solution and is defined as follows:

$$w_p = qM_c/S_0V_0 = (S_0 - S_e)/S_0 \tag{2.1}$$

q: equilibrium expressed in solid phase in mass of adsorbate per mass of the adsorbent (M_s/M_c),
M_c: mass of the adsorbent (M_c),
S_0: initial adsorbate concentration in the fluid phase (M_s/L^3),
S_e: equilibrium adsorbate concentration (M_s/L^3),
V_0: volume of the liquid phase (L^3).

In this equation the subscripts 's' and 'c' denote the 'adsorbate' and 'activated carbon,' respectively.

The adsorption isotherm represents the distribution of the adsorbed material between the adsorbed phase and the solution phase at equilibrium. An isotherm is characteristic of a specific system at a particular temperature. The adsorption isotherm is the primary source of information on the adsorption process. The relation between amount adsorbed, q, and the equilibrium concentration in the liquid phase, S_e, at temperature, T, is called the adsorption isotherm at T [Eq. (2.2)].

$$q = q(S_e) \text{ at temperature } T \tag{2.2}$$

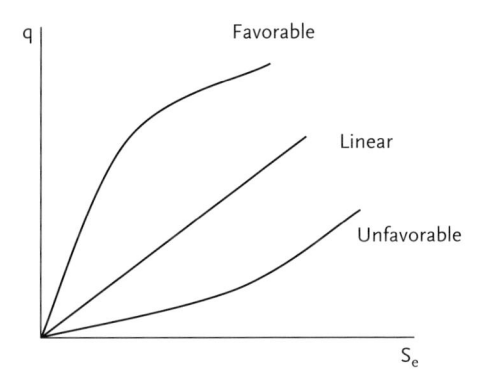

Figure 2.3 Adsorption isotherm curves.

The amount of adsorbed material per unit weight of adsorbent increases with increasing concentration, but not in direct proportion [3]. Generally, an isotherm is favorable if its shape is convex upward, and unfavorable if it is concave upward (Figure 2.3). Any point on an isotherm curve gives the amount of adsorbed contaminant per unit weight of carbon, in other words the adsorption capacity at a particular concentration.

For practical purposes, isotherms can be used to select the appropriate activated carbon type, to estimate the carbon service life, and to test the remaining adsorption capacity of a carbon adsorber in continuous-flow operation. Due to structural and energetic heterogeneity of solid surfaces, adsorption isotherms are characterized by complex analytical equations. This heterogenous characteristic is valid for most of the adsorbents used in practice, including activated carbons. Adsorption isotherms are described in many mathematical forms, some of which are based on a simplified physical picture of adsorption, while others are purely empirical and intended to correlate the experimental data in simple equations. The most widely used mathematical expressions of adsorption equilibrium are the Freundlich, Langmuir, and BET (Brunauer, Emmett, and Teller) isotherms.

The Freundlich isotherm can be used to determine the adsorption capacity of activated carbon over a wide range of concentrations. Adsorbents that follow the Freundlich isotherm equation are assumed to have a heterogeneous surface consisting of sites with different adsorption potentials, and each site is assumed to adsorb molecules. The Freundlich equation as shown below is very widely used for activated carbon applications.

$$q = K_F S_e^{1/n} \tag{2.3}$$

q: adsorption capacity (mg adsorbate adsorbed per g adsorbent), (M_s/M_c)
S_e: equilibrium adsorbate concentration, (M_s/L^3)
K_F: Freundlich exponent
$1/n$: Freundlich slope

The K_F value is an indicator of adsorption capacity. The adsorption intensity, $1/n$, is shown by the slope of the adsorption isotherm curve. An increase in activated carbon dose is more effective at low $1/n$ values than in the case of steeper isotherm curves with high $1/n$ values.

The theoretical Langmuir equation, defined as in Eq. (2.4), is also very widely used for the adsorption isotherms. The Langmuir equation often does not describe adsorption data as accurately as the Freundlich equation. However, the Langmuir expression introduced a clear concept of the monomolecular adsorption on energetically homogeneous surfaces. Adsorbents that exhibit the Langmuir isotherm behavior are supposed to contain fixed individual sites, each of which equally adsorbs only one molecule, forming thus a monolayer, namely, a layer with the thickness of a molecule. The constants in the Langmuir equation have a strictly defined physical meaning in contrast to the ones in the empirical Freundlich equation.

$$q = \frac{q_{max} b S_e}{(1 + b S_e)} \quad (2.4)$$

q_{max}: solute adsorbed per unit weight of adsorbent in forming a complete monolayer on the surface, (M_s/M_c)
b: a constant related to the energy of adsorption.

The linear form of the Langmuir equation is given in Eq. (2.5). From the slope and intercept, b and q_{max} can be calculated.

$$\frac{1}{q} = \frac{1}{q_{max}} + \left(\frac{1}{b q_{max}}\right)\left(\frac{1}{S_e}\right) \quad (2.5)$$

The essential characteristics of the Langmuir equation can be expressed in terms of a dimensionless separation factor R_L, which is defined by the following equation [25]:

$$R_L = 1/(1 + q_{max} \times b \times S_o) \quad (2.6)$$

where S_o is the initial solute concentration in M_s/L^3.

The value of R_L indicates the shape of the isotherm to be either unfavorable ($R_L > 1$), linear ($R_L = 1$), favorable ($0 < R_L < 1$) or irreversible ($R_L = 0$). By using the dimensionless separation factor R_L from the Langmuir equation, adsorption isotherms can be successfully used to have an idea on the reversibility of adsorption.

The Freundlich equation is a special case of heterogeneous surface energies in which the energy term (b) in the Langmuir equation Eq. (2.4) varies as a function of surface coverage (q), strictly due to variations in the heat of adsorption.

The Freundlich equation cannot apply to all equilibrium concentrations. It agrees quite well with the Langmuir equation at moderate concentration ranges. However, unlike the Langmuir equation, it does not reduce to a linear adsorption expression at very low concentrations and does not agree well with the Langmuir equation at very high concentrations. As seen in Eq. (2.4), in the Langmuir equation as S_e increases q increases until the adsorbent approaches saturation. At saturation, q is a constant, independent of further increases in S_e, and the Freundlich equation no longer applies. Also, the Freundlich equation may not apply to concentrations less than the saturation concentration.

With the BET theory, the monolayer adsorption assumption of Langmuir theory was extended to cover multilayer adsorption. The BET model assumes that a number of layers of adsorbate molecules form at the surface. The principal assumption of the BET theory is that the Langmuir equation applies to every adsorption layer. This theory was the first attempt to create a universal explanation of physical adsorption despite many restrictions. The BET equation given below describes the entire course of the isotherm including the areas of monomolecular adsorption, polymolecular adsorption, and capillary condensation. A further assumption of the BET model is that a given layer does not need to be completely formed prior to initiation of subsequent layers. With the additional assumption that layers beyond the first have equal energies of adsorption for adsorption from solution, the BET equation takes the following simplified form:

$$q = \frac{BS_e q_{max}}{(S_s - S_e)[1 + (B-1)(S_e/S_s)]} \quad (2.7)$$

S_S: saturation coefficient of the solute, (M_s/L^3)
B: constant expressing the energy of interaction with the surface.

The linearized form of the BET equation is given below. From the slope and intercept of this equation B and q_{max} can be obtained.

$$\frac{S_e}{(S_s - S_e)q_e} = \frac{1}{Bq_{max}} + \left(\frac{B-1}{Bq_{max}}\right)\left(\frac{S_e}{S_s}\right) \quad (2.8)$$

The BET theory assumes that only the first adsorbed layer is strongly attracted by the surface. The second layer is adsorbed essentially by the first adsorbed layer, and the adsorption propagates from layer to layer to result in a multilayer adsorption. The adsorptive forces are assumed to result from the total of forces usually designated as van der Waals forces, among which are the dispersion forces of London [10]. The BET theory allows the correct calculation of the specific surface area of macroporous adsorbents lacking a great number of micropores. However, incorrect results are obtained in the presence of micropores if the BET theory is applied.

2.2.5
Single- and Multisolute Adsorption

2.2.5.1
Single Solute Adsorption

The simplest description of adsorption refers to single solute adsorption. Assuming that the solution is highly diluted, the interactions between the molecules of the dissolved substance and the solvent can be neglected, and the process can be described as in the case of a single gas adsorption. The above-mentioned adsorption theories are based primarily on single solute adsorption. However, in real applications the presence of a single solute is not common, and a combination of solutes is more commonly encountered in real water and wastewater streams. Therefore, the description of multisolute adsorption is more complex, as summarized below.

2.2.5.2
Multisolute Adsorption

The competitive adsorption of aromatic compounds from a multicomponent mixture on activated carbon has been extensively investigated in the literature [7, 19, 26–30]. The outcome of the competition on activated carbon depends upon the strength of adsorption of the competing solutes, their concentration, and the type of activated carbon.

For the prediction of multisolute adsorption equilibria some important expressions were developed making use of single solute adsorption parameters. In the case of competitive adsorption, the loadings on activated carbon (q) can be modeled by using the single solute adsorption parameters. Several theoretical approaches have been suggested for predicting multicomponent equilibria. Perhaps the simplest is the Ideal Adsorbed Solution Theory (IAST). As the name implies, the adsorbed materials are assumed to form an 'ideal solution' in the solid adsorbent. The theory provides satisfactory predictions under different conditions, and it is therefore more universally applicable as compared to other models [21]. The Ideal Adsorbed Solution (IAS) model, based on the thermodynamics of adsorption, was originally developed to describe competitive adsorption of gases and was later extended by Radke and Prausnitz to aqueous mixtures of solutes [31]. Others workers developed a Simplified Ideal Adsorbed Solution (SIAS) model, which utilizes single solute Freundlich isotherm parameters to simplify the handling of the mathematical equations [26]. According to the SIAS model, the individual loadings in the mixture can be calculated from the equation below:

$$q_i = K_{F'}^{\left(\frac{n'-1}{n'}\right)} [K_{F_i} S_{e_i}^{n_i}]^{1/n'} \left[\sum_N \left(\frac{K_{F_i}}{K_{F'}} S_{e_i}^{n_i}\right)^{1/n'}\right]^{(n'-1)} \qquad (2.9)$$

q_i: solid phase equilibrium concentration of solute i, (M_s/M_c)
K_{F_i}, n_i: single solute empirical Freundlich constants for solute i,
S_{e_i}: liquid phase equilibrium concentration of solute i (M_s/L^3),
n': average value of n_i,
K_F': average value of K_{Fi}

The SIAS model was modified by other researchers and the Improved Simplified Ideal Adsorbed Solution (ISIAS) model was developed in order to account for competition for adsorption sites by adding a competition factor defined as 'a' [7]. This model equation can be seen below.

$$q_i = K_{F'}^{\left(\frac{n'-1}{n'}\right)} \left[\frac{K_{F_i}}{a_i} S_{e_i}^{n_i}\right]^{1/n'} \left[\sum_N \left(\frac{K_{F_i}/a_i}{K_{F'}} S_{e_i}^{n_i}\right)^{1/n'}\right]^{(n'-1)} \qquad (2.10)$$

where

$$K_{F'} = \frac{\sum(K_{F_i}/a_i)}{N} \qquad (2.11)$$

The single solute Freundlich parameters can be used to predict the competitive bisolute loadings using both the SIAS and ISIAS models (Figure 2.4). The SIAS model and its improved form were investigated for phenol and 2-chlorophenol (2-CP) using different activated carbon types: Norit SA4 (thermally activated) and Norit CA1 (chemically activated) [32, 33]. The SIAS model underestimated the 2-CP loadings and overestimated the phenol loadings, particularly those on the thermally activated carbon SA4. Irreversible adsorption and unequal competition for adsorption sites were considered to be the main reasons for the failure of the SIAS model. On the other hand, the ISIAS model, accounting for nonideal competition for adsorption sites, provided a reasonable fit for bisolute data. A modified ISIAS model considering solubilities of competing phenol and 2-CP (when $a_2 = 1$) was not successful in describing the bisolute data for the chemically activated carbon, although it could reasonably predict data for the thermally activated one. It was shown that the carbon activation type was very important for prediction of the adsorption behavior in multisolute competitive adsorption [33]. According to Sontheimer and co-workers, the better adsorbable compound influences the adsorbability of the weakly adsorbable one rather than the reverse [34]. In such a case, adsorption data of the weakly adsorbable compound takes a curved shape [34], as seen in the case of phenol in another study [33]. The shape of 2-CP data was not curved, which exhibited preferential adsorption compared to phenol (Figure 2.4).

The adsorption behavior of multiple substrates is also important, in particular from the aspect of bioregeneration, when activated carbon adsorption is combined with biological processes. For example, it is known that preferentially adsorbed substrates will be irreversible adsorbed and therefore will not be available for

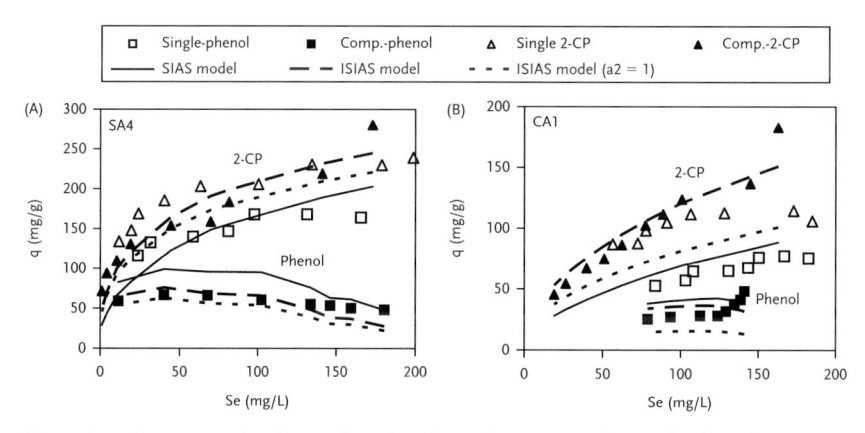

Figure 2.4 Competitive bisolute and single solute adsorption isotherms for phenol and 2-CP with their Simplified Ideal Adsorbed Solution (SIAS) and Improved SIAS (ISIAS) model fits for two different activated carbon types [32].

further biological degradation, whereas the nonpreferred adsorbate may be available for bioregeneration (Chapter 7). Hence, it is very important to determine the adsorption characteristics of pollutant molecules in multisolute adsorption, which actually takes place in real wastewater treatment applications.

2.3
Activated Carbon Reactors in Water and Wastewater Treatment

2.3.1
PAC Adsorbers

PAC is applied in continuous-flow drinking water systems in one of the following process variations: single stage concurrent flow, multistage concurrent flow, and multistage countercurrent flow [21]. The application of PAC involves the contact of the adsorbent with the solution for a prescribed time. Then, a proper separation is needed to obtain a treated effluent.

Also in wastewater treatment PAC may be used, mainly for the removal of nonbiodegradable substances after biological treatment. However, in both water and wastewater treatment the use of PAC contactors proves to be mostly impractical compared to GAC adsorbers. Therefore, in water treatment PAC is temporarily added to existing water treatment units, as described in Chapter 8. In wastewater treatment, PAC is largely incorporated into biological treatment, as described in Chapter 3.

2.3.2
GAC Adsorbers

2.3.2.1
Purpose of Use

Activated carbon adsorbers are applied for surface and groundwater treatment, mainly for the removal of Natural Organic Matter (NOM) and Synthetic Organic Compounds (SOCs). GAC adsorption is also employed for tertiary treatment of municipal and industrial wastewaters (physicochemical treatment following secondary treatment) or as a step in the physicochemical treatment train (coagulation, settling, filtration, GAC adsorption) in place of biological treatment. If applied for tertiary treatment, GAC is used mainly to adsorb organic molecules that are not removed in biological treatment. Pretreatment is usually required before GAC application, such as lime precipitation followed by rapid filtration [35].

In industrial wastewater treatment, GAC may be used to comply with pretreatment standards for discharge to municipal sewage systems or to meet the discharge standards into receiving waters [35].

2.3.2.2
Types of GAC Adsorbers

Continuous-flow GAC systems are generally composed of carbon adsorbers, virgin and spent carbon storage, carbon transport systems, and carbon regeneration systems. The carbon adsorber consists of a lined steel column or a steel or concrete rectangular tank in which the carbon is placed to form a 'filter' bed.

A fixed-bed downflow GAC adsorber is often used to contact water or wastewater with activated carbon. Fixed-bed carbon adsorbers may be operated under pressure or gravity flow. Wastewater is applied at the top of the carbon column, flows downward through the carbon bed, and is withdrawn at the bottom of the column. As the wastewater flows through the column, the contaminants are adsorbed. The carbon is held in place with a drain system at the bottom of the contactor. Provisions for backwash and surface wash of the carbon bed are required to prevent buildup of excessive head loss due to accumulation of solids and to prevent the bed surface from clogging.

The main reactor configurations for GAC adsorption systems are the fixed (packed), expanded, and fluidized beds. A downflow adsorber with low fluid velocities essentially functions as a fixed-bed reactor. Fixed-bed operation provides filtration as well as adsorption. The fixed-bed configuration is suitable in the case of low-strength wastewaters containing little or no suspended solids. Fixed-bed adsorption is commonly applied in the treatment of drinking water, which contains a comparatively lower amount of pollutants. In such cases, fixed-bed GAC adsorbers are operated mainly in downflow mode and under gravity.

On the other hand, the suspended solids present in municipal and industrial wastewaters and the potential for biological growth can lead to problems if fixed beds are used. Passing wastewater upwards at velocities sufficient to expand the bed is a way of minimizing the problems of fouling, plugging, and increasing pressure drop in municipal and industrial wastewater treatment. For this purpose, expanded-, moving-, or fluidized-bed systems may be used. In an expanded-bed system, water or wastewater introduced at the bottom expands the bed when it flows upward. In the moving-bed system, spent carbon is continuously replaced so that no head loss builds up [36]. Another option in water and wastewater treatment is to use fluidized-bed reactors (FBRs), in which the upflowing liquid suspends the fine solid particles, which remain in the reactor. Fluidized beds minimize clogging and unintentional filtration. The carbon can be continuously removed from the reactor while fresh carbon can be added to the top at the same rate. In contrast to fixed-bed operation, this process will eliminate the need to shut down the contactor after exhaustion occurs.

2.3.2.3
Operation of GAC Adsorbers

In a carbon adsorption system, the principal factors which must be taken into consideration are the characteristics of the activated carbon, the operating conditions such as the flow rate and contact time, and the mode of operation (fixed-, expanded-, or fluidized-bed, pumped or gravity flow). Figure 2.5 illustrates a typical fixed-bed GAC adsorber. As shown in Table 2.5, in fixed-bed GAC operation, process variables can be divided into two groups as independent and dependent.

In sizing of GAC columns, the following factors are mainly taken into consideration: the empty-bed contact time (EBCT), the hydraulic loading rate, the carbon depth, and the number of contactors [36]. GAC filters may be designed as

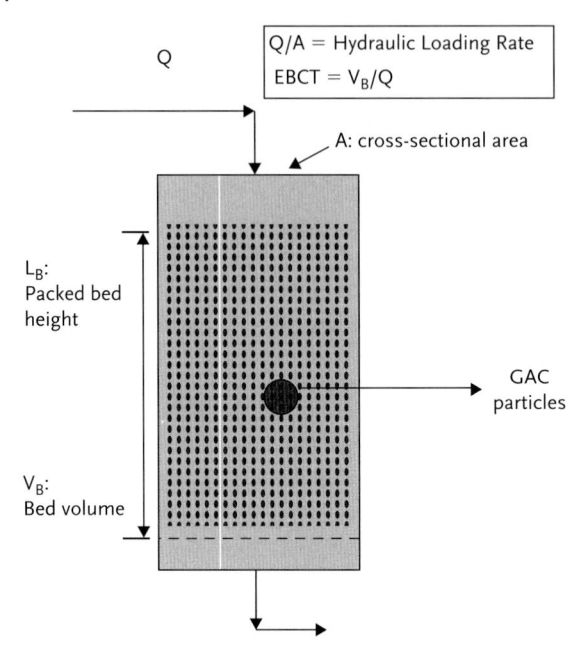

Figure 2.5 Downflow fixed-bed GAC adsorber.

up- or downflow systems consisting of one or more vessels in series or in parallel. The main parameters characterizing the GAC operation are as follows:

2.3.2.3.1 Empty-Bed Contact Time (EBCT)
The EBCT is calculated as the total volume of the carbon bed divided by the flow rate of water:

$$\text{EBCT} = \frac{V_B}{Q} \qquad (2.12)$$

where

V_B = volume of the GAC bed (L^3)
Q = flow rate (L^3/T)

EBCT can also be expressed as follows:

$$\text{EBCT} = \frac{L_B}{Q/A} \qquad (2.13)$$

where

L_B = depth of GAC bed, (L)
A = cross-sectional area of bed, (L^2)
Q/A = hydraulic loading rate (HLR), ($L^3/L^2 \cdot T$)

The EBCT represents the theoretical residence time in the filter in the absence of packing media. Therefore, it is actually a fictive parameter whose meaningfulness has been discussed in water treatment [35].

Table 2.5 Process variables for a fixed-bed reactor (adapted from [35]).

Independent process variables	
Adsorbent Variables	Adsorbent Type, Shape, Size, Porosity, Particle Density, Apparent Density, Surface area, Molasses number, Iodine number, Abrasion number, Dispersion coefficient, Intrinsic permeability
Design Variables	Hydraulic Loading Rate (HLR), Bed Area, Length of Reactor Bed, Contact Time
Operating Variables	Influent concentration of adsorbate, S_0
	Flow
	Adsorbent capacity as a function of S_0 (in terms of isotherm coefficients)
	Hydraulic gradient (head loss per unit length)
	Backwash rate
	Mode of operation (upflow or downflow)
	Carbon Usage Rate (CUR)
Dependent Process Variables	Column performance: $S(z, t)$
	Wave front: $S(z)_t$
	Breakthrough curve: $S(t)_{z=L_B}$

As seen in Table 2.5, in water and wastewater treatment the EBCT varies over a wide range depending on the application, water/wastewater constituents, and desired effluent quality. In practice, the depth of the GAC bed or hydraulic loading can be changed to increase the EBCT. EBCT and system configuration have in turn an effect on carbon usage rate.

2.3.2.3.2 Effective Contact Time Since the GAC bed has a void fraction, the effective contact time (τ) of the fluid is as follows in an adsorber:

$$\tau = \frac{V_B \cdot \varepsilon_B}{Q} \quad (2.14)$$

where

ε_B = void fraction in the carbon bed.

This time is of interest for mass transfer and is responsible for the elimination of organic pollutants from the bulk liquid phase.

2.3.2.3.3 Filter Velocities In water and wastewater treatment the superficial velocity shows the velocity of liquid in an empty bed with a cross-sectional area A and corresponds to the hydraulic loading rate (HLR). It is also termed the surface loading rate [21]:

$$v_F = \frac{Q}{A} \quad (2.15)$$

On the other hand, the interstitial velocity in an adsorber is a more meaningful parameter, and is defined as follows:

$$v_i = \frac{Q}{A\epsilon_B} \tag{2.16}$$

2.3.2.3.4 Filter Operation Time
The filter operation time (t_F) is the time until the replacement of activated carbon with regenerated or new activated carbon.

2.3.2.3.5 Throughput Volume
The throughput volume is the water volume which passes through the reactor during the filter operation time (t_F):

$$V_L = Q\,t_F \tag{2.17}$$

2.3.2.3.6 Bed Volume
The bed volume (BV) shows the normalization of throughput volume to the carbon bed volume V_B. As such, it allows a comparison of removal efficiencies of different adsorbers [21].

$$BV = \frac{V_L}{V_B} = \frac{t_F}{EBCT} \tag{2.18}$$

2.3.2.3.7 Carbon Usage Rate
The breakthrough curve can be used to determine the activated carbon usage rate (CUR), which is defined as the mass of activated carbon required per unit volume of water treated until breakthrough [20]:

$$\text{CUR}\left(\frac{\text{mass}}{\text{volume}}\right) = \frac{\text{mass of GAC in column}}{\text{volume treated to breakthrough}} \tag{2.19}$$

As stated in the literature there are no institutionalized guidelines for design of GAC reactors. The relevant process variables for sizing are HLR and L_B [35]. Further, a common design parameter is EBCT, a derivative of these two variables. The common values of various operational and design parameters are shown in Table 2.6 for a wide range of GAC applications.

2.3.2.4
Breakthrough Curves

As the water or wastewater flows down in a GAC adsorber, contaminants are removed by adsorption. Adsorption of contaminants takes place in Zone A, called the mass transfer zone (MTZ) (Figure 2.6). As a GAC adsorber continues to operate, this zone moves downwards. Above this lies another zone (Zone-Sat) which is saturated with contaminants and is in equilibrium with the influent concentration S_0.

Compared to adsorption kinetics, adsorption dynamics is a more general term that indicates the time evolution of adsorption processes in adsorbers. Inside an adsorber the concentrations of contaminants change with respect to both time and reactor length. The region of rapid change in concentration with the length z is

Table 2.6 Typical values in the design and operation of GAC filters.

	Range	Unit	Reference
Flow rate, Q			
Water treatment	1200–9600	$m^3\ d^{-1}$	[21]
Tertiary treatment of municipal wastewaters	5000–380000	$m^3\ d^{-1}$	[11]
Physicochemical treatment of municipal wastewaters	2200–19000	$m^3\ d^{-1}$	[11]
Industrial wastewater treatment	19–16000	$m^3\ d^{-1}$	[11]
Empty-Bed Contact Time (EBCT)			
General	10–50	min	[35]
Typical	30	min	[35]
Water treatment	5–30	min	[21]
Tertiary treatment of municipal wastewaters	17–50	min	[11]
Physicochemical treatment of municipal wastewaters	20–66	min	[11]
Industrial wastewater treatment	30–540	min	[11]
Hydraulic Loading Rate, HLR			
General	5–25	$m\ h^{-1}$	[35]
Typical	12	$m\ h^{-1}$	[35]
Water treatment	5–15	$m\ h^{-1}$	[21]
Tertiary treatment of municipal wastewaters	7–16	$m\ h^{-1}$	[11]
Physicochemical treatment of municipal wastewaters	6–15	$m\ h^{-1}$	[11]
Carbon Usage Rate (CUR)			
Tertiary treatment of municipal wastewaters	0.12–0.23	kg GAC/m^3	[35]
Physicochemical treatment of municipal wastewaters	0.29–1.04	kg GAC/m^3	[35]
Typical values of the GAC bed			
Bed volume, V_B	10–50	m^3	[21]
Cross-sectional area, A	5–30	m^2	[21]
Length, L_B			
General	3–9	m	[35]
Water treatment	1.8–4	m	[21]
Tertiary treatment of municipal wastewaters	3–10	m	[11]
Physicochemical treatment of municipal wastewaters	2.7–11	m	[11]
Void fraction in GAC bed, ε_B	0.3–0.6	m^3/m^3	[35]
Apparent filter density, ρ_B	430–480	kg solids/m^3 bed	[35]
GAC particle density, ρ_P	0.92–1.5	kg solids/m^3 solid	[35]

termed the wave front. However, in a real operation, mainly the concentration at the exit of the adsorber or the effluent concentration S would be of interest rather than that inside the bed. The plot of effluent concentration divided by influent concentration (S/S_0) as a function of elapsed time, processed volume, or bed volume generates the 'breakthrough curve.' The point at which a predetermined concentration (S_B) appears is defined as the 'breakthrough point.' This point is rather arbitrary and is selected according to purity requirements [37].

Basically, in a GAC adsorber treating water or wastewater, the breakthrough profiles depend on adsorption equilibrium conditions and mass transfer limitations [21]. In the ideal case, when mass transport occurs at infinite rate, meaning that the external and internal mass transfer shown in Figure 2.1 are very fast, the height of the mass transfer zone would approach zero. Under this condition, the ideal breakthrough curve would take the form shown in Figure 2.6 and could be easily estimated using equilibrium data and process conditions only. In such a case the breakthrough time would be identical with the saturation time of the adsorber.

However, as shown in Figure 2.6, in reality, dispersed breakthrough curves are generally observed that deviate from the ideal one. This occurs because the height of the adsorption zone (Zone-A) and the steepness of the breakthrough curves are determined by adsorption kinetics [23]. As explained in Section 2.2.3, adsorption kinetics depends in turn on the rate of external diffusion and intraparticle diffusion (pore or surface diffusion). The mathematical description of external diffusion, intraparticle diffusion, and adsorption rate is presented in detail in Chapter 6. If the mass transport to the inner surface of adsorbent particles in a GAC adsorber is slow, a flat breakthrough curve is observed that deviates more from the ideal breakthrough. In that case, the amount of mass adsorbed until breakthrough differs more from the saturation capacity of an adsorber, and a higher proportion of the GAC in the adsorber remains unused.

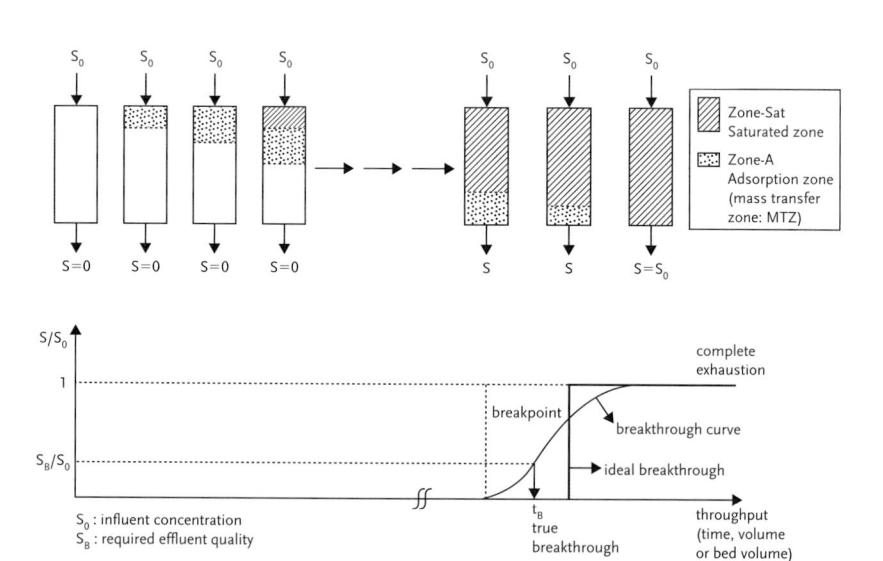

Figure 2.6 Breakthrough characteristics in a fixed-bed GAC adsorber.

The independent variables listed in Table 2.5 influence both the concentration profile inside the adsorber and the shape of the breakthrough curve. For example, compound properties influence the size of the adsorption Zone-A or the so-called mass transfer zone (MTZ). Easily adsorbed compounds show a shorter MTZ than compounds that are difficult to adsorb. Also, the concentration of a compound is important in that higher concentrations result in stronger diffusion gradients and enhanced mass transfer [38]. In addition, the shape of an equilibrium isotherm has a great influence on the breakthrough curve. More detailed information about the inluence of such factors on breakthrough can be acquired from the literature [20, 21, 35].

In a GAC adsorber, the prediction of breakthrough profiles is relatively easy when single solutes are of interest. However, in reality single solute adsorption is seldom if ever the case. In contrast to this, when multiple substances are removed from water or wastewater by adsorption, competition takes place for adsorption sites on activated carbon, as exemplified by the bisolute isotherm curves in Figure 2.4. In a GAC adsorber where adsorption is the only removal mechanism, the shape of breakthrough curves is strongly affected by the presence of such compounds of different adsorbability. In a fixed-bed operation this leads to the separation of weakly and strongly adsorbing components into their respective mass transfer zones. Compared to strongly adsorbing compounds, weakly adsorbing ones are held in deeper parts of the column. In accordance with this, nonadsorbable compounds that cannot be held on activated carbon migrate rapidly through the filter bed and emerge in the effluent stream [21]. Therefore, in reality, breakthrough profiles may strongly deviate from the profile shown in Figure 2.6.

If biodegradation is the additional removal mechanism in a GAC adsorber, the shape of the breakthrough curve can change entirely. For the case of wastewater treatment, this type of information is basically presented in Chapter 6, which addresses the modeling of bioactive GAC adsorbers (BAC reactors). However, also in drinking water treatment the biological activity inside GAC adsorbers alters the breakthrough profiles. This type of information is mainly presented in Chapters 9 and 11, which focus on substrate removal in BAC reactors and their modeling, respectively.

2.4
Activated Carbon Regeneration and Reactivation

The adsorptive capacity of carbons gradually deteriorates upon usage. When the carbon adsorber effluent quality reaches minimum water quality standards, the exhausted carbon has to be regenerated, reactivated, or disposed of. Activated carbon involves a large capital investment and high operating costs both in batch and column operation, mainly due to the need for regeneration. The regeneration of spent adsorbents is the most difficult and expensive part of adsorption technology. It accounts for about 75% of total operating and maintenance cost for a fixed-bed GAC operation. Regeneration involves removal of the contaminants from the carbon without destroying the contaminants. Reactivation means destroying the contaminants and reactivating the carbon, which usually occurs at very high temperatures. If regeneration of the spent carbon is not feasible, or the carbon is irreversibly contaminated by the adsorbed substance, the spent carbon should be disposed of. The increased use of activated carbons has been accompanied by the development and broader application of a number of new regeneration processes. The most commonly used conventional techniques are thermal, chemical, and electrochemical regeneration.

Carbon regeneration is accomplished primarily by thermal means. Thermal volatilization usually refers to the process of drying, thermal desorption, and high temperature heat treatment in the presence of a limited amount of oxidizing gases such as water vapor, flue gas, and oxygen. Organic matter within the pores of the carbon is oxidized and thus removed from the carbon surface. The two most widely used regeneration methods involve the use of rotary kiln and multiple hearth furnaces. Approximately 5–10% of the carbon is destroyed in the regeneration process or lost during transport and must be replaced with virgin carbon. The capacity of the regenerated carbon is slightly less than that of virgin carbon. Repeated regeneration degrades the carbon particles until equilibrium is eventually reached providing predictable long term system performance.

Pressure swing adsorption is a common regeneration process, involving the application of low pressure to desorb the gaseous contaminants from the solid phase. Other common processes involve the use of steam, microwaves, embedded heaters, or a hot inert gas for volatilization of adsorbed contaminants. These processes, also called thermal swing regeneration, are most convenient when the adsorbed species are volatile organic compounds (VOCs) having low vapor pressures. The contaminants are then recovered in liquid form after a condensation step. During the regeneration process, the contaminants are desorbed and a waste stream is produced. The regeneration process should therefore be followed by waste treatment [5].

Activated carbon regeneration (or reactivation) is achieved using thermal destruction/scrubbing systems in most cases. Under these conditions, the organic contaminants are destroyed during the regeneration process at a high temperature (typically in excess of 800°C). Carbon losses during reactivation processes can be held at 3–15% [5]. Regeneration by wet air oxidation (WAO) is a liquid-phase reaction in water using dissolved oxygen to oxidize sorbed contaminants in a spent carbon slurry. Regeneration is conducted at moderate temperatures of 205–260°C and at pressures between 50 and 70 bars. The process converts organic contaminants to CO_2, water, and short-chain organic acids. Sorbed inorganic constituents such as heavy metals are converted to stable, nonleaching forms that can be separated from the regenerated carbon.

Regeneration of GAC is also carried out in electrochemical batch reactors consisting of two electrodes, anode and cathode, and a reference electrode submerged in electrolyte. Other physicochemical regeneration methods include aqueous solution extraction using NaOH, extraction with conventional organic solvents, and extraction with supercritical carbon dioxide [21].

GAC regeneration has several advantages including reduced solid waste handling problems due to spent carbon and consequent reduction in carbon costs. GAC regeneration also has some disadvantages associated with air pollution. Air emissions from the furnace of a thermal regeneration unit may contain particulate matter and volatile organics. In addition, carbon monoxide may be formed as a result of incomplete combustion. Therefore, afterburners and scrubbers are usually needed to treat exhaust gases [36]. Moreover, thermal regeneration is energy intensive, that is, it requires large amounts of energy.

Activated carbon adsorbers must be equipped with proper mechanisms for carbon removal and replacement of spent carbon with virgin or regenerated carbon. Spent, regenerated, and virgin carbon is typically transported hydraulically by pumping as a slurry.

Another method for the regeneration of activated carbon is biologically induced regeneration, termed the bioregeneration, which is extensively discussed in Chapter 7.

References

1 Teng, H., Ho, J.A., and Hsu, Y.F. (1997) Preparation of activated carbons from bituminous coals with CO_2 activation – influence of coal oxidation. *Carbon*, **35**, 275–283.

2 Khalili, N.R., Campbell, M., Sandi, G., and Golas, J. (2000) Production of micro- and mesoporous activated carbon from paper mill sludge I. Effect of zinc chloride activation. *Carbon*, **38**, 1905–1915.

3 Suzuki, M. (1990) *Adsorption Engineering* Elsevier, Tokyo, Japan.

4 Cheremisinoff, N.P. (2002) *Handbook of Water and Wastewater Treatment Technologies*, Butterworth-Heinemann, Woburn, MA, USA.

5 Inglezakis, V.J. and Poulopoulos, S.G. (eds) (2006) *Adsorption, Ion Exchange and Catalysis: Design of Operations and Environmental Applications*, Elsevier, Amsterdam, The Netherlands.

6 Yonge, D.R., Keinath, T.M., Poznanska, K., and Jiang, Z.P. (1985) Single-solute irreversible adsorption on granular activated carbon. *Environ. Sci. Technol.*, **19**, 690–694.

7 Yonge, D.R. and Keinath, T.M. (1986) The effects of non-ideal competition on multi-component adsorption equilibria. *J. Water Pollut. Control Fed.* **58**, 77–81.

8 Julien, F., Baudu, M., and Mazet, M. (1998) Relationship between chemical and physical surface properties of activated carbon. *Water Res.*, **32** (11), 3414–3424.

9 Weber, W.J. (1972) *Physicochemical Processes for Water Quality Control*, Wiley-Interscience, New York, USA.

10 Hassler, J.W. (1963) *Activated Carbon*, Chemical Publishing Company, Inc., New York, USA.

11 Hung, Y., Lo, H.H., Wang, L.K., Taricska, J.R., and Li, K.H. (2005) Granular activated carbon adsorption, in *Handbook of Environmental Engineering – vol. 3: Physicochemical Treatment Processes* (eds L. K. Wang, Y. Hung, and N.H. Shammas), The Humana Press Inc., Totowa, New Jersey, USA.

12 Karanfil, T. and Kilduff, J.E. (1999) Role of granular activated carbon surface chemistry on the adsorption of organic

compounds. 1. Priority pollutants. *Environ. Sci. Technol.*, , **33**, 3217–3224.
13 Ahnert, F., Arafat, H.A., and Pinto, N.G. (2003) A study of the influence of hydrophobicity of activated carbon on the adsorption equilibrium of aromatics in non-aqueous media. *Adsorption*, **9**, 311–319.
14 Dabrowski, A. (2001) Adsorption – from theory to practice. *Advances in Colloid and Interface Science*, **93**, 135–224.
15 Aktaş, Ö. and Çeçen, F. (2006) Effect of type of carbon activation on adsorption and its reversibility. *J. Chem. Technol. Biotechnol.*, **81** (1), 94–101.
16 Jonge, R.J. de, Breure, A.M., and van Andel, J.G. (1996) Reversibility of adsorption of aromatic compounds onto powdered activated carbon. *Water Res.*, **30** (4), 883–892.
17 Knappe, D.R.U., Li, L., Quinlivan, T.B., and Wagner, T.B. (2003) *Effects of Activated Carbon Characteristics on Organic Contaminant Removal*, AWWA Res. Foundation, Denver, USA.
18 Puri, B.R. (1980) Carbon adsorption of pure compounds and mixtures from solution phase, in *Activated Carbon Adsorption of Organics from the Aqueous Phase*, vol. **1** (eds I.H. Suffet and M.J. McGuire), Ann Arbor Science Publishers, Inc., Michigan, pp. 353–376.
19 Garcia-Araya, J.F., Beltran, P. Alvare, P. and Masa, F.J. (2003) Activated carbon adsorption of some phenolic compounds present in agroindustrial wastewater. *Adsorption*, **9**, 107–115.
20 Snoeyink, V. L. and Summers, R.S. (1999) Adsorption of organic compounds, in *Water Quality and Treatment*, 5th edn (ed. R.D. Letterman), McGraw-Hill, New York, NY.
21 Sontheimer, H., Crittenden, J.C. and Summers, S. (1988) *Activated Carbon for Water Treatment*, DVGW-Forschungsstelle am Engler-Bunte Institut der Universität Karlsruhe (TH), G.Braun GmbH, Karlsruhe, Germany.
22 Thomas,W.J. and Crittenden, B. (1998) *Adsorption Technology and Design*, Elsevier Science & Technology Books, ISBN 0750619597.
23 Worch, E. (2009) Adsorptionsverfahren in der Trinkwasseraufbereitung. *Wien. Mitt.*, Band **212**, pp. 207–230.
24 Worch, E. (2008) Fixed-bed adsorption in drinking water treatment: a critical review on models and parameter estimation. *J. Water Supply: Res. Technol.-AQUA*, **57** (3), 171–183.
25 Al-Degs, Y., Khraisheh, A.M., Allen, S.J., and Ahmad, M.N. (2000) Effect of carbon surface chemistry on the removal of reactive dyes from textile effluent. *Water Res.*, **34** (3), 927–935.
26 DiGiano, F.A., Baldauf, G., Frick, B., and Sontheimer, H. (1980) Simplifying the description of competitive adsorption for practical application in water treatment, in *Activated Carbon Adsorption of Organics from the Aqueous Phase*, vol. **1** (eds I.H. Suffet and M.J. McGuire), Ann Arbor Science Publishers, Inc., Michigan, USA, pp. 213–228.
27 Fritz, W., Merk, W., Schlünder, E.U., and Sontheimer, H. (1980) Competitive adsorption of dissolved organics on activated carbon, in *Activated Carbon Adsorption of Organics from the Aqueous Phase*, vol. **1** (eds I. H. Suffet and M.J. McGuire), Ann Arbor Science Publishers, Inc., Michigan, USA, pp. 193–211.
28 Singer, P.C. and Yen, C.Y. (1980) Adsorption of alkyl phenols by activated carbon, in *Activated Carbon Adsorption of Organics from the Aqueous Phase*, vol. **1** (eds I.H. Suffet and M.J. McGuire), Ann Arbor Science Publishers, Inc., Michigan, USA, pp. 167–189.
29 Knettig, E., Thomson, B.M. and Hrudey, S.E. (1986) Competitive activated carbon adsorption of phenolic compounds. *Environ. Pollut.B*, **12**, 281–299.
30 Garner, I.A., Watson-Craik, R. Kirkwood, E. and Senior, E. (2001) Dual solute adsorption of 2,4,6-trichlorophenol and N-(2-(2,4,6-trichlorophenoxy) propyl amine on to activated carbon. *J. Chem. Technol. Biotechnol.*, **76**, 932–940.
31 Radke, C.J. and Prausnitz, J.M. (1972) Thermodynamics of multi-solute adsorption in dilute liquid solution. *AIChE J.*, **18**, 761–768.

32 Aktaş, Ö. (2006) Bioregeneration of Activated Carbon in the Treatment of Phenolic Compounds. Ph.D. Thesis, Bogazici University, Istanbul, Turkey.
33 Aktaş, Ö. and Çeçen, F. (2007) Competitive adsorption and desorption of a bi-solute mixture: effect of activated carbon type. *Adsorption*, **13** (2), 159–169.
34 Sontheimer, H., Frick, B.R., Fettig, J., Hörner, G., Hubele, C., and Zimmer, G. (1985) *Adsorptionsverhalten zur Abwasserreinigung*, DVGW-Forschungsstelle am Engler-Bunte Institut der Universität Karlsruhe (TH), G. Braun GmbH, Karlsruhe, Germany.
35 Hendricks, D. (2006) *Water Treatment Unit Processes: Physical and Chemical*, CRC Press, USA.
36 EPA (2000) Wastewater technology fact sheet – GAC adsorption and regeneration, EPA 832-F-00–017.
37 EPA (1991) Engineering Bulletin – granular activated carbon treatment, EPA/540/2–91/024.
38 Suthersan, S.S. (1999) Pump and treat systems, in *Remediation Engineering: Design Concepts* (eds S. S. Suthersan and B. Raton), CRC Press LLC.

3
Integration of Activated Carbon Adsorption and Biological Processes in Wastewater Treatment

Ferhan Çeçen and Özgür Aktaş

3.1
Secondary and Tertiary Treatment: Progression from Separate Biological Removal and Adsorption to Integrated Systems

Increasingly, concerns are being raised about the presence of various pollutants and micropollutants (pollutants detected in the concentration range of $ng\,L^{-1}$ up to $\mu g\,L^{-1}$ in the environment) in domestic wastewaters, industrial wastewaters, landfill leachates, and municipal wastewaters from domestic and other sources. Correspondingly, stringent discharge requirements are currently in place for such wastewaters. Among the various types, organic micropollutants have received special attention in the last decades. Organic micropollutants, including hormones, pharmaceuticals, personal care products (PCPs), pesticides, brominated flame retardants, industrial products, household products, disinfectants, and antiseptics, are often alien to the biota; they are therefore also referred to as xenobiotic organic compounds (XOCs). Conventional treatment processes often lead to inadequate removal of these compounds, and the reuse of treated wastewater or its discharge into water bodies often raises concern about direct or indirect risks to human health and the environment. Therefore, elaborate wastewater treatment methods have to be applied for an adequate elimination of these compounds from the water phase. For this pupose, the use of activated carbon adsorption, both in powdered (PAC) and granular (GAC) form is becoming more widespread. The decision when and where to use activated carbon depends on the characteristics and flow rate of a wastewater. As well as providing effective adsorption of organic micropollutants, activated carbon can bring about many positive effects in wastewater treatment, as discussed in later sections of this chapter.

Activated carbon is widely utilized by a number of industries for the treatment of their effluents and sanitary and hazardous landfill leachates. In each case, the main aim is the removal of specific compounds down to the level acceptable for discharge into Publicly Owned Treatment Works (POTWs) or into receiving waters. Activated carbon adsorption also finds application in sewage treatment

Activated Carbon for Water and Wastewater Treatment: Integration of Adsorption and Biological Treatment.
First Edition. Ferhan Çeçen and Özgür Aktaş
© 2011 WILEY-VCH Verlag GmbH & Co. KGaA, Weinheim.
Published 2011 by WILEY-VCH Verlag GmbH & Co. KGaA

plants (STPs), particularly in those that receive a significant input of industrial wastewater.

The idea of integrated adsorption and biological removal has its roots in both secondary and tertiary treatment. This approach has received great attention since the 1970s as a more effective pollutant removal method. When activated carbon was initially introduced, mainly in tertiary treatment of municipal or industrial wastewaters, the focus was on the adsorptive properties of this material. However, it has been recognized that biological activity could develop inevitably in GAC adsorbers receiving secondary effluents. As a result of this biological activity, GAC adsorbers were gradually converted into attached-growth (biofilm) reactors, often referred to as biological activated carbon (BAC) reactors. Similarly, it was recognized that in secondary treatment of industrial wastewaters, substrate removal could be enhanced by the addition of PAC to suspended-growth reactors.

Integrated adsorption and biological removal can be realized using both PAC or GAC (Figures 3.1 and 3.2). At the current state of development, integrated systems are basically divided into two based on the configuration of the biomass:

- Suspended-growth biological systems receiving activated carbon dosage, these being the Powdered Activated Carbon Treatment (PACT) process and the PAC added membrane bioreactor (PAC-MBR) process.
- Attached-growth (biofilm) biological systems containing GAC media, namely the Biological Activated Carbon (BAC) process.

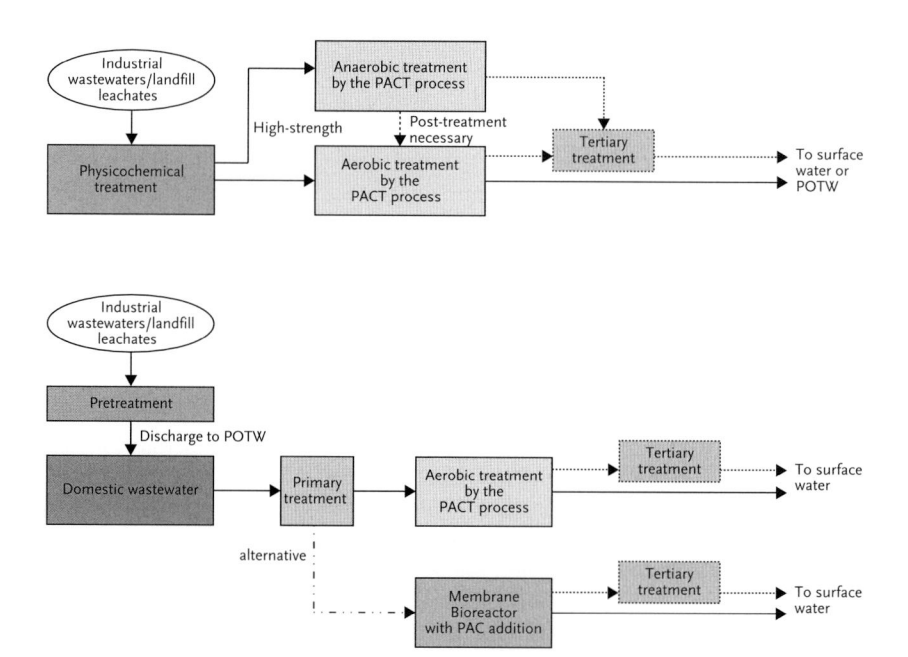

Figure 3.1 Use of PAC in secondary treatment of wastewaters.

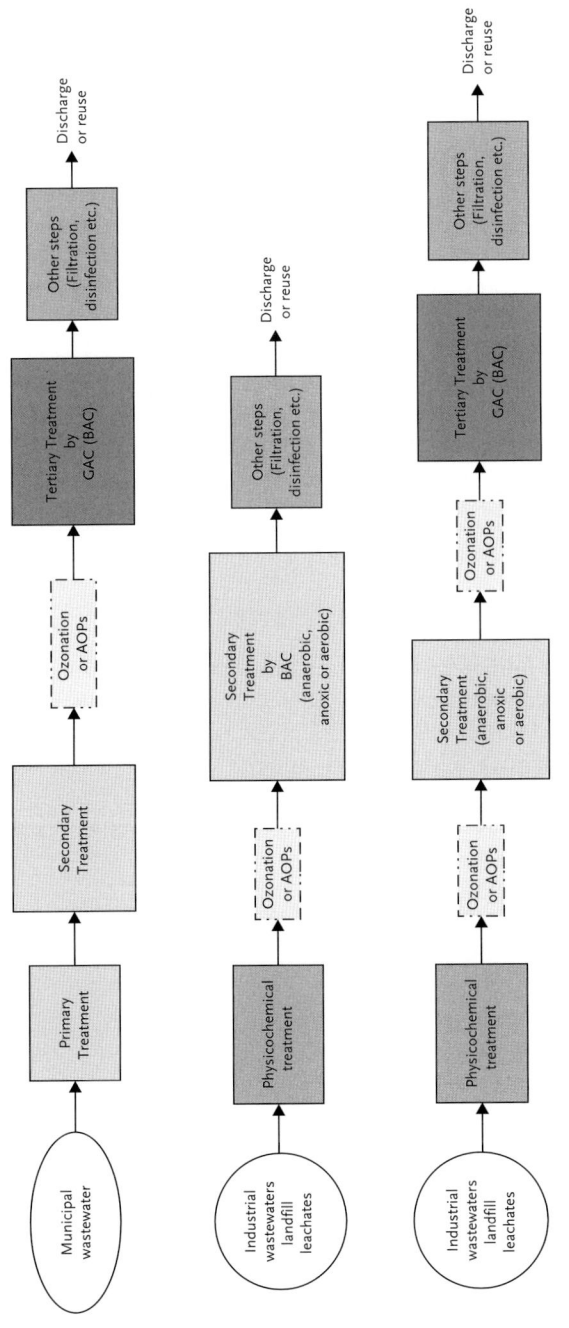

Figure 3.2 Use of GAC in tertiary and secondary treatment of wastewaters.

3.1.1
Activated Carbon in Secondary Treatment

3.1.1.1
PAC

If microorganism growth is appreciable, which is often the case in the treatment of wastewaters that contain a high concentration of biodegradable organics, technologies in which the activated carbon is held in suspension are generally more favorable than those in which activated carbon is in stationary form. In that regard, PACT is the major suspended-growth process, in which PAC is directly added into a biological tank for simultaneous adsorption and biological removal.

The PACT process is often employed in secondary treatment of industrial wastewaters and landfill leachates. PACT varieties are aerobic PACT, anaerobic PACT, and sequential anaerobic–aerobic PACT (Figure 3.1). Aerobic PACT is mainly applied to relatively dilute wastewaters whereas anaerobic PACT is introduced in the treatment of highly concentrated ones. Depending on the strength of the wastewater, the anaerobic PACT may sometimes precede the aerobic PACT. Detailed information about the PACT process is provided in Section 3.4.

In more recent MBR applications, PAC is also used as a material assisting the operation (Figure 3.1). Detailed information about the positive effects of PAC on MBR operation are discussed in Section 3.6.7.

3.1.1.2
GAC

GAC adsorbers can also be introduced in secondary treatment as a substitute for conventional biological processes such as activated sludge. This application is mainly chosen for industrial wastewaters and landfill leachates. Since the effluent of primary treatment contains a relatively high amount of biodegradable substances, the exposure of GAC surface to such effluents leads to rapid colonization by microorganisms. Finally, a GAC adsorber is transformed into a biofilm reactor, referred to as a BAC reactor (Figure 3.2).

Since biomass growth is often gradual, the transition from GAC to BAC operation cannot be strictly defined. Therefore, other descriptions such as GAC/BAC or B(GAC) are also commonly used.

3.1.2
Activated Carbon in Tertiary Treatment

3.1.2.1
PAC

PAC adsorption can also be introduced as a tertiary treatment step for the removal of organics that are still present in municipal or industrial wastewaters after secondary treatment. However, the separation of PAC from the treated wastewater is regarded as a challenging task. This separation is conducted either by settling followed by sand filtration or membrane filtration, which requires additional energy [1]. Moreover, elimination of organics is slower in PAC adsorption compared to other tertiary treatment alternatives such as ozonation and Advanced Oxidation Processes (AOPs). Hence, in PAC units having a hydraulic retention time (HRT) of the order of minutes, equilibrium is usually not established between PAC surface and pollutants, leading to an inefficient use of PAC capacity.

In order to overcome these disadvantages, a better choice would be the concurrent addition of PAC in secondary treatment, namely the PACT process (Section 3.4). In this process, PAC is retained in the system for a period of time equal to the sludge retention time (SRT), also called the sludge age. This high retention in the system allows the full exploitation of PAC capacity. In addition, as discussed in the following sections, the simultaneous presence of microorganisms and PAC in the reactor brings about synergistic effects. Further, in a PACT process PAC leaves the system with the wasted sludge and often no separation procedures are needed.

3.1.2.2
GAC

GAC adsorption is often introduced in tertiary treatment of municipal and industrial wastewaters and landfill leachates where a smaller fraction of organic material has to be removed by biodegradation and clogging problems are not encountered (Figure 3.2). The main purpose is generally the removal of trace organics from secondary effluents. If a GAC adsorber receives secondary effluents containing also microorganisms, the conversion of a GAC adsorber into a BAC reactor is very probable.

In tertiary GAC adsorption systems, wastewater flows through either fixed, expanded, or fluidized beds. Using such GAC beds, activated carbon separation problems encountered in PAC adsorption are overcome. At the same time, the passage of wastewater through GAC beds allows the full exploitation of carbon capacity.

Following secondary treatment of municipal or industrial wastewaters, ozonation and AOPs are the common alternatives for the elimination of residual pollutants. GAC adsorption is often preceded by such oxidative techniques, most frequently by ozonation. These techniques usually have the effect of increasing the biodegradable fraction in wastewater. Consequently, they enhance the conversion of GAC into BAC.

3.2
Fundamental Mechanisms in Integrated Adsorption and Biological Removal

3.2.1
Main Removal Mechanisms for Organic Substrates

In a biological system involving activated carbon, removal of bulk organic matter as well as organic micropollutants can take place principally by the mechanisms biodegradation/biotransformation, sorption onto sludge (biosorption), adsorption onto activated carbon, and others such as chemical transformation and volatilization. These processes are described in the following sections.

3.2.1.1
Biodegradation/Biotransformation
3.2.1.1.1 Fractionation of Organic Matter in Wastewater

Bulk organic matter Wastewaters are composed of different types of organic compounds such as carbohydrates, proteins, and lipids that constitute the bulk organic matter characterized by the sum parameters COD, BOD, and TOC.

As shown in Figure 3.3 the organic matter has two major components: the nonbiodegradable or inert organics, and the biodegradable organics [2]. In a biological system, such as activated sludge, nonbiodegradable soluble organic matter in the influent passes unchanged in form. On the other hand, inert particulate organic matter (nonbiodegradable particulates) may be removed from the system through wasting of sludge. The biodegradable organic matter can be further differentiated as readily or slowly biodegradable. The readily biodegradable fraction is mainly composed of soluble compounds such as volatile fatty acids, simple carbohydrates, alcohols, amino acids which can be directly utilized by microorganisms. However, some of the biodegradable organics need to undergo extracellular hydrolysis before they can be utilized by microorganisms. Such organics amenable to hydrolysis are further classified as rapidly and slowly hydrolysable organics.

Organic micropollutants In addition to bulk organic matter, many organic micropollutants are present in wastewaters that cannot be identified by the sum parameters BOD, COD, and TOC since they concentrations are at ng L^{-1} or µg L^{-1} levels. In the last decades, significant efforts have been directed toward detection and removal of micropollutants since they are known or suspected to have adverse effects on human health as well as on the environment.

The biodegradation potential of any compound depends largely on its inherent physicochemical properties. A pollutant may often not be amenable to biodegradation because of its molecular structure. Biodegradability of a pollutant is also related to its solubility in the aqueous phase. A high solubility is a factor increasing the bioavailability of the pollutant to microorganisms. Environmental conditions,

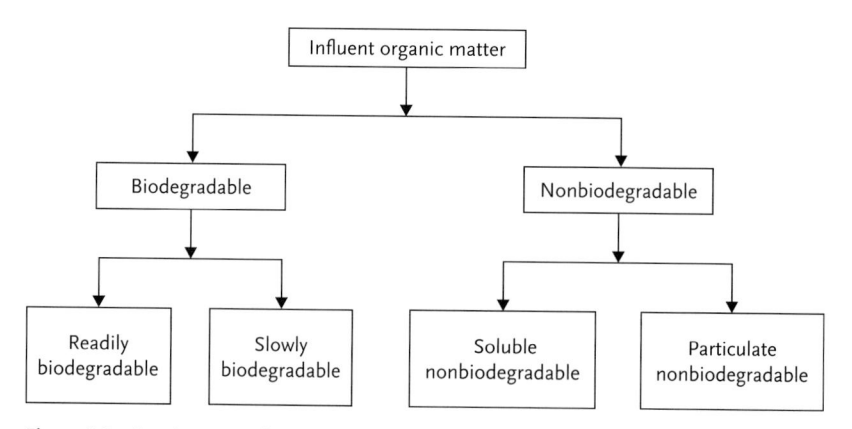

Figure 3.3 Fractionation of organic matter in wastewaters.

such as lack of nutrients, minerals, oxygen may also limit the biodegradation potential of a pollutant.

Biodegradation of pollutants depends also on the presence, type, and activity of microorganisms. Biodegradation of XOCs is often hindered since there may be an insufficient number of microorganisms capable of their biodegradation. Acclimation of biomass to specific pollutants has usually a positive influence on biodegradation. However, even upon acclimation some xenobiotics are often incompletely mineralized, meaning that they are transformed into organic metabolites only.

3.2.1.1.2 Main Pathways for Biological Removal of Organic Substrates

Removal as a primary substrate The bulk organic matter in a wastewater is often removed by normal metabolism where the different types of substrates serve the energy and growth requirements of microorganisms. Also, some of the micropollutants may be utilized as growth and energy substrates.

Oxygen is essential for decomposition of organic matter in aerobic biological treatment processes. As shown in Eq. (3.1), the heterotrophic bacterial culture carries out the conversion of organic substrates (CHONS), and decomposes them mainly into carbon dioxide and water.

$$CHONS + O_2 + \text{nutrients} \rightarrow CO_2 + H_2O + NH_3 + C_5H_7NO_2 + \text{other end products} \quad (3.1)$$

The above reaction shows that also new bacterial cells ($C_5H_7NO_2$) are produced. However, when the concentration of available organic substrate is at a minimum, the microorganisms are forced to metabolize their own protoplasm. This phenomenon, termed endogenous respiration, is shown as follows:

$$C_5H_7NO_2 + 5O_2 \rightarrow 5CO_2 + NH_3 + 2H_2O + \text{energy} \quad (3.2)$$

Biodegradation of organic matter can take place also under anoxic and anaerobic conditions. Under the anoxic conditions, the main interest relies on the reduction of nitrate into nitrogen gas through denitrification whereby organic matter acts as an electron donor. In anaerobic processes, organic matter is converted into a variety of end products, consisting mainly of methane and carbon dioxide.

Removal as a secondary substrate If the concentration of a substrate (pollutant) is very low in biological treatment systems, no net growth of biomass can occur using this substrate since the decay rate of microorganisms just balances the growth rate. The bulk substrate concentration at which this phenomenon is observed is defined as S_{min} (Eq. (3.3)).

$$S_{min} = K_s \frac{b}{Y k_{max} - b} \quad (3.3)$$

where

K_s = half-velocity constant (M_s/L^3),
b = decay rate (1/T),

Y = yield coefficient (M_x/M_s),
k_{max} = maximum specific substrate utilization rate (μ_{max}/Y), ($M_s/M_x \cdot T$).

In the above notation, the subscripts s and x represent substrate and biomass, respectively.

Below the S_{min} concentration no biomass can be kept in the system [3]. The S_{min} concept is relevant in water and wastewater treatment and in soil and groundwater bioremediation.

However, bacteria can still consume a pollutant if there is another substrate at a concentration high enough to support the growth of existing population. In such cases, the compound at the higher concentration is denoted as the primary substrate whereas the pollutant at the lower concentration becomes the secondary substrate. In contrast to cometabolism, the secondary substrate serves as a source of carbon and energy.

Cometabolic removal of substrate If a substrate (pollutant) is entirely resistant to biodegradation under the specific condition, there is still a possibility of removal by cometabolism. Cometabolism is the biological transformation of a nongrowth (cometabolic) substrate by bacteria through enzymes which can only be induced in the presence of a growth substrate (primary substrate) providing energy for cell growth and maintenance. Since the cometabolic transformation of a substrate yields no carbon and energy benefits to the cells and cannot induce production of enzymes, a growth substrate must be supplied to the microorganism, at least periodically, for the growth of new cells that produce the necessary enzymes.

Cometabolic degradation of various chlorinated organics by phenol, toluene or propane oxidizers is widely reported (Figure 3.4). For example, phenol serves as an ideal growth substrate in cometabolic transformation of chlorinated phenols. Phenol is used by phenol oxidizers to induce the necessary enzymes for the degradation of chlorinated phenols that have a structural analogy with phenol. As exemplified in Figure 3.5, in the presence of phenol (bisolute case), 2-chlorophenol (2-CP), which is normally nonbiodegradable under aerobic conditions (single solute case), acts as a cometabolic substrate and is efficiently dechlorinated [4].

Also, easily biodegradable compounds such as glucose and dextrose can serve as primary or growth substrates in cometabolism. Aerobic cometabolism of chlorinated compounds is also possible using the inorganic ammonia as primary substrate [5]. Cometabolism is also commonly encountered under anaerobic conditions.

Cometabolic degradation of (micro)pollutants is often encountered in the treatment of domestic, municipal, and industrial wastewaters, sanitary and hazardous landfill leachates, in bioremediation of polluted soils and groundwaters, and in drinking water treatment.

3.2.1.1.3 Biological Removal of Micropollutants

Due to the relatively low concentrations of organic micropollutants in wastewater treatment systems, biodegradation is often modeled using a pseudo first order degradation expression [6].

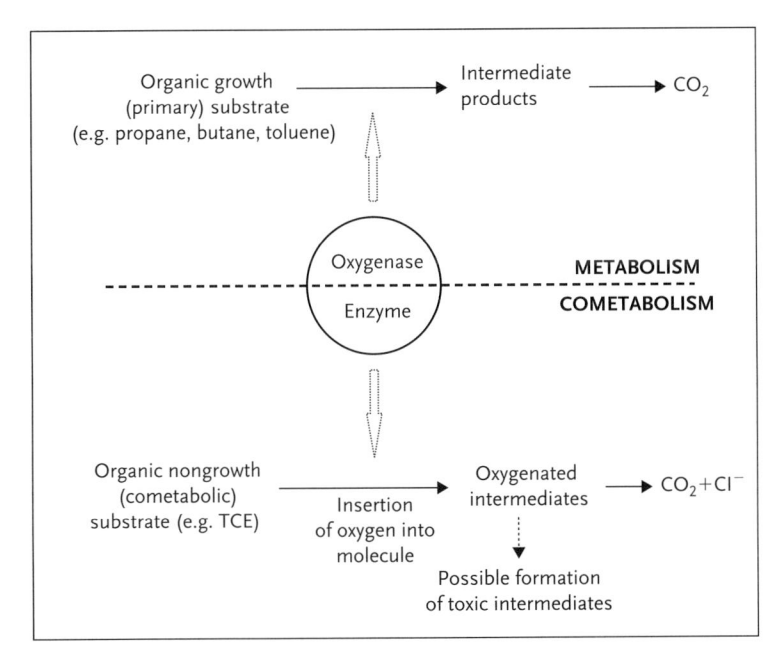

Figure 3.4 Cometabolic removal of nongrowth substrates in the presence of an organic growth substrate.

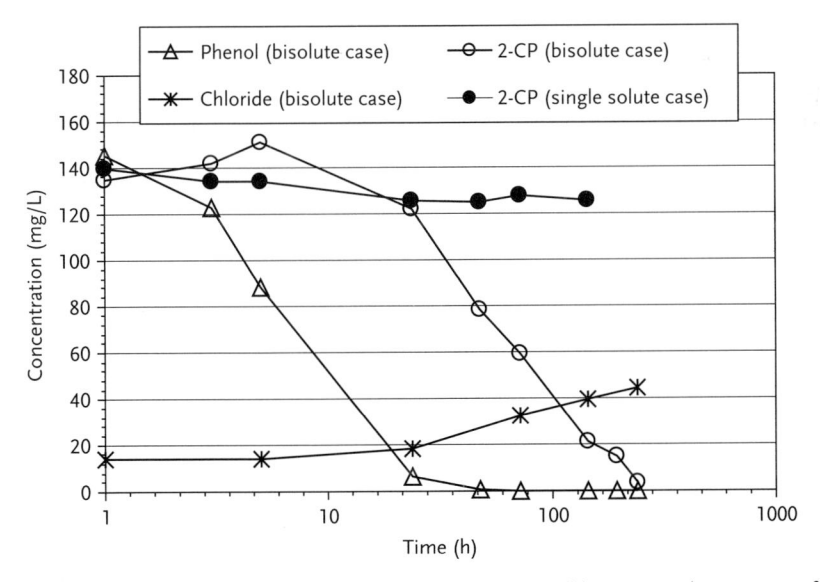

Figure 3.5 Cometabolic degradation of 2-CP by an acclimated biomass in the presence of phenol [4].

$$r_s = \frac{k_{max}S}{K_s}X_{ss} = k_{biol}SX_{ss} \qquad (3.4)$$

where

r_s = rate of soluble substrate removal, $(M_s/L^3.T)$,
S = soluble substrate concentration, (M_s/L^3),
k_{max} = maximum specific substrate utilization rate (μ_{max}/Y), $(M_s/M.T)$,
k_{biol} = biodegradation rate constant, $(L^3/M.T)$,
X_{ss} = suspended solids concentration, (M/L^3),
K_s = half-velocity constant, (M_s/L^3).

Organic compounds with a biodegradation rate constant (k_{biol}) below $0.1\,m^3$/kg SS.d are regarded as hardly biodegradable. Compounds with $0.1 > k_{biol} > 10\,m^3$/kg SS.d and $k_{biol} > 10\,m^3$/kg SS.d are considered moderately and highly biodegradable, respectively [7]. Some of the organic micropollutants of interest today, such as the pharmaceuticals diclofenac and carbamazepine (Table 3.2) are very persistent chemicals whose removal is very limited or nil in large or small wastewater treatment plants (WWTPs). On the other hand, many other micropollutants can be effectively eliminated. For example, 98% and 84% removal is reported for the anti-inflammatories ibuprofen and naproxen, respectively. The biodegradability of such substances depends also on the redox potential of the system. For example, the pharmaceuticals ibuprofen, bezafibrate, iopromide, and estrone were found to be biodegradable only under aerobic or anoxic conditions [10].

It is widely accepted that the removal percentage of micropollutants increases at high sludge retention time (SRT). This is attributed to the establishment of slowly growing microorganisms and to higher biodiversity of the sludge at increased SRTs.

Biodegradation of a biodegradable pollutant is likely to occur if the concentration is above the S_{min} value defined by Eq. (3.3). However, even if the concentration is below this value, bacteria may still be able to utilize the pollutant as a secondary substrate. Generally, in a wastewater matrix composed of different compounds at variable concentrations, it is difficult to determine whether a (micro)pollutant is removed as a growth, secondary, or cometabolic substrate.

3.2.1.2
Sorption onto Sludge

The sorption potential of a pollutant onto sludge may be predicted based on its octanol–water partition coefficient, K_{ow}. Micropollutants with a high K_{ow} are less soluble in water and would have a high affinity for the solid phase. In general, low, medium, and high sorption potentials are indicated by log $K_{ow} < 2.5$, $2.5 <$ log $K_{ow} < 4$ and log $K_{ow} > 4$, respectively [14]. More specifically, the solid–water partition coefficient (K_d) can be used in characterizing the uptake potential of the pollutant onto sludge. For some of the micropollutants, such as the 17α-ethinylestradiol (EE2), often encountered in wastewater treatment systems, sorption onto biological sludge is an important removal mechanism (Table 3.2). Yet, for others, such as some pharmaceuticals, biological transformation is the main elimination mechanism.

Although no distinct limits have been set for biosorption potential, often substances with $K_d < 0.3$ L g^{-1} SS are regarded to have low affinity to activated sludge [9].

3.2.1.3
Sorption onto Activated Carbon

3.2.1.3.1 Adsorption The adsorption potential of any pollutant onto PAC or GAC is usually expressed by adsorption isotherm constants. If adsorption is expressed by the Freundlich equation as shown in Chapter 2, the adsorption parameters K_F and $1/n$ are taken as indicators of adsorbability (Table 3.2).

The nonpolar surface of activated carbon preferably adsorbs hydrophobic compounds that have high log K_{ow} values. On the other hand, ionic or polar soluble compounds are hardly adsorbed on this surface. The log K_{ow} value of a compound may be regarded as an initial indicator of the sorption onto activated carbon. Some of the pollutants found in wastewaters have a high affinity for sorption. For example, the endocrine disrupting compounds bisphenol A and 17 α-ethinylestradiol are more hydrophobic than cytarabine and 5-fluorouracil, as indicated by their respective log K_{ow} values. In accordance with this, the adsorption constants belonging to bisphenol A and 17 α-ethinylestradiol indicate that the possibility to retain these compounds by adsorption onto PAC is greater than in the case of cytarabine and 5-fluorouracil (Table 3.2) [12].

Correspondingly, compounds with low k_{biol} values but high log K_{ow} values would tend to sorb to the activated carbon in a biological system where they will be retained for a long period of time. As discussed in the next sections, this long retention facilitates their biological removal. On the other hand, compounds with low k_{biol} and log K_{ow} values are neither biodegraded nor adsorbed onto biological sludge or activated carbon. Therefore, such nonbiodegradable and nonadsorbable compounds often leave the treatment plant with the effluent stream unless they are removed by other methods.

3.2.1.3.2 Desorption Adsorbed pollutants may also be desorbed from activated carbon if equilibrium conditions are disturbed. Many organic substrates tend to desorb from activated carbon as a result of the reversal of concentration gradient between carbon surface and bulk medium. It is also possible that adsorbed pollutants are displaced by more adsorbable ones. In biological systems the desorption of organic substrates brings about some consequences. Desorption of biodegradable organics from activated carbon is advantageous since it paves the way for bioregeneration of activated carbon, as discussed in Chapter 7. However, desorption of adsorbable and nonbiodegradable organics is not desired since this would deteriorate the effluent quality.

3.2.1.4
Abiotic Degradation/Removal

Besides biodegradation, abiotic mechanisms such as photodegradation, settling, chemical degradation, and volatilization can be effective in the removal of pollutants. While chemical transformation often occurs, photodegradation of a compound is not very probable under the conditions of biological wastewater treatment. For particulate pollutants, settling may be an effective elimination mechanism.

Volatilization may be an important mechanism leading to the elimination of a pollutant from the water phase. The volatilization potential of organic compounds in sewage treatment is predicted based on both the Henry constant and the octanol–water partition coefficient [14]. For example, trichloroethylene (TCE) is considered a volatile chemical. In STPs, volatilization and air stripping are not relevant processes for the removal of most of the micropollutants. However, they may be important elimination mechanisms in industrial wastewater treatment.

Table 3.1 shows some parameters that are typically used in predicting the biodegradability, biosorption, volatility, and adsorbability of compounds in wastewater treatment studies. Table 3.2 exemplifies the biodegradation, biosorption, volatility, and adsorbability of some selected micropollutants.

3.2.2
Main Interactions between Organic Substrates, Biomass, and Activated Carbon

In all biological systems involving activated carbon, the synergism between carbon and biomass is attributed to the adsorption, desorption, and biodegradation mechanisms that are going on concurrently or sequentially. In that regard, biodegradability of the substrate is the main criterion. If the wastewater is entirely composed of nonbiodegradable organics, an integrated system will yield no benefits. Similarly, no benefits would be expected if the wastewater is composed of biodegradable but nonadsorbable organics. For example, for some compounds, which are rather resistant to biodegradation, such as aniline, simultaneous adsorption and biological treatment is shown to be a simple combination of adsorption and biodegradation, with no synergistic effect. On the other hand, for

Table 3.1 Characterization of biodegradability, biosorption, volatility, and adsorbability of organic substances.

	Characteristic parameter
Biodegradation	Biodegradation rate constant (k_{biol})
Sorption onto biological sludge	Sludge–water partition coefficient (K_d)
	or
	Octanol–water partition coefficient (K_{ow})
Volatility	Henry constant (H_c)
Adsorbability	Adsorption isotherm constants
	(If Freundlich isotherm is valid:
	Freundlich equation: K_F and $1/n$)

Table 3.2 Characteristics of some micropollutants found in wastewaters.

Micropollutants	Biodegradation k_{biol} (m³/kg SS.d)	Sorption to sludge log K_{ow}	K_d (L/gSS)	Adsorbability onto PAC K_F and $1/n$	Reference
Pharmaceuticals					
Diazepam (DZP), (psychoactive)	≤ 0.03	2.5–3	0.02		[8]
	<0.02				[6]
Naproxen (NPX), (analgesic)	0.08 ± 0.016		0.01 ± 0.01		[9]
	0.4–0.8, 1.0–1.9		0.013		[6]
		3.2			[11]
Diclofenac (DCF), (anti-inflammatory)	≤ 0.1		0.016 ± 0.003		[6]
	≤ 0.1	4.5–4.8	0.016		[8]
	≤ 0.02				[9]
		4.51			[10]
Carbamazepine (CBZ), (antiepileptic)	<0.005, <0.008		<0.008, <0.075		[9]
	<0.01	2.3–2.5	0.00125–0.005		[7]
	<0.01				[6]
Ibuprofen (IBP), (analgesic)	1.33 ± 0.02, >3		0.05 ± 0.018		[9]
			0.006 ± 0.004		
	9–35	3.5–4.5	0.008–0.025		[7]
	9–22, 21–35		0.007 ± 0.002		[6]
Cyctostatics (anti-cancer drugs)					
Cytarabine (CytR)	–[b]	–2.2		1.12 and 0.31[a]	[12]
5-fluorouracil (5-Fu)	–[b]	–0.9		0.35 and 0.36[a]	[12]
Endocrine Disrupters					
17 β-Estradiol (E2)		4.01			[11]
	>100				[6]
	300–800	3.9–4.0	0.24–0.63		[8]
Estriol (E3)		2.45			[11]
Bisphenol A (BPA)	–[b]	3.2		4.1 and 0.35[a]	[12]
	–[b]	3.1–3.3			[7]
			0.07		[13]
17 α-ethinylestradiol (EE2)		4.1		9 and 0.41[a]	[12]
	7–9	2.8–4.2	0.31–0.63		[8]

[a] Micropollutant concentration: 200 µg L^{-1}, K_F in (µg/mg)(L/µg)$^{1/n}$, PAC concentration: 4.5–180 mg L^{-1}
[b] not available

others, like phenol, which have a relatively high biodegradability and adsorbability, the synergism created by simultaneous treatment is obvious [15].

The main interactions taking place in a carbon environment with biological activity are explained below. Figure 3.6 illustrates some of these interactions.

3.2.2.1
Retention of Slowly Biodegradable and Nonbiodegradable Organics on the Surface of Activated Carbon

The presence of activated carbon in a biological system provides a vast surface area for adsorption of organics. Hydrophobic compounds with log $K_{ow} > 2$ would be preferably adsorbed onto carbon. Adsorption is particularly beneficial for biodegradation of slowly biodegradable and apparently nonbiodegradable organics. A number of compounds found in domestic, municipal, industrial wastewaters, and in landfill leachates fall into these categories. Adsorption of such compounds onto carbon would decrease the probable inhibitory effects in the bulk solution. In addition to the initial organics, activated carbon can also effectively adsorb the intermediate and/or end products of biodegradation.

The presence of bulk organic matter in a wastewater would decrease the adsorption capacity of activated carbon for a specific pollutant. Furthermore, it should be recognized that a pollutant never exists alone in a wastewater. Therefore, competition may occur between different types of specific pollutants for adsorption sites on activated carbon.

Upon adsorption onto activated carbon, the retention time of an organic substance in a biological system is considerably increased. For example, in the PACT process (Section 3.4), dissolved organics adsorbed on activated carbon are retained in the system for a time equal to the SRT of about 10–50 days, while in conventional biological systems they will be kept for a period equal to the hydraulic retention time (HRT), typically 6–36 h [16]. This arises because PAC acts in the same way as the biomass in the system; in other words, the 'age' of PAC in the system is equal to that of biological sludge. In comparison to PACT, the retention of organics may even be longer in BAC reactors since the GAC is kept for a very long period of time in the reactor. The long retention on carbon surface enables the acclimation of attached and suspended microorganisms to these organics, eventually leading to their efficient biodegradation (Figure 3.6a).

3.2.2.2
Retention of Toxic and Inhibitory Substances on the Surface of Activated Carbon

Some organic and inorganic compounds may exert inhibitory or toxic effects on biological processes. Such adverse effects are usually encountered during industrial wastewater treatment, during leachate treatment, or in POTWs receiving a significant input of these wastewaters.

Generally, a pollutant may exert either substrate inhibition (inhibition of its own removal) or competitive and noncompetitive inhibitory effects on other substances. The inhibitory characteristics of a pollutant in a biological system are not easily determined by a single parameter, but tests have to be conducted to predict

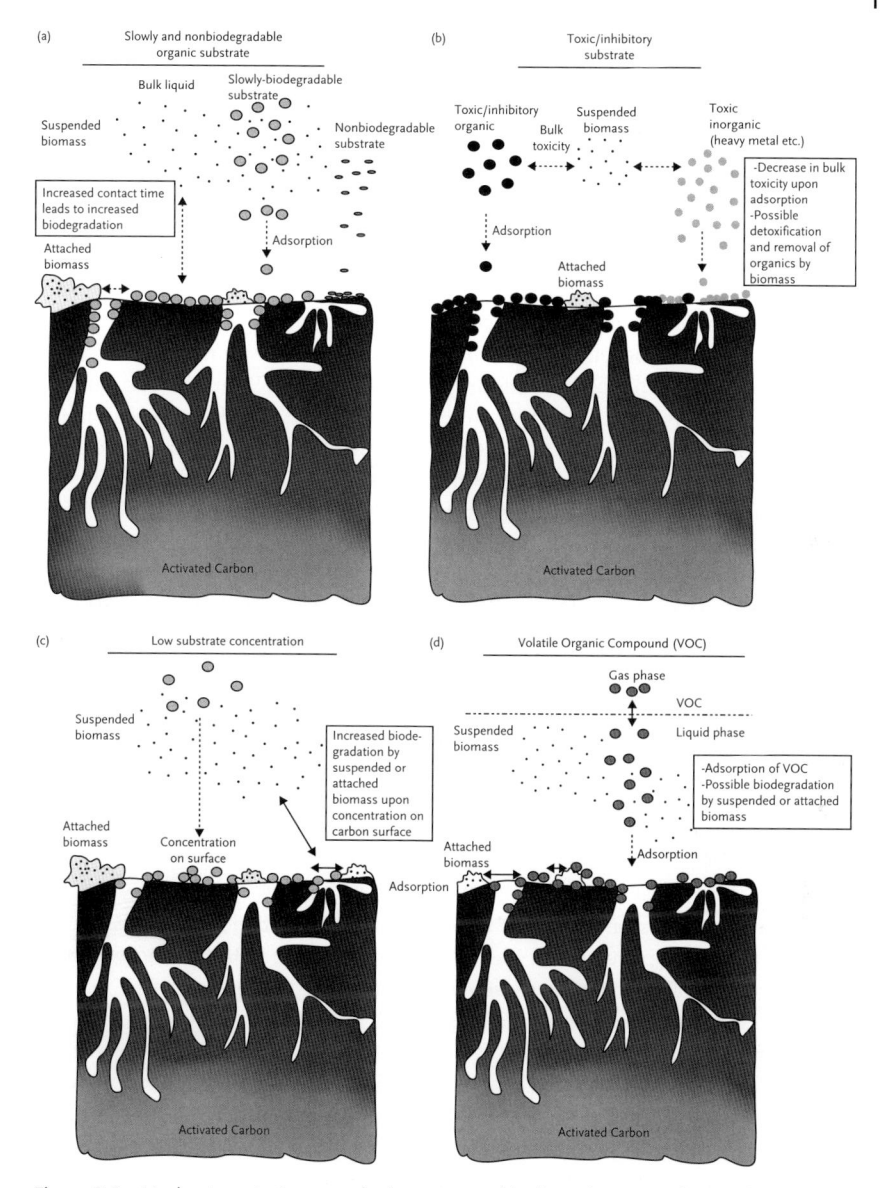

Figure 3.6 Mechanisms in integrated adsorption and biological systems for (a) slowly and nonbiodegradable substrate, (b) toxic/inhibitory substrate, (c) low substrate concentration, and (d) volatile organic compound (VOC).

the type of inhibition and the inhibitory range. Further, the effect of a pollutant differs from one system to another. For example, most organic pollutants are known to influence nitrification more severely than organic carbon removal.

Specific organic substances can exert a toxic or inhibitory effect on the biomass if they are present above threshold concentrations and sufficient acclimation of biomass is not provided. This results then in the inhibition of the removal of bulk organic matter. For example, 2-nitrophenol may inhibit the removal of other organics denoted by sum parameters such as COD, BOD_5, or TOC while its own removal is nil or very limited under aerobic conditions. Thus, it is regarded as nonbiodegradable and inhibitory.

Fortunately, most of the inhibitory or toxic organic pollutants are amenable to adsorption onto activated carbon (Figure 3.6b). Their adsorption from bulk water onto activated carbon leads to the protection of biomass, which is retained in the biological reactor in suspended or attached form. Moreover, microbial processes can take place on the surface of activated carbon that convert a specific toxic organic pollutant into an innocuous one. This beneficial effect of activated carbon is even more enhanced in BAC systems where microorganisms are attached to the GAC surface (Section 3.3).

Various methods can be used to test the inhibitory behavior of a substance in the absence and presence of activated carbon. Monitoring of the Specific Oxygen Uptake Rate (SOUR) is a common procedure in the assessment of the inhibitory response of a wastewater due to presence of organic and/or inorganic substances [17]. Phenol is well known for its inhibitory effects on biological processes, and has therefore often been employed as a model compound in testing the effect of activated carbon on biological processes. For example, PAC addition to activated sludge enabled continuous operation with inlet concentrations of phenol up to 1000 mg L^{-1} [18].

Heavy metals constitute the major group of inorganics that have adverse effects on biological processes. For example, copper, zinc, nickel, lead, cadmium, and chromium can react with microbial enzymes to retard or completely inhibit metabolism. On the other hand, as shown in Figure 3.6, heavy metals can also be adsorbed on activated carbon. The affinity of activated carbon for toxic heavy metals may even eliminate the need for metal precipitation prior to biological treatment [19].

3.2.2.3
Concentration of Substrates on the Surface of Activated Carbon

Most of the organic pollutants in wastewater treatment are found nowadays at very reduced concentrations. Under these conditions, most substrates can be used as secondary substrates only since microorganisms are unable to utilize them as the main energy and carbon sources. In that respect, the presence of activated carbon in a biological system is advantageous because it provides a tremendous surface on which substrates become concentrated upon adsorption (Figure 3.6c). Accordingly, the use of activated carbon is particularly advantageous for dilute wastewater streams. Additionally, depending on the type of activated carbon and environment, activated carbon has the ability to concentrate other substrates on its surface such as oxygen and nutrients. This concentration of various substrates further enhances biodegradation.

3.2.2.4
Retention of Volatile Organic Compounds (VOCs) on the Surface of Activated Carbon

Depending on the extent of adsorbability, the retention of a VOC on the carbon surface prevents its emission into the air. Additionally, if a VOC is amenable to biodegradation, the retention on the carbon surface increases the probability of biodegradation by suspended or attached biomass (Figure 3.6d).

3.2.2.5
Attachment and Growth of Microorganisms on the Surface of Activated Carbon

One of the most important features of activated carbon is that it also provides a vast surface for attachment of microorganisms. The activated carbon surface contains functional groups which enhance attachment of microorganisms. It also provides shelter for microorganisms from fluid shear forces [20]. The enrichment of substrates, nutrients, oxygen, and so on favors growth of microorganisms.

If sufficient time is allowed for immobilization of bacteria, a biofilm is formed on the carbon surface as shown in Figure 3.6. Biofilm formation is particularly pronounced in BAC reactors. To some extent also in suspended-growth systems the PAC surface can be occupied by microorganisms.

3.2.2.6
Biological Regeneration (Bioregeneration) of Activated Carbon

In early studies, removal of inhibitory compounds and bioregeneration of activated carbon were proposed as the two most important phenomena leading to the synergism observed in integrated systems [21].

As shown in Figure 3.6, activated carbon may serve as a depot for pollutant molecules. However, it is also probable that adsorbed molecules desorb out of carbon upon the reversal of concentration gradient between carbon surface and bulk medium. Upon desorption, the molecules may be removed by biological action, a process known as bioregeneration. Also, microbial enzymes secreted from microorganisms into carbon pores, can bring about the extracellular biodegradation of adsorbed organics and bioregeneration of activated carbon. Thus, bioregeneration of activated carbon is based on the three fundamental processes, adsorption, desorption, and biological removal. Conditions leading to bioregeneration of activated carbon are discussed in detail in Chapter 7.

3.2.3
Behavior and Removal of Substrates in Dependence of their Properties

The behavior of a substance in an integrated system depends on its biodegradability, adsorbability, volatility, and inhibitory properties. A substrate present in a carbon-biomass environment is often characterized by one or more properties (Table 3.3). For example, phenol is an adsorbable, biodegradable, and semi-volatile substance. It can be biodegraded at relatively low concentrations whereas it exerts self-inhibition at high concentrations. Phenol may also exert an inhibitory effect on others.

3.3
Integration of Granular Activated Carbon (GAC) into Biological Wastewater Treatment

3.3.1
Positioning of GAC Reactors in Wastewater Treatment

Figure 3.7 illustrates the options for the positioning of GAC adsorption in wastewater treatment. The GAC adsorption can be used as a stand-alone unit process

Table 3.3 Behavior and removal of substrates in integrated adsorption and biological wastewater treatment.

Item of interest		Examples	Behavior/removal in integrated adsorption and biological treatment
Biodegradability	biodegradable and noninhibitory	Glucose, sucrose	Compounds are completely removed by biodegradation. Their adsorption is usually limited.
	biodegradable and inhibitory	Phenol	Compounds can be removed by biodegradation. However, they may exert self-inhibition and/or inhibition to biodegradation of others. They may be removed by adsorption.
	nonbiodegradable and noninhibitory	Humic substances	Compounds can be removed by adsorption. Otherwise, they may be found in the effluent.
	nonbiodegradable and inhibitory	Chlorophenols, nitrophenols	Compounds exert inhibitory effects on biodegradation of others. They may be removed by adsorption. Otherwise, they may be found in the effluent.
Pathway of Biological Removal	primary substrate	Glucose, phenol	Compounds can be removed by biodegradation. Some may also be removed by adsorption.
	secondary substrate	Chlorobenzene, 2,4-dichlorophenoxyacetate (2,4-D)	Compounds at low concentrations can be biodegraded during biodegradation of primary substrates. They may also be removed by adsorption.
	cometabolic substrate	Chlorinated organics; e.g., 2-chlorophenol (2-CP), 2-nitrophenol (2-NP)	Compounds are nonbiodegradable themselves and can be removed only in the presence of primary substrates. They may also be removed by adsorption.
Inhibitory Characteristics	substrate inhibition	Phenol (at high concentrations)	Compounds can be removed by biodegradation at reduced bulk concentrations. At high concentrations, they inhibit their own removal (self-inhibition). They may also be removed by adsorption.

Adsorbability	competitive inhibition	Trichloroethylene (TCE), 1,2-Dichloroethane (1,2-DCA)	Compounds compete with others and inhibit their biodegradation. The nonbiodegradable ones may undergo biodegradation by cometabolism in the presence of primary substrates. They may also be removed by adsorption.
	noncompetitive inhibition	Heavy metals such as Ni, Cu, Cr(VI), Cr(III).	Substances inhibit the biodegradation of other compounds. To some extent they may also be removed by adsorption. In biological systems, they be be transformed into other forms (i.e., Cr(VI) into Cr(III)).
	adsorbable	Phenol, chlorophenols	Compounds can be removed by adsorption and possibly by biodegradation. Such compounds may also inhibit the biodegradation of others.
	moderately adsorbable	Dichloroethylene (DCE), Tetrachloretylene (PCE)	Compounds can be removed by adsorption and possibly by biodegradation. Such compounds may also inhibit the biodegradation of others.
	nonadsorbable	Methanol, ethanol	Compounds can be removed by biodegradation.
Volatility	volatile	Tetrachloretylene (PCE), Trichloroethylene (TCE), Dichloromethane (DCM, methylene chloride), Toluene, Benzene, Chloroform, Trichlorobenzene (1,2,4-TCB)	Compounds can be eliminated by volatilization. To some extent, they can be adsorbed on carbon. Depending on the type of compound, adsorption onto carbon decreases volatilization. Such compounds may inhibit the biodegradation of others. For nonchlorinated VOCs, biodegradation under aerobic conditions is possible.
	semi-volatile/nonvolatile	Phenol, Lindane	Compounds such as phenol are removed by adsorption and biodegradation. Nonbiodegradable compounds such as lindane can be removed by adsorption only. Such compounds are hardly eliminated by volatilization.

in physicochemical treatment of wastewaters which are not amenable to biological treatment (Figure 3.7a). Another option is to place GAC adsorption ahead of biological treatment if the wastewater is likely to contain substances that exert toxic or inhibitory effects on microorganisms (Figure 3.7b). The schemes shown in Figure 3.7a and b mainly apply to the treatment of industrial wastewaters.

However, typically, GAC adsorption is placed in tertiary wastewater treatment when the secondary effluent contains nonbiodegradable and/or toxic organics which are amenable to adsorption (Figure 3.7c). Tertiary stage GAC adsorption is applied as a polishing step in the treatment of wastewaters from industries that have to comply with stringent toxicity reduction protocols. For example, this configuration is often utilized for the reduction of residual COD, DOC, color, Adsorbable Organic Xenobiotics (AOX), and so on in industrial wastewaters such as pulp bleaching effluents [22].

In sewage treatment plants, tertiary stage GAC adsorption is mainly employed for the removal of micropollutants such as pharmaceuticals and hormones which are inadequately eliminated in secondary treatment. Reduction in the emission of such compounds helps protect the aquatic environment and/or enables the reuse of treated wastewater.

Ozonation is the most frequently used oxidative alternative for the degradation of residual pollutants since many compounds are susceptible to destruction with ozone. In the treatment of industrial wastewaters and landfill leachates, the sequential combination of ozonation and GAC filtration is often the choice. However, it is frequently observed that ozonation decreases the adsorption potential of organic compounds in subsequent steps. Simultaneously, however, ozonation can increase the biodegradability of organic compounds and enhance their biological removal. Therefore, biological activity develops inevitably in GAC

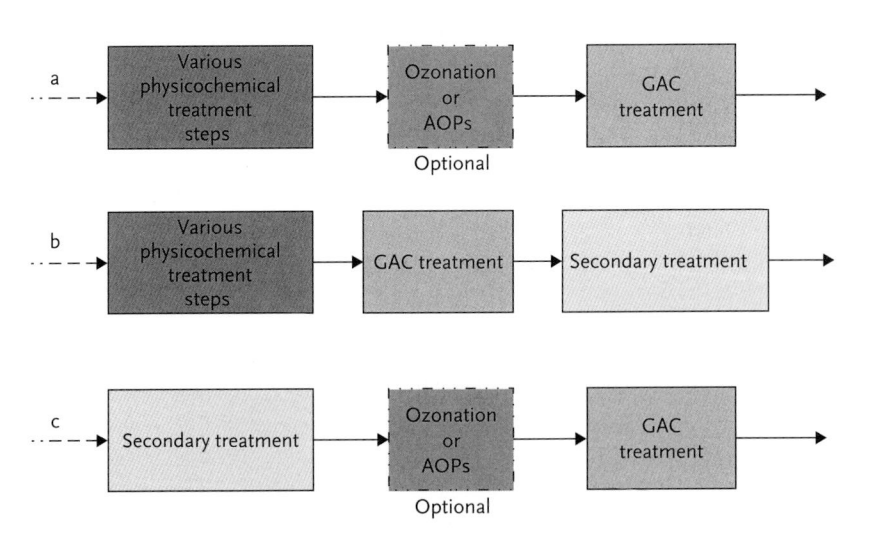

Figure 3.7 Positioning of GAC adsorption in wastewater treatment.

reactors if these receive ozonated wastewaters containing biodegradable organics along with microorganisms.

3.3.2
Recognition of Biological Activity in GAC Reactors

GAC filtration was first employed at around 1930 for adsorption of specific compounds from specific industrial wastewaters [23]. The installation of tertiary GAC units dates back to circa the 1960s. In those years, it was noticed that long-term operation without regeneration was possible for GAC filters which received secondary effluents. This long-term operation was attributed to biological activity in these filters.

In the 1970s Weber and co-workers carried out extensive experiments on physicochemical treatment of municipal wastewater in fixed-and expanded-bed GAC filters after flocculation and sedimentation [23, 24]. The main evidence for biological activity in GAC filters was that attained effluent levels could not be achieved with adsorption alone. Moreover, the frequent clogging of fixed-bed filters was indicative of microorganism growth. Further, the researchers examined the treatment of pretreated municipal wastewater in expanded-bed GAC filters and concluded that aerobic operation (inlet DO: 8–10 $mg\,L^{-1}$) yielded better TOC removal than anaerobic operation (inlet DO $<2\,mg\,L^{-1}$) [24].

Ying and Weber noted that if simultaneous adsorption and biodegradation were provided in the same reactor, expected capital costs were lower, regeneration of activated carbon was less frequent, and therefore, energy requirements and operating expenditure were reduced [25]. They also recognized that a BAC bed was able to remove slowly biodegradable compounds to a higher extent compared to a bioactive but nonadsorbing media. They suggested that acclimation of bacteria to less biodegradable compounds was the result of an adsorption-induced mechanism.

In spite of evidenced biological activity, GAC filters were not initially used for secondary treatment of domestic wastewater, but rather for industrial wastewaters, such as those from the textile industry, that were regarded as more difficult to treat. For example, the KATOX process incorporated the advantages of biological activity and adsorption and was taking an intermediate position between PAC and GAC application [23].

In the 1960s and 1970s GAC adsorbers were mainly introduced at the tertiary stage for wastewater reuse purposes. The enhancement of biological activity in GAC adsorbers by preozonation was demonstrated in the early 1980s [23].

Although biological activity was recognized in GAC adsorbers in the 1970s and 1980s, the extent of biodegradation was rarely investigated in detail. Tests carried out on Zürich domestic wastewater in pilot-scale GAC columns revealed that about 35% of DOC removal took place due to direct biological oxidation at the carbon surface. Biological regeneration was estimated to account for not more than 10% of the overall removal, while the remaining removal was due to adsorption [23].

3.3.3
Conversion of GAC into Biological Activated Carbon (BAC)

In the 1970s, Weber and co-workers carried out experiments on a completely mixed batch reactor and on an expanded-bed reactor, these being fed with coagulated settled sewage and humic acid, respectively [20]. The physical and structural characteristics of biological growth on GAC were investigated by Scanning Electron Microscopy (SEM). After exposure to sewage, the surface of the activated carbon was rapidly colonized by bacteria and protozoa. Similarly, in the expanded-bed reactor exposed to 5 mg L^{-1} humic acid, bacteria attached themselves on the surface of the carbon; a nonhomogeneous coverage of carbon surface and production of a slime layer were evident.

In another study, investigation of a carbon surface by SEM revealed that biological material did not penetrate into the interior of carbon particles [26]. This is a significant finding, which also receives acceptance nowadays. As discussed in Chapters 6 and 11, modeling of integrated adsorption and biological removal is mostly based on the assumption that there is no biological activity inside the carbon particles.

Table 3.4 Removal of different organic fractions in BAC reactors.

Type of organic compound	Removal mechanism	Ideal condition	Remarks
Biodegradable and adsorbable	1. Biodegradation (preferred) 2. Adsorption	High desorption of adsorbed matter followed by biodegradation (bioregeneration)	Bioregeneration possibility extends the life of GAC bed.
Biodegradable and nonadsorbable	Biodegradation	Complete biodegradation	Biological removal takes place in intermediate or later operational stages. The compounds do not appear in the effluent after initiation of biological activity.
Adsorbable and nonbiodegradable	Adsorption	Nondesorbability of adsorbed compounds	No removal after exhaustion of adsorption capacity. Probable emergence of desorbable compounds in the effluent.
Nonadsorbable and nonbiodegradable	No removal	No competition with removable compounds, nontoxic properties	Apparent in the effluent from the beginning of operation.

3.3.3.1
Removal Mechanisms and Biofilm Formation in BAC Operation

In BAC operation, adsorption and biological removal of contaminants take place simultaneously in the same reactor. Moreover, the relative importance of adsorption and biodegradation changes with respect to operation time. The reader may refer to Chapter 6 for a mathematical representation and analysis of GAC/BAC filtration.

Table 3.4 gives an overview of the behavior of different fractions in influent organic matter in BAC reactors.

3.3.3.1.1 Initial Stage of Operation – Removal by Adsorption At this stage, the organic matter is primarily removed by adsorption onto the GAC surface since a microorganism layer has not yet been developed. The extent of adsorption depends on a number of factors such as the charge, size, and polarity of the adsorbate, and the relationship between the adsorbate and activated carbon surface. This period may last for several weeks or months, depending on operating conditions and the characteristics of the wastewater. The effluent contains mainly nonadsorbable compounds, which can be nonbiodegradable or biodegradable.

3.3.3.1.2 Intermediate and Later Stages of Operation The formation of a biofilm layer on the GAC surface is crucial for biodegradation. In wastewater treatment, the high microorganism and substrate concentrations in the influent wastewater usually results in rapid colonization of the GAC surface. The extent of colonization and biofilm growth depends on whether the GAC reactors receive a wastewater containing biodegradable substances or a secondary effluent that has a low fraction of these compounds. If the wastewater does not contain any microorganisms and the GAC adsorber is placed before secondary treatment, the adsorbers are usually seeded with microorganisms for startup purposes. The high organic load to the filter then ensures a rapid growth of microorganisms.

At an intermediate stage of operation, organics are both adsorbed and biodegraded by the biomass present in the filter. During this period, most of the adsorbable and biodegradable fraction will be removed from the influent. In BAC reactors, biodegradation is the preferred removal mechanism rather than adsorption since biodegradation would lower the organic loading on the GAC surface, a factor increasing the service life of the activated carbon (Table 3.4).

During the course of operation, desorption of biodegradable compounds from the GAC surface would result in additional adsorption capacity for nonbiodegradable compounds. Therefore, the right option is to choose a GAC grade providing a high desorption of biodegradable organics.

In the intermediate stage of GAC operation, adsorption gradually decreases whereas biodegradation gains importance. At later stages, biological processes prevail since the adsorption capacity of the GAC is completely exhausted in most cases. Then, the GAC surface loaded with organic matter may eventually be renewed by biological activity. This phenomenon, which is referred to as bioregeneration, is the main subject of Chapter 7.

3.3.3.2
Advantages of BAC Over GAC Operation

The capital costs of a BAC system are lower than they are when two separate reactors are used. However, the merits of the BAC system extend beyond the simple combination of adsorption and biodegradation in the same unit.

Combination of GAC adsorption and biofilm processes in a BAC reactor creates ideal conditions for the removal of substrates that otherwise would be not removed in attached-growth systems. The presence of GAC has a dampening function in the case of shock inputs since it rapidly adsorbs a number of organic and inorganic substances. Moreover, under shock load conditions, the biomass is less sensitive to toxic and inhibitory compounds because of the adsorptive ability of GAC. In that respect, BAC operation is well suited to the treatment of wastewaters which are highly variable in concentration and composition.

Since BAC reactors are operated at high SRTs like other biofilm reactors, another advantage is that target pollutants can be decreased to much lower levels compared to suspended-growth reactors. The considerable age of the sludge allows for the selection of slowly growing bacterial populations that are specific for the removal of target pollutants. As discussed in Chapter 7, in BAC operation bioregeneration of activated carbon is possible, leading to an effective use of GAC.

3.3.3.3
Advantages of the BAC Process Compared to Other Attached-Growth Processes

In comparison to other attached-growth processes, the main advantages of the BAC process derive from the adsorption and desorption properties of GAC. GAC is a very adsorptive medium having a very high surface-to-volume ratio. The superiority of activated carbon over nonadsorptive media, such as sand and anthracite coal, for the removal of organic matter was recognized very early [24]. In contrast to GAC, nonadsorptive media such as sand, plastic rings, or stone serve for attachment of biomass only and exhibit lower substrate removal.

Activated carbon also offers a relatively rough surface. As such, it supports significantly more biomass than a synthetic carbonaceous adsorbent, an anion exchange resin, a natural sand, or glass beads [27]. Thus, for example, microorganisms immobilized on GAC and on a plastic medium removed 100% and 91% of trichlorophenol, respectively [28]. Other studies indicate that a GAC biofilter is more stable under shock and/or intermittent loading and starvation conditions compared to nonadsorbing media [29].

3.3.4
Main Processes in BAC Reactors

Figure 3.8 illustrates the processes occurring in BAC systems. In such systems, the surface of the GAC is covered with a biofilm. In many cases, a thin stationary layer, known as the liquid film or diffusion layer, exists over the surface of the biofilm. Diffusion of substrates from the bulk into the biofilm occurs through this layer. Substrates diffusing into the biofilm are removed by biological mechanisms. The products of bacterial activity also diffuse out from the biofilm. In addition, the

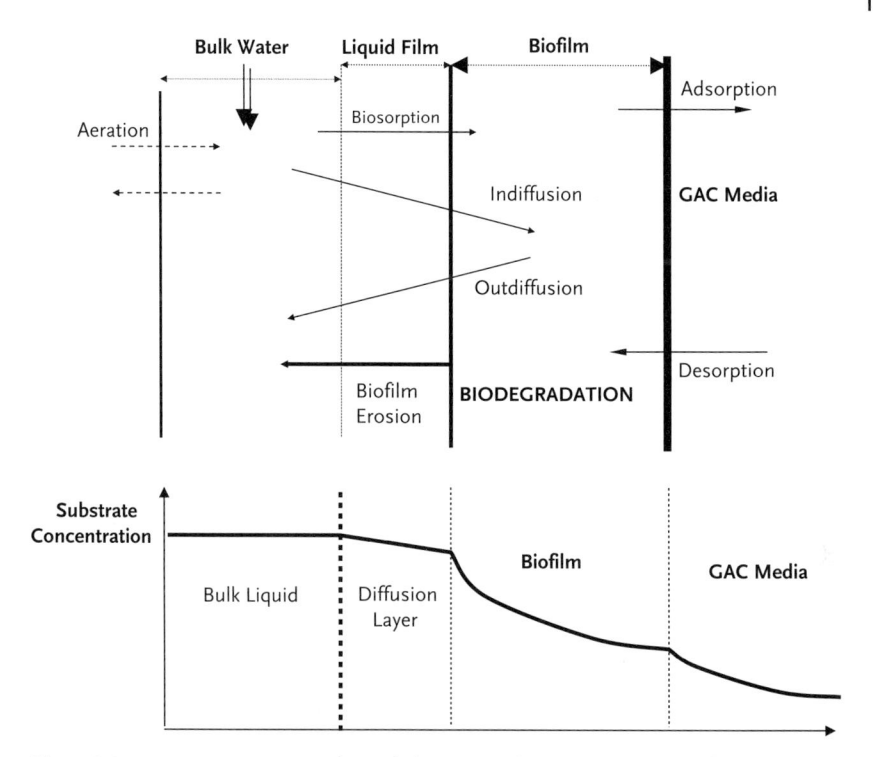

Figure 3.8 Processes on GAC media and change in substrate concentration from the bulk liquid to the GAC surface.

biofilm surface may adsorb and desorb substrates. Particulate organic matter adsorbed on biofilm may be utilized by microorganisms upon hydrolysis [30]. At the biofilm–carbon boundary, substrate diffuses further into the activated carbon, where it is adsorbed. Desorption from activated carbon may be another important mechanism. The change in substrate concentration through the bulk liquid, the liquid film, the biofilm, and activated carbon is illustrated in Figure 3.8. All these processes are further analyzed mathematically in Chapter 6.

3.3.5
Types of GAC Reactors with Biological Activity (BAC Reactors)

With regard to the mobility of GAC media, BAC reactors can be mainly divided into fixed-, expanded-, or fluidized-bed reactors. GAC operation with biological activity provides the advantage that the SRT is higher compared to a PACT system. Also, the attachment efficiency of microorganisms is much higher in a BAC reactor compared to PACT (Section 3.4) [31].

3.3.5.1
Fixed-Bed BAC Reactors
Although most fixed-bed GAC reactors are operated in the continuous-flow mode, they may also be operated in the sequencing batch mode; such a

configuration is referred to as the GAC-Sequencing Batch Biofilm Reactor (GAC-SBBR).

Fixed-bed reactors packed with GAC can be operated under aerobic, anoxic or anaerobic conditions. A well known type of a fixed-bed reactor operated under aerobic conditions is the Biocarbone, which is a biological aerated filter (BAF) packed with GAC aiming at carbonaceous BOD removal and nitrification [32].

For municipal and industrial wastewaters, the presence of suspended solids as well as biomass growth may present problems in fixed-bed operation. Clogging problems often occur when the wastewater is inadequately pretreated.

3.3.5.2
Expanded- and Fluidized-Bed BAC Reactors

In a conventional fluidized-bed reactor (FBR), water or wastewater is pumped upwards through the bed. The velocity of the pumped water should be sufficient to expand the bed against gravity resulting in fluidization of media. The difference between expanded- and fluidized-bed reactors is that the former are operated with smaller upflow velocities resulting in an incomplete fluidization of the biofilm support medium [33].

The development of systems in which the biomass was kept in fluidized form in water and wastewater treatment is traced back to the 1940s. However, in such reactors the use of a carrying media began only in the early 1970s. Moreover, nonadsorptive media such as sand and anthracite were employed in early applications rather than GAC [34].

Nowadays, GAC is often incorporated into the operation of an FBR. The configuration is designated as the GAC-FBR, or the BAC-FBR if biological activity is evident. The development of full-scale GAC-FBR technologies began in about 1986 for the treatment of contaminated surface waters and groundwaters [34]. Since then several patented GAC-FBR technologies have been developed, such as the Envirex® FBR, capable of removing a number of organic and inorganic target contaminants from water or wastewater under aerobic, anoxic, or anaerobic conditions [32]. For the removal of refractory organic matter and nitrogen from high-strength wastewaters such as landfill leachates, anaerobic and aerobic GAC-FBR reactors can also be used in series [35].

Fluidized GAC media provide a large surface area for biological growth which leads to biomass concentrations 5–10 times higher (about 10–50 g MLVSS/L) compared to conventional activated sludge [34, 36]. The size of GAC particles used in GAC-FBR reactors may affect the process performance. Medium-sized GAC particles (0.40–0.59 mm) were suggested to achieve higher efficiency and biomass attachment compared to other sizes [37].

High SRTs further enhance acclimation of bacteria to refractory compounds and degradation of slowly biodegradable organics. High-quality effluent is achieved in GAC-FBR operation in terms of suspended solids and organic matter [36].

The benefits of expanded- or fluidized-bed operation over fixed-bed are summarized below:

1. Many problems in operation are minimized if wastewater is passed upward at sufficiently high velocities to expand the bed. Effective operation can be maintained for a long time without significant pressure loss since in expanded or fluidized beds the dependence of pressure loss on particle size is small.
2. Fouling problems can be minimized. Clogging problems are not encountered in fluidized-bed filters due to the higher void volume. Also, due to the high velocity of fluid in a FBR, the sloughing of biofilm is easier to maintain, preventing the clogging problems caused by a thick biofilm.
3. Operational capacity of the activated carbon is higher in expanded and fluidized beds than predicted from the adsorption isotherms or from sorption capacities obtained in fixed-bed systems. In the early 1970s Weber and co-workers postulated that the enhancement of effective capacity was attributable to microbial activity on the carbon surface. This was particularly enhanced in expanded-bed adsorbers, whereas biological growth led to fouling in fixed-bed adsorbers [24]. Because expanded beds require little maintenance, extended periods of undisturbed operation facilitate the development and continuous growth of bacteria on carbon surfaces.
4. Higher removal efficiencies of carbon and nitrogen are achieved at low hydraulic retention times.

Application of the GAC-FBR configuration has also limitations such as the following:

1. High operating costs are required for the pumping of recycle streams in order to provide fluidization of the reactor. Compared to fluidized-beds, systems involving fixed-bed configurations incur lower operating costs.
2. GAC-FBR systems are not designed for suspended solids removal. The inlet distribution system of FBRs has a high sensitivity to clogging [38]. Therefore, pretreatment by lime or alum addition is usually required.
3. Control of fluidization and bed height is difficult in FBR reactors [38].

3.4
Integration of Powdered Activated Carbon (PAC) into Biological Wastewater Treatment

The process in which PAC is directly added to activated sludge is known as Powdered Activated Carbon Treatment (PACT). The PACT process can be effectively applied in the treatment of the following water/wastewaters:

   a. Municipal wastewaters
   b. Combined municipal and industrial wastewaters
   c. Industrial wastewaters
   d. Sanitary and hazardous landfill leachates
   e. Contaminated groundwater.

The primary advantages of using PAC are the low capital investment costs and the possibility of changing the influent PAC dose in response to incoming wastewater.

However, at high PAC doses the operating costs may be relatively high. The PACT process can be adapted to continuous-flow and sequencing batch conditions.

3.4.1
Single-Stage Continuous-Flow Aerobic PACT® Process

3.4.1.1
Development of the PACT Process

Starting from the late 1960s, a variety of modifications and alternatives to conventional biological treatment have been evaluated for removal of organic compounds designated as priority pollutants, and, among these, the addition of PAC to an activated sludge tank has proved to be an attractive integrated process [39].

The PACT process was originally developed in the 1970s by DuPont as a result of efforts to find alternative treatment procedures for DuPont's Chamber Works Wastewater [40]. Zimpro Environmental purchased the patent rights to this process, named the PACT® Process, from DuPont and extended the technology [41]. DuPont's Chamber Works was a chemical plant manufacturing various products [42]. The mostly batch type processes operated in the plant resulted in a highly variable wastewater. Operation of the PACT plant led to BOD, COD, and color removal that exceeded expectations. The PACT system was very effective in removing the organics that were at that time designated as priority pollutants by EPA, and it was recognized that the PACT process protected the biological system from upsets caused by shock loads of organic compounds. In that respect, it was noted that the carbon dose was a key parameter for efficient removal. Regarding solid residuals, the most significant result was that PAC could be regenerated from PACT sludge in a multiple-hearth regeneration furnace.

3.4.1.2
Basic Features of the Activated Sludge Process

The activated sludge process involves the growth of an activated mass of microorganisms capable of stabilizing the wastewater aerobically. The major objective of the activated sludge process is the removal of organic matter, and in many cases the removal of the nutrients nitrogen and phosphorus. Trace concentrations of inorganics may also be removed due to precipitation or adsorption onto biomass during biological treatment [43].

The methods for optimizing the biological degradation include controlling the dissolved oxygen (DO) level, nutrient addition, increasing the concentration of microorganisms, and maintaining many environmental factors such as pH, temperature, and mixing. In principle, from the point of hydraulic mixing, the operation of the aeration tank ranges from plug flow reactor (PFR) to continuous-flow stirred tank reactor (CSTR). The design parameters include reactor volume, desired effluent quality, sludge return rates, food to microorganism (F/M) ratio, HRT, and SRT [43].

The mixture of the activated sludge and wastewater in the aeration basin is called the mixed liquor suspended solids (MLSS), whereas the volatile organic fraction is referred to as the mixed liquor volatile suspended solids (MLVSS).

3.4 Integration of Powdered Activated Carbon (PAC) into Biological Wastewater Treatment

Biological reactions, namely biodegradation and/or biotransformation of organic compounds, are assumed to take place mainly in the aeration basin. The biological sludge continuously flows from the aeration basin into a secondary clarifier where it is settled. A portion of the settled sludge is returned to the aeration basin to maintain the proper F/M ratio. Generally, a surplus of activated sludge is produced. Therefore, some of the activated sludge is wasted from the aeration tank or from the sludge return line. Waste activated sludge is introduced to the sludge handling systems for further treatment and disposal.

3.4.1.3
Characteristics of the PACT Process

The PACT process is simply a modification of the activated sludge process. In this aerobic system the influent enters an aeration tank where activated carbon in the form of PAC is added to the mixed liquor. PAC is dosed in dry form or as a slurry. Typically, PAC is added directly to the aeration tank or the sludge return line. The dosage rate of carbon depends on the type and concentration of pollutants. Following aeration, the mixed liquor enters the secondary settling tank, where the slurry composed of biomass and carbon particles is allowed to settle. A small dose of cationic polyelectrolyte may be added ahead of the secondary settling tank to enhance solids-liquid separation and to ensure maximum overflow clarity. After settling, filtration is optional and is usually recommended if a suspended solids level of less than 20 mg L^{-1} is to be consistently obtained in the effluent [44]. Following treatment, a portion of the sludge is daily wasted from the system. Since a portion of the PAC is also wasted with this sludge, continuous addition of virgin or regenerated PAC is required to maintain the desired carbon concentration in the aeration tank. Waste solids can be disposed of as slurry, dewatered to a compact stable cake, or pumped to a Wet Air Regeneration (WAR) unit for regeneration of carbon (Section 3.7) (Figure 3.9).

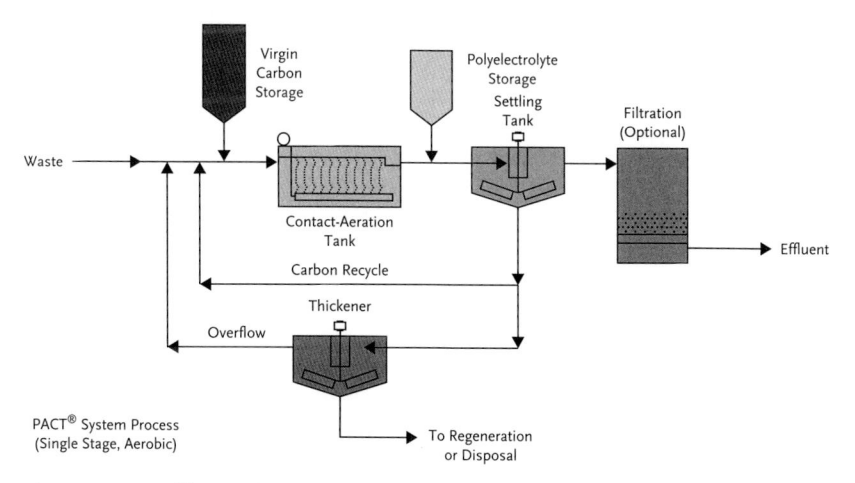

Figure 3.9 PACT® general process diagram [45] (permission received).

3.4.1.4
Process Parameters in PACT Operation

Depending on the characteristics of wastewater and the requirements set for effluent quality, the process parameters vary over a wide range. Operation is mainly controlled by the regulation of the following parameters:

- Hydraulic Retention Time (HRT)
- Sludge Retention Time (SRT)
- Carbon Dose
- Carbon Type.

In contrast to GAC filters, which are filled with GAC and undergo colonization with respect to operation time (Section 3.3), in the PACT process biomass already exists in the aeration tank. It is the PAC whose dosage to an aeration tank can be adjusted, depending on operating conditions. Consequently, the process has a high flexibility for wastewaters fluctuating in composition and concentration. As shown in Section 6.2 of Chapter 6, which discusses the modeling of the PACT process, the PAC concentration in the mixed liquor is a function of influent PAC dose, HRT, and SRT. The concentration of carbon in the aeration tank is of utmost importance because of the beneficial effect of carbon. For a proper design and operation of PACT systems, the optimum carbon dose should be initially established by isotherms in batch experiments. However, in the real treatment system, the adsorption capacity of PAC can differ from that predicted by isotherms. The type of carbon is also important since it determines the adsorption and bioregeneration properties.

The typical conditions in PACT systems are presented in Table 3.5. The hydraulic loading rates (HLRs) in the PACT system are comparable to those in conventional activated sludge systems. On the other hand, the addition of PAC, a solid, to activated sludge increases the MLSS concentration in the aeration tank to very high values. It is reported that the aeration systems and settling tanks are designed to handle high levels of mixed liquor solids ($>15\,000\,\text{mg}\,\text{L}^{-1}$) to ensure that sufficient carbon is maintained therein to enable efficient removal. A high proportion of the MLSS can be made up of carbon (55%), while the rest consists of biomass and inorganics [47].

Table 3.5 Typical conditions in a PACT process.

Parameter	Typical range	References
Hydraulic Retention Time, HRT	4–48 h	[16, 41]
Sludge Retention Time, SRT	10–50 d	[41, 46]
Influent PAC Dose	20–500 mg L^{-1}	[19, 23, 41, 46]
MLSS in aeration tank	7000–22000 mg L^{-1}	[23, 41, 47]
PAC concentration in aeration tank	1000–12000 mg L^{-1}	[23, 41]
Polyelectrolyte dose rates	0.5–1.5 mg L^{-1}	[44]
Clarifier hydraulic loading rates	16–32 m^3/m^2.d	[16, 44]
Solids loading rates	490–610 kg m^{-2}	[44]

Due to the high MLSS concentration in aeration tanks upon PAC addition, the solids loading rates in settling tanks are much higher, and larger amounts of waste PACT sludge are produced than in conventional activated sludge operation. However, dewatering of the PACT sludge will result in less sludge volume compared with an activated sludge due to the formation of a drier cake (50% solids vs <20% solids) [19, 48]. Instead of settling, for clarification purposes the effluent of the PACT system can also be led to a dissolved air flotation (DAF) unit or a membrane filter [49].

In continuous-flow PACT operation, an important decision to be made is the point of PAC dosing. Correspondingly, in a batch or sequencing batch operation, the time of PAC addition during the course of operation would be of interest [44].

There is also a strong relationship between biomass concentration and pollutant removal in a PACT system. In the PACT system an optimum biomass concentration exists above which removal of organic matter may decrease [50]. The concentration of biomass is particularly important in the case of adsorbable and nonbiodegradable organics such as dyes. Above the optimum biomass concentration, the carbon particles are trapped within the floc matrix, resulting in a mass transfer resistance between the pores and the bulk liquid which eventually leads to a decrease in adsorption properties. For a better understanding of the mass transfer in integrated adsorption and biological systems, the reader may refer to Chapter 6.

3.4.2
Sequencing Batch PACT Reactors

The aerobic sequencing batch reactor (SBR) operation is a modification of an activated sludge system operating on a fill- and draw basis. The SBR method of operation lies in between batch and continuous-flow operation. Also, the PACT process can be adapted to the SBR mode (Figure 3.10).

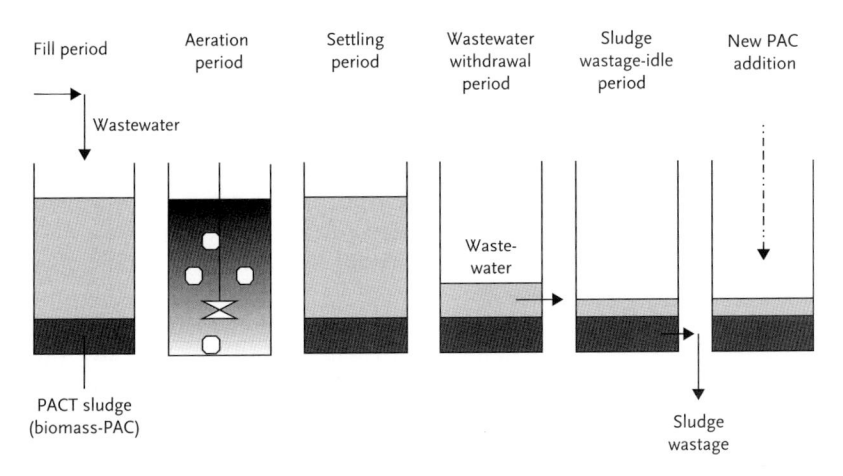

Figure 3.10 Operation of an aerobic SBR Reactor with PAC addition.

The SBR-PACT system is primarily introduced in the case of wastewaters with a small flow rate. It has frequently been applied in the treatment of industrial wastewaters and leachates prior to discharge into receiving waters or as a pretreatment alternative before discharge into the sewer system.

Each cycle of a typical SBR operation consists of five discrete periods: fill, react, settle, draw, and idle. The minimum number of cycles in full-scale operation is reported to be five cycles per day [51]. In the fill mode of operation, the wastewater enters the tank containing the sludge composed of biomass and PAC. If the operation is aerobic, the reactor is aerated and mixed during the react period. The duration of aeration depends on the character of the wastewater and the required level of treatment. In the settling period, aeration and mixing is stopped and the MLSS is allowed to settle. One of the advantages of the SBR process is that no separate clarifier is required. In the draw period the clarified effluent is withdrawn from the tank. The sludge, composed of biomass and PAC, is retained in the tank for the next cycle. The surplus sludge is wasted as in other biological systems. Since some portion of PAC is also wasted with this sludge, makeup PAC is directly added to the tank. The time period between draw and fill is named 'idle.'

3.4.3
Anaerobic PACT Process

In anaerobic PACT systems, the same synergistic mechanisms apply as in the aerobic PACT process. As in other anaerobic systems, the methane gas emerging from the reactor can be recovered and used for energy purposes. Following treatment, a portion of carbon and biomass slurry is wasted to solids handling. The choice of solids handling method will depend upon the amount of solids, disposal costs, and carbon use. The PACT process can also be adapted to anoxic conditions [52].

3.5
Biomembrane Operation Assisted by PAC and GAC

3.5.1
Membrane Bioreactors (MBRs)

In the advanced treatment of wastewaters, the use of membrane processes has been increasing. The main advantage of an MBR, an alternative to the conventional activated sludge process, is that no separate settling tank is required, since the MBR combines biotreatment with membrane separation by microfiltration (MF) or ultrafiltration (UF) [53].

Initially, MBRs involved cross-flow membrane modules through which the mixed liquor of the biological unit is continuously circulated. However, this reactor configuration requires high energy costs related to pumping for recirculation of mixed liquor. Moreover, the imposed shear stress can cause breakage of microbial flocs. These drawbacks necessitated the use of submerged membrane bioreactors (sMBRs), also called immersed MBRs (iMBRs), in which the membrane module is immersed in activated sludge [54]. Existing activated sludge systems can easily be upgraded with a submerged membrane without significant modifications [55].

MBRs typically operate at high SRTs, in the range of 20–100 d, whereas in conventional centralized WWTPs the SRT is in the range of 5–20 d [9, 56]. Consequently, MBRs are operated at higher MLSS concentrations (typically 8000–15 000 mg L^{-1}) than activated sludge systems. Therefore, the required reactor size is reduced.

The operation of MBRs at high SRTs favors the growth of slowly growing microorganisms such as nitrifiers. It also allows the selection of other bacterial populations by which the degradation of organic micropollutants becomes possible [54]. Microbial degradation of complex organics and detoxification of toxic organics are favored in systems operating at high SRTs. Also, enhanced metal removal is reported in MBR operation compared to conventional activated sludge [57]. Therefore, MBRs are a good choice in the treatment of complex industrial wastewaters requiring long SRTs. MBRs produce a better quality effluent free of suspended solids and containing less residual organics. The quality of the effluent is equivalent to tertiary filtration. Due to the high SRT, the production of excess sludge is potentially lower in MBRs.

MBRs have become a feasible option for aerobic treatment of municipal and industrial wastewaters and potentially for anaerobic treatment of municipal and/ or low- to medium-strength industrial wastewaters. They are often preferred when sludge settling and clarification problems are frequently observed [58].

Microfiltration and ultrafiltration membranes are able to retain compounds with a molecular weight (MW) of 200 000 and 20 000 g mol^{-1}, respectively. The retention of most micropollutants, such as the pharmaceutical carbamazapine and others, would only be possible by nanofiltration, which does not allow the passage of compounds with a molecular weight greater than 200 g mol^{-1} [59]. Therefore, the main removal mechanism in MBRs using MF and UF is not filtration, but increased retention of pollutants on biological sludge and/or increased biodegradation at high SRTs. The disadvantages of MBRs involve membrane fouling and operational costs related to membrane renewals [60].

3.5.2
The PAC-MBR Process

The accumulation of rejected constituents on the membrane surface leads to a reduction in water flux (flow per unit area) at a given transmembrane pressure (TMP). The TMP is the difference in pressure between the filtrate side and the permeate side of the membrane. The rise in the TMP for a given flux results in the reduction of permeability which is defined as the ratio of flux to TMP. These phenomena are collectively referred to as fouling [54]. Membrane fouling is a frequently encountered problem in MBR operation. Biofouling, on the other hand, is defined as a special type of membrane fouling caused by microbial activity. The mechanisms leading to biofouling are not fully understood yet.

PAC may also be added to the mixed liquor of an MBR, a process designated as PAC-MBR. Compared to the MBR operation alone, this type of treatment has the advantage of reducing biofouling, increasing membrane permeability, stabilizing membrane operation, and reducing the organic matter to very low levels in the effluent. Particularly, elimination of micropollutants can be further enhanced by PAC addition. Like other suspended-growth systems, MBRs involving PAC addition can also be operated under sequencing batch conditions. Details of the effect of PAC in MBRs are discussed in Section 3.6.7.

3.5.3
Membrane-Assisted Biological GAC Filtration – the BioMAC Process

Full-scale plants involving MBR technology for municipal and industrial wastewater treatment are already in operation in several countries. MBRs can be introduced at different locations of the treatment train, for example, after primary treatment of wastewater for secondary treatment purposes. Then, adsorption in GAC columns is introduced as a tertiary treatment step. However, this combination would not prevent fouling of membranes.

MBRs are also often used in the polishing of wastewater treatment effluents. When MBRs are combined with GAC adsorption, biofouling can be prevented through immobilization of bacteria on activated carbon instead of a membrane surface, since attached bacteria cannot reach the membrane.

Membrane-assisted BAC is used for the removal of priority pollutants from secondary effluents as well as from drinking water. In coupling MBRs with GAC adsorption, one of the major aims is usually the enhanced removal of micropollutants. For example, the patented Biological Membrane-Assisted Carbon Filtration (BioMAC) technology uses a combination of BAC filtration and micro/ultrafiltration for enhanced treatment of WWTP effluent or for the treatment of concentrated streams. It is effective, for example, in the removal of antibiotics, pharmaceuticals, pathogens, and endocrine disrupting compounds from a wastewater. The permeate of a BioMAC reactor treating secondary effluents contains very low levels of organic matter, nutrients, odor, and indicator organisms, and can be suitable for direct industrial reuse or groundwater recharge [61].

3.6
Observed Benefits of Integrated Systems

The observed benefits of using PAC or GAC in conjunction with biological treatment can be summarized as follows:

1. Improvement of organic carbon removal (reduction in COD, BOD, and TOC)
2. Promotion of nitrification
3. Stabilization of the treatment system against upsets and shock loads
4. Adsorption of inhibitory and toxic compounds

5. Control of color and odor
6. Reduction in effluent toxicity
7. Reduction in VOC emissions to the atmosphere
8. Improvement of sludge settling and dewatering properties (in the PACT process)
9. Adsorption of biologically generated substances (soluble microbial products (SMP) and extracellular polymeric substances (EPS)).

Table 3.6 gives an overview of the underlying mechanisms leading to these observed benefits.

Table 3.6 Mechanisms leading to observed beneficial effects.

Mechanisms	Importance in PACT and/or BAC Operation	Observation
Adsorption of inhibitory compounds	both	• Alleviation of toxicity to biological system • Protection of suspended and attached biomass • Improvement in organic carbon removal • Improvement of effluent quality • Improved nitrification
Adsorption of nonbiodegradable compounds	both	• Improvement in organic carbon removal
Enhanced biodegradation of slowly biodegradable compounds	both	• Improvement in organic carbon removal
Adsorption of biodegradable and nonbiodegradable compounds under shock loads	both	• Stabilization of reactor performance
Adsorption of SMP and EPS	both PACT	• Improvement of effluent quality • Improvement of sludge dewaterability (filterability)
Adsorption of color- and odor-causing compounds	both	• Alleviation of color • Alleviation of odor • Alleviation of toxicity • Improvement of aesthetic quality
Adsorption and biodegradation of Volatile Organic Compounds (VOCs)	both	• Reduction in VOC stripping into the atmosphere • Reduction of VOC in the water phase
Adsorption of metals	both	• Improvement in organic carbon removal • Improved nitrification
Weighting effect of activated carbon	PACT	• Improvement in settling

3.6.1
Enhancement of Organic Carbon Removal by Activated Carbon

The most striking effect of activated carbon in integrated systems is observed in the case of organic carbon removal that takes place as a result of mechanisms discussed in Section 3.2. The benefits of enhanced organics removal are observed directly or indirectly. Direct benefits include mainly the decrease in effluent concentration, alleviation of toxicity, removal of odor, color, reduction in VOC emissions, and so on. Furthermore, the decrease in SMP and EPS is related to enhanced organics removal. Indirect effects include improved nitrification as a result of enhanced biological removal and/or adsorption of inhibitory or toxic compounds. Also, improvement in sludge settling and dewaterability is to a certain extent related to enhanced organics removal.

The enhancement of organic carbon removal in integrated systems is extensively discussed in Chapters 4 and 5, for the case of small- and full-scale systems, respectively.

3.6.2
Enhancement of Nitrification by Activated Carbon

Nitrogeneous matter in wastewater, like carbonaceous matter, can be divided into two categories, nonbiodegradable and biodegradable. The nonbiodegradable fraction is associated with nonbiodegradable organic matter while the biodegradable fraction contains ammonia and soluble and particulate organic nitrogen. Particulate organic nitrogen is hydrolyzed to soluble organic nitrogen in parallel to the hydrolysis of slowly biodegradable organic matter. The soluble organic nitrogen can be converted to ammonia by ammonification [62].

Nitrification is an autotrophic process in which energy for bacterial growth is derived from the oxidation of ammonia. At neutral pH, most of the nitrogen is present in the form of the ammonium ion rather than ammonia. Therefore, equations delineating nitrification involve the ammonium ion as the initial substrate. Nitrification bacteria, or nitrifiers, use inorganic carbon instead of organic carbon for cell synthesis. Nitrification is a two-step process carried out by Ammonia Oxidizing Bacteria (AOB) and Nitrite Oxidizing Bacteria (NOB), converting ammonium into nitrite and nitrite into nitrate, respectively. The two steps of oxidation reactions, which yield the energy required for the synthesis of new nitrifier cells, are shown as follows.

$$NH_4^+ + \frac{3}{2}O_2 \xrightarrow{AOB} NO_2^- + 2H^+ + H_2O \quad (3.5)$$

$$NO_2^- + \frac{1}{2}O_2 \xrightarrow{NOB} NO_3^- \quad (3.6)$$

3.6.2.1
Inhibition of Nitrification

Several factors such as reduced temperature, limiting oxygen, and elevated pH may retard nitrification. Inorganic (e.g., heavy metals) and organic pollutants (e.g., chlorinated aliphatics and aromatics) generally exert a greater inhibitory effect on nitrifiers than heterotrophs, which are responsible for carbonaceous removal. The effects of some pollutants on organic carbon removal and nitrification are comparatively presented in Table 3.7.

Heavy metals are commonly found in industrial effluents and leachates and are known for their adverse effects on nitrification. However, recent studies show that the total concentration of a heavy metal entering a system is not the critical factor for inhibition, but rather the speciation properties [63, 64].

In many cases, the most important compounds leading to inhibition of nitrification are the two substrates in nitrification: ammonium (NH_4^+) and nitrite (NO_2^-). Depending on the concentration of these ions, and the pH and temperature, unionized nitrogen forms, namely free ammonia (FA) and free nitrous acid (FNA) may reach levels that are inhibitory to nitrification [65]. At high ammonium concentrations in wastewater, both ammonium oxidation and nitrite oxidation steps may be inhibited as a result of increased FA.

3.6.2.2
Nitrification in the Presence of PAC

In many cases, PAC addition to activated sludge has been shown to enhance nitrification [66–71]. This observation is actually contrary to expectations since adsorption of the ammonium ion onto activated carbon at the pH and concentrations found in wastewater treatment is negligibly small. Obviously, other mechanisms play a role in the enhancement of nitrification.

PACT for nitrification purposes is usually considered in the treatment of leachate or industrial wastewaters. It has also been applied at municipal wastewater treatment plants where industrial sources significantly contribute to the incoming wastewater [49].

Table 3.7 Threshold concentrations of some selected pollutants inhibitory to the activated sludge process (adapted from [43]).

Pollutant	Concentration (mg L^{-1})	
	Organic carbon removal	Nitrification
Chromium (hexavalent)	1–10	0.25
Copper	1.0	0.005–0.5
Cyanide	0.1–5	0.34
Nickel	1.0–2.5	0.25
Zinc	0.8–10	0.08–0.5
Phenol	200	4–10

In general, the beneficial effect of PAC on the nitrification process is higher than that on organic carbon removal. The main mechanism leading to enhanced nitrification is the adsorption of toxic/inhibitory organics and inorganics onto activated carbon upon which the bulk water concentrations are reduced (Figure 3.6). However, nitrifiers can only be protected from such compounds if these are adsorbable on activated carbon. Further, in the presence of PAC, slowly biodegradable inhibitory compounds may be more effectively removed by heterotrophic bacteria upon acclimation. This enhanced elimination of organics is also a factor improving nitrification.

Addition of PAC may improve the nitrification potential of an industrial effluent. In an early study, low PAC dosages were shown to permit nitrite formation. However, full oxidation of nitrite into nitrate became possible only at high PAC dosages. With 50 mg L^{-1} PAC in the influent stream, ammonium concentrations in the effluent of a coke plant were reduced from 80 to 1 mg L^{-1}. At this dose almost complete nitrification was achieved with only 1 mg L^{-1} nitrite formation [72]. As discussed in detail in Chapter 4, also in the co-treatment of sanitary landfill leachate and domestic wastewater in laboratory-scale activated sludge reactors, inhibition of nitrification could be largely overcome by the addition of PAC to activated sludge [71].

3.6.3
Enhancement of Denitrification by Activated Carbon

GAC serves as a support medium for denitrifiers in anoxic reactors such as the FBR reactors [73]. As in the case of aerobic biological processes, use of GAC as an adsorptive and bioactive medium enhances denitrification. For example, the GAC-FBR process could achieve complete denitrification at an influent nitrate concentration of 330 mg $NO_3^- - N$ L^{-1}, without any nitrite accumulation [73]. Since nitrate is a nonadsorbable ion, the main role of GAC in such biological systems is attributed to the retention of pollutants that could be detrimental to anoxic removal.

3.6.4
Effect of Activated Carbon Addition on Inorganic Species

Several studies have revealed that activated carbon also has an adsorptive capacity for inhibitory or toxic substances such as heavy metals. For example, it was shown that the adsorptive capacity of PAC for hexavalent chromium (Cr(VI)) was much greater than that of activated sludge flocs. The addition of PAC to a Cr(VI)-containing stream improved biological removal, as evidenced from increased oxygen uptake rate (OUR) and COD removal [74].

The combination of adsorption and biodegradation in the PACT process partially enables the removal of heavy metals such as nickel, copper, and cadmium from the solution phase. This can possibly be attributed to the adsorption of metals upon complexation with an organic molecule or to the precipitation of metals on the carbon surface by the sulfides present in the carbon [16].

In suspended- and attached-growth systems, the behavior of inorganics substrates such as heavy metal ions is quite different from that of organic substrates. Heavy metal ions have a much lower capability for adsorption onto biological sludge and PACT sludge than organic substrates. Some studies showed that heavy metals such as zinc have a lower inhibitory effect on PACT sludge than on activated sludge because of adsorption onto PAC [17]. The PACT process was considered to be a good alternative in resisting the inhibition caused by copper; it also allowed the rapid recovery of biological activity [75]. Removal of less biodegradable organics was predominantly influenced by the presence of heavy metal ions. The biodegradability index (BOD_5/COD ratio) could be an indicator for the evaluation of inhibition caused by heavy metals [76]. Low BOD_5/COD ratios in a wastewater suggest that heavy metals will possibly exhibit a higher adverse effect on microorganisms. Therefore, addition of activated carbon would be more beneficial in such cases.

The effects of heavy metals were also studied in activated sludges operated in the SBR mode. In one study, the biomass was exposed to high concentrations of Cd(II) and Cu(II). A kinetic study conducted in the react period of SBR operation showed that PAC minimized the inhibitory effects of these heavy metals on bioactivity. As a result of PAC dosing to the influent at 143 mg L^{-1}, the average COD removal efficiency was increased from about 60% to 85% [77].

3.6.5
Enhancing Effects of Activated Carbon in Anaerobic Treatment

Anaerobic biological treatment is also combined with PAC and GAC adsorption. Addition of PAC to an anaerobic digester could stimulate methane production and improve sludge filterability and supernatant clarity. A hybrid anaerobic PACT reactor was developed for treating high-strength wastewaters with COD ranging from 10 000 to 20 000 mg L^{-1} [41]. Anaerobic biological treatment and activated carbon adsorption are often integrated for the treatment of high-strength, refractory, or toxic wastewaters such as phenolic [78] or textile wastewaters [79]. The GAC-FBR configuration is widely employed in the anaerobic treatment of such wastewaters [78, 80]. In another configuration, where GAC is added to an Upflow Anaerobic Sludge Blanket reactor (GAC-UASB), wastewater first passes through the GAC bed at the bottom layer before reaching the sludge blanket. Thus, the GAC bed provides an adsorption layer and protects the granular sludge in the UASB from toxicants [79].

3.6.6
Properties of Biological Sludge in the Presence of Activated Carbon

3.6.6.1
Improvement of Sludge Settling and Thickening

The main reason for enhanced settling upon PAC addition is the weighting effect of activated carbon particles. Production of heavier and well-structured flocs results in an increased sludge settling velocity and a more efficient solid–liquid separation [81]. In particular, coarse-ground activated carbons, which are mesoporous carbons produced from lignite, are reported to promote rapid settling and enhance the dewatering of sludge [82].

Very early studies with the PACT system at DuPont's Chamber Works demonstrated excellent capability of the clarifier in terms of thickening. It was argued at that time that this good performance even raised the possibility of eliminating a separate sludge thickener [83].

The Sludge Volume Index (SVI) is a widely used parameter for a rough characterization of sludge settleability. Low SVI values indicate good settleability, whereas high values normally show poor settling. In membrane operation, PAC addition was shown to improve sludge settling properties as indicated by the decrease in SVI [70, 84].

3.6.6.2
Improvement in the Dewaterability of Sludge

The effect of PAC addition on substrate removal and sludge dewaterability was also demonstrated in the co-treatment of sanitary landfill leachate and domestic wastewater with activated sludge [81, 85]. The sludge dewaterability was expressed in terms of filterability using the Specific Resistance to Filtration (SRF). One of the indications of enhanced dewaterability is the decrease in SRF. The ratio of leachate in the total wastewater varied from 5% to 25% on volumetric basis. Coinciding with the removal of readily biodegradable COD, SRF first peaked and then declined when mainly nonbiodegradable matter was left over (Figure 3.11). PAC addition at a dose of 2500 mg L^{-1} decreased the nonbiodegradable COD in the leachate to very low levels. When leachate was treated alone, SRF of sludge was much higher than in the treatment of domestic wastewater alone or in the co-treatment with leachate (leachate fraction: 10%) (Figure 3.11). In any case, at high leachate inputs, sludge dewaterability and sludge settling were negatively affected. PAC addition to activated sludge could decrease the residual COD and suppress the peak SRF values. PAC also had a considerable positive impact on sludge dewaterability, particularly at high leachate fractions. Therefore, when the leachate input is at a high level, the addition of PAC to activated sludge may be an effective solution in improving effluent quality and increasing the dewaterability of sludge.

3.6.7
Effect of PAC on Membrane Bioreactors (MBRs)

3.6.7.1
Importance of Microbial Products

Microorganisms are embedded in a matrix of extracellular polymeric substances (EPS). It is largely accepted that membrane fouling in MBRs can primarily be attributed to the presence of EPS. EPS are composed of a variety of organic substances such as carbohydrates, proteins, nucleic acids, (phospo)lipids, and other

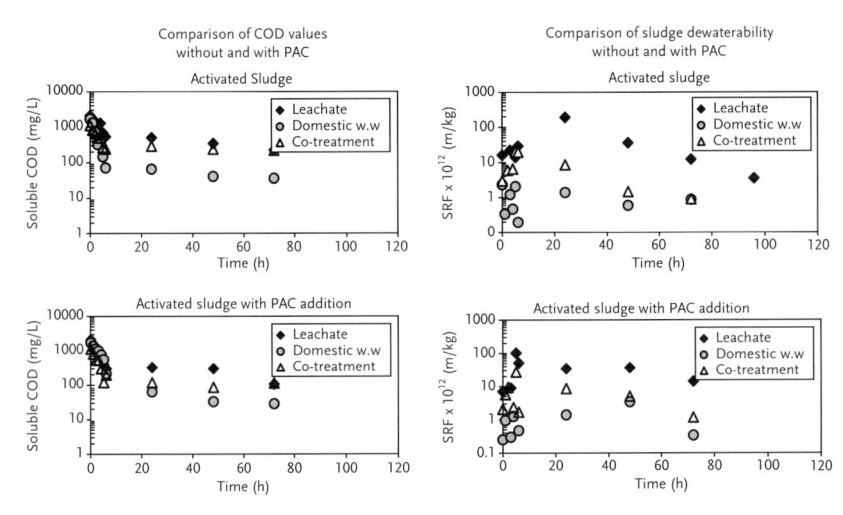

Figure 3.11 Domestic wastewater treatment, leachate treatment, and co-treatment of domestic wastewater and leachate: change in soluble COD and dewaterability (expressed as SRF) without and with PAC (redrawn after [81]).

substances found at or outside the cell surface and in the intercellular space of microbial aggregates [54]. In general, EPS are known to be mainly dominated by proteins and carbohydrates. EPS play an essential role in wastewater treatment processes. They are involved in the formation of activated sludge flocs and the development of fixed biofilms. Additionally, they have an influence on the dewaterability of sludge [86].

There are two kinds of EPS: soluble and bound EPS. Bound EPS include microbially produced bound polymers, but also lysis and hydrolysis products as well as absorbed or attached matter. Soluble EPS include microbially produced soluble polymers, hydrolysis products of attached organic matter, and organic molecules released by cell lysis.

Another group of microbial substances is the SMP. SMP are defined as soluble cellular components that are released during cell lysis, diffuse through the cell membrane, are lost during synthesis, or are excreted for some purpose. Compared to the residual substrate, it is mostly the SMP that constitutes the majority of organic matter found in the effluent of biological treatment units [3]. It has been shown that EPS and SMP intersect in many cases and that soluble EPS was equivalent to loosely bound EPS, which in turn, was the same as SMP [87].

3.6.7.2
Effect of Activated Carbon on Microbial Products

In early studies activated carbon was considered as one of the most promising adsorbents for the removal of biologically generated residual organics [88]. Further, it was suggested that biodegradation of metabolic end products (MEP) was possible. About 20% of microbial products were removed in the PACT system.

However, removal of MEP took place mainly by adsorption, whereas 50% of MEP were irreversibly adsorbed. Only 4% could be metabolized through bioregeneration [89].

Several additives have been tested in MBR operation to overcome the problem of fouling. In MBR operation the positive effect of PAC on sludge filterability is attributed to several factors. One of them is the enhanced membrane scouring by PAC. The solid and sharp-edged PAC may work as a scouring agent that removes deposited foulants from membrane surface [90]. Another effect of PAC is the adsorption of membrane foulants which tend to deposit on the membrane surface [55, 91]. Since the foulants are kept within the PAC sludge they cannot reach the membrane, resulting in improved filterability. The adsorption effect of activated carbon for organic foulants and fine colloids is considered more important than the scouring effect [60]. Another study demonstrated that the positive effect of PAC was due to formation of stronger flocs and therefore a higher shear resistance and lower release of foulants [90].

The major components of EPS, namely the proteins and carbohydrates, are large molecules which are mostly held to be responsible for biofouling. Thus, the decrease in EPS allows a reduction in membrane fouling [92]. Some authors have reported that it is the soluble (free) fraction of EPS, referred to as SMP that causes membrane fouling [54]. However, the relative importance of EPS and SMP components in causing fouling is not yet clear.

A sludge incorporating PAC had a lower content of EPS inside the floc than conventional activated sludge. This comparatively increased the porosity of the cake layer and thus enhanced membrane flux [70]. It was also shown that EPS deposition on the membrane surface was reduced by PAC addition. This suggested that the permeable PAC particulate layer filtered out microbial cells and colloids and prevented them from reaching the membrane surface [93].

PAC addition may also bring about a change in EPS composition. PAC addition was shown to alter the protein-to-carbohydrate (polysaccharide) ratio in EPS, whereas this ratio increased with PAC dose. The reduction in the amount of polysaccharides by PAC addition was attributed to (i) the attached growth of biomass upon PAC addition resulting in reduced synthesis of polysaccharides and (ii) the adsorption of polysaccharides onto PAC. Large amounts of polysaccharides in EPS cause severe fouling. Therefore, their reduction due to PAC addition lowers the fouling potential of an MBR. In the presence of PAC the protein to polysaccharide ratio in the MBR reactor was high, implying better settleability due to the hydrophobic nature of protein. This argument correlated well with the consistently lower SVI obtained in the case of PAC addition to MBR [84].

The effectiveness of PAC in suppressing membrane biofouling and enhancing permeability relies on its continuous addition to an MBR system. Since PAC adsorbs EPS and SMP, its adsorption capacity for other organic compounds may be lowered. For instance, the adsorption capacities of PACs produced from coal, soft coal, and sawdust for the target compound 3,5-dichlorophenol (3,5-DCP) were reduced to 29%, 34%, and 17%, respectively, when placed in an activated sludge compared to PAC placed into an abiotic environment. This reduced capacity was attributed to the loading of PAC with SMPs that were characterized by Dissolved Organic Matter (DOM) having an MW greater than 50 kDa [94]. Similarly, 'aged PAC' loses its adsorption capacity due to pore blocking by fine foulants, microbial products, and dead cells. This may also explain the poor performance of the PACT sludge in MBR operation at 'infinite' SRT. Aged PAC may even increase the fouling potential of a membrane by increasing the MLSS concentration and viscosity of the suspension in contact with the membrane [60].

3.6.7.3
Effect of Activated Carbon on Membrane Filtration

The addition of PAC may slightly reduce sludge production [92], a factor improving filterability. However, the most striking effect of PAC in MBRs is the reduction in membrane fouling and extension of the operation interval [95].

An important parameter considered in membrane operation is the Specific Cake Resistance (SCR). According to the Carman–Kozeny equation, the specific cake resistance (SCR), represented by α, was expressed as follows in a study [60]:

$$\alpha = \frac{180(1-\varepsilon)}{\rho d_p^2 \varepsilon^3} \tag{3.7}$$

where α is the SCR (L M^{-1}), ε is the porosity of the cake layer, ρ is the particle density (M L^{-3}), and d_p is the effective particle diameter (L).

Equation (3.7) indicates that a relatively small change in porosity (ε) can bring about a large change in SCR. Moreover, SCR is strongly dependent on the effective particle diameter (d_p). Consequently, the floc size distributions in conventional and in PAC-added MBR operations have an impact on filtration characteristics.

Another important parameter in MBR operation is the irreversible fouling resistance which indicates the cake resistance caused by irreversible pore plugging and restriction. In fixed-pressure tests, the flux decline profile is estimated as an indicator of filtration performance. In one case, the sustainable flux was described as the maximum flux at which the transmembrane pressure (TMP) did not rise over a period of 15 min [60]. Filtration tests including these measurements indicated superior filtration of the PAC-added MBR sludge compared to conventional sludge [60]. Further, it was shown that the TOC in the MBR permeate went down with PAC addition.

The same study related the SCR to the mean floc size (D_{50}) in biological sludge. At the PAC dose of 1 g L^{-1}, the biomass floc was marginally bigger compared to that in activated sludge alone. However, at 3 and 5 g L^{-1} PAC, the reverse was observed and the mean particle size in the PACT sludge (D_{50}) shifted to somewhat lower values [60]. Similar trends were noted by other researchers. For example, in ultrafiltration of sludges, the particle size distribution of PAC was considerably lower than that of activated sludge. Thus, the mean size of activated sludge decreased from 100 to 82 μm by PAC addition [70]. In another study, addition of PAC at a low dosage (2 g L^{-1}) increased the particle size, whereas the opposite was observed at higher concentrations [84]. These observations were attributed to the dominant particles present. At low PAC concentrations, suspended biomass growth is dominant. The mean particle size of sludge increases since PAC merely attaches onto biomass flocs. However, at higher PAC concentrations, PAC is dominant in the sludge and attached biomass growth on PAC particles prevails. Since PAC particles have a smaller size than activated sludge flocs, the mean particle size of PACT sludge shifts to lower values [60, 84].

Although Eq. (3.7) indicates that the PACT sludge would have a higher SCR because of the smaller d_p, PACT sludges with higher PAC concentrations are reported to have a lower SCR, higher membrane fluxes and higher permeabilities compared to activated sludge alone [60, 70]. This increased performance is attributed to the higher porosity (ε) of the cake upon PAC addition [70]. This improved porosity is believed to result from the reduction of EPS in interstitial voids and/or reduced cake compressibility [60]. PAC addition to an MBR also decreased the rise in TMP since flocs on the membrane surface and organic matter in the inner pores of the membrane were remarkably reduced [96].

The importance of PAC dosage has also been shown in other cases [93]. The cake resistance decreased with increase in PAC dosage. The irreversible fouling resistance was effectively reduced at a PAC dosage of $0.75\,g\,L^{-1}$; however, this effect weakened significantly at $1.5\,g\,L^{-1}$. The degree of irreversible fouling was equal at PAC dosages of 0 and $1.5\,g\,L^{-1}$. A PAC dosage of $0.75\,g\,L^{-1}$ was the optimum in terms of organic removal and filtration flux [93]. Another study also showed that low PAC dose ($0.5\,g\,L^{-1}$) was better in reducing membrane fouling in MBR reactors [97]. It is apparent that there exists an optimum PAC dose above which filtration characteristics may diminish. However, the optimum dose seems to be case-specific. For example, Munz and co-workers showed that fouling rate continuously decreased with an increase in PAC dose up to $3\,g\,L^{-1}$ [98].

3.7
Regeneration of PACT and BAC Sludges

Management of residuals is an important aspect of any activated carbon operation. The residuals of integrated carbon adsorption and biological treatment, or the waste solids, are composed of biological solids (biosolids), spent carbon, organics, and inorganics that are adsorbed onto carbon. The typical disposal methods for these waste solids are landfilling, incineration, and regeneration.

Regeneration processes are discussed in Chapter 2 for activated carbon adsorption applications without biological activity. The major advantage of the GAC process is the possibility to regenerate the spent carbon. This is one of the main reasons why in wastewater treatment GAC is used more frequently than PAC.

Wet Air Oxidation (WAO), as explained in Chapter 2, is a liquid-phase oxidation reaction using dissolved oxygen. WAO is used in industry for the treatment of spent caustic wastewater streams generated by ethylene plants and refineries and oxidation of reduced sulfur species and complex organic contaminants such as phenols [99].

Wet Air Regeneration (WAR) is a form of WAO. The advantage of the WAR process is that it can be used to regenerate activated carbon in the sludges emerging from PACT and BAC operations.

Because of the liquid phase operation, WAR is more frequently used for PACT systems where regeneration of activated carbon is difficult by the methods described in Chapter 2. At large-scale plants, operators of PACT® systems can regenerate PAC using the WAR technology while at the same time achieving the destruction of biological sludge and adsorbed organics.

In the WAR technology, the biological sludge-PAC mixture is heated up to 205–260 °C at a pressure of 500–1000 psi in the presence of excess oxygen. The system operates autothermally, requiring no outside source of fuel other than that supplied by the biomass/adsorbed organics in the spent carbon. At this point it is crucial that the quality of regenerated carbon has not deteriorated to the extent that upon reuse the effluent quality is affected.

Nowadays, the patented Zimpro® WAR operates at moderate temperatures so that the spent carbon is regenerated without damaging the surface [100]. According to this process, the carbon is regenerated in slurry form without labor-intensive dewatering steps. Therefore, WAR is the preferred method for regeneration of PACT sludges. Carbon recovery rates of 90% or more are common. At the same time, the process oxidizes the biological sludge, leaving behind a small amount of sterile ash. Therefore, expensive sludge disposal is not needed [47]. If the wastewater contains priority pollutants, destruction of such pollutants would also be of interest during regeneration. The PACT/WAR system was shown to effectively oxidize nonbiodegradable priority pollutants adsorbed on PAC. For most of the adsorbable priority pollutants such as benzene, chlorobenzenes, phenol, chlorophenols, and nitrophenols, removal efficiencies exceeded 99%. For example, in the case of 2-chlorophenol, destruction exceeded 98.4% during WAR whereas the residual 2-chlorophenol was irreversibly adsorbed and did not appear in the liquid phase [101]. During WAR, no formation of priority pollutants was observed. Only low-molecular-weight organic molecules appeared in the effluent of WAR as a result of destruction [101].

WAR is also effective for the stabilization of activated sludge. In the WAR process, at temperatures and pressures of <220 °C and <500 psi, respectively, biological sludge is broken down and the dewaterability of the sludge is enhanced. When higher temperatures and pressures are applied, destruction of volatile solids occurs, and wet oxidation can also be used for sludge destruction [99]. However, a problem that may be encountered in regeneration of PACT sludge by WAR at higher temperature (260 °C) and pressure (750 psi) is that the organic and ammonia content of the return liquor may be very high (COD: 3000–8000 mg L^{-1}, NH$_4^+$–N > 1000 mg L^{-1}) due to destruction. This brings about an additional load on the PACT process upon recycling [16].

Another possibility in the regeneration of loaded activated carbon (both GAC and PAC) is the regeneration by microorganisms, known as bioregeneration. Bioregeneration, a challenging process depending on various conditions, is extensively discussed in Chapter 7.

References

1 Abegglen, C. (2009) Spurenstoffe eliminieren: Kläranlagentechnik. *Eawag News 67d/Juni*, **2009**, 25–27.
2 Orhon, D. and Artan, N. (1994) *Modelling of Activated Sludge Systems*, Technomic Publishing Co., Lancaster.
3 Rittmann, B.E. and McCarty, P.L. (2001) *Environmental Biotechnology: Principles and Applications*, McGraw-Hill.
4 Aktaş, Ö. (2006) Bioregeneration of activated carbon in the treatment of phenolic compounds. Bogazici University, Institute of Environmental Sciences, Ph.D. Thesis, Istanbul, Turkey.
5 Çeçen, F., Kocamemi, B.A., and Aktaş, Ö. (2010) Chapter 9: Metabolic and co-metabolic degradation of industrially important chlorinated organics under aerobic conditions, in *Xenobiotics in the Urban Water Cycle: Mass Flows, Environmental Processes, Mitigation and Treatment Strategies* (eds D. Kasinos, K. Bester and K. Kümmerer), Book Series: Environmental Pollution 16, Springer Science+Business Media, pp. 161–178.
6 Joss, A., Zabczynski, S., Göbel, A., Hoffmann, B., Löffler, D., McArdell, C.S., Ternes, T.A., Thomsen, A., and Siegrist, H. (2006) Biological degradation of pharmaceuticals in municipal wastewater treatment: proposing a classification scheme. *Water Research*, **40**, 1686–1696.
7 Omil, F., Suarez, S., Carballa, M., Reif, R., and Lema, J.M. (2010) Chapter 16: Criteria for designing sewage treatment plants for enhanced removal of organic micropollutants, in *Xenobiotics in the Urban Water Cycle: Mass Flows, Environmental Processes, Mitigation and Treatment Strategies* (eds D. Kasinos, K. Bester, and K. Kümmerer), Book Series: Environmental Pollution 16, Springer Science+Business Media, pp. 283–306.
8 Suarez, S., Carballa, M., Omil, F., and Lema, J.M. (2008) How are pharmaceutical and personal care products (PPCPs) removed from urban wastewaters? *Reviews in Environmental Science and Biotechnology*, **7**, 125–138.
9 Abegglen, C., Joss, A., McArdell, C.S., Fink, G., Schlüsener, M.P., Ternes, T.A., and Siegrist, H. (2009) The fate of selected micropollutants in a single-house MBR. *Water Research*, **43**, 2036–2046.
10 Aga, D.S. (ed.) (2008) *Fate of Pharmaceuticals in the Environment and in Water Treatment Systems*, CRC Press Taylor & Francis Group.
11 Virkutyte, J., Rajender S., Varma, R.S., and Jegatheesan, V. (2010) *Treatment of Micropollutants in Water and Wastewater*, IWA Publishing, ISBN: 9781843393160.
12 Lehnberg, K., Kovalova, L., Kazner, C., Wintgens, T., Schettgen, T., Melin, T., Juliane Hollender, J., and Dott, W. (2009) Removal of selected organic micropollutants from WWTP effluent with powdered activated carbon and retention by nanofiltration, in *Atmospheric and Biological Environmental Monitoring* (eds Y.J. Kim et al.), DOI 10.1007/978-1-4020-9674-7 10, Springer Science+Business Media B.V.
13 Press-Kristensen, K. (2007) Biodegradation of xenobiotic organic compounds in wastewater treatment plants, Ph.D. Thesis, Institute of Environment & Resources, Technical University of Denmark.
14 van Beelen, E.S.E. (2007) Municipal waste water treatment plant (WWTP) effluents: a concise overview of the occurrence of organic substances. *Association of River Waterworks – RIWA*, ISBN: 978-90-6683-124-7.
15 Orshansky F. and Narkis, N. (1997) Characteristics of organics removal by PACT simultaneous adsorption and biodegradation. *Water Research*, **31**, 391–398.
16 Lankford, P.W. and Eckenfelder, W.W. (1990) *Toxicity Reduction in Industrial Effluents*, Van Nostrand Reinhold, New York, USA.

17 Sher, M.I., Arbuckle, Wm. B., and Shen, Z. (2000) Oxygen uptake rate inhibiton with PACT^{TM} sludge. *Journal of Hazardous Materials*, **73** (2), 129–142.

18 Sundstrom, D.W., Klei, H.E., Tsui, T., and Nayar, S. (1979) Response of biological reactors to the addition of powdered activated carbon. *Water Research*, **13**, 1225–1231.

19 Meidl, J.A. (1997) Responding to changing conditions: how powdered activated carbon systems can provide the operational flexibility necessary to treat contaminated groundwater and industrial wastes. *Carbon*, **35** (9), 1207–1216.

20 Weber, W.J., Pirbazari, M., and Melson, G.L. (1978) Biological growth on activated carbon: an investigation by scanning electron microscopy. *Environmental Science and Technology*, **12** (7), 817–819.

21 Sublette, K.L., Snider, E.H., and Sylvester, N.D. (1982) A review of the mechanism of powdered activated carbon enhancement of activated sludge treatment. *Water Research*, **16**, 1075–1082.

22 Çeçen, F., Urban, W., and Haberl, R. (1992) Biological and advanced treatment of sulfate pulp bleaching effluents. *Water Science and Technology*, **26** (1/2), 435–444.

23 Sontheimer, H., Crittenden, J., and Summers, R.S. (1988) *Activated Carbon for Water Treatment*, 2nd edn, Forschungstelle Engler-Bunte Institute, Universität Karlsruhe, Karlsruhe, Germany.

24 Weber, W.J., Friedman, L.D., and Bloom, R. (1972) Biologically-extended physicochemical treatment. Proceedings of the Sixth Conference of the International Association on Water Pollution Research, Jerusalem, Pergamon: Oxford.

25 Ying, W.C. and Weber, W.J. Jr. (1979) Bio-physicochemical adsorption model systems for wastewater treatment. *Journal of the Water Pollution Control Federation*, **55** (11), 2661–2677.

26 Peel, R.G. and Benedek, A. (1983) Biodegradation and adsorption within activated carbon adsorbers. *Journal WPCF*, **55** (9), 1168–1173.

27 Pirbazari, M., Voice, T.C. and Weber, W.J. (1990) Evaluation of biofilm development on various natural and synthetic media. *Hazardous Waste and Hazardous Materials*, **7** (3), 239–250.

28 Zhou, L., Li, G., An, T., Fu, J., and Sheng, G. (2008) Recent patents on immobilized microorganism technology and its engineering application in wastewater treatment. *Recent Patents on Engineering*, **2**, 28–35.

29 Emanuelsson, M.A.E., Henriques, I.S., Ruben, M., Ferreira, J., and Castro, P.M.L. (2006) Biodegradation of 2-fluorobenzoate in upflow fixed bed bioreactors operated with different growth support materials. *Journal of Chemical Technology and Biotechnology*, **81**, 1577–1585.

30 Arvin, E. and Harremoes, P. (1990) Concepts and models for biofilm reactor performance. *Water Science and Technology*, **22**, 171–192,

31 Imai, A., Onuma, K., Inamory, Y., and Sudo, R. (1995) Biodegradation and adsorption in refractory leachate treatment by the biological activated carbon fluidized bed process. *Water Research*, **29**, 687–694.

32 Shammas, N.K. and Wang, L.K. (2009) Chapter 17: Emerging attached-growth biological processes, in *Handbook of Environmental Engineering-Vol. 9: Advanced Biological Treatment Processes* (eds L.K. Wang, Y. Hung, and N.H. Shammas), Humana Press, a part of Springer Science+Business media, USA, pp. 649–682.

33 Morgenroth, E. (2008) *Biofilm reactors, Biological Wastewater Treatment, Principles, Modelling and Design* (eds M. Henze, M.C.M. van Loosdrecht, G.A. Ekama, and D. Brdjanovic), IWA Publishing, pp. 493–511.

34 Sutton, P.M. and Mishra, P.N. (1994) Activated carbon based biological fluidized beds for contaminated water and wastewater treatment: a state-of-the-art review. *Water Science and Technology*, **29** (10–11), 309–317.

35 Imai, A., Iwami, N., Matsushige, K., Inamori, Y., and Sudo, R. (1993) Removal of refractory organics and nitrogen from landfill leachate by the microorganism-attached activated carbon fluidized bed process. *Water Research*, **27** (1), 143–145.

36 Verma, M., Brar, S.K., Blais, J.F., Tyagi, R.D., and Surampalli, R.Y. (2006) Aerobic biofiltration processes – advances in wastewater treatment. *Practice Periodical of Hazardous, Toxic and Radioactive Waste Management*, **10** (4), 264–276.

37 Ouyang, C.F. and Liaw, C.M. (1994) The optimum medium of the suspended bio-medium aeration contactor process. *Water Science and Technology*, **29** (10–11), 183–188.

38 Lessard, P. and Le Bihan, Y. (2003) Fixed film processes, in *The Handbook of Water and Wastewater Microbiology* (eds D. Mara and N. Horan), Academic Press, ISBN 0–12–470100–0.

39 Weber, W.J. Jr., Jones, B.E., and Katz, L.E. (1987) Fate of toxic compounds in activated sludge and integrated PAC systems. *Water Science and Technology*, **19**, 471–782.

40 Hutton, D.G. and Robertaccio, F.L. (1975) Waste water treatment. US Patent, 3,904,518, assigned to E. I. Du Pont de Nemours and Co., Wilmington, DE, 9 pages, 9 September 1975.

41 Hutton, D.G., Meidl, J.A., and O'Brien, G.J. (1996) Chapter 6: The PACT® system for wastewater treatment, in *Environmental Chemistry of Dyes and Pigments* (eds A. Reife and H.S. Freeman), John Wiley & Sons Inc., pp. 105–131.

42 Heath, H.W. (1981) Full-scale demonstration of industrial wastewater treatment utilizing Du Pont's PACT Process. Project Summary, Report No: EPA-600/S2–81–159, U.S. EPA, Robert S. Kerr Environmental Research Laboratory, Ada, Oklohama, USA.

43 Tchobanoglus, G. and Burton, F.L. (1991) *Wastewater Engineering Treatment Disposal, Reuse, Metcalf and Eddy Inc.*, 3rd edn, McGraw-Hill Book Company.

44 Meidl, J.A. (1990) *PACT® Systems for Industrial Wastewater Treatment in Physical Chemical Processes, Innovative Hazardous Waste Treatment Technology Series*, vol. **2**, Technomic Publishing Company, Pennsylvania, USA.

45 Siemens Water Technologies Corp. (2006) Brochure: ZP-PACT-BR-1206, http://www.siemens.com/water.

46 Eckenfelder, W.W. and Grau, P. (eds) (1992) *Activated Sludge process Design and Control Theory and Practice*, Technomic Publishing Co. Inc. Printed in the USA.

47 Meidl, J. and Lu, C.C. (2006) Wastewater treatment: raising the standard of industrial wastewater PACT system operation. *Filtration +Separation*, December 2006, Application 32.

48 Meidl, J.A. (1999) Use of the PACT System to Treat Industrial Wastewaters for Direct Discharge or Reuse, Technical Report No: 097, US Filter Zimpro systems Products and Services, Rothschild, Wisconsin, USA, presented at Jubilee XX Conference Science & Technology, 14–15 October 1999, Katowice, Poland.

49 Shammas, N.K. and Wang, L.K. (2009) Chapter 16: Emerging suspended-growth biological processes, in *Handbook of Environmental Engineering – Vol. 9: Advanced Biological Treatment Processes* (eds L.K. Wang, Y. Hung, and N.H. Shammas), Humana Press, a part of Springer Science+Business media, USA, pp. 619–648.

50 Marquez, M.C. and Costa, C. (1996) Biomass concentration in PACT process. *Water Research*, **30**, 2079–2085.

51 Siemens Water Technologies Corp. (2007) Brochure: ZP-BTCHdr-DS-0507, http://www.siemens.com/water.

52 Siemens Water Technologies Corp. (2010) Brochure: OGS-PACTWAR-BR-0610, http://www.siemens.com/water.

53 Judd, S. (2006) *The MBR Book: Principles and Applications of Membrane Bioreactors in Water and Wastewater Treatment*, 1st edn, Printed in Great Britain, Elsevier Ltd.

54 Judd, S., Kim, B.G., and Amy, G. (2008) Membrane Bio-reactors, in *Biological Wastewater Treatment, Principles, Modelling and Design* (eds M. Henze, M.C.M. van Loosdrecht, G.A. Ekama, and D. Brdjanovic), IWA Publishing, pp. 335–360.

55 Kim, J.S. and Lee, C.H. (2003) Effect of powdered activated carbon on the performance of an anaerobic membrane bioreactor: comparison between cross-flow and submerged membrane systems. *Water Environment Research*, **75** (4), 300–307.

56 Schultz, T.E. (2005) Biological wastewater treatment. Reprinted from the October 2005 Chemical Engineering Magazine. © 2005 Access Intelligence, LLC.

57 Fatone, F. (2010) Chapter 18: Membrane bio-reactors: A cost-effective solution to enhance the removal of xenobiotics from urban wastewaters, in *Xenobiotics in the Urban Water Cycle: Mass Flows, Environmental Processes, Mitigation and Treatment Strategies* (eds D. Kasinos, K. Bester, and K. Kümmerer), Book Series: Environmental Pollution 16, Springer Science+Business Media, pp. 339–354.

58 Sutton, P.M. (2006) Membrane Bioreactors for Industrial Wastewater Treatment: Applicability and Selection of Optimal System Configuration. Proceedings of the Water Environment Federation, WEFTEC 2006: Session 41 through Session, pp. 3233–3248.

59 Pinnekamp, J. and Merkel, W. (2008) Abschlussbericht zu den Forschungsvorhaben: "Senkung des Anteils organischer Spurenstoffe in der Ruhr durch zusätzliche Behandlungsstufen auf kommunalen Kläranlagen - Gütebetrachtungen" Vergabe-Nr. 07/111.1 (IV-7–042 1 D 7) und "Senkung des Anteils organischer Spurenstoffe in der Ruhr durch zusätzliche Behandlungsstufen auf kommunalen Kläranlagen - Kostenbetrachtungen"Vergabe-Nr. 07/111.2 (IV-7–042 1 D 6). ISA, RWTH Aachen IWW, Mülheim an der Ruhr.

60 Ng, C.A., Sun, D., and Fane, A.G. (2006) Operation of membrane bioreactor with powdered activated carbon addition. *Separation Science and Technology*, **41**, 1447–1466.

61 van Hege, K., Dewettinck, T., Claeys, T., de Smedt, G., and Verstraete, W. (2002) Reclamation of domestic wastewater using biological membrane assisted carbon filtration (BioMAC). *Environmental Technology*, **23**, 971–980.

62 Henze, M., Grady, C.P.L., Gujer, W., Marais, G.V.R., and Matsuo, T. (1987) A general model for single-sludge wastewater treatment systems. *Water Research*, **21** (5), 505–515.

63 Çeçen, F., Semerci, N., and Geyik, A.G. (2010) Inhibition of respiration and distribution of Cd, Pb, Hg, Ag and Cr species in a nitrifying sludge. *Journal of Hazardous Materials*, **178**, 619–627.

64 Çeçen, F., Semerci, N., and Geyik, A.G. (2010) Inhibitory effects of Cu, Zn, Ni and Co on nitrification and relevance of speciation. *Journal of Chemical Technology and Biotechnology*, **85**, 520–528.

65 Anthonisen, A., Loehr, R.C., Prakasam, T.B.S., and Srinath, E.G. (1976) Inhibition of nitrification by ammonia and nitrous acid. *Journal of the Water Pollution and Control Federation*, **48**, 835–852.

66 Lee, J.S. and Johnson, W.K. (1979) Carbon-slurry activated sludge for nitrification-denitrification. *Journal of the Water Pollution Control Federation*, **51** (1), 11–126.

67 Ng, A., Stenstrom, K., and Marrs, D.R. (1987) Nitrification enhancement in the powdered activated carbon-activated sludge process for the treatment of petroleum refinery wastewaters. *Journal of the Water Pollution and Control Federation*, **59**, 199–211.

68 Ng, A. and Stenstrom, M.K. (1987) Nitrification in powdered activated carbon-activated sludge process. *Journal of Environmental Engineering*, **113**, 1285–1301.

69 Specchia, V. and Gianetto, A. (1984) Powdered activated carbon in an

activated sludge treatment plant. *Water Research*, **18**, 133–137.
70 Kim, J.S., Lee, C.H. and Chun, H.D. (1998) Comparison of ultrafiltration characteristics between activated sludge and BAC sludge. *Water Research*, **32** (11), 3443–3451.
71 Aktaş, Ö. and Çeçen, F. (2001) Nitrification inhibition in landfill leachate treatment and impact of activated carbon addition. *Biotechnology Letters*, **23**, 1607–1611.
72 Eckenfelder, W.W. (1980) Recent developments and future trends in industrial wastewater technology. *Pure and Applied Chemistry*, **52**, 1973–1983.
73 Ersever, I., Ravindran, V., and Pirbazari, M. (2007) Biological denitrification of reverse osmosis brine concentrates: II. Fluidized bed adsorber reactor studies. *Journal of Environmental Engineering and Science*, **6**, 519–532.
74 Lee, S.E., Shin, H.S., and Paik, B.C. (1989) Treatment of Cr(VI)-containing wastewater by addition of powdered activated carbon to the activated sludge process. *Water Research*, **23** (1), 67–72.
75 Bing, X. and Kyoung, S. (2003) Uptake of copper ion by activated sludge and its bacterial community variation analyzed by 16S rDNA. *Journal of Environmental Science*, **15** (3), 328–333.
76 Mochidzuki, K. and Takeuchi, Y. (1999) The effects of some inhibitory components on biological activated carbon process. *Water Research*, **33** (11), 2609–2616.
77 Lim, P.E., Ong, S.A., and Seng, C.E. (2002) Simultaneous adsorption and biodegradation processes in sequencing batch reactor (SBR) for treating copper- and cadmium-containing wastewater. *Water Research*, **36**, 667–675.
78 Chen, J., Wu, J., Tseng, I., Liang, T., and Liu, W. (2009) Characterization of active microbes in a full-scale anaerobic fluidized bed reactor treating phenolic wastewater. *Microbes and Environment*, **24** (2), 144–153.
79 Kuai, L., De Vreeze, I., Vandevivere, P., and Verstraete, W. (1998) GAC-ammended UASB reactor for the stable treatment of toxic textile wastewater. *Environmental Technology*, **19**, 1111–1117.
80 Fox, P. and Suidan, M.T. (1991) A fed-batch technique to evaluate biodegradation rates of inhibitory compounds with anaerobic biofilms attached to granular activated carbon. *Water Science and Technology*, **23**, 1337–1346.
81 Çeçen, F., Erdinçler, A., and Kiliç, E. (2003) Effect of powdered activated carbon addition on sludge dewaterability and substrate removal in landfill leachate treatment. *Advances in Environmental Research*, **7**, 707–713.
82 Norit Americas Inc. (2001) Wastewater Treatment and Remediation with Norit Activated Carbon, www.norit-americas.com (access on Sep.14, 2010).
83 Flynn, B.P. and Stadnik, J.G. (1979) Start-up of a powdered activated carbon-activated sludge treatment system. *Journal of Water Pollution Control Federation*, **51** (2), 358–369.
84 Satyawali, Y. and Balakrishnan, M. (2009) Effect of PAC addition on sludge properties in an MBR treating high strength wastewater. *Water Research*, **43**, 1577–1588.
85 Kiliç, E. (2001) The effect of PAC addition on the substrate removal and sludge characteristics in leachate treatment. Bogazici University, Institute of Environmental Sciences, M.Sc. Thesis, Istanbul, Turkey.
86 Flemming, H.C. and Wingender, J. (2003) The crucial role of extracellular polymeric substances in biofilms, in *Biofilms in Wastewater Treatment: A Multidisciplinary Approach* (eds S. Würtz, P. Bishop, and P. Wilderer), IWA Publishing.
87 Laspidou, C.S. and Rittmann, B.E., (2002) A unified theory for extracellular polymeric substances, soluble microbial products, and active and inert biomass. *Water Research*, **36**, 2711–2720.
88 Tsezos, M. and Benedek, A. (1980) Removal of organic substances by biologically activated carbon in a fluidized-bed reactor. *Journal of the*

Water Pollution Control Federation, **52** (3 I), 578–586.
89 Schultz, J.R. and Keinath, T.M. (1984) Powdered activated carbon treatment process mechanisms. *Journal of the Water Pollution Control Federation*, **56** (2), 143–151.
90 Remy, M., Potier, V., Temmink, H., and Rulkens, V. (2010) Why low powdered activated carbon addition reduces membrane fouling in MBRs. *Water Research*, **44**, 861–867.
91 Fang, H.H.P., Shi, X., and Zhang, T. (2006) Effect of activated carbon on fouling of activated sludge filtration. *Desalination*, **189**, 193–199.
92 Lesage, N., Sperandio, M., and Cabassud, C. (2008) Study of a hybrid process: adsorption on activated carbon/membrane bioreactor for the treatment of an industrial wastewater, Chemical Engineering and Processing: Process Intensification 47(3), 10th French Congress on Chemical Engineering, March 2008, pp. 303–307.
93 Ying, Z. and Ping, G. (2006) Effect of powdered activated carbon dosage on retarding membrane fouling in MBR. *Seperation and Purification Technology*, **52**, 154–160.
94 Widjaja, T., Miyata, T., Nakano, Y., Nishijima, W., and Okada, M. (2004) Adsorption capacity of powdered activated carbon for 3,5-dichlorophenol in activated sludge. *Chemosphere*, **57**, 1219–1224.
95 Li, Y., He, Y., Liu, Y., Yang, S., and Zhang, G. (2005) Comparison of filtration characteristics between biological powdered activated carbon sludge and activated sludge in submerged membrane bioreactors. *Desalination*, **174**, 305–314.
96 Vyrides, I. and Stuckey, D.C. (2009) Saline sewage treatment using a submerged anaerobic membrane reactor (SAMBR): effects of activated carbon addition and biogas-sparging time. *Water Research*, **43**, 933–942.
97 Remy, M., van der Marel, P., Zwijnenburg, A., Rulkens, W., and Temmink, H. (2009) Low dose powdered activated carbon addition at high sludge retention times to reduce fouling in membrane bioreactors. *Water Research*, **43**, 345–350.
98 Munz, G., Gori, R., Mori, G., and Lubello, C. (2007) Powdered activated carbon and membrane bioreactors (MBRPAC) for tannery wastewater treatment: long term effect on biological and filtration process performances. *Desalination*, **207**, 349–360.
99 Siemens Water Technologies Corp. (2006) Brochure:ZP-WAO-BR-0906, http://www.siemens.com/water.
100 Siemens Water Technologies Corp. (2009) Brochure: ZP-WARdr-CS-0909, http://www.siemens.com/water.
101 Randall, T.L. (1983) Wet oxidation of PACT process carbon loaded with toxic compounds. 38th Purdue Industrial Waste Conference Proceedings of Purdue University, West Lafayette, Lewis Publishers, Inc., Chelsea, Michigan, USA, 323–338.

4
Effect of Activated Carbon on Biological Treatment of Specific Pollutants and Wastewaters: Laboratory- and Pilot-Scale Studies

Özgür Aktaş and Ferhan Çeçen

After highlighting the main mechanisms in integrated adsorption and biological removal in Chapter 3, Chapter 4 delineates the extent of removal achieved in such integrated processes. This chapter addresses the impacts of powdered (PAC) and granular activated carbon (GAC) in biological treatment of industrial wastewaters and landfill leachates, co-treatment of domestic wastewaters with other wastewaters, and in the removal of specific compounds. Chapter 4 basically provides an overview of laboratory- and pilot-scale studies, while results from full-scale operation are covered in Chapter 5.

4.1
Treatment of Industrial Wastewaters

Industrial wastewaters are very variable in composition and may contain a wide range of organic contaminants including chlorinated, phenolic, aromatic, polyaromatic, aliphatic, and volatile organics. The main reason for integration of activated carbon into biological processes in industrial wastewater treatment is to meet discharge standards and/or to reuse the treated wastewater.

4.1.1
Pharmaceutical Wastewaters

Various pharmaceutical industries synthesize raw chemicals for the pharmaceutical sector. In consequence, pharmaceutical wastewaters are very variable in both quality and quantity. Depending on the type of raw chemicals and finished products, the COD of these wastewaters often ranges from 1000 to 10 000 mg L^{-1} [1]. If organic solvents used in the production of pharmaceuticals are not properly recovered through in-plant control, they appear in the effluent and increase the organic load. Many compounds found in pharmaceutical wastewaters are considered to be nonbiodegradable, while others exhibit toxic and/or inhibitory

Activated Carbon for Water and Wastewater Treatment: Integration of Adsorption and Biological Treatment.
First Edition. Ferhan Çeçen and Özgür Aktaş
© 2011 WILEY-VCH Verlag GmbH & Co. KGaA, Weinheim.
Published 2011 by WILEY-VCH Verlag GmbH & Co. KGaA

effects. Examples of toxic compounds include the solvents chloroform, isopropyl alcohol, methanol, and methylene chloride [2]. Moreover, some pharmaceuticals and their metabolites are known to be partially or completely nonbiodegradable [3]. Some solvents also have a high potential for stripping during treatment. Therefore, for industries such as pharmaceuticals manufacturing and chemical synthesis, PAC addition to biological systems can be regarded as an option for the elimination of toxic and/or nonbiodegradable matter and for the prevention of the stripping of volatile organic compounds (VOCs) into the atmosphere.

In a study on pharmaceutical effluents, mainly two treatment options were considered: (i) pretreatment of the pharmaceutical wastewater with PAC before its co-treatment with the domestic wastewater of the facility or (ii) direct addition of PAC to activated sludge in the co-treatment. The latter simulated the PACT process [2]. The wastewater samples investigated in this study were taken from a factory which produced various active pharmaceutical ingredients that were mainly used in antibiotic formulations. Solvents such as isopropyl alcohol, acetone, ethyl acetate, isopropyl ether, methylene chloride, N,N-dimethyl formamide, N,N-dimethyl acetamide, ethanol, methanol, chloroform, toluene, and tetrahydrofurane were employed in production. After the recovery of solvents, the wastewater first underwent physicochemical treatment and then entered an activated sludge tank for co-treatment with the domestic wastewater of the facility. In the samples taken from the plant, after physicochemical treatment, the industrial wastewater was analyzed as follows: COD: $25\,890\,\text{mg}\,\text{L}^{-1}$, BOD_5: $13\,300\,\text{mg}\,\text{L}^{-1}$, Total Kjeldahl Nitrogen (TKN): $306\,\text{mg}\,\text{N}\text{L}^{-1}$, NH_4^+-N: $206\,\text{mg}\,\text{L}^{-1}$, pH: 9.52. When this wastewater was mixed in laboratory studies in 1 : 1 ratio (v/v) with domestic wastewater as in actual plant operation, the initial COD was about $12\,500\,\text{mg}\,\text{L}^{-1}$.

In the above-mentioned study, inhibition was monitored by oxygen uptake rate (OUR) measurements in both activated sludge systems receiving wastewater pretreated with PAC or operated with simultaneous PAC addition [2]. OUR mainly indicated the organic substrate removal. In both cases, PAC addition relieved the inhibition, as evidenced from the rise in OUR of activated sludge. Also, the nonbiodegradable fraction in the wastewater could be decreased by PAC addition. The adsorption capacity of PAC in terms of soluble COD was relatively low. On the other hand, PAC addition to activated sludge proved to be very effective in the reduction of UV_{254} and UV_{280}. These two parameters indicate the presence of specific organic pollutants with aromatic structures. Therefore, it was concluded that such compounds were mainly susceptible to removal by adsorption. PAC addition was also particularly effective in the reduction of color that was expressed in terms of absorbances at 400 and 436 nm. On the other hand, the second option, namely pretreatment of wastewater with PAC before activated sludge treatment, did not bring about an additional advantage. It was recognized that such pretreatment would require additional units whereas direct PAC addition to activated sludge was a promising option in the upgrading of existing systems. However, PAC addition to activated sludge did not improve nitrification, indicating the high sensitivity of nitrifiers to a number of compounds in these effluents [2].

Table 4.1 Effect of PAC addition to activated sludge in the treatment of a pharmaceutical wastewater (adapted from [5]).

Parameter	Influent	Control effluent	PACT effluent
COD (mg L^{-1})	10207	1992	920
BOD$_5$ (mg L^{-1})	5734	57	43
MLSS (mg L^{-1})	–	2430	6070
SVI (mL g^{-1})	–	157	65

Inhibition of nitrification is also reported by others when PACT is applied to pharmaceutical wastewaters [4]. PAC addition also did not lead to complete adsorption of residual COD. Destructive techniques such as ozonation and advanced oxidation processes (AOPs) were suggested for an effective removal of residual inhibitory and noninhibitory compounds before and/or after PACT.

Data from another study on pharmaceutical wastewaters demonstrated much better improvement in effluent quality by PAC addition (1500 mg L^{-1}) in a laboratory-scale activated sludge reactor (Table 4.1) [5]. In comparison to BOD, particularly COD removal was improved (from 80% to 91%), indicating that PAC addition was more plausible for the removal of nonbiodegradable organics than the biodegradable ones. Also, lower Sludge Volume Index (SVI) values indicated the positive impact of PAC on sludge settling. This finding may be important in activated sludge systems which are prone to bulking and foaming in the treatment of these wastewaters.

Pilot continuous-flow activated sludge reactors (5.5–6 m^3) with and without PAC addition were also tested for the treatment of a pharmaceutical wastewater [1]. In the raw wastewater at the inlet of a full-scale activated sludge reactor COD was in the range of 2260–12 000 mg L^{-1} with an average value of 7030 mg L^{-1}, whereas BOD$_5$ was in the range of 1700–4400 mg L^{-1} with an average value of 2830 mg L^{-1}. Dosing of PAC at 208 and 827 mg L^{-1} to activated sludge decreased COD from 825 to 459 mg L^{-1} and 265 mg L^{-1}, respectively. In both reactors with and without PAC, effluent BOD$_5$ concentrations were below 10 mg L^{-1}. PAC addition also resulted in improvement of sludge settleability.

Also, PAC addition to a membrane bioreactor (MBR) was shown to lead to the removal of persistent pharmaceutical compounds such as ciprofloxacin, enrofloxacin, moxifloxacin, and their precursors with the efficiency increasing with PAC dosage [3].

4.1.2
Paper and Pulp Wastewaters

Effluents from the pulp and paper industry are complex and difficult to treat. The high organic load and toxicity in these wastewaters pose a hazard to aquatic organisms. The anaerobic GAC process was applied to bleaching effluents from the pulp and paper industry. At an initial COD of 1400 mg L^{-1}, the removal efficiencies in terms of COD and color were higher than 50% [6]. Toxicity bioassays showed that PAC addition to activated sludge slightly improved the treatment of the highly toxic bleached kraft pulp mill effluent (BKME) [7]. In another study, the PACT process was considered for the treatment of paper coating industrial wastewater. However, physical/chemical treatment coupled with a two-stage activated sludge performed better than the PACT process [8]. Nonbiodegradable COD constitutes an important proportion of total COD in these wastewaters. In one study, the nonbiodegradable COD amounted to 40–60% of the initial COD in a kraft pulp bleaching wastewater [9]. Batch and continuous-flow activated sludge treatment studies showed that PAC addition led to the removal of the nonbiodegradable COD and color from the wastewater, but did not enhance biodegradation [9].

4.1.3
Petroleum Refinery and Petrochemical Wastewaters

Refinery and petrochemical facilities produce complex wastewaters. Wastewaters produced in these facilities exert toxicity due to a number of constituents [10]. The PACT process has great potential for controlling the effluent toxicity in refinery wastewaters. Therefore, in several cases, PAC is added to the activated sludge of petroleum refineries. A pilot study conducted at the Amoco refinery in Texas City showed that 60% of soluble COD, 98% of NH_4^+–N and most phenolics were removed after PAC dosing at 50 mg L^{-1} to activated sludge [10]. According to the results of the pilot plant study at the Amoco refinery, the effluent qualities of the PACT system (4000 mg PAC L^{-1} in the reactor) and the GAC reactors were comparable to each other, both leading to better results than activated sludge (Table 4.2) [11]. For the reduction of toxicity in the West Coast refinery wastewater a pilot-scale comparison was made between the PACT (70 mg PAC L^{-1}) and the extended aeration process. Although both performed similarly with regard to COD removal, only the effluent of the PACT met the discharge requirements set for whole-effluent toxicity. As a result, a full-scale PACT system was installed at this refinery [10].

Table 4.2 Pilot-plant effluent quality comparison of Amoco Refinery wastewater treated in activated sludge (AS), PACT, and GAC systems (adapted from [11]).

Parameter	Influent	AS control effluent	PACT effluent	GAC effluent
TOC (mg L^{-1})	59	21	13	10
COD (mg L^{-1})	214	68	44	31
Phenols (mg L^{-1})	0.98	0.050	0.034	0.034
Oil and grease (mg L^{-1})	15	3	1	Not measured

Table 4.3 PAC addition to activated sludge for the treatment of a refinery wastewater (adapted from [12]).

Parameter	Influent (mg L^{-1})	Effluent (mg L^{-1})
COD	300	140
BOD$_5$	90	1.6
Phenolics	1.6	0.02
Oil and Grease	54	<5

In a refinery in the United States, addition of PAC to an activated sludge system was tested at laboratory-scale with the purpose of reusing the wastewater in the cooling tower [12]. The low BOD$_5$/COD ratio in the wastewater was a factor favoring the use of activated carbon. The PACT system met the requirements of reuse in the cooling tower (Table 4.3).

Another study examined the BAC treatment of a wastewater contaminated with petroleum products [13]. Special emphasis was put on polycyclic aromatic hydrocarbons (PAHs). The study showed that bacteria-covered coconut shell-based activated carbon removed 99.5–99.6% of total petroleum hydrocarbons and PAHs. The BAC reactors performed much better than activated sludge. Because of the higher adsorption capacity of activated carbon, the BAC process was also superior compared to other attached-growth processes involving sorbents such as zeolite and anthracite.

Petrochemical industrial wastewater containing acrylonitrile butadiene styrene (ABS) was treated by a moving bed BAC reactor following ozone treatment [14]. Depending on the organic loading rate, COD removal efficiencies ranged between 70% and 95%. The service life of the BAC bed was four to five times longer than the GAC bed.

In the treatment of an oilfield wastewater in a two-stage Granular Activated Carbon-Fluidized-Bed Reactor (GAC-FBR) system (aerobic and anaerobic), removal efficiencies were 99% for BTEX (Benzene, Toluene, Ethylbenzene, Xylene), 74% for COD, 94% for total volatile hydrocarbons (TVH) and 94% for oil at initial COD, BTEX, TVH, and oil concentrations of 588, 17.7, 64, and 72 mg L^{-1}, respectively [15]. Another two-stage GAC-FBR system, anoxic and aerobic in series, also exhibited >98% removal for BTEX, >99% removal for COD, and 81% removal for oil and grease in the case of a gas well-produced wastewater at initial BTEX, oil, and TOC of 15.9, 53, and 211 mg L^{-1}, respectively [15]. In the treatment of an oily wastewater emerging from the reprocessing of used emulsions or suspensions, PAC addition to an activated sludge reactor increased the TOC removal from 70% to 96% [16].

4.1.4
Textile Wastewaters

Textile wastewaters are generally strongly colored with a high organic content. They are difficult to treat by conventional biological treatment processes because of the presence of refractory dyes and complex ingredients. The high color of these wastewaters results from a number of dyestuffs. The BOD_5/COD ratio in these wastewaters is generally too low (<0.2) to support biological treatment. In such cases, activated sludge can be effectively combined with PAC to decolorize textile effluents since PAC serves as a deposit for dye molecules. For example, it was shown that in the treatment of a wool-dyeing wastewater the decolorization yield reached 86% upon the addition of PAC to a batch activated sludge reactor [17].

Also, BAC filters can be used for dye removal. Simulated aqueous discharge from a carpet printing plant comprising a mixture of acid dyes was treated for color removal in an aerobic batch BAC reactor dominated by the bacteria *Pseudomonas putida* [18]. This BAC system outperformed the combined application of conventional GAC adsorption and biological treatment. Enhanced color removal was achieved in the BAC system due to the high utilization of biodegradable anthraquinone dyes. The main factor underlying the enhanced utilization was the accumulation of dye at the carbon surface, as discussed in Chapter 3 in Section 3.2.2.3. For nonbiodegradable azo dyes, biosorption was an important removal mechanism. In any case, in BAC systems biosorption was higher than that in conventional immobilized systems.

An upflow biological aerated filter packed with two different layers consisting of GAC and ceramsite (artificial foundry sand) was employed for tertiary treatment of a secondary textile effluent [19]. The reactor performed well under steady-state conditions. The average COD, NH_4^+-N, and total nitrogen (TN) in the effluent were 31, 2, and $8\,mg\,L^{-1}$, respectively, satisfying the standards for domestic reuse in China. At a constant dissolved oxygen (DO) concentration, the increase in hydraulic loading rate (HLR) from 0.13 to 0.78 $m^3\,m^{-2}\,h^{-1}$ decreased COD, NH_4^+-N, and TN removal efficiencies from 52% to 38%, 90% to 68%, and 45% to 33%, respectively. When the DO concentration was raised from 2.4 to $6.1\,mg\,L^{-1}$, the efficiency of COD and NH_4^+-N removal increased from 39% to 53% and 64% to 88%, respectively.

Anaerobic processes combined with activated carbon were also considered for the treatment of textile effluents. For example, the wastewater of a carpet dyeing factory with COD and BOD_5 concentrations up to 9000 and $1500\,mg\,L^{-1}$, respectively, was anaerobically treated in a GAC-UASB (upflow anaerobic sludge blanket) reactor with about 80% efficiency [20].

4.1.5
Other Industrial Wastewaters

The efficiency of the PACT system was tested in an industrial district producing fossil fuel products. Estrogenic and oxidative hepatotoxic substances were detected in the wastewater that was pretreated via filtration and pH adjustment. *In-vitro* bioassays showed that in the effluent of the PACT system the concentrations of such substances were reduced [21].

The PACT was also applied for a wastewater containing organic pesticides. Compared with the activated sludge system, in the presence of a special bacterial culture, the PACT system led to up to 20% higher removal in COD in plug flow activated sludge reactors [22]. Other researchers showed the successful treatment of a pretreated bactericide wastewater in the PAC-MBR configuration [23]. The bactericide wastewater containing a biocide, named isothiazolinones, along with toxic and nonbiodegradable organics, was pretreated by chemical coagulation and electrochemical oxidation. This pretreatment decreased the COD to about 9000 mg L^{-1}. The PAC-MBR configuration further decreased the COD below 100 mg L^{-1} with an efficiency of 99.6%, whereas the COD in the effluent of a conventional MBR was about 1000 mg L^{-1}.

The PAC-MBR configuration was also used in the treatment of a tannery wastewater. The system decreased the COD from about 4000 to 800 mg L^{-1}, NH_4^+-N from 220 to 82 mg L^{-1} and phenols from 225 to 32 mg L^{-1}. PAC addition also enhanced the filtration characteristics of sludge [24].

Caustic hydrolysate wastewater from a chemical agent destruction facility was studied for its treatability in the PACT process. The caustic waste stream mainly contained sodium-2-(diisopropylamino) ethylthiolate and sodium ethylmethyl phophonate, while COD ranged from 50 to 200 mg L^{-1}. Using the actual wastewater and activated sludge, in laboratory-scale bioreactors the conditions at the WWTP were simulated. The PACT effluent met the discharge limits in terms of parameters such as BOD_5 and toxicity. Therefore, PAC-assisted activated sludge operation was proposed for full-scale application at the DuPont WWTP in Newport, Indiana [25].

Several studies point out that the use of PAC controls the toxicity inside biological reactors as well as the effluent toxicity. In a PACT system receiving chemical manufacturing wastewater, increasing the PAC dose from 100 to 500 mg L^{-1}, significantly decreased the effluent concentrations of BOD, TOC, color, and the heavy metals Cu, Cr, and Ni. Simultaneously, an apparent reduction was recorded in effluent toxicity, as evidenced from LC_{50} tests [26].

Phenolic wastewaters can also be treated by the integration of activated carbon into biological systems. Treatment of wastewaters from phenolic resin manufacturing in an anaerobic GAC-FBR reactor yielded 100% removal of phenols at an initial concentration of less than 250 mg L^{-1} within 15 days of aeration. The system yielded partial removal at phenol concentrations exceeding 250 mg L^{-1} [27].

Patoczka and Johnson studied the performance of PACT in the case of a chemical process wastewater generated at a large facility for formulating and packaging finished products consisting mostly of industrial and commercial cleaners [28]. The wastewater was fairly concentrated, with average COD in the range of 10 000–15 000 mg L^{-1}, whereas a significant part of the organics consisted of nonylphenol-based ethoxylated surfactants (NPEs) at a concentration of about 3000 mg L^{-1}. Moreover, NPEs, which biodegrade slowly, and their by-products of degradation are toxic to aquatic life. The toxicity of the wastewater due to the presence of these compounds, as measured by the Microtox® test, raised concerns for the recipient publicly owned treatment works (POTW). The target was to reduce toxicity to 30% of the EC_{50} value. Even at a sludge age of 20 days, activated sludge treatment alone was not capable of achieving this level of reduction.

Addition of 2000 mg L^{-1} PAC to batch biological reactors achieved the targets set for toxicity and effectively stabilized the performance. On the other hand, in abiotic PAC treatment in batch vessels, an average carbon concentration of 31 000 mg L^{-1} was required for the reduction of toxicity to the target value. While the PACT system required a larger capital expenditure, the PAC dosage was lower compared to abiotic PAC treatment. Therefore, the performance of the PACT system was considered to be substantially better considering the overall economics over a 15-years period [28].

Wastewater of a dyes and pigments processing plant, having a high organic content (mean COD: 2590 mg L^{-1}, BOD$_5$: 960 mg L^{-1}), as well as metals and color, was treated in a pilot-scale PAC-added (70–1800 mg PAC L^{-1}) activated sludge reactor. PAC addition raised COD and color removal efficiencies from 63% to 96% and 0% to 96%, respectively. The impact of PAC was less pronounced in the case of BOD$_5$ since already 99% reduction was achieved in this parameter in the conventional activated sludge process [29].

Paint wastewater was treated by four alternative schemes: (i) activated sludge followed by activated carbon, (ii) hybrid activated sludge–activated carbon reactor followed by activated carbon, (iii) activated carbon followed by activated sludge, and (iv) activated carbon followed by hybrid activated sludge–activated carbon reactor [30]. The best results were obtained with activated carbon followed by a hybrid reactor. The TOC was reduced from 1600 to 70 mg L^{-1} at an activated carbon dose of 500 mg L^{-1}, and further to 35 mg L^{-1} at a dose of 1000 mg L^{-1}. Higher carbon doses were considered more economical, since they also lowered SVI, which decreased from 160 mL g^{-1} at the carbon dose 500 mg L^{-1} to 120 mL g^{-1} at a dose of 1000 mg L^{-1}.

Automotive plants emit significant amounts of volatile organic compounds (VOCs) due to the use of solvents in painting operations. The VOC control process, based on vapor-phase adsorption followed by thermal oxidation, is costly to install and operate. In comparison, PAC dosing to activated sludge was considered technically feasible and cost effective. Pilot reactors effectively removed hydrophilic paint solvents (methyl ethyl ketone, n-butanol, and butyl cellosolve) and a hydrophobic solvent (toluene) with efficiencies up to 93% [31].

Wastewater from a steel mill coke plant was examined for its treatability in continuous-flow activated sludge reactors receiving a PAC dose of 200–1000 mg L^{-1} [32]. In coke plant wastewaters the concentrations of complex hydrocarbons, ammonia, phenolic compounds, cyanide, thiocyanate, and sulfides are high. Conventional activated sludge systems require pretreatment such as ammonia and cyanide stripping, specifically for this type of wastewater. For the reduction of toxicity and the removal of nonbiodegradable organics, cyanide and thiocyanate wastewaters in these wastewaters, another alternative was the addition of PAC to activated sludge [32]. While the total phenols initially amounted to 80–160 mg L^{-1}, they could be reduced to below 1 mg L^{-1}. The initial COD value was reduced from 600–1400 mg L^{-1} to below the discharge limit of 250 mg L^{-1} with removal efficiencies of 81–86%. On the other hand, the control reactor without PAC addition never did meet the discharge limits set in Taiwan. The reduction in COD was mainly due to the removal of nonbiodegradable organics since OUR was comparable in PAC-added and control reactors. On the other hand, through adsorption of toxic compounds, PAC addition also enhanced the biological removal of cyanide and thiocyanate by autotrophic microorganisms which can utilize cyanide as a carbon and nitrogen source. With PAC addition, influent cyanide at 40–80 mg L^{-1} and influent thiocyanate at 90–180 mg L^{-1} were reduced to below 4 and 20 mg L^{-1}, respectively. The control reactor could achieve less (and insufficient) cyanide removal (82%) and poor thiocyanate removal. However, the cyanide metabolism produced ammonia which could not be nitrified at a sludge age of 15 days, even in PAC-added cases.

Coke oven plant wastewater was simulated by a synthetic mixture containing the inhibitory chemicals, phenol, thiocyanide, and cyanide at concentrations up to 1400, 270, and 100 mg L^{-1}, respectively. For comparison purposes, PAC was dosed both to a biological continuous-flow stirred tank reactor (CSTR) and to a Sequencing Batch Reactor (SBR). Although efficient removal was observed in both, the performance of SBR was better than that of CSTR. Toxicity reduction amounted to up to 90% in the SBR, whereas it was limited to about 60% in the CSTR. In the SBR reactor, up to 93% of COD was removed, corresponding to an effluent COD below 400 mg L^{-1}, while the COD in the CSTR effluent was higher than 600 mg L^{-1} [33].

Wastewater from an abondoned toxic waste storage site clean-up was tested for its treatability by a pilot-scale PACT system [34]. The wastewater contained chlorinated aliphatic, aromatic, and polyaromatic hydrocarbons, nitrobenzenes, phenols, oils, solvents, thinners, pesticide production wastes, and so on. Pretreatment by wet air oxidation (WAO) aimed to produce biodegradable organics for subsequent removal in the PACT system. In pretreated and diluted wastewaters, BOD$_5$ and COD values were about 9000–10 000 and 17 000–19 000 mg L^{-1}, respectively. In the PACT system, BOD$_5$ decreased to below 1 mg L^{-1} with an efficiency exceeding 99%, while COD decreased to below 900 mg L^{-1} with efficiencies above 95%. The aim was also to decrease the extractable organic chlorine (EOCl) to below 0.1 mg L^{-1}. At influent EOCl concentrations as high as 150 mg L^{-1}, the removal efficiency exceeded 99%. Based on the pilot study results, the PACT system was selected and designed for full-scale treatment.

The nature of pollutants and the high salinity had an unfavorable effect in activated sludge treatment of a wastewater produced in an oilfield. However, PAC addition successfully stabilized the process. The positive effect of PAC was attributed to the adsorption of pollutants and immobilization of microorganisms on PAC surface. Immobilization prevented bacteria from washout at high hydraulic loads. Addition of PAC also improved sludge settling [35].

High-strength wastewater from an acrylonitrile manufacturing wastewater in which COD and cyanide concentrations were 25 000 mg L^{-1} and 15–20 mg L^{-1}, respectively, was treated in a laboratory-scale activated sludge system. PAC addition raised the COD removal in activated sludge from about 70% to 80%. OUR was significantly increased in the system whereas the SVI decreased from about 90 to 60 mL g^{-1}, indicating enhancement of substrate removal and sludge settling, respectively [36].

Another high-strength wastewater from a sugar cane molasses-based alcohol distillery was treated in a PAC-MBR. The influent COD ranging from 29 000 to 48 000 mg L^{-1} was decreased by 41% in the PAC-MBR, whereas in the absence of PAC the reduction in the MBR was only 27%. The presence of PAC also enhanced the filtration of sludge [37]. The positive impacts of PAC on the filterability of sludges are extensively discussed in Chapter 3.

4.2
Removal of Specific Chemicals

Biological treatment in combination with activated carbon is shown to be effective in the removal of various hazardous chemicals, in particular those that are strongly adsorbable or readily biodegradable. Most studies on specific pollutants deal with relatively high concentrations of pollutants. However, as discussed in Chapter 3, in recent years the elimination of micropollutants is receiving special attention. In this regard, activated carbon can play a significant role in assisting biological treatment. Compared to the removal of pollutants at high concentration, the removal of micropollutants at ng L^{-1} or low µg L^{-1} levels is harder to achieve since the extremely low concentration is a factor complicating both adsorption and biological removal. Moreover, in real wastewater treatment systems pollutants do not exist alone, but in combination with others.

The adsorbability of (micro)pollutants is roughly predicted from their octanol-water partitioning coefficient K_{ow}. This value is also indicative of the sorption potential of the pollutant onto sludge. However, in order to test the biodegradability of such pollutants, extensive studies are needed. As discussed in Chapter 3, the biodegradation potential of organic pollutants is often expressed by the k_{biol} value.

4.2.1
Volatile Organic Compounds (VOCs)

The volatility of compounds is evaluated based on Henry's constant H_c (Chapter 3). The ability of PAC or GAC to adsorb VOCs and odor-causing compounds may sometimes be so high in a biological system that the addition of a separate emission control system becomes unnecessary.

For instance, a pilot-scale GAC-FBR removed more than 99% of total BTEX with no loss of these volatile compounds to the atmosphere [38]. This BAC system allowed a tenfold stepwise increase in the BTEX loading rate up to 3 kg m^{-3} d^{-1}. Such high loadings were possible since activated carbon in the reactor first captured BTEX and was subsequently bioregenerated by the biofilm.

In another case, the GAC-FBR configuration significantly increased the removal of BTX (Benzene, Toluene, p-Xylene) compared to a nonactivated carbon FBR with no significant adsorption [39]. GAC-FBR improved the removal of benzene, toluene, and xylene from 25% to 64%, 34% to 85%, and 58% to 91%, respectively. Additionally, GAC-FBR significantly removed intermediate products of BTX biodegradation, whereas 29–47% of total BTX removed appeared as intermediates in the effluent of the nonadsorptive FBR. In another study, the GAC-FBR was used to treat groundwater contaminated with toluene [40]. Toluene at an initial concentration about 2.7 mg L^{-1} was removed with 99.4% efficiency.

Upflow fixed-bed reactors packed with GAC (GAC-UFBR) and expanded clay (EC) as support medium were operated under dynamic conditions of shock loads and starvation periods for 8 months. The reactors were fed with a synthetic wastewater containing 2-fluorobenzoate (2-FB) and the volatile compound dichloromethane (DCM) [41]. GAC had the capacity to adsorb 390 mg DCM g^{-1}, whereas the expanded clay did not have any adsorptive capacity. Hence, DCM was never detected in the effluent of the GAC reactor, which received an organic load up to 250 mg L^{-1} d^{-1} and was inoculated with DCM-degrading bacteria. On the other hand, in the effluent of the EC reactor, small amounts of DCM were observed. Through adsorption of DCM, operation was stabilized in the GAC-UFBR under high organic loading rates and shock loadings. The microbial community analysis showed that the presence of GAC in the reactor also provided better microbial stability. Compared to nonadsorbing EC, the use of GAC was also advantageous during the starvation period due to the adsorption–desorption characteristics of this material. Adsorbed 2-FB and DCM were kept within the GAC reactor. Therefore, the microbial community did not experience the same starvation as in EC.

In another study, the BAC system was found to achieve total removal of the biodegradable compounds toluene and benzene Their initial concentrations were about 600 and 400 µg L^{-1}, respectively [42]. On the other hand, the concentration of tetrachloroethylene (PCE), which is nonbiodegradable under aerobic conditions, decreased from about 900 to about 600 µg L^{-1} in 10 days of operation. However, this combined biological–adsorptive process was not effective in the treatment of carbon tetrachloride, which is a poorly adsorbed and nonbiodegradable compound.

The BAC technology was also used under anaerobic conditions for the reductive dechlorination of highly chlorinated ethylenes such as tetrachloroethylene (PCE) and trichloroethylene (TCE). Reductive dechlorination is the biological degradation of chlorinated compounds by reduction under anaerobic conditions resulting in the release of chloride ions. A study on reductive dechlorination showed that PCE (initial concentration: 152.2 µg L^{-1}) was removed with an efficiency of 91.5% in 12 days in the anaerobic BAC reactor, whereas the removal efficiency by abiotic adsorption and by biodegradation only was limited to 74.5% and 73.3%, respectively. Additionally, about 60 µg L^{-1} dichloroethene (1,2-DCE), which accumulated in the biodegradation reactor, was totally eliminated in the BAC reactor through adsorption [43].

An aerobic BAC reactor inoculated with a special culture able to degrade 1,2-dichloroethane (1,2-DCA) removed over 95% of 5 mg L^{-1} of this substance [44]. By biological activation the service life of GAC was increased from 20 days to 170 days. Although 1,2-DCA was not efficiently adsorbed on activated carbon, its biodegradability allowed efficient removal in BAC media.

As discussed in Chapter 3, PAC addition to activated sludge increases the retention of VOCs and eliminates their stripping. For example, a mass balance for

Table 4.4 Comparative effluent VOCs (in terms of % of influent) in PACT and activated sludge systems (adapted from [45]).

	% of Influent			
	Activated sludge		PACT (100 mg PAC L^{-1})	
Compound	Effluent	Off-gas	Effluent	Off-gas
Benzene	<1	16	<1	14
Toluene	<1	17	<1	0
o-Xylene	<1	25	<1	0
1,2-Dichlorobenzene	6	59	<1	6
1,2,4-Trichlorobenzene	10	90	<1	6

benzene in a PACT system showed that most of the benzene was retained in the waste PACT sludge while only a minor fraction of it was found in the off-gas [45]. In this case, PAC had a greater positive impact on off-gas quality than the liquid effluent as shown in Table 4.4.

4.2.2
Phenols

Phenol is a biodegradable compound, but it is also inhibitory at high concentrations. The buffering effect of GAC was also studied for phenol in an activated sludge operated as an SBR [46]. The reactor was subjected to three types of phenol shock loadings, namely step-up shock load, short-term fluctuation, and stepwise augmentation. The SBR configuration receiving a GAC dosage of 5 g L^{-1} was superior to conventional SBR for all cases of shock loading in the range of 500–3500 mg L^{-1} phenol. For example, in the case of stepwise augmentation, shock phenol loading was applied stepwise at every cycle by increasing the phenol concentration from 500 to 3000 mg L^{-1}. Under these conditions, the SBR dosed with GAC removed 99.9% of phenol whereas the conventional SBR removed only 70%.

In another case, PAC addition to activated sludge was reported to increase phenol removal from 58% to 98.7% at an initial phenol concentration of 100 mg L^{-1} [47]. Another study reveals that phenol at 500 mg L^{-1} was reduced to 0.15 mg L^{-1} as a result of PACT application [48]. Inhibition tests based on OUR measurements showed that PAC addition to activated sludge increased the IC_{50} value for phenol to 855–947 mg L^{-1} whereas this value was 637–862 mg L^{-1} in a conventional activated sludge [49]. In PAC added cases the IC_{50} value for 3,5-dichlorophenol (3,5-DCP) was 24–28 mg L^{-1} whereas this was 14–16 mg L^{-1} in activated sludge, indicating that in the presence of PAC higher concentrations of this compound were tolerated in activated sludge. The tests were performed with sludges taken from systems at sludge ages of 4 and 12 days [49].

Table 4.5 Improvement of removal efficiency for halogenated phenols by the PACT process (adapted from [52]).

	Removal via adsorption (%)	Removal via biodegradation (%)
3,5-DCP, PACT	88	0
3,5-DCP, control AS	9	0
3-BP, PACT	50	37
3-BP, Control AS	13	25

Nitrophenols are used in the production of dyes, photochemicals, pesticides, wood preservatives, explosives, and in leather treatment. Industries like textile and pesticide manufacturing generate large quantities of effluents containing nitrophenols like *m*-nitrophenol (MNP) and *p*-nitrophenol (PNP). Due to the toxic character of these compounds, their concentrations must be reduced to specified limits before discharge. In one study, the batch type BAC process utilizing a specific bacterial community, Pseudomonas putida ATCC 70047, could almost completely remove $100\,mg\,L^{-1}$ MNP and $150\,mg\,L^{-1}$ PNP through adsorption and biodegradation in about 16 h [50]. In another study, the removal of PNP as the sole organic carbon source was investigated in an SBR with PAC dosing. Without the addition of PAC, at a PNP concentration of $200\,mg\,L^{-1}$, the removal efficiency was poor. However, when 1.0 g PAC was applied per cycle of the SBR, the COD removal efficiency was 95%. Also, at an influent PNP concentration of $300\,mg\,L^{-1}$ almost complete PNP removal was achieved [51]. At PAC doses of 100 and $300\,mg\,L^{-1}$, the respective 2,4-dinitrophenol (2,4-DNP) concentrations in the PACT effluent were 3.9 and below $0.1\,mg\,L^{-1}$. On the other hand, in the effluent of activated sludge, the concentration was $86\,mg\,L^{-1}$. The effluent concentrations of 4-nitrophenol (4-NP) were 100 and $29\,mg\,L^{-1}$ at carbon doses of 100 and $300\,mg\,L^{-1}$, respectively. In the activated sludge effluent this value was as high as $830\,mg\,L^{-1}$ [45].

Halogenated phenols can exert toxicity on biological processes. PACT and BAC processes are widely used for the removal of these compounds, particularly for chlorophenols. In one study, for the halogenated phenols 3,5-dichlorophenol (3,5-DCP) and 3-bromophenol (3-BP), which have high adsorbabilities, the removal achieved in PACT was higher than that in the control activated sludge [52]. For the nonbiodegradable 3,5-DCP, the removal mechanism consisted of adsorption, whereas 3-BP was removed by both adsorption and biodegradation (Table 4.5). Biodegradation occurred continuously after shock loading of 3-BP, which is a highly desorbable and biodegradable compound. This indicated that activated carbon was bioregenerated. In another study, at carbon doses of 100 and $300\,mg\,L^{-1}$ in a PACT reactor, the effluent concentrations of 2,4-dichlorophenol (2,4-DCP) were 3.1 and $1.3\,mg\,L^{-1}$, respectively. Compared to that, the effluent concentration in activated sludge was $22\,mg\,L^{-1}$ [45].

A BAC biofilm reactor was established using a bacterial consortium capable of degrading 4-chlorophenol (4-CP) obtained from the rhizosphere of *Phragmites australis* [53]. The degradation of 4-CP was investigated under continuous-flow operation at a feed concentration of 20–50 mg L^{-1}. In the reactor, 4-CP removal efficiencies were in the range of 69–100%. In another study, in batch biodegradation of phenol and 4-CP, the presence of GAC minimized the substrate inhibition caused by the growth substrate phenol, reduced the toxicity of the cometabolic substrate 4-CP by adsorption, and minimized the competitive inhibition between the two substrates. In general, the presence of GAC reduced the overall degradation time and increased degradation efficiency [54]. Combination of biofilm removal with GAC adsorption also gave satisfactory results for higher chlorinated phenols such as pentachlorophenol (PCP). Batchwise operation of BAC removed 90% of 100 mg L^{-1} PCP, whereas nonbiological GAC was able to remove 64% only. In a continuous-flow BAC filter, the PCP removal efficiency exceeded 98% when a PCP-adapted culture was used [55].

The influence of PAC on activated sludge was also shown in the case of complex phenolic effluents emerging from an integrated oil refinery and gasoline washery. The average phenolic concentration in the wastewater was in the range of 90–130 mg L^{-1} consisting of 20% xylenols, 40% cresols, 35% phenol, and other phenolic compounds. The pilot study showed that 17% of organic matter was adsorbed on PAC and 28% was biodegraded. The inhibitory compounds methylphenols, xylenols, and cresols were primarily removed via adsorption without further desorption and bioregeneration. However, their adsorption enabled biological removal of other biodegradable phenolic compounds [56]. As discussed in Chapter 3, this is one of the main advantages of activated carbon in integrated treatment.

In biological removal of inhibitory compounds such as phenols, acclimation of biomass is a critical issue. Phenol-acclimated biomass could remove 99% of 500 mg L^{-1} phenol and 95% of 100 mg L^{-1} 2,4-DCP in a BAC reactor coupled with membrane filtration [57]. Another way of obtaining high removal efficiencies is to use special cultures for phenolic compounds. At initial phenol, chlorophenol, and o-cresol concentrations in the range of 100–1700 mg L^{-1} in batch reactors, a biofilm consisting of *Arthrobacter viscosus* was grown on the GAC surface, and these substances were removed with efficiencies of 99.5% to 93.4%, 99.3% to 61.6%, and 98.7% to 73.5%, respectively [58]. The addition of PAC to an MBR enabled maintenance of biological activity when 2,4-dimethylphenol, a toxic compound, was fed to the reactor at 40 mg L^{-1}. The compound could be biodegraded after an acclimation pediod [59]. Acclimation of biomass to a growth substrate such as phenol also led to the cometabolic degradation of nongrowth (cometabolic) substrates such as 2-chlorophenol (2-CP) (up to 110 mg L^{-1}) and 2-nitrophenol (2-NP) (up to 60 mg L^{-1}). These compounds were otherwise not removed by nonacclimated biomass in PAC and GAC amended reactors, even in the presence of phenol [60].

For highly chlorinated phenols, which are more difficult to break down biologically, anaerobic and aerobic BAC processes can be applied sequentially. In the presence of ethanol as the primary substrate, anaerobic GAC-FBRs degraded 100–200 mg L^{-1} PCP to monochlorophenols with an efficiency exceeding 99%. The aerobic GAC-FBR led then to complete mineralization of residual chlorophenols and phenol [61].

4.2.3
Pharmaceuticals and Endocrine Disrupting Compounds (EDCs)

In the presence of activated carbon, two different strains of *Sphingomonas sp.* effectively biodegraded the endocrine disrupting compound bisphenol A at a high concentration of 300 mg L^{-1} [62]. The presence of activated carbon also prevented the release of 4-hydroxyacetophenone, which is the major intermediate product of bisphenol A degradation, into the receiving medium. A pilot-scale study at the Utrecht wastewater treatment plant showed that BAC filtration, when applied as a tertiary treatment step under anoxic conditions, could remove 99.9% of 17β-estradiol and more than 90% of bisphenol A and nonylphenol and decrease them to ng L^{-1} levels from several μg L^{-1} [63].

GAC was added to an activated sludge reactor to improve the removal of organic micropollutants such as those within the scope of Pharmaceutical and Personal Care Products (PPCPs). Removal efficiencies of PPCPs varied depending on compound characteristics. Some substances such as the acidic pharmaceuticals naproxen and ibuprofen and more hydrophobic organic substances, like musk fragrances, were almost completely removed (90%) in activated sludge reactors with and without GAC. The main difference between these two reactors was observed in the case of more recalcitrant compounds like diazepam, carbamazepine, and diclofenac. Via adsorption in GAC added activated sludge, the former two could be removed up to 40%, whereas the latter was removed up to 85% [64].

PAC addition to activated sludge may bring about the elimination of various micropollutants. However, most studies done so far center around pharmaceuticals. For example, dosing of PAC to activated sludge was shown to reduce the persistent carbamazepine very effectively [65]. However, the removal efficiencies recorded in various studies still show discrepancies. Moreover, for various micropollutants, such as flame retardants, complexing agents (EDTA/DTPA) and perfluorinated tensides, information about the impact of PAC is still limited.

4.2.4
Pesticides and Polychlorinated Biphenyls (PCBs)

PAC addition to activated sludge improves the removal of some pesticides and PCBs [66]. By PAC addition, the concentrations of chlorinated pesticides, organosulfur pesticides, organophosphate pesticides, and PCBs were decreased from 0.35 to 0.017 mg L^{-1}, 15 to 0 mg L^{-1}, 3.03 to 1.23 mg L^{-1}, and 0.13 to 0.008 mg L^{-1}, respectively.

4.2.5
Priority Pollutants

A PACT system can also be useful when the removal of priority pollutants is of interest [5]. As shown in Table 4.6, regarding the removal of a number of priority pollutants, a PACT system can outperform activated sludge operation. The

Table 4.6 Comparison of PACT and activated sludge in removal of priority pollutants (adapted from [5]).

Pollutant	Influent (µg L^{-1})	Removal in AS (%)	Removal in PACT (%)
Benzene	81	98.5	99.6
Chlorobenzene	3660	99.1	99.8
Methylene Chloride	138	98.5	>99.7
1,2-DCE	18	90.6	>99
2,4-Dinitrotoluene	1000	31	90
2,6-Dinitrotoluene	1100	14	95
Nitrobenzene	330	94.5	99.9
2,4-Dichlorophenol	19	0	93
2,4-Dinitrophenol	140	39	>99
4-Nitrophenol	1100	25	97

advantage of PACT was most pronounced for compounds such as dinitrotoluene, dinitrophenol, nitrophenol, and dichlorophenol. Another PACT system also showed enhanced elimination of such compounds. The effluent concentrations of 1,2-dichlorobenzene, nitrobenzene, and 2,4-dinitrotoluene were 0.4, 1.7, and 70 mg L^{-1}, respectively. On the other hand, the concentrations of the same compounds in the effluent of an activated sludge were 1.7, 18, and 950 mg L^{-1}, respectively [45].

PACT was also effective in the removal of lindane at influent concentrations up to 100 µg L^{-1} with efficiencies above 96% at a PAC dosage of 30 mg L^{-1} [67]. Lindane is a compound having a high octanol–water partition coefficient. Other priority pollutants with physicochemical properties similar to lindane are expected to exhibit comparable behavior in a PACT system.

The anaerobic GAC-FBR configuration was shown to be effective in the removal of hazardous chemicals, namely phenol (500–1000 mg L^{-1}), pentachlorophenol (PCP) (100–400 mg L^{-1}), tetrachloroethylene (PCE) (2.7–16 mg L^{-1}), and chloroacetaldehyde (CAA) (90 mg L^{-1}). Stable and high removal was achieved for these compounds. Even under startup and shock load conditions, the first three compounds and CAA were removed to the extent of 99.9% and 98%, respectively [68].

4.2.6
Dyes

Specific textile azo dyes such as Orange II and Reactive Black 5 were reduced anaerobically in upflow fixed-bed BAC reactors [69]. Physical properties of activated carbon had a significant impact on the decolorization rate of azo dyes whereas chemical surface characteristics had a moderate effect. The researchers suggested that specific sites of activated carbon play a role in the catalytic reduction of azo dyes in the absence of surface oxygen groups [69]. The importance of surface oxgen groups in integrated treatment is discussed in Chapter 7. In another study, a synthetic high-strength wastewater containing methylene blue up to 1350 mg L^{-1} could be treated in an anaerobic fixed-bed BAC reactor with removal efficiencies of over 96% [70]. A fixed-bed BAC column was used in another study for the treatment of a wastewater containing the azo dye Acid Orange 7 [71]. At a high loading rate of 2.1 g L^{-1} d^{-1} the extent of removal of this dye achieved was 100%. The decolorization rate of the dye increased up an optimum concentration of 1150 mg L^{-1} whereas a decline was observed beyond this concentration. The same dye at a concentration of 20 mg L^{-1} was removed with an efficieny of 97 ± 1.6% in a PACT reactor [72]. Compared to other media such as kaolin and bentonite, PAC was regarded as a better attachment medium for bacterial cultures that are able to degrade several textile dyes [73]. The sequential anaerobic–aerobic mechanism proposed for the degradation of the azo dye Acid Orange 7 on PAC media involved cleavage of the azo bond in anaerobic microniches and oxidation of amines in aerobic microniches within the porous structure of activated carbon.

Submerged membrane batch reactors colonized by fungi or a mixed microbial community were used to treat a textile wastewater. The synthetic wastewater was prepared to contain one or both of the following azo dyes: Acid Orange II and a polymeric dye (Poly S119). The former has a simpler structure than the latter. Following the addition of PAC to the MBR, excellent and stable removal (>99%) was observed both in cases when Acid Orange II was the only substrate and in cases where both dyes were present concurrently [74].

4.2.7
Organic Pollutants in Secondary Sewage Effluents

BAC reactors are often used in the polishing of secondary sewage effluents. For example, in one case, BAC operation was shown to remove about 97% of DOC from a secondary effluent [75]. However, secondary effluents also contain specific pollutants which cannot be removed by the conventional activated sludge process. In one study, the treatment of secondary effluent containing dibutyl phthalate, bis (2-ethylhexyl) phthalate, 4-bromo-3-chloroaniline, and other phenol derivatives was examined [76]. Upon pretreatment by TiO$_2$/UV/O$_3$, some aromatic compounds including 2,4-dichloro-benzenamine, 4-bromo-3-chloroaniline, and 3,5-dimethoxy-acetophenone disappeared and some easily biodegradable small molecules 1,3-cyclohexanediamine, 9-octadecenamide, *tert*-butyldimethylsilanol, and phenol were formed. The subsequent BAC unit greatly reduced the concentrations of these organic pollutants.

In order to remove the refractory organic matter from secondary sewage effluent, PAC may be incorporated with membrane bioreactors (PAC-MBR) involving microfiltration. In one case, the secondary effluent contained nonbiodegradable organics such as humic acid, lignin sulfonate, tannic acid, and gum arabic powder [77]. The TOC removal at a mixed liquor PAC concentration of 20 g L^{-1} was higher than at 0.5~2 g PAC L^{-1} (83% and 66~68%, respectively). More than 90% of nonbiodegradable compounds (detected as UV$_{280}$) were removed in the PAC-MBR.

4.2.8
Other Chemicals

A fixed-bed BAC reactor operated in the SBR mode, the so-called GAC-SBR or BAC-SBR, removed 95% of 3-chlorobenzoate, thioglycolic acid, or a combination of both at concentrations above 100 mg L^{-1} [78]. The performance of the reactor in the SBR mode was better than that in continuous-flow operation, especially when the toxic substance, thioglycolic acid, was fed to the reactor. The GAC-SBR operation also provided a higher flexibility in the case of shock loads.

A high-strength synthetic industrial wastewater containing 1,1,1-trichloroethane (TCA), acetic acid, and phenol was treated in an expanded-bed anaerobic BAC reactor [79]. 20–430 mg L^{-1} of TCA was removed at an efficiency exceeding 99.4%. Adsorption of TCA on the GAC surface relieved the inhibitory effects of this substance on biodegradation of others, namely acetic acid and phenol. Particularly, the activity of methanogens was improved upon the adsorption of TCA. The treatment system tolerated fluctuations in TCA and removed more than 93% of acetic acid and 99% of phenol at high concentrations.

Halogenated organic compounds are characterized by the parameter Adsorbable Organic Xenobiotics (AOX). Often, this parameter is used to indicate the concentration of chlorine in chlorinated organics. The principal sources of AOX are solvents, pesticides, disinfectants, and so on. In one case, PAC addition to activated sludge at a dose above 150 mg L^{-1} raised the removal of about 120 µg L^{-1} AOX from 24% to 44% [80].

Cyanide (CN^-) emerges as a pollutant at varying concentrations in effluents from industries like metal finishing, mining (extraction of gold, silver, and so on), coke plants, paint, and ink formulation, petroleum refining, explosives manufacture, automobile manufacture, pesticide production, and synthetic fiber production. Cyanide is often found in the form of metal complexes. Due to their toxic effects, metal cyanide complexes are strictly regulated worldwide. Cyanide-containing effluents cannot be discharged to the environment without treatment. In one study, a special culture, *Pseudomonas fluorescens* immobilized on GAC was used for the removal of ferrocyanide. In this case, ferrocyanide acted as a secondary substrate in the presence of glucose, which served as the primary substrate [81]. At cyanide concentrations of 50–300 mg L^{-1}, the combined effect of adsorption and biodegradation resulted in 70–99% removal. Concurrent adsorption and biodegradation was more effective than biodegradation (69.3–96.4% removal) and adsorption (50.2–85.6% removal) performed separately. The main role of GAC in this reactor was to provide a surface on which cyanide could be adsorbed and easily biodegraded by bacteria.

Hydrogen sulfide (H_2S) is a major odorous substance in sewage air. To investigate the removal of H_2S from the gas phase, a laboratory-scale BAC filter was set up by immobilizing the isolated H_2S-degrading bacterium *Thiomonas* sp. on GAC surface. High H_2S removal efficiencies of 97–99.9% could be achieved [82]. In another study on a BAC system colonized by sulfide-oxidizing bacteria, at an influent H_2S concentration of 200–4000 mg L^{-1} the removal exceeded 98% [83].

Chromium compounds are usually released from steelworks, chromium electroplating, leather tanning, and chemical manufacturing. In the environment, chromium is usually encountered in the form of Cr(III) and Cr(VI). The solubility of Cr(VI) salts is high. In general, Cr(VI) is regarded to be more toxic than Cr(III).

Under the aerobic conditions of activated sludge, the reduction of Cr(VI) to the less mobile form Cr(III) has been reported. Cr(VI) removal (10–100 mg L^{-1}) in the combined activated carbon-activated sludge system (1–8 g PAC L^{-1} in the aeration tank) was faster than in the case of PAC or activated sludge treatment alone. Complete removal was observed below 25 mg L^{-1}, and considerable removal was obtained above 25 mg L^{-1}, while efficiencies decreased at increased concentrations [84]. A previous study had shown that PAC addition to a continuous-flow activated sludge increased Cr(VI) removal efficiency from 9% to 41% at a PAC dose of 200 mg L^{-1} to obtain a PAC concentration of 4000 mg L^{-1} in the aeration tank [85]. Since in such cases PAC will act primarily as an adsorbent for chromium, it is essential to consider the surface charges of PAC and chromium species formed under specific conditions. Moreover, also in evaluation of toxicity, the speciation of chromium, either in Cr(VI) or Cr(III) form, with the inorganics and organics in the medium, should be taken into consideration [86].

4.3
Landfill Leachate Treatment

Leachates from sanitary landfills can constitute a large pollution potential for receiving waters. Leachates usually exhibit very high concentrations in terms of BOD, COD, TKN, and other parameeers. The concentration of organic matter is high in landfill leachates during the active decomposition stage in the landfill and decreases as the landfill stabilizes. For a young landfill (less than 2 years), the COD is in the range of 3000–60 000 mg L^{-1}, with a typical value of 18 000 mg L^{-1}. BOD_5 varies in the range of 2000–30 000 mg L^{-1}, with a typical value of 10 000 mg L^{-1} [87]. The high BOD_5/COD ratio in a young landfill leachate is a factor favoring biological treatment.

Many small-scale experimental studies demonstrated that leachates from sanitary landfills could be treated aerobically. In general, all studies show that BOD and COD can be substantially reduced by aerobic treatment, particularly in cases where readily degradable substances such as volatile fatty acids comprise a high proportion of the organic matter [88].

However, in leachates from mature landfills, the BOD_5 is much lower (100–200 mg L^{-1} for landfills over 10 years old) than in young landfills, since the wastewater is mainly composed of nonbiodegradable organics. Therefore, such leachates have a very low BOD_5/COD ratio, implying that biological treatment itself cannot be a good option. Hence, advanced treatment methods such as activated carbon adsorption should be employed in order to achieve an efficient removal of organics. Moreover, most leachates from old landfills also have high levels of ammonia that require additional treatment.

Landfill leachates contain significant amounts of refractory, inhibitory, or toxic organics. Additionally, they contain heavy metals, which may decrease the efficiency of biological treatment. At high concentrations some pollutants (like phenol and cyanide) become toxic and reduce the biological activity of microorganisms [89]. Biological systems are often prone to shock loadings, since leachate composition is highly variable. Compared to organic carbon removal, nitrification is more easily inhibited by the chemicals present in leachates, since the threshold levels of inhibitors is often lower for nitrifiers than for heterotrophs, as discussed in Chapter 3.

Activated carbon treatment is usually not effective in the case of leachates having a high BOD, because this parameter represents rather low-molecular-weight (LMW) compounds that have a low adsorption potential onto activated carbon. Activated carbon is usually more suitable for the removal of the nonbiodegradable fraction, which often consist of high-molecular-weight (HMW) organics. In leachate, the majority of refractory organics consist of humic substances (about 30%) that have molecular weights less than 10 000 g mol^{-1} [90]. Activated carbon can hardly adsorb humic acids with molecular weights higher than 10 000 g mol^{-1}.

4.3.1
Leachate Treatment in PAC-added Activated Sludge Systems

PAC addition to an activated sludge reactor was shown to enhance the removal of COD, color, and ammonia from raw leachate. For an initial COD of 1800 mg L^{-1}, PAC addition raised the COD removal efficiency from 8.2–11% to 23–31%. However, in order to meet discharge standards, further treatment would still be required [91].

In another study, the anaerobic PACT was also applied to highly concentrated leachates [45]. The initial COD and BOD$_5$ in the leachate, 10 500 and 4500 mg L^{-1}, respectively, were reduced in the effluent to 2400 and 200 mg L^{-1}, respectively. When a packed reactor with nonadsorbing media was used instead of PAC, effluent COD and BOD$_5$ were 3800 and 510 mg L^{-1}, respectively.

PAC addition to activated sludge systems is an effective solution for the treatment of landfill leachate. However, in the case of high-strength leachates, the activated sludge process may be severely inhibited. Therefore, a convenient way is to discharge high-strength leachates to domestic sewage treatment plants for co-treatment. The resultant dilution of leachate with domestic wastewater can enable biological treatability. Supplement of PAC to the system further increases the effectiveness of biological treatment.

EXAMPLE 4.1: Leachate Treatment in Laboratory-Scale PACT reactors The effectiveness of PAC was investigated experimentally in the co-treatment of sanitary landfill leachate and domestic wastewater [92–96]. The main aim was to study organic carbon and nitrogen removal. For this purpose, laboratory-scale batch reactors and a continuous-flow activated sludge with sludge recycling were operated with and without PAC supplement. The landfill leachate sample taken from a

Table 4.7 Improvement of effluent quality in leachate treatment by PAC addition to batch reactors [96].

Leachate ratio in the total wastewater (%)	Reactor configuration	PAC dosage (mg L^{-1})	Initial SCOD (mg L^{-1})	Final SCOD after 72 h/(mg L^{-1})
5	AS	0	670	95
5	AS+PAC	2500	598	29
10	AS	0	1492	200
10	AS+PAC	2500	1467	68
15	AS	0	1575	280
15	AS+PAC	2500	1538	99
20	AS	0	2466	548
20	AS+PAC	2500	2556	112
25	AS	0	3135	399
25	AS+PAC	2500	3128	109

AS: activated sludge reactor
AS + PAC: activated sludge reactor with powdered activated carbon (PAC) addition

sanitary landfill had the following characteristics: pH: 8.2, Total COD (TCOD): 10750 mg L^{-1}, Soluble COD (SCOD): 9070 mg L^{-1}, BOD$_5$: 6380 mg L^{-1}, TKN: 2031 mg L^{-1}, NH$_4^+$−N : 2002 mg L^{-1}, NO$_x$-N: 128 mg L^{-1}, Total P: 6.8 mg L^{-1}, Alkalinity: 10 600 mg CaCO$_3$ L^{-1}. Leachate and domestic wastewater were mixed such that the volumetric ratio of leachate in the total wastewater varied from 5% to 20%.

Biological treatability of the leachate was first studied in batch activated sludge reactors. Activated sludge (AS) reactors received a feed composed of leachate and domestic wastewater. The abbreviation AS+PAC indicates activated sludge reactors which also contained PAC at a concentration ranging from 100 to 3500 mg L^{-1}. The performances of both systems were compared with each other. Batch studies showed that the leachate contained a nonbiodegradable COD fraction of about 30%. Thus, the nonbiodegradable COD was as high as 3225 mg L^{-1}. The effect of PAC addition was not significant in the initial periods of aeration where biodegradable matter was mainly removed. The main effect of PAC was observed in later periods when PAC adsorbed nonbiodegradable matter and considerably lowered the final residual COD (Table 4.7). Moreover, PAC addition considerably improved nitrification and accelerated NO$_x$-N production because of adsorption of toxic/inhibitory compounds. However, in all cases, high nitrite accumulation was observed in the reactors, indicating that the activity of nitrite-oxidizing bacteria was still inhibited.

Experiments were also conducted in Semi-Continuously Fed Batch (SCFB) and Continuous-Flow (CF) activated sludges with recycling. SCFB operation involved daily feeding and wastage of wastewater in batch reactors. In both types of operations, the positive effect of PAC on organic carbon removal and nitrification became more apparent at high leachate inputs. Organic carbon removal and nitrification were significantly enhanced by PAC addition in SCFB operation, as

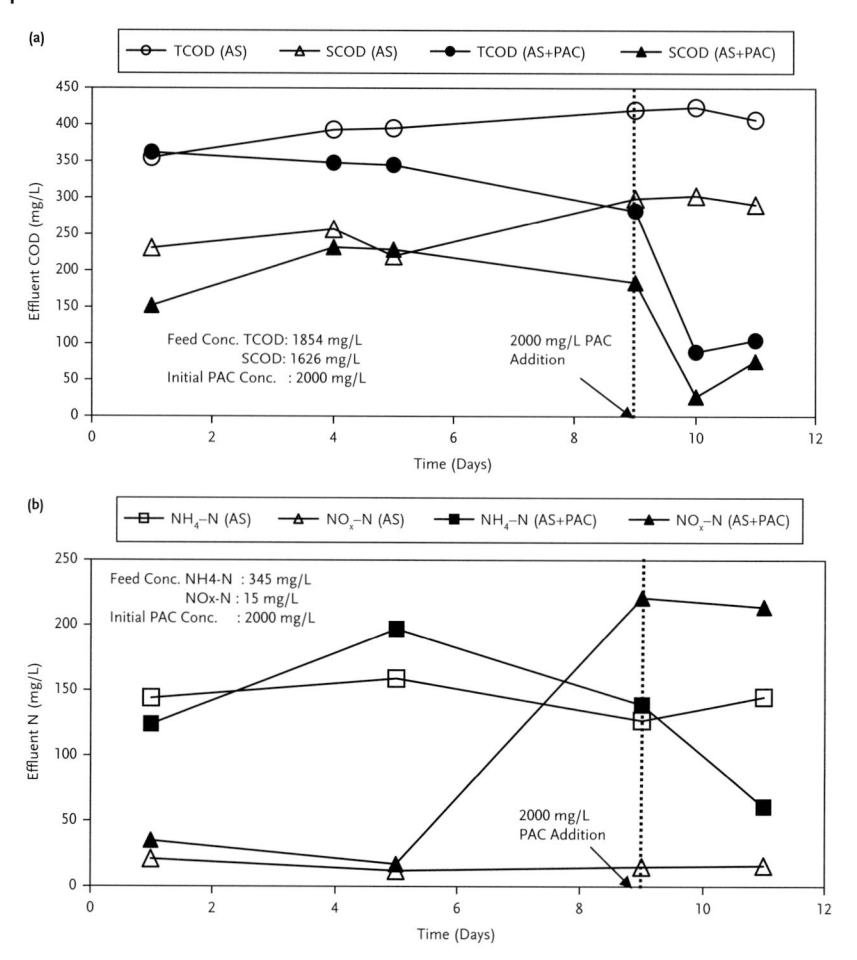

Figure 4.1 Effect of PAC addition to semi-continuously fed batch activated sludge reactors in terms of (a) organic matter and (b) ammonia removal (redrawn after [95]).

seen in Figure 4.1 a and b, respectively. Generally, at high leachate inputs, nitrification was more inhibited. With PAC addition this inhibition could be relieved to some extent.

The results pointed to differences between SCFB and CF operations. Figure 4.2a and b illustrate the performance of the CF reactor in organic carbon removal and nitrification, respectively. The positive effect of PAC on COD reduction and nitrification was more striking in CF operation than with SCFB operation. The reason was that the reactor operation was stabilized, and relatively lower COD concentrations could be attained with CF at steady-state, whereas in SCFB operation microorganisms were subject to high concentrations at each feeding step, a factor which led to inhibition of nitrification in SCFB operation. For practical purposes, the results with SCFB operation imply that the positive impact

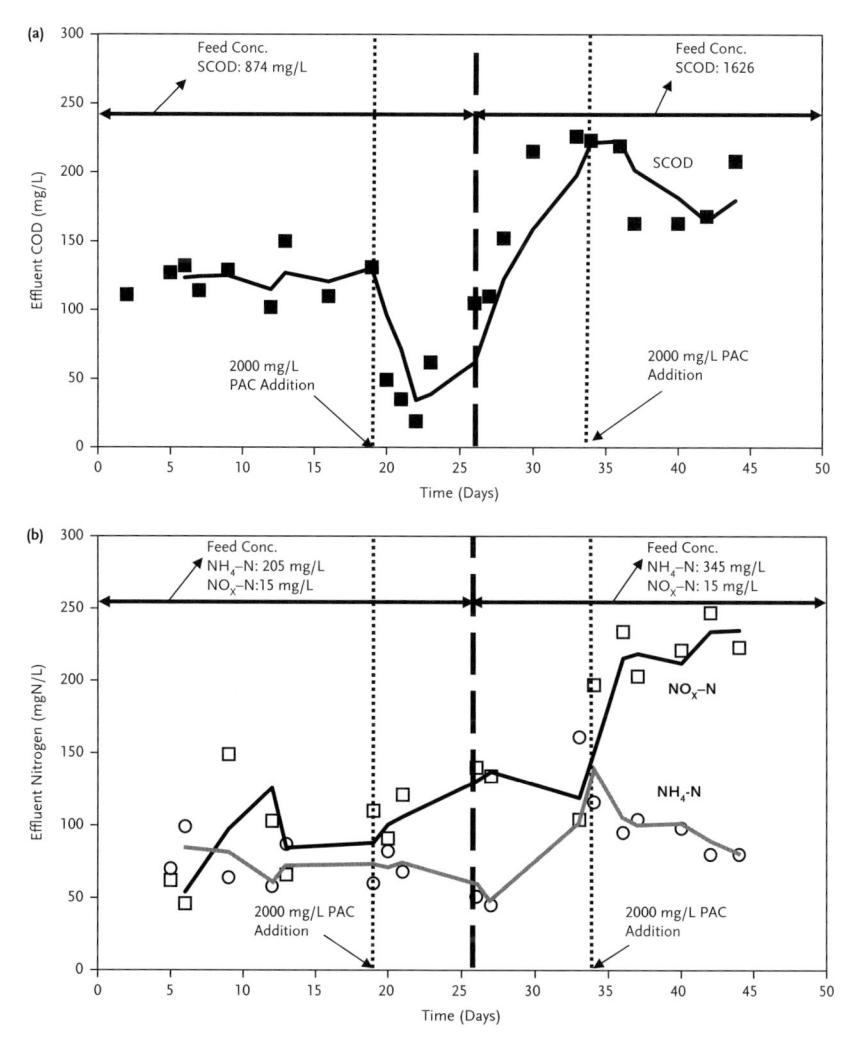

Figure 4.2 Effect of PAC addition to a continuous-flow activated sludge reactor in terms of (a) organic matter and (b) ammonia removal (redrawn after [95]).

of PAC would be reduced under intermittent substrate loadings compared to steady continuous-flow operation. The process were successfully monitored by OUR measurements, which normally corresponded directly to depletion of soluble organic substrate. Thus, OUR can be regarded as a rapid tool for estimating the changes in substrate.

The results of this study suggested that the combination of an activated sludge system with PAC adsorption was additive rather than synergistic when organic carbon removal was taken into consideration. PAC was seemingly not bioregenerated. However, PAC addition prevented nitrification inhibition and significantly enhanced nitrification. In general, PAC addition had a more pronounced effect on nitrification than on organic carbon removal.

4.3.2
Leachate Treatment Using the PAC-MBR Process

As discussed in Chapter 3, one alternative for the treatment of landfill leachates is the PAC-MBR combination. For example, in one study the aim was to investigate the impact of PAC in an MBR consisting of ultrafiltration [97]. A tubular cross-flow ultrafiltration membrane separated colloids and microorganisms to obtain a high quality permeate. In two leachate samples used in the experiments, the COD and BOD_5 were within the ranges 3000–10 000 mg L^{-1} and 1500–7000 mg L^{-1}, respectively. In terms of organic carbon removal, the performance of the PAC-MBR was similar to that of PACT and SBR; however, PAC-MBR performed better with regard to suspended solids. The process achieved 95–98% TOC removal, more than 97% removal of specific organics, and 100% removal of suspended solids. The effluent concentrations of benzene, toluene, dichloroethane, trichloroethane, methylene chloride, and dichloropropane were below 0.05 mg L^{-1}, whereas those of phenol, 2,4-dimethylphenol, naphthalene, and isopropane were below 0.005 mg L^{-1}. The average removal of phenol and benzoic acids exceeded 99.7%, whereas the removal in chlorobenzoic acids was within the range 97.5–99%.

4.3.3
Leachate Treatment in Biological Activated Carbon (BAC) Media

The GAC-FBR configuration is widely used for the removal of bulk organic matter expressed by BOD and COD and for the removal of specific aromatic compounds, aliphatics, and halogenated aliphatics. Although GAC is widely used for leachate treatment, the application of the GAC-FBR combination is rather limited. The GAC-FBR is normally not applied as a treatment step alone; advanced treatment techniques would be needed, especially for leachates containing very HMW nonadsorbable and nonbiodegradable organic compounds.

Pretreatment of leachate is almost always required before GAC-FBR operation because most leachates have high iron levels which create plugging problems above 20 mg L^{-1}. Pretreatment for suspended solids removal facilitates the operation of a GAC-FBR. Introducing a backwash phase to the GAC-FBR system may remove the suspended solids from the media, and the backwashing water can be recirculated to the pretreatment step for further removal of suspended solids. Ozonation and alum coagulation can also be successfully used as pretreatment steps before GAC filtration. Advanced oxidation processes are particularly promising pretreatment techniques, because they usually increase the BOD_5/COD ratio in a wastewater and thus favor biological treatability. For example, pretreatment of leachate with Fenton reagent was shown to increase the organic carbon removal efficiency from 40% to 75% in subsequent BAC filtration [98].

Laboratory-scale continuous-flow anaerobic and aerobic GAC-FBRs in series were found to remove 60% of refractory organics and 70% of ammonium nitrogen from the landfill leachate steadily over more than 700 days of operation. The leachate emerged from an old co-disposal site of municipal and industrial wastes and had a BOD$_5$/COD ratio lower than 0.1, a COD of 131 mg L^{-1}, and an NH$_4^+$−N content of 181 mg L^{-1} [99]. GAC-FBR was also shown to treat leachate from a hazardous waste landfill following an adjustment of pH to neutral by aeration and settling [100].

EXAMPLE 4.2: Leachate Treatment in Pilot-Scale GAC-FBR The GAC-FBR process was used to treat landfill leachate with COD and ammonia in the ranges of 800–2700 and 220–800 mg L^{-1}, respectively [101]. For this purpose, two identical pilot-scale GAC-FBRs were used. The first reactor was introduced for organic carbon removal, while the second was intended for nitrification.

The leachate taken from an old landfill had an average COD, BOD, and NH$_4^+$−N of 2450, 185, and 744 mg L^{-1}, respectively. As implied by these values, it had a high nonbiodegradable organic fraction and a high ammonia concentration. Before biological treatment, pretreatment was applied to remove heavy metals that could inhibit the growth of attached microorganisms. This pretreatment was achieved by adding lime and increasing the pH to around 9.0.

The startup of the GAC-FBR reactors was as follows: Reactors having an effective bed volume of 1500 cm^3 were filled with approximately 500 g of GAC. They were seeded with 4 L of return activated sludge from a treatment plant, such that a biofilm formed on the GAC particles. The reactors were operated at room temperature (20 °C). Instead of air, pure oxygen was utilized in the experiments to keep the DO above 5 mg L^{-1} such that COD and ammonia removal were not limited. The pH within the reactor was maintained initially at about 6.5–7. Alkalinity was also monitored during the process to ensure that nitrification was not adversely affected by pH drops. For this purpose, soda ash or sodium bicarbonate were added to the second reactor. The experimental scheme is shown in Figure 4.3.

The overall removal achieved in the two-stage GAC-FBR system for COD and NH$_4^+$−N was 54% and 70%, respectively. The corresponding effluent BOD and COD concentrations were about 2 and 900 mg L^{-1}, respectively. The high effluent COD was due to the presence of nonbiodegradable organics that had to be further reduced by advanced treatment methods. Nitrification removed 70% of the total ammonia removed in the system; the remaining 30% was removed by assimilation by heterotrophs.

Two parallel activated sludge units were operated in order to compare the results with the two-stage GAC-FBR system [101]. The sludge age in activated sludge units was kept above 20 days in order to achieve effective nitrification and BOD removal. In each reactor, BOD removal exceeded 80%. However, this value was still low for an activated sludge operating at high sludge ages. The low removal was attributed to the inhibitory and toxic effects of leachate organics. The effluent COD from activated sludge reactors was much higher (1687 mg L^{-1}) than the GAC-FBRs

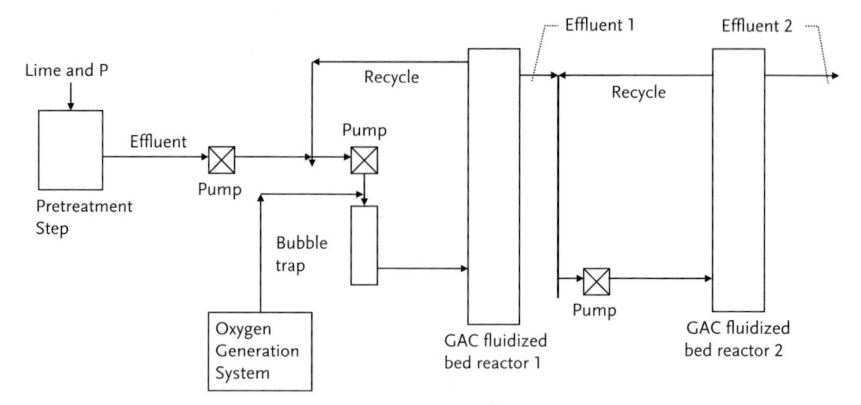

Figure 4.3 Schematic Diagram of the two-stage GAC-FBR for the treatment of landfill leachate (adapted from [101]).

(911 mg L^{-1}). Most of the ammonia remained in the effluent; nitrite and nitrate concentrations were also low, indicating that nitrification was almost completely inhibited due to the presence of leachate constituents. On the other hand, the GAC-FBR system exhibited much better organic carbon removal than that achieved using activated sludge. In the GAC-FBR, nitrification was also achieved successfully.

References

1 Osantowski, R.A., Dempsey, C.R., and Dostal, K.A. (1986) Enhanced COD removal from pharmaceutical wastewater using powdered activated carbon addition to an activated sludge system. *40th Industrial Waste Conference Proceedings*, Purdue University, West Lafayette, Indiana, USA, Lewis Publishers, Inc., Chelsea, Michigan, USA, pp. 719–727.

2 Çeçen, F. and Aktaş, Ö. (2001) Powdered activated carbon assisted biotreatment of a chemical synthesis wastewater. *J. Chem. Technol. Biotechnol.*, **76**, 1249–1259.

3 Baumgarten, S., Schroeder, H.F., Charwath, C., Lange, M., Beier, S., and Pinnekamp, J. (2007) Evaluation of advanced treatment technologies for the elimination of pharmaceutical compounds. *Water Sci. Technol.*, **56** (5), 1–8.

4 Xu, H.J., Webster, J.B., and Merchant, B.E. (1994) Biotreatment of pharmaceutical wastewater using powdered activated carbon treatment PACT system. *48th Industrial Waste Conference Proceedings*, Purdue University, West Layafette, Indiana, USA, Lewis Publishers, Inc., Chelsea, Michigan, USA, pp. 589–596.

5 Eckenfelder, W.W., Argaman, Y., and Miller, E. (1989) Process selection criteria for the biological treatment of industrial wastewaters. *Environ. Prog.*, **8** (1), 40–45.

6 Jackson-Moss, C.A., Maree, J.P., and Wotton, S.C. (1992) Treatment of bleach plant effluent with the biological granular activated carbon process. *Water Sci. Technol.*, **26** (1–2), 427–434.

7 Kennedy, K.J., Graham, B., Drostes, R. L., Fernandes L., and Narbaitz R. (2000) Microtox™ and Ceriodaphnia

dubia toxicity of BKME with powdered activated carbon treatment™. *Water SA*, **26**, 205–216.

8   Obayashi, A.W., Stover, E.L., Thomas, J.A., and Pereira, J.A. (1990) Comparison of PACT process to coupled physical/chemical biological treatment. *44th Purdue Industrial Waste Conference Proceedings*, West Layafette, Indiana, USA, Lewis Publishers, Inc., Chelsea, Michigan, USA.

9   Çeçen, F. (1994) Activated carbon addition to activated sludge in the treatment of kraft pulp bleaching wastes. *Water Sci. Technol.*, **30** (3), 183–192.

10   Wong, J.M. and Hung, Y. (2004) Oilfield and refinery wastes, in *Handbook of Industrial and Hazardous Wastes Treatment*, 2nd edn (eds L.K Wang, Y. Hung, H.H. Lo, and C. Yapijakis), Marcel Dekker, Inc., New York, USA, pp. 235–306.

11   Grieves, G., Crame, L.W., Venardos, D. G., and Ying, S. (1980) Powdered versus granular carbon for oil refinery wastewater treatment. *J. WPCF (Water Pollution Control Federation)*, **52** (3), 483–497.

12   Meidl, J.A. (1999) *Use of the PACT System to Treat Industrial Wastewaters for Direct Discharge or Reuse, Technical Report No:097*, US Filter Zimpro Systems Products and Services, Rothschild, Wisconsin, USA, presented at Jubilee XX Conference Science & Technology, 14–15 October 1999, Katowice, Poland.

13   Augulyte, L., Kliaugaite, D., Racys, V., Jankunaite, D., Zaliauskiene, A., Andersson, P.L., and Bergqvist, P.A. (2008) Chemical and ecotoxicological assessment of selected biologically activated sorbents for treating wastewater polluted with petroleum products with special emphasis on polycyclic aromatic hydrocarbons. *Water Air Soil Pollut.* **195**, 243–256.

14   Lin, C.K., Tsai, T.Y., Liu, J.C., and Chen, M.C. (2001) Enhanced biodegradation of petrochemical wastewater using ozonation and BAC advanced treatment system. *Water Res.*, **35** (3), 699–704.

15   Allen, E.W. (2008) Process water treatment in Canada's oil sands industry: II. A review of emerging technologies. *J. Environ. Eng. Sci.*, **7**, 499–524.

16   Martienssen, M. and Simon, J. (1996) Effect of activated carbon on the biological treatment of oil–water emulsions. *Acta Biotechnologica*, **16** (4), 247–255.

17   Chen, M., Wang, J., and Wang, Y. (2000) The combination of carbon adsorption with activated sludge for the decolorization of wool-dyeing wastewater. *J. Environ. Sci. Health*, **A35** (10), 1789–1795.

18   Walker, G.M. and Weatherley, L.R. (1999) Biological activated carbon treatment of industrial wastewater in stirred tank reactors. *Chem. Eng. J.*, **75**, 201–206.

19   Liu, F., Zhao, C., Zhao, D., and Liu, G. (2008) Tertiary treatment of textile wastewater with combined media biological aerated filter (CMBAF) at different hydraulic loadings and dissolved oxygen concentrations. *J. Hazardous Mater.*, **160**, 161–167.

20   Kuai, L., De Vreeze, I., Vandevivere, P., and Verstraete, W. (1998) GAC-ammended UASB reactor for the stable treatment of toxic textile wastewater. *Environ. Technol.*, **19**, 1111–1117.

21   Chen, F.A., Shue, M.F., Chen, T.C. (2004) Evaluation on estrogenicity and oxidative hepatotoxicity of fossil fuel industrial wastewater before and after the powdered activated carbon treatment. *Chemosphere*, **55**, 1377–1385.

22   Xiao-yun, Z., Ming, C. (1999) Kinetics of continuous biodegradation of pesticide organic wastewater by activated carbon-activated sludge. *J. Environ. Sci.*, **11** (1), 118–123.

23   Han, W., Wang, L., Sun, X., and Li, J. (2008) Treatment of bactericide wastewater by combined process chemical coagulation, electrochemical oxidation and membrane bioreactor. *J. Hazardous Mater.*, **151**, 306–315.

24 Munz, G., Gori, R., Mori, G., and Lubello, C. (2007) Powdered activated carbon and membrane bioreactors (MBRPAC) for tannery wastewater treatment: long term effect on biological and filtration process performances. *Desalination*, **207**, 349–360.

25 DuPont (2004) DuPont Technical Assessment on U.S. Army Newport (Indiana) Project, E.I. Du Pont de Nemours and Company Wilmington, Delaware, U.S.A.

26 Meidl, J.A. (1997) Responding to changing conditions: how powdered activated carbon systems can provide the operational flexibility necessary to treat contaminated groundwater and industrial wastes. *Carbon*, **35** (9), 1207–1216.

27 Chen, J., Wu, J., Tseng, I., Liang, T., and Liu, W. (2009) Characterization of active microbes in a full-scale anaerobic fluidized bed reactor treating phenolic wastewater. *Microbes Environ.*, **24** (2), 144–153.

28 Patoczka, J. and Johnson, R.K. (1995) Pretreatment of chemical wastewater for toxicity reduction. Presented at 68th Annual Water Environment Federation Conference & Exposition. http://www.patoczka.net/Jurek%20Pages/Papers/Pretreatment%20of%20Chemical%20%28Microtox%29.pdf (22.5.2010).

29 Shaul, G.M., Barnett, M.W, Neiheisel, T.W., and Dostal, K.A. (1983) Activated Sludge with Powdered Activated Carbon Treatment of a Dyes and Pigments Processing Wastewater. Technical report EPA-600/D-83-, Industrial Environmental Research Lab., Cincinnati, Edison, New Jersey, USA.

30 Wu, Y.C. and Mahmud, Z. (1980) Biological treatment of paint industry wastewater with activated carbon. *35th Industrial Waste Conference Proceedings*, Purdue University, West Lafayette, Indiana, USA, Lewis Publishers, Inc., Chelsea, Michigan, USA, pp. 200–210.

31 Kim, B.R., Adams, J.A., Klaver, P.R., Kalis, E.M., Contrera, M., Griffin, M., Davidson, J., and Pastick, T. (2000) Biological removal of gaseous VOCs from automotive painting operations. *J. Environ. Eng.*, **126** (8), 745–753

32 Chao, Y.M. and Shieh, W.K. (1985) PAC-activated sludge treatment of coke-plant wastewater. *40th Industrial Waste Conference Proceedings*, Purdue University, West Lafayette, Indiana, USA, Lewis Publishers, Inc., Chelsea, Michigan, USA. pp. 33–41.

33 Papadimitrou, C.A., Samaras, P., and Sakellaropoulos, G.P. (2009) Comparative study of phenol and cyanide containing wastewater in CSTR and SBR activated sludge reactors. *Bioresource Technol.*, **100**, 31–37.

34 Canney, P.J., Mordorski, C.J., Rollins, R.M., and Berndt, C.L. (1984) PACT wastewater treatment for toxic waste cleanup, *39th Industrial Waste Conference Proceedings*, Purdue University, West Lafayette, Indiana, USA, Lewis Publishers, Inc., Chelsea, Michigan, USA, pp. 413–428.

35 Dalmajica, B., Karlovic, E., Tamas, Z., and Miskovic, D. (1996) Purification of high salinity wastewater by activated sludge process. *Water Res.*, **30** (2), 295–298.

36 Ramakrishna, C., Kar, D., and Desai, J.D. (1989) Biotreatment of acrylonitrile plant effluent by powdered activated carbon-activated sludge process. *J. Ferment. Bioeng.*, **67** (6), 430–432.

37 Satyawali, Y. and Balakrishnan, M. (2009) Performance enhancement with powdered activated carbon (PAC) addition in a membrane bioreactor (MBR) treating distillery effluent. *J. Hazardous Mater.*, **170**, 457–465.

38 Hickey, R.F., Wagner, D., and Mazewski, G. (1990) Combined biological fluid-bed carbon adsorption system for BTEX contaminated groundwater remediation. *Proceedings of the Fourth National Outdoor Action Conference on Aquifer Restorating, Groundwater Modelling and Geophysical Methods*, NWWA, Las Vegas, NV, USA, pp. 813–827.

39 Zhao, X., Doh, K., Criddle, C.S., and Voice T.C. (1997) Accumulation of metabolic intermediates during shock

loads in biological fluidized bed reactors. *J. Environ. Eng.*, **123** (12), 1185–1193.
40 Shi, J., Zhao, X., Hickey, R.F., and Voice, T.C. (1995) Role of adsorption in granular activated carbon-fluidized bed reactors. *Water Environ. Res.*, **67** (3), 302–309.
41 Emanuelsson, M.A.E, Osuna, M.B., Sipma, J., and Castro, P.M.L. (2008) Treatment of halogenated organic compounds and monitoring of microbial dynamics in up-flow fixed bed reactors under sequentially alternating pollutant scenarios. *Biotechnol. Bioeng.*, **99** (4), 800–810.
42 Putz, A.R.H., Losh, D.E., and Speitel, G.E. Jr. (2005) Removal of nonbiodegradable chemicals from mixtures during granular activated carbon bioregeneration. *J. Environ. Eng.*, **131** (2), 196–205.
43 Wu, Y., Tatsumoto, H., and Aikawa, M. (2000) Modeling of tetrachloroethylene degradation by anaerobic granular biological activated carbon. *J. Health Sci.*, **46** (6), 434–440.
44 Stucki, G. and Thüer, M. (1994) Increased removal capacity for 1,2-dichloroethane by biological modification of the granular activated carbon process. *Appl. Microbiol. Biotechnol.*, **42**, 167–172.
45 Meidl, J.A. (1990) PACT systems for industrial wastewater treatment, in *Physical Chemical Processes*, Innovative Hazardous Waste Treatment Series – vol. 2 (ed. H. Freeman), Technomic Publishing Company, Pennsylvania, USA.
46 Vinitnantharat, S., Woo-Suk, C., Ishibasi, Y., and Ha, S.R. (2001) Stability of biological activated carbon-sequencing batch reactor (BAC-SBR) to phenol shock loading. *Thammasat Internat. J. Sci. Technol.*, **6** (2), 27–32.
47 Martin, M.J., Artola, A., Balaguer, M. D., and Rigola, M. (2002) Enhancement of the activated sludge process by activated carbon produced from surplus biological sludge. *Biotechnol. Lett.*, **24**, 163–168.
48 Orshansky F. and Narkis, N. (1997) Characteristics of organics removal by PACT simultaneous adsorption and biodegradation. *Water Res.*, **31**, 391–398.
49 Sher, M.I., Arbuckle, W.B., and Shen, Z. (2000) Oxygen uptake rate inhibition with PACT^{TM} sludge. *J. Hazardous Mater.*, **B73**, 129–142.
50 Escobar, E., Moreno-Pirajan, J.C., Perez, M., and Sanchez, O. (2008) Evaluation of m-Nitrophenol and p-Nitrophenol degradation with biological activated carbon by immobilization of *Pseudomona putida* ATCC 700447. Webpaper of poster presented at Industrial Biotechnology Conference, 8 –11 June 2008, Naples, Italy, http://www.aidic.it/IBIC2008/webpapers (22 April 2010).
51 Loo, Y.M., Lim, P.E., and Seng, C.E. (2010) Treatment of p-nitrophenol in an adsorbent supplemented sequencing batch reactor. *Environ. Technol.*, **31** (5), 479–487.
52 Nakano, Y., Widjaja, T., Miyata, T., Nishijima, W., and Okada, M. (2007) Effect of different adsorption/desorption properties and biodegradability of chemicals on the performance of powdered activated carbon (PACT) processes. *J. Water Environ. Technol.*, **5** (2), 71–78.
53 Carvalho, M.F., Vasconcelos, L., Bull, A.T., and Castro, P.M.L. (2001) A GAC biofilm reactor for the continuous degradation of 4-chlorophenol: treatment efficiency and microbial analysis. *Appl. Microbiol. Biotechnol.*, **57**, 419–426.
54 Loh, K.C. and Wang, Y. (2006) Enhanced cometabolic transformation of 4-chlorophenol in the presence of phenol by granular activated carbon adsorption. *The Canadian J. Chem. Eng.*, **84**, 248–255.
55 Rahman, R.A. and Anuar, N. (2009) Pentachlorophenol removal via adsorption and biodegradation. *World Acad. Sci. Eng. Technol.*, **55**, 194–199.
56 Galil, N. and Rebhun, M. (1992) Powdered activated carbon combined

with biological treatment to remove organic matter containing cresols and xylenols. *J. Environ. Sci. Health A*, **27** (1), 199–218.

57 Thuy, Q.T. and Visvanathan, C. (2006) Removal of inhibitory phenolic compounds by biological activated carbon coupled membrane bioreactor. *Water Sci. Technol.*, **53** (11), 89–97.

58 Quintelas, C., Silva, B., Figueiredo, H., and Tavares, T. (2010) Removal of organic compounds by a biofilm supported on GAC: modelling of batch and column data. *Biodegradation*, **21**, 379–392.

59 Lesage, N., Sperandio, M., and Cabassud, C. (2008) Study of a hybrid process: adsorption on activated carbon/membrane bioreactor for the treatment of an industrial wastewater. *Chem. Eng. Process.* **47** (3), 10th French Congress on Chemical Engineering, March 2008, pp. 303–307.

60 Aktaş, Ö. (2006) Bioregeneration of activated carbon in the treatment of phenolic compounds. Ph.D. Dissertation, Boğaziçi University, Institute of Environmental Sciences, Turkey.

61 Wilson, G.J., Khodadoust, A.P., Suidan, M., and Brenner, R.C. (1997) Anaerobic/aerobic biodegradation of pentachlorophenol using GAC fluidized bed reactors:optimization of the empty bed contact time. *Water Sci. Technol.*, **36** (6–7), 107–115.

62 Yamanaka, H., Moriyoshi, K., Ohmoto, T., Ohe, T., and Sakai, K. (2008) Efficient microbial degradation of bisphenol-A in the presence of activated carbon. *J. Biosci. Bioeng.*, **105** (2), 157–160.

63 Miska, V., Menkveld, H.W.H., Kuijer, L., Boersen, M., and van der Graaf, J.H.J.M (2006) Removal of nitrogen, phosphorus and other priority (hazardous) substances from WWTP effluent. *Water Sci. Technol.*, **54** (8), 189–195.

64 Serrano, D., Lema, J. M., and Omil. F. (2010) Influence of the employment of adsorption and coprecipitation agents for the removal of PPCPs in conventional activated sludge (CAS) systems. *Water Sci. Technol.*, **62** (3), 728–735.

65 Pinnekamp, J. and Merkel, W. (2008) Abschlussbericht zu den Forschungsvorhaben: "Senkung des Anteils organischer Spurenstoffe in der Ruhr durch zusätzliche Behandlungsstufen auf kommunalen Kläranlagen - Gütebetrachtungen" Vergabe-Nr. 07/111.1 (IV-7–042 1 D 7) und "Senkung des Anteils organischer Spurenstoffe in der Ruhr durch zusätzliche Behandlungsstufen auf kommunalen Kläranlagen - Kostenbetrachtungen" Vergabe-Nr. 07/111.2 (IV-7–042 1 D 6). ISA, RWTH Aachen IWW, Mülheim an der Ruhr.

66 Eckenfelder, W.W. (1980) Recent developments and future trends in industrial wastewater technology. *Pure Appl. Chem.*, **52**, 1973–1983.

67 Weber, W.J. Jr., Corfis, N.H., and Jones, B.E. (1983) Removal of priority pollutants in integrated activated sludge-activated carbon treatment systems. *J. WPCF*, **55** (4), 369–376.

68 Tsuno, H., Kawamura, W., and Oya, T. (2006) Application of biological activated carbon anaerobic reactor for treatment of hazardous chemicals. *Water Sci. Technol.*, **53** (11), 251–260.

69 Mezohegyi, G., Goncalvez, F., Orfao, J. J.M., Fabregat, A., Fortuny, A., Font, J., Bengoa, J., and Stuber, F. (2010). Tailored activated carbons as catalysts in biodecolourisation of textile azo dyes. *Appl. Catalysis B: Environ.*, **94**, 179–185.

70 Ong, S., Toorisaka, E., Hirata, M., and Hano, T. (2007) Treatment of methylene blue-containing wastewater using microorganisms supported on granular activated carbon under packed column operation. *Environ. Chem. Lett.*, **5**, 95–99.

71 Ong, S., Toorisaka, E., Hirata, M., and Hano, T. (2008) Combination of adsorption and biodegradation processes for textile effluent treatment using a granular activated carbon-biofilm configured packed column system. *J. Environ. Sci.*, **20**, 952–956.

72 Marquez, M.C. and Costa, C. (1996) Biomass concentration in PACT process. *Water Res.*, **30**, 2079–2085.
73 Barragan, B.E., Costa, C., and Marquez, M.C. (2007) Biodegradation of azo dyes by bacteria inoculated on solid media. *Dyes Pig.* **75**, 73–81.
74 Hai, F.I., Yamamato, K., Nakajima, K., and Fukushi, K. (2008) Removal of structurally different dyes in submerged membrane fungi reactor—Biosorption/PAC-adsorption, membrane retention and biodegradation. *J. Membrane Sci.*, **325**, 395–403.
75 Xing, W., Ngo, H.H., Kim, S.H., Guo, W.S., and Hagare, P. (2008) Adsorption and bioadsorption of granular activated carbon (GAC) for dissolved organic carbon (DOC) removal in wastewater. *Bioresource Technol.*, **99**, 8674–8678.
76 Li, L., Zhang, P., Zhu, W., Han, W., and Zhang, Z. (2005) Comparison of O_3-BAC, UV/O_3-BAC and TiO_2/UV/O_3-BAC processes for removing organic pollutants in secondary effluents. *J. Photochem. Photobiol. A: Chem.*, **171**, 145–151.
77 Seo, G.T. and Ohgaki, S. (2001) Evaluation of refractory organic removal in combined biological powdered activated carbon – microfiltration for advanced wastewater treatment. *Water Sci. Technol.*, **43** (11), 67–74.
78 Jaar, M.A.A. and Wilderer, P.A. (1992) Granular activated carbon sequencing batch biofilm reactor to treat problematic wastewaters. *Water Sci. Technol.*, **26** (5–6), 1195–1203.
79 Suidan, M.T., Wuellner, A.M., and Boyer, T.K. (1991) Anaerobic treatment of a high-strength industrial waste bearing inhibitory concentrations of 1,1,1-trichloroethane. *Water Sci. Technol.*, **23**, 1385–1395.
80 Bornhardt, C., Drewes, J.E., and Jekel, M. (1997) Removal of organic halogens (AOX) from municipal wastewater by powdered activated carbon (PAC)/ activated sludge (AS) treatment. *Water Sci. Technol.*, **35** (10), 147–153.
81 Dash, R.R., Balomajumder, C., and Kumar, A. (2008) Treatment of metal cyanide bearing wastewater by simultaneous adsorption and biodegradation (SAB). *J. Hazardous Mater.*, **152**, 387–396.
82 Koe, L. and Tong, D. (2005) Feasibility of using biologically activated carbon for treatment of gaseous H_2S. *J. Inst. Eng.-Singapor*, **45** (4), 15–23.
83 Rattanapan, C., Boonsawang, P., and Kantachote, D. (2009) Removal of H_2S in down-flow GAC biofiltration using sulfide oxidizing bacteria from concentrated latex wastewater. *Bioresource Technol.*, **100**, 125–130.
84 Orozco, A.M.F., Contreras, E.M., and Zaritzky, N.E. (2008) Modelling Cr(VI) removal by a combined carbon-activated sludge system. *J. Hazardous Mater.*, **150**(1), 46–52.
85 Lee, S.E., Shin, H.S., and Paik B.C. (1989) Treatment of Cr(VI) containing wastewater by addition of powdered activated carbon to the activated sludge process. *Water Res.*, **23** (1), 67–72.
86 Çeçen, F., Semerci, N., and Geyik, A.G. (2010) Inhibition of respiration and distribution of Cd, Pb, Hg, Ag and Cr species in a nitrifying sludge. *J. Hazardous Mater.*, **178**, 619–627.
87 Quasim, S.R. and Chiang, W. (1994) *Sanitary Landfill Leachate: Generation, Control and Treatment*, Technomic Publishing Co., Lancaster.
88 Robinson, H.D. and Maris, P.J. (1985) The treatment of leachates from domestic waste in landfill sites. *J. WPCF*, **57** (1), 30–38.
89 Kang, S.J., Englert, C.J., Astfalk, T.J., and Young, M.A. (1990) Treatment of leachate from a hazardous waste landfill. *44th Industrial Waste Conference Proceedings*, Purdue University, West Layafette, Indiana, USA, Lewis Publishers, Inc., Chelsea, Michigan, USA, pp. 573–579.
90 Imai, A., Onuma, K., Inamory, Y., and Sudo, R. (1995) Biodegradation and adsorption in refractory leachate treatment by the biological activated carbon fluidized bed process. *Water Res.*, **29**, 687–694.

91 Aghamohammadi, N., Aziz, H.B., and Isa, M.H. (2007) Powdered activated carbon augmented activated sludge process for treatment of semi-aerobic landfill leachate using response surface methodology. *Bioresource Technol.*, **98** (18), 3570–3578.

92 Aktaş, Ö. (1999) Powdered activated carbon addition to activated sludge in the treatment of landfill leachate. M.Sc. Thesis, Boğaziçi University, Institute of Environmental Sciences, Turkey.

93 Aktaş, Ö. and Çeçen, F. (2001) Activated carbon addition to batch activated sludge reactors in the treatment of landfill leachate and domestic wastewater. *J. Chem. Technol. Biotechnol.*, **76**, 793–802.

94 Aktaş, Ö. and Çeçen, F. (2001) Nitrification inhibition in landfill leachate treatment and impact of activated carbon addition. *Biotechnol. Lett.*, **23**, 1607–1611.

95 Çeçen, F. and Aktaş, Ö. (2001) Effect of PAC addition in combined treatment of landfill leachate and domestic wastewater in semi-continuously fed batch and continuous-flow reactors. *Water SA*, **27**, 177–188.

96 Çeçen, F. and Aktaş, Ö. (2004) Aerobic co-treatment of landfill leachate with domestic wastewater. *Environ. Eng. Sci.*, **21**, 303–312.

97 Pirbazari, M., Ravindran, V., Badriyha, B.N., and Kim, S.H. (1996) Hybrid membrane filtration process for leachate treatment. *Water Res.*, **30** (11), 2691–2706.

98 Abdul, J.M., Vigneswaran, S., Ngo, H. H., Kandasamy, J., and Goteti, P. (2007) Investigation of granular activated carbon as a biofilter for the treatment of landfill leachate. Proceedings of the International Conference on Sustainable Solid Waste Management, 5–7 September 2007, Chennai, India, pp. 210–217.

99 Imai, A., Iwami, N., Matsushige, K., Inamori, Y., and Sudo, R. (1993) Removal of refractory organics and nitrogen from landfill leachate by the microorganism-attached activated carbon fluidized bed process. *Water Res.*, **27** (1), 143–145.

100 Pirbazari, M., Badriyha, B.N., Ravindran, V., and Kim, S.H. (1990) Treatment of landfill leachate by biologically active carbon adsorbers, 44th *Industrial Waste Conference Proceedings*, Purdue University, West Lafayette, Indiana, USA, Lewis Publishers, Inc., Chelsea, Michigan, USA, pp. 555–563.

101 Horan, N.J., Gohar, H., and Hill, B. (1997) Application of a granular activated carbon-biological fluidized bed for the treatment of landfill leachates containing high concentrations of ammonia. *Water Sci. Technol.*, **36** (2–3), 369–375.

5
Combination of Activated Carbon with Biological Wastewater Treatment at Full-Scale

Özgür Aktaş and Ferhan Çeçen

Activated carbon adsorption and biological processes are widely integrated in full-scale treatment of industrial and municipal wastewaters, leachates from sanitary and hazardous landfills, and contaminated groundwaters. The PACT process, which is basically a modification of suspended-growth biological processes, is commonly applied in secondary treatment of industrial wastewaters or landfill leachates. On the other hand, BAC filtration, which is based on development of biological activity in GAC adsorbers, mostly finds application in tertiary treatment and serves as a polisher of relatively treated wastewaters. Nowadays, BAC filtration is often considered for tertiary treatment of domestic and industrial wastewaters for reuse purposes. In some cases, BAC filtration can also be introduced in secondary treatment. The aim of this chapter is to present the results of some full-scale PACT and BAC applications.

5.1
Full-Scale PACT Systems

PACT systems date back more than 30 years, and PACT® systems, as a registered trade mark, have now been installed at several places of the world. The main object of these facilities is compliance with regulations set for industries producing organic chemicals, plastics, and synthetic fiber (OCPSF), compliance with pre-treatment regulations controlling industrial discharges to Publicly Owned Treatment Works (POTWs), and compliance with stringent standards for direct effluent discharges. PACT systems are also used for treatment of landfill leachates, and may also be introduced as a pretreatment step for downstream membrane systems that are used for reuse purposes. PACT systems have lower operation costs than those incurred by separate biological and physicochemical systems [1].

PACT systems may be designed from the beginning or existing biological systems can be converted into PACT systems. Factory-built units treat about 2–380 $m^3 d^{-1}$, whereas in field-erected units and custom-designed systems flow rates up to 4000–200 000 $m^3\ d^{-1}$ can be handled. Both single and two-stage

Activated Carbon for Water and Wastewater Treatment: Integration of Adsorption and Biological Treatment.
First Edition. Ferhan Çeçen and Özgür Aktaş
© 2011 WILEY-VCH Verlag GmbH & Co. KGaA, Weinheim.
Published 2011 by WILEY-VCH Verlag GmbH & Co. KGaA

PACT® systems are available in continuous or batch flow configurations [1]. A two-stage PACT system is often used where the treated effluent has to comply with stringent environmental regulations. In such a configuration, the first stage PACT system has high concentrations of biomass and PAC and removes most of the contaminants whereas the second-stage acts as a polisher of first-stage effluent. Excess solids removed from the first and second stages are treated as in a single PACT® system.

The removal efficiency of the PACT® system depends primarily on the characteristics of wastewater. The most important ones are biodegradability, adsorbability, and the concentration of toxic inorganic and organic compounds. In general, the PACT® system can treat wastewaters with COD of up to 60 000 mg L^{-1}, including volatile organic compounds (VOCs) up to 1000 mg L^{-1}.

PACT® systems are applied to various wastewaters such as municipal, joint municipal-industrial, industrial, landfill leachates, and to contaminated groundwaters [1]. The largest applications of the PACT process are found in the treatment of wastewaters from the refinery and petrochemical industries, and in the treatment of leachates and highly contaminated groundwaters [2]. The spent PAC from some of the full-scale PACT facilities is regenerated by incorporating wet air regeneration (WAR). After regeneration of carbon, the only waste is the ash that is disposed of to a drying bed and then landfilled [3].

5.1.1
Full-Scale PACT for Industrial Effluents

5.1.1.1
Organic Chemicals Production Industry
PACT systems are used at many organic chemicals, plastics, synthetic fibers, solvents, and dye and pesticide manufacturing sites, both for pretreatment and direct discharge purposes. At a specialty chemical plant in Louisiana, USA, a two-stage aerobic PACT system was reported to meet the effluent organic matter and toxicity requirement for discharge to the Mississippi River. Also, another wastewater containing 19 pesticides at concentrations exceeding 3400 mg L^{-1} was treated in the PACT system in which COD reduction exceeded 99% and pesticide removal amounted to 99.8% [1].

One of the first applications of the PACT process was in DuPont's Chambers Works site at Deepwater, New Jersey. Chambers Works was an organic chemical plant producing freon fluorocarbons, general organic intermediates, dyes and dye intermediates, petroleum additives, isocyanates and elastomeric products, aliphatic surfactants, and chlorinated and amminated derivatives of benzenoid chemicals [4, 5]. Batch-type production resulted in highly variable organic loadings and highly colored wastewater. Following technical tests, the PACT process was selected in order to meet discharge standards, particularly those relating to organic matter. In the initial wastewater, DOC and color were about 171 mg L^{-1} and 1690 APHA units, respectively. Laboratory experiments showed that the PACT process at a carbon dose of 160 mg L^{-1} reduced DOC and color to 26 mg L^{-1} and 310

APHA units, respectively. In an activated sludge system the corresponding values were as 57 mg L^{-1} and 1020 APHA units. At the design PAC dose of 180 mg L^{-1}, a full-scale PACT system achieved 31.2 mg DOC/L, 6.7 mg BOD$_5$/L and 483 APHA color units in the effluent with removal efficiencies of 82%, 96%, and 66%, respectively [4]. For further details on the startup and steady-state operation at this plant the reader may refer to literature [6].

Introducing a more biodegradable wastewater to an existing PACT plant treating organic chemical wastewaters may be a convenient way of diluting a high-strength wastewater and increasing the biodegradation potential. In that respect, screening tests were performed by mixing various types of industrial effluents with the original wastewater of Chambers Works in laboratory scale continuous-flow PACT reactors [5]. When agricultural wastewater (40% of total BOD) was added to the original wastewater, BOD and DOC removal efficiencies increased from 95.2% to 96.7% and from 81.6% to 86.7%, respectively. The test showed that addition of a more biodegradable organic waste slightly increased the removal efficiency in the PACT reactor [5].

The wastewater from an organic chemical factory was treated at full-scale in an activated sludge system that was upgraded by PAC addition [7]. The factory was producing several kinds of organic chemicals such as substituted phenols, benzene derivatives, nitrated aromatic substances, aniline derivatives, chlorinated hydrocarbons, solvents, plastics, and pesticides. After neutralization, settling, and equalization of wastewater, the activated sludge units received a wastewater with a high organic content (COD: 1700–4000 mg L^{-1}, BOD$_5$: 1100–2700 mg L^{-1}). Operational difficulties were encountered since biological units suffered from overloading. Despite equalization and pretreatment, the quality of wastewater was very variable. Sudden changes in the composition of wastewater led to decreases in sludge activity, and to sludge bulking and foaming. This observation is typical for industries that utilize batch-type productions and change their processes frequently to produce different kinds of chemicals. Two cylindrical activated sludge units were operated with aeration volumes of about 1500 m^3 each. One of the biological units was operated as a control, without PAC addition, while the other received different PAC doses. Under equal loading rates, in the first month, the organic content of the effluent at the PAC dose of 70 mg L^{-1} was 30–40% lower than the control unit. The treatment efficiency reliably approached 85%, while that of the control unit went down to 67%. The PAC addition decreased the sludge volume index (SVI) from 35 to 21 mL g^{-1} and improved sludge thickening. It also enabled higher hydraulic loading to be applied. PAC addition also decreased sludge foaming to a minimum. An important result was also that 20–50% decrease was achieved in the concentration of micropollutants compared to the control unit. The optimal PAC dose was relatively low (<100 mg L^{-1}) despite the high organic content of the influent.

Another example showed that the PACT® system was able to reduce organic compounds in a chemicals production wastewater to much lower levels than a conventional activated sludge, although the sludge age in the latter was more than twice that in the PACT® system (Table 5.1) [8]. The effluent concentrations in Table 5.1

Table 5.1 PACT® system performance for the treatment of an organic chemical manufacturing wastewater (adapted from [8]).

Parameter	Influent (mg L^{-1})	AS effluent (mg L^{-1})	PACT® effluent (mg L^{-1})
Chlorinated aromatics	5.08	0.91	0.1
Phenol	8.1	0.22	0.01
COD	10 230	296	102
BOD$_5$	4035	17	11
Color (APHA units)	–	820	94

show that the PACT® system was more effective in the removal of COD and color compared to BOD. Both systems were effective in removing biodegradable compounds. However, nonbiodegradable organic compounds could be successfully removed through activated carbon adsorption only. The most important advantage of PAC addition was the enhanced removal of specific organics such as phenols and chlorinated aromatics, which were decreased to µg L^{-1} levels. A similar result was obtained for phenol manufacturing wastewater containing 150 mg L^{-1} phenol. The PACT® system decreased phenol concentrations to below 10 µg L^{-1}, whereas in activated sludge effluent phenol stayed above 100 µg L^{-1} [8].

A manufacturer of herbicides and organic chemicals used a PACT system following wet air oxidation (WAO) [9]. WAO acted as a pretreatment step before PACT and destroyed 99.8% of toxicity. Spent PAC was also regenerated by WAO, which led to the destruction of adsorbed toxic organics. Within the PACT system the toxic process wastewater of the facility having a flow rate in excess of 5700 m^3 d^{-1} was co-treated with contaminated groundwater. Contaminated groundwater did not undergo WAO pretreatment. Overall, for an initial COD of 70 000–80 000 mg L^{-1}, the removal exceeded 99.5%.

5.1.1.2
Synthetic Fiber Manufacturing Industry

Wastewater of a synthetic fiber manufacturing plant contained high amounts of 2-methyl-1-dioxolane and the resistant compound 1,4-dioxane, which was not removable by biological treatment, GAC, or BAC filtration [8]. However, proper application of the PACT system led to full removal of these compounds (Table 5.2).

5.1.1.3
Propylene Oxide/Styrene Monomer (PO/SM) Production Wastewater

In 1999 a sophisticated industrial wastewater treatment plant was constructed in Tarragona, Spain, to remove aromatic and polyol compounds from the high-strength PO/SM wastewater. These compounds were unaffected by conventional biological treatment processes. The goal was to achieve a liquid effluent that was directly dischargeable to the Mediterranean Sea [10]. WAO was employed for pretreatment of the high-strength PO/SM wastewater. This was followed by a two-stage

Table 5.2 PACT® system performance for the treatment of a synthetic fiber manufacturing wastewater (adapted from [8]).

Parameter	Influent (mg L^{-1})	Effluent (mg L^{-1})
1,4-dioxane	542	1
2-methyl-1-dioxolane	2540	1.4
COD	11 950	216
BOD$_5$	2560	<6

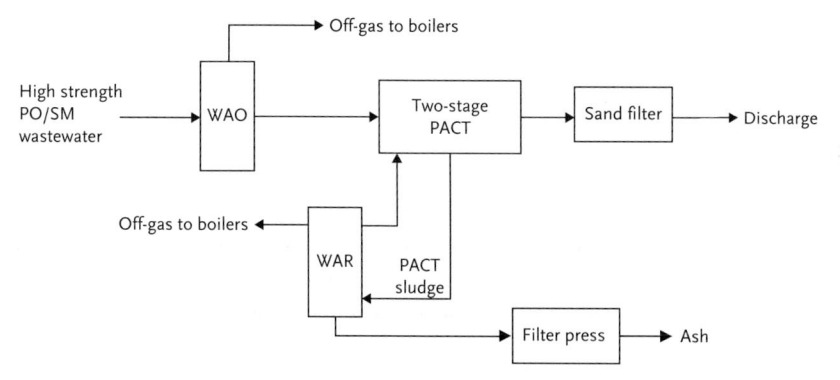

Figure 5.1 Repsol Tarragona wastewater treatment plant process flow diagram (redrawn after [10]).

PACT system (Figure 5.1). The purpose of WAO was to oxidize nonbiodegradable high-molecular-weight (HMW) branched molecules and to reduce the organic load to the PACT system. The effluent quality was excellent and the average effluent COD amounted to only 15% of the effluent specification limit. The two-stage PACT system was designed to treat a flow rate of 2160 m^3 d^{-1} with an average feed COD of 26 000 mg L^{-1}. The design effluent COD was 615 mg L^{-1}, representing 97.6% removal in the PACT system. The WAR was used to regenerate the spent carbon from the PACT system. While the bulk of activated carbon was regenerated, there was approximately 10% loss of activated carbon via oxidation in WAR and through the effluent stream. This was compensated by the addition of fresh carbon to the second PACT aeration tank.

5.1.1.4
Refinery and Petrochemical Wastewaters

The PACT® system has been used to treat wastewaters from petroleum refining, and contaminated surface runoff at refineries. At several refineries and petrochemical plants, the PACT® system is used to meet a number of regulatory requirements. For example, at a chemical terminal in New Jersey, a factory-built

PACT® unit cleans up contaminated surface runoff and meets the stringent National Pollutant Discharge Elimination System (NPDES) permit in the USA [1].

A full-scale test in which carbon was added to an extended aeration treatment system at the Sun Oil Refinery in Texas indicated substantial improvement in BOD, COD, and total suspended solids at even very small carbon dosages (9–24 mg L^{-1}). Carbon dosing also stabilized the effluent quality and led to a clearer effluent; it also eliminated foam formation. Another field-scale test at the Exxon refinery, for aeration tank PAC concentrations in excess of 1000 mg L^{-1}, showed significantly improved effluent quality. It was noted that under such operating conditions the process stability was improved under shock loading conditions [11].

The wastewater of a refinery plant at a flow rate of 5760 $m^3 d^{-1}$, was treated by PACT in combination with WAR. The system achieved sufficient reduction in COD, phenols, and sulfides, and met all discharge standards (Table 5.3). The plant was able to withstand shock loadings of up to 4000 mg L^{-1} COD or 900 mg L^{-1} phenols [8]. The PACT/WAR process produced a high-quality effluent with minimal residual solids to landfill. Also, no concerns were raised about air pollution or the escape of toxics, odors, and color to the environment [2].

A full-scale PACT® system treated 280–380 $m^3 d^{-1}$ of a high-strength petrochemical industry wastewater emerging from the production of organic acids, solvents, and aromatics [2]. Preliminary pilot tests showed that PACT decreased COD from 3600 to 100 mg L^{-1}. In comparison, the effluent from the conventional activated sludge reactor had a COD of 285 mg L^{-1}, thus exceeding the discharge limit of 250 mg L^{-1}. The PACT system was selected over conventional biological treatment both for environmental reasons and cost considerations. At full-scale PACT the effluent COD was about 135 mg L^{-1} whereas BOD_5 was below 30 mg L^{-1}. The capital costs of the PACT system were 10% less than those of conventional biological treatment, primarily because the PACT reactors were more compact and had a size of about two-thirds that of an activated sludge reactor. Operating costs were also about 10% less than those of conventional biological treatment despite the additional expenditure for PAC. The introduction of the PACT provided significant savings in man-hours to run the system and in the amount of dewatered sludge generated (0.5 tons per day with PACT compared to 1.4 tons per day with activated sludge).

Table 5.3 PACT® system performance for the treatment of a refinery wastewater (adapted from [8]).

Parameter	Influent (mg L^{-1})	Effluent (mg L^{-1})
COD	1494	78
BOD_5	718	7
Phenolics	70	0
Sulfides	142	0

When the petrochemical wastewaters emerge mainly from the production of a single compound such as vinyl chloride monomer, high concentrations of that chemical are often observed in the wastewater. Vinyl chloride-containing wastewater from a South African petrochemical company was successfully treated in a PACT unit. High-quality wastewater was produced for reuse in production [2]. Pilot tests with a two-stage PACT system showed that the TOC of the wastewater went down from 910 to 160 mg L^{-1} in the first stage and further to 60 mg L^{-1} in the second stage. The aim was to decrease TOC below 5 mg L^{-1} in order to comply with reuse standards. For this purpose, a full-scale PACT system was constructed along with UV oxidation as a post-treatment step. Although the standards for reuse could not be achieved, the full-scale PACT system performed better than the pilot-scale system and decreased effluent TOC to less than 20 mg L^{-1}.

In the industrial park Renda in TaSheh Taiwan, a single-stage PACT system was combined with WAR to treat chemical and petrochemical wastewaters. The aim was to meet effluent standards of COD (<80 mg L^{-1}) (Figure 5.2). The wastewater had a BOD$_5$/COD ratio less than 0.1, indicating very low biodegradability. In the treatment system, the aeration tank and clarifiers were designed to handle high levels of mixed liquor suspended solids (MLSS). The MLSS concentration in the PACT sludge was about 22 000 mg L^{-1}. The MLSS was made up of approximately 15% biomass, 55% carbon, and 30% ash. It was ensured that sufficient carbon was maintained in the system to meet the stringent effluent COD and color limits. Effluent COD values could be decreased to below 90 mg L^{-1} while negligibly small BOD$_5$ values (1–3 mg L^{-1}) were recorded [3].

Recycling of wastewater was investigated at a coal gasification facility in South Africa [12]. Wastewaters at the facility stem from various oily and organic wastes that are generated in coal gasification. For the additional removal of organic matter, nitrification of nearly 500 mg L^{-1} ammonia, and adsorption of recalcitrant compounds, an activated sludge system with PAC addition was introduced

Figure 5.2 Full-scale PACT at Renda Industrial Park (RIP) in TaSheh, Taiwan [3]. (permission received)

downstream of a biofilm reactor. In the PAC reactor, COD went down from about 2000 mg L^{-1} to less then 600 mg L^{-1}, achieving the target. Also, residual BOD$_5$ values approached insignificant levels, and almost full nitrification took place. Based on these positive results, the full-scale system was designed for an average flow rate of 550 m^3 d^{-1}. The average influent COD concentration to the PAC system was 2893 mg L^{-1}, whereas the effluent COD from the secondary clarifier was 1042 mg L^{-1}. The combination of the biofilm reactor with the PACT system reduced the concentration of organics by 95% on average, provided complete nitrification, and met the overall treatment requirements.

5.1.1.5
Treatment of Priority Pollutants

An extensive test at DuPont's Chambers Works Plant [8] showed that priority aromatic pollutants could be more successfully removed with PACT® compared to activated sludge (Table 5.4). The effluent concentrations of priority pollutants had to be kept at levels of a few µg L^{-1}. However, in general, in conventional activated sludge systems it is very difficult to achieve such low target values. Fortunately, PAC addition enabled the removal of these compounds to the level required in discharge standards. The removal of some aromatic compounds exceeded 90% after PAC addition, although biodegradation efficiencies of these compounds were much lower in most cases. The off-gas from activated sludge may contain volatile priority organic chemicals such as chlorobenzenes, toluene, and xylene. PAC addition eliminated the stripping of these volatiles almost totally, whereas up to 90% of the influent could appear in the off-gas of a conventional system [8]. In other work, the removal efficiencies for priority pollutants at the Chambers Works PACT® system were reported to vary in the range of 81–99% for volatiles (benzene, carbon tetrachloride, chlorobenzene, chloroform, ethylbenzene, methyl chloride, PCE, toluene, TCE, and so on), 44–99% for base-neutral extractables (dichlorobenzene, dinitrotoluene, nitrobenzene, and trichlorobenzene), 94–98% for acid extractables (phenol, 2-chlorophenol, 4-nitrophenol, 2,4-dinitrophenol), and up to 52% for heavy metals [4].

Table 5.4 PACT® System performance for the treatment of priority pollutants (adapted from [8]).

Parameter	Influent concentration, (µg L^{-1})	Activated sludge (AS) removal (%)	PACT® removal (%)
1,2-Dichlorobenzene	18	90.6	>99
2,4-Dinitrotoluene	1000	31	99
2,6-Dinitrotoluene	1100	14	95
Nitrobenzene	330	94.5	>99
1,4-Dichlorophenol	19	0	93
2,4-Dinitrophenol	140	39	>99
4-Nitrophenol	1100	25	97

5.1.1.6
Treatment of Pharmaceutical Wastewaters

As shown in Chapter 3, the PACT system can also be operated in the sequencing batch reactor (SBR) mode. Synthetech, Inc. in Albany, Oregon made use of the SBR-PACT system for the treatment of a pharmaceutical chemicals production effluent [13]. The company produced peptide building blocks and specialty amino acids, alcohols, esters, amides, and chiral intermediates that are distributed to pharmaceutical companies. A wide range of chemicals were used in the synthesis of these products. This resulted in large fluctuations in the type and concentration of contaminants in the wastewater. The wastewater was finally discharged to the Publicly Owned Treatment Works (POTW) of the City of Albany. In 2001, the Albany POTW became subject to Federal Pharmaceutical Pretreatment Standards imposed by the USEPA. These treatment standards include limitations for 23 VOCs, pH, and ammonia. The POTW also levies a toll charge on industrial wastewater dischargers tied to BOD and total suspended solids (TSS) loadings.

In bench-scale testing in biological reactors, an equal mixture of process wastes (COD: 30 000–90 000 mg L^{-1}) and non-process wastewaters (COD: 1000–2000 mg L^{-1}) was used as a feed. In such cases, VOCs were removed without a problem. Also, efficient BOD and COD removal was observed. Some fraction of ammonia was removed by assimilation into biomass. However, the removal of ammonia via nitrification was problematic because of the inhibitory effect of sulfate [13].

Using the results of the bench-scale testing program, the company decided on full-scale application of PACT in order to meet the standards and decrease the costs of discharge. The treatment system consisted of an evaporator to pretreat the process stream and a PACT unit to further treat the evaporator distillate after combining it with the non-process wastewater (Figure 5.3). Evaporation was chosen as a pretreatment step to eliminate total dissolved solids (TDS) and to control ammonia. The evaporator is a factory-built system that exploits the combined effect of vacuum and heat pump technology to carry out distillation at low temperatures. The PACT system is a factory-built SBR unit consisting of aeration/settling tank, aeration blower, air distribution system, and delivery systems. The SBR-PACT system was designed to handle approximately 13.25 $m^3 d^{-1}$ of non-process water and 13.25 $m^3 d^{-1}$ of evaporator condensate from the process water [13].

For the system shown in Figure 5.3, effluent COD was 261 mg L^{-1} with a removal efficiency of 98%. The effluent BOD to the sewer averaged about 0.9 kg d^{-1} or 44 mg L^{-1}. Thus, these values met the target BOD of Albany City, specified as 88.5 kg d^{-1}. Prior to the installation of the PACT system, effluent discharges frequently surpassed 450 kg d^{-1}, which necessitated that the process wastewater be treated offsite. However, after the installation of the PACT system no wastewater had to be treated offsite. Moreover, VOCs were consistently eliminated, and their levels easily complied with Federal Pretreatment Standards. The wastewater was rich in nitrogenous compounds such as amino acids. In the PACT system, nitrification and denitrification were also achieved by switching aeration on and off during the SBR cycle [13].

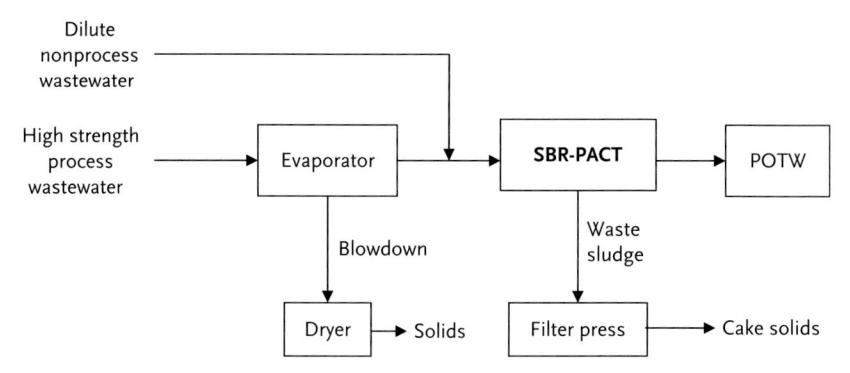

Figure 5.3 PACT system of Synthetech in Albany Oregon for the treatment of pharmaceutical wastewater. (redrawn after [13])

Table 5.5 PACT® system performance for the treatment of a domestic wastewater receiving industrial effluents (adapted from [8]).

Parameter	Influent concentration (mg L^{-1})	Effluent concentration (mg L^{-1})
COD	~700	<90
BOD$_5$	~250	<5
Suspended Solids	~400	<10
$NH_4^+ - N$	~20	<0.5

5.1.2
PACT for Co-treatment of Domestic and Industrial Wastewaters

PACT was also applied to a city's municipal wastewater at a flow rate of 200 000 m^3 d^{-1}. The municipal wastewater in question also received input from pharmaceutical, pulp and paper, and organic chemicals industries [8]. The PACT system was combined with WAR to reuse the loaded PAC. Following the upgrading of the system from activated sludge to PACT/WAR, excellent removals were immediately recorded (Table 5.5). PAC addition enabled the achievement of river discharge standards; it also eliminated fish toxicity, the unpleasant color of the receiving river, and the odor problems in the treatment plant located near the city.

5.1.3
PACT for Landfill Leachates

Leachates collected from solid waste disposal sites contain high levels of organic contaminants. GAC treatment results in poor efficiency if leachates emerge from young landfills and have a BOD$_5$/COD ratio and COD concentration greater than

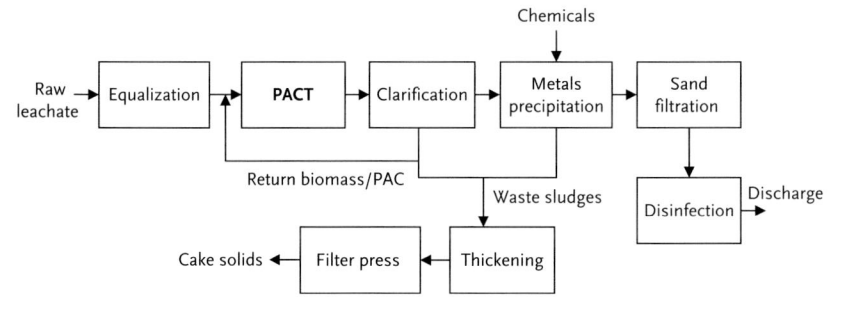

Figure 5.4 Typical leachate treatment including PACT system. (redrawn after [14])

0.5 and 10 000 mg L^{-1}, respectively [14]. Therefore, for such wastewaters, biological processes are typically followed by GAC adsorption used as a polishing step to remove resistant organics. Tightening of landfill regulations has led to the introduction of the PACT® system for the treatment of leachates from sanitary and hazardous waste landfills. The PACT system was reported to be the only technology that provides good treatability of organic matter in all cases, namely in young (<5 years), medium (5–10 years), and old (>10 years) landfills [14]. Figure 5.4 shows the positioning of PACT for landfill leachates before discharge into surface waters or a POTW.

PACT systems are in operation at dozens of landfills in the USA [14]. For example, at a combined hazardous and municipal landfill site near Los Angeles, CA, a PACT® system was installed in 1988 after an evaluation showed it to be the lowest in cost and land usage and superior in treatment stability when compared with other systems [1]. The Reichs Ford Road landfill in Frederick County, MD, has used a PACT® system since 1995 to treat landfill leachate (COD: 2250 mg L^{-1}, BOD$_5$:1700 mg L^{-1}, TSS: 820 mg L^{-1}) for direct discharge to the receiving stream. Average flow rate through the system is 300 m^3 d^{-1}, with a future design value of 530 m^3 d^{-1}. The system includes five SBR-PACT® reactors, each 170 m^3, a sludge storage tank, a carbon addition system, and a recessed plate and frame filter press for dewatering solids. The aerobic SBR-PACT® configuration enabled operators to treat influent, to clarify and decant the treated liquid, and to store sludge in the same tank [15].

The Zernel Road Municipal Solid Waste Landfill serving Charlotte County in Florida receives the solid waste of about 125 000 people and has to meet new standards for deep-well injection of treated leachate into groundwater [16]. A new technology had to be applied to handle the variation in contaminant concentrations, whereby design influent concentrations were 1000 mg L^{-1} COD, 500 mg L^{-1} BOD$_5$, and 500 mg L^{-1} suspended solids. Three SBR-PACT reactors, each with a volume of 180 m^3, were constructed for this purpose. Each batch unit had one cycle per day. Finally, the effluent was filtered to remove remaining suspended solids. It was then chlorinated before deep-well injection into an aquifer for reuse purposes. The system was cost-effective and met the requirements, with $COD < 33\,mg\,L^{-1}$, $BOD_5 < 6\,mg\,L^{-1}$, and suspended solids $< 3\,mg\,L^{-1}$ at a wastewater flow rate of $300–380\,m^3\,d^{-1}$.

The leachate treatment plant of the Stewartby Landfill Site in England also uses an activated sludge system with PAC addition. At that facility, the leachate was heated before the PACT process in order to achieve a temperature of $20°C$ in the aeration tank and increase nitrification efficiency. The system also involved dissolved air flotation (DAF) and sand filtration for removal of suspended solids and phosphorus. In the leachate, with a flow rate of $300–400\,m^3\,d^{-1}$, the average COD and $NH_4^+–N$ were about 2800 and $800\,mg\,L^{-1}$, respectively. The full-scale system removed 70% of COD and almost 100% of ammonia. The effluent concentrations, $867\,mg\,L^{-1}$ COD and $< 1\,mg\,L^{-1}$ $NH_4^+–N$, satisfied sewer discharge standards [17].

5.1.4
PACT for Contaminated Groundwaters

Groundwaters may be contaminated with a wide range of organics including chlorinated solvents, polyaromatic hydrocarbons (PAHs), phenols, mineral oils, benzene, toluene, and inorganics such as cyanides. Traditionally, highly contaminated groundwaters are treated *ex-situ* with GAC according to the pump-and-treat method. However, PACT® systems can also be effective in the treatment of contaminated groundwaters. Near Los Angeles, CA, a batch-operated PACT® system achieved more than 99% COD and BOD removal from groundwater that was contaminated with wastes from a mobile home products and paint manufacturing plant [1].

A PACT system at a PAC dose of $667\,mg\,L^{-1}$ was used to treat Stringfellow Quarry site's contaminated groundwater, having a flow rate of $12.96\,m^3\,d^{-1}$ [9]. For an influent COD of $1788\,mg\,L^{-1}$, the PACT system accomplished more than 70% removal. Organic priority pollutants such as benzene, chlorobenzene, dichloromethane, chloroform, 1,2-dichloroethylene, trichloroethylene, tetrachloroethylene, ethyl benzene, and toluene were completely removed. The cost of the PACT system was compared with that of a GAC system which achieves the same level of effluent COD. Although capital costs were higher in PACT, fivefold savings in operation and maintenance costs favored the selection of PACT [9].

An engineering study to compare PACT and GAC systems revealed that a batch PACT system incurred 10–15% less capital and operating costs in the treatment of a BTEX (Benzene, Toluene, Ethylbenzene, Xylene) contaminated groundwater compared to a GAC system followed by air stripping. The PACT system decreased BTEX concentrations from $4434\,\mu g\,L^{-1}$ to less than $5\,\mu g\,L^{-1}$, while it reduced COD from 45 to $22\,mg\,L^{-1}$ [2].

Groundwater contaminated by halogenated and aromatic compounds from a former chemical manufacturing plant in Michigan had COD and $NH_4^+–N$ values as high as 1500 and $80\,mg\,L^{-1}$, respectively. The Granular Activated Carbon-Fluidized Bed Reactor (GAC-FBR) and two-stage PACT system were tested for bioremediation of this site. Although the GAC-FBR system achieved high removal of organic compounds, the PACT system performed better and exhibited almost complete

Table 5.6 Comparison of removal in GAC-FBR and PACT systems in the treatment of a highly contaminated groundwater (adapted from [2]).

Parameter	GAC-FBR (%)	PACT (%)
BOD	>95	>99
COD	80–90	>90
NH_4^+-N	Unstable	>99

nitrification (Table 5.6). Based on performance and costs, the PACT system was chosen and applied for the treatment of this contaminated groundwater [2].

5.1.5
PACT for Reuse of Domestic Wastewaters

The Hueco Bolson aquifer in Texas supplied 65% of the city's total water demand, but was being consumed at a rate 20 times faster than its natural recharge capacity. In order to halt the drop in water table, a plan was made to recycle wastewater and to supply the public with water in the long term at the least cost. In 1986, 16 650 $m^3\,d^{-1}$ of wastewater of El Paso in Texas was converted to a potable water resource by injecting the water into the Hueco Bolson aquifer following advanced treatment [18]. Among 20 different treatment steps, the two-stage PACT process was the one mainly responsible for the removal of organic pollutants. The effluent of primary sedimentation had a BOD_5 concentration of 85 mg L^{-1}, which was decreased first to 3 mg L^{-1} and then to 1 mg L^{-1} after the first and second stages of PACT, respectively. COD was reduced from 155 to 31 mg L^{-1}, whereas Total Kjeldahl Nitrogen (TKN) went down from 25.4 to 0.6 mg L^{-1} in the PACT system. The second stage anoxic PACT reactor served as a denitrification unit and reduced nitrate concentrations from 5–22 to 1.6 mg L^{-1}. Regeneration of the PACT sludge was achieved by WAO. The regeneration unit achieved 81% oxidation; this efficiency was regarded as acceptable for a PACT/WAR system [18]. The effluent from the PACT systems was further filtered and disinfected. Full-scale operation was started in 1985 and has been in service since then without violation of discharge limits. The treated water had the following characteristics: COD < 10 mg L^{-1}, BOD_5 < 3 mg L^{-1}, TKN < 1 mg L^{-1}, NH_4^+-N < 0.2 mg L^{-1}, and THM < 0.1 mg L^{-1}. These concentrations met US EPA water standards and drinking water quality standards [16].

5.1.6
PACT for Contaminated Surface Runoff Waters

A PACT system was also used to treat surface runoff waters from the Bayonne terminal in New Jersey and to meet discharge limits prior to discharge into New York Harbor [9]. At the terminal, surface water was being collected in order to prevent the contamination of groundwater. The terminal served ocean-crossing chemical parcel tankers and stored a variety of chemicals. Therefore, the surface runoff water was highly contaminated. In this water, COD concentration exceeded 600 mg L^{-1}, BOD$_5$ was 230 mg L^{-1}, and oil and grease was 40 mg L^{-1}. Pilot studies showed that COD, BOD$_5$, and oil and grease could be decreased to below 20, 6, and 1 mg L^{-1}, respectively. These values were in compliance with discharge limits. The full-scale PACT system also performed well. The effluent COD, BOD$_5$, and oil and grease concentrations were about 44, 18, and 2.7 mg L^{-1}, respectively.

5.2
Biological Activated Carbon (BAC) Filtration at Full Scale

5.2.1
BAC Filtration for Reuse Purposes

5.2.1.1
Treatment of Sewage for Reuse in Agriculture

BAC filtration has been introduced in recent years for the treatment of sewage for reuse purposes. For example, a wastewater treatment plant was designed in Australia for a population of 11 000 such that the effluent could be reused in agriculture for various purposes such as dairy farming [19]. As shown in Figure 5.5, raw sewage first undergoes treatment consisting of removal of organic carbon, biological nitrogen, and phosphorus according to the BioDenipho process. This is then followed by sand filtration. Then, advanced treatment is conducted consisting of ozonation, BAC filtration, microfiltration, and disinfection. Ozonation is introduced for the reduction of COD and organic nitrogen that remain after biological treatment. As a result of ozonation, BAC filtration receives a water which is more amenable to biological removal. The main idea in this treatment scheme is to reduce the concentrations of pesticides and endocrine disrupting compounds (EDCs) before the reuse of water.

5.2.1.2
Reclamation of Domestic Wastewater for Drinking Purposes

Water shortages in arid areas can necessitate the reclamation of wastewater for drinking purposes. A reclamation plant with a capacity of 21 000 m^3 d^{-1} was constructed in 2002 in Windhoek, Namibia [20]. The treatment plant contains BAC and GAC unit in series as well as an activated sludge maturation pond, preliminary physicochemical sedimentation, rapid sand filtration, ozonation, and final disinfection (Figure 5.6). Ozonation oxidizes HMW organics. In particular,

Municipal wastewater → BioDenipho® process → Sand filtration → Ozonation → BAC filtration → Microfiltration → UV disinfection → Reuse by dairy farmer

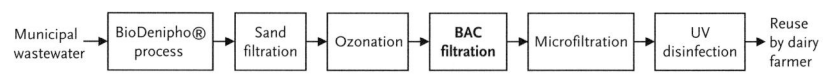

Figure 5.5 Process flow diagram of Gerroa sewage treatment plant for water reuse. (redrawn after [19])

refractory organics are oxidized and converted into more readily biodegradable organics that are amenable to removal in BAC filtration. In this treatment scheme, preliminary ozonation did not lead to a decrease in the sum parameters COD and DOC, but reduced 36% of UV_{254}. This indicated that large aromatic compounds were degraded by ozonation, but total mineralization was not achieved. The succeeding BAC and GAC units removed 29% and 13% of COD, respectively. These two units decreased COD to about $10\,mg\,L^{-1}$ in total.

5.2.1.3
Reclamation of Domestic and Industrial Wastewaters for Nonpotable Uses

In a recent study aiming at reuse of treated wastewater, BAC filtration could significantly reduce organic carbon and total nitrogen from a secondary effluent. Continuous-flow laboratory-scale BAC columns were operated for 320 days for the treatment of the secondary effluent of Pasakoy Advanced Wastewater Treatment Plant in Istanbul, Turkey. BAC filters removed about 65–81% of DOC with fresh activated carbon during the initial stage of operation. After breakthrough was reached, DOC removal efficiencies dropped to about 38–46%. This removal took place primarily by biodegradation. Nitrification and denitrification were also achieved within the BAC column, resulting in total nitrogen removal of about 52–54%. For reclamation purposes, the recommendation was made to place BAC filters ahead of disinfection units at full-scale domestic wastewater treatment plants [21].

A pilot study was performed in Beishiqiao wastewater purification center of Xi'an municipality in China to obtain high-quality water from domestic wastewater in order to meet the water reuse standards for various purposes [22]. In order to achive this aim, secondary domestic effluent was ozonated and then treated by BAC. Ozonation removed only 12% of the initial TOC, which was about 10–15 $mg\,L^{-1}$. On the other hand, ozonation oxidized nonbiodegradable substances and raised the ratio of biodegradable DOC (BDOC) to total DOC from 0.28 to 0.6. This increase then led to an additional 35% removal in succeeding BAC filtration. While the COD in secondary effluent was at about $20\,mg\,L^{-1}$, it was reduced to about $5\,mg\,L^{-1}$ after BAC filtration.

The South Caboolture Water Reclamation Plant in Queensland in Australia was designed to reduce river pollution and to provide water at a rate of $10\,000\,m^3\,d^{-1}$ to industry and community consumers for nonpotable uses. In fact, the plant was also capable of providing water that met drinking water standards [23]. The physicochemical/biological treatment scheme incorporated biological denitrification, preozonation, coagulation/flocculation, dissolved air flotation/sand filtration, ozonation, and BAC treatment. Preozonation converted refractory organic

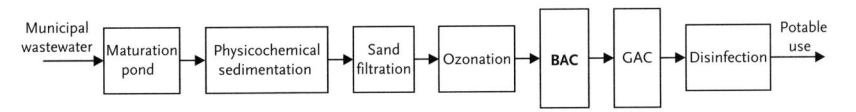

Figure 5.6 New Goreangab reclamation plant in Windhoek, Namibia. (redrawn after [20])

Figure 5.7 Four parallel BAC reactors in the South Caboolture Water Reclamation Plant [23]. (permission received)

compounds into biodegradable ones. Thereby, the service life of activated carbon was prolonged. Water could be supplied well below the cost of water from conventional sources. Four open gravity BAC beds were used in parallel with a depth of 2.2 m, an empty-bed contact time (EBCT) of 18 min and a total bed surface area of 13 m^2 (Figure 5.7). After 36% removal was achieved in the BAC filter, the mean COD in the effluent was around 12.8 mg L^{-1}.

BAC filters can also be used for recycling or reuse of industrial wastewaters within an industrial plant. In five textile factories in Iraq, two types of BAC systems were evaluated to recycle 80% of treated wastewater in the factory and reuse the remaining 20% for secondary uses [24]. BAC filters removed 63–87% of COD and more than 80% of color.

References

1 Siemens Water Technologies Corp. (2009) Brochure: ZP-PACT-BR-0409, http://www.siemens.com/water.
2 Meidl, J.A. (1997) Responding to changing conditions: how powdered activated carbon systems can provide the operational flexibility necessary to treat contaminated groundwater and industrial wastes. *Carbon*, **35** (9), 1207–1216.
3 Meidl, J.A. and Lu, C.C. (2006) Wastewater treatment: raising the standard of industrial wastewater PACT system operation. *Filtration & Separation*, **43** (10), 32–33.
4 Heath, H.W. (1981) Full-scale demonstration of industrial wastewater treatment utilizing Du Pont's PACT Process. Project Summary, Report No: EPA-600/S2–81–159, U.S. EPA, Robert S. Kerr Environmental Research Laboratory, Ada, Oklohama, USA.
5 Soderberg, R.W. and Bockrath, R.E. (1984) Treatability of diverse waste streams in the PACT activated carbon-biological process. 39th Industrial Waste Conference Proceedings, Purdue University, West Lafayette, Lewis Publishers, Inc., Chelsea, Michigan, USA, pp. 121–127.

6. Flynn, B.P. and Stadnik, J.G. (1979) Start-up of a powdered activated carbon-activated sludge treatment system. *Journal of Water Pollution Control Federation*, **51** (2), 358–369.

7. Benedek, P., Takacs, I., and Vallai, I. (1986) Upgrading of a wastewater treatment plant in a chemical factory. *Water Science and Technology*, **18**, 75–82.

8. Meidl, J.A. (1999) Use of the PACT System to Treat Industrial Wastewaters for Direct Discharge or Reuse, Technical Report No:097, US Filter Zimpro systems Products and Services, Rothschild, Wisconsin, USA, presented at Jubilee XX Conference Science & Technology, 14–15 October 1999, Katowice, Poland.

9. Meidl, J.A. (1990) PACT systems for industrial wastewater treatment, in *Physical Chemical Processes, Innovative Hazardous Waste Treatment Series-Vol. 2* (ed. H. Freeman), Technomic Publishing Company, Pennsylvania, USA.

10. Gallego, J.P., Lopez, S.R., and Maugans, C.B. (2002) The Use of Wet Oxidation and PACT® for the Treatment of Propylene Oxide/Styrene Monomer (PO/SM) Industrial Wastewaters at the Repsol PO/SM Plant in Tarragona-Spain. U.S. Filter Technical Report No. 429, Presented at: CHISA 2002, Praha, The Czech Republic, August 25–29, 2002.

11. Wong, J.M. and Hung, Y. (2004) Oilfield and refinery wastes, in *Handbook of Industrial and Hazardous Wastes Treatment*, 2nd edn (eds L.K. Wang, Y. Hung, H.H. Lo, and C. Yapijakis), Marcel Dekker, Inc., New York, USA, pp. 235–306.

12. Ratcliffe, M., Rogers, C., Merdinger, M., Prince, J., Mabuza, T., and Johnson, C.H. (2006) Treatment of high strength chemical industry wastewater using moving bed biofilm reactor (MBBR) and powdered activated carbon technology. Proceedings of the Water Environment Federation, WEFTEC 06, pp. 1677–1694.

13. Virnig, T. and Melka, J. (2004) The use of evaporation and biological treatment to meet pharmaceutical pre-treatment standards. Proceedings of the Water Environment Federation, WEF/A&WMA Industrial Wastes 2004, pp. 548–555. http://www.synthetech.com/download/Wastewater.pdf (22 May 2010).

14. Siemens Water Technologies Corp. (2006) Brochure: ZP-LCH-BR-1206, http://www.siemens.com/water.

15. Siemens Water Technologies Corp. (2007) Brochure: ZP-FRCO-CS-0207, http://www.siemens.com/water.

16. Siemens Water Technologies Corp. (2010) www.water.siemens.com/en/applications/wastewater_treatment (15 February 2010).

17. Toddington, R., Pankhania, M., and Clark, E.C. (2005) Advanced leachate treatment at the Stewartby landfill site. *Water and Environment Journal*, **19** (3), 264–271.

18. Knorr, D.B. (1986) *City of El Paso Ground Water Recharge Project*, Parkhill, Smith & Cooper, Inc., El Paso, Texas, USA.

19. Boake, M.J. (2005) Recycled water-case study: Gerringong gerroa, in *Integrated Concepts in Water Recycling* (eds S.J. Khan, A.I. Schäfer, and M.H. Muston) ISBN 1 74128 082 6.

20. Menge, J. (2010) Treatment of wastewater for re-use in the drinking water system of Windhoek, http://www.wastewater.co.za (8 May 2010).

21. Kalkan, Ç., Yapsakli, K., Mertoglu, B., Tufan, D., and Saatci, A. (2011) Evaluation of Biological Activated Carbon (BAC) process in wastewater treatment secondary effluent for reclamation purposes. *Desalination*, **265**, 266–273.

22. Wang, X.C., Wang, S., and Kin, P.K. (2002) Tertiary treatment of domestic wastewater by low-concentration ozonation and biofiltration for water reclamation. Proceedings of the International Conference Advances in Ozone Science and Engineering, April 15–16, 2002, Hong Kong, China, http://water.xauatst.com/pages/resource.

23 van Leeuwen, J.H., Pipe-Martin, C., and Lehmann, R.M. (2003) Water reclamation at South Caboolture, Queensland, Australia. *Ozone Science and Engineering*, **25**, 107–120.
24 Abud, D.W., Talab, S.A., and Hamed, S.L. (2006). Water recycling in textile industry using biological activated carbon treatment system. Proceedings of the Conference "Towards Harvesting the Success of Science Output in the Arab World", December 11–14, 2006, Damascus, Syria. http://www.astf.net/sro/sro4.

6
Modeling the Integration of Adsorption with Biological Processes in Wastewater Treatment

Ferhan Çeçen

Modeling gives insight into the processes taking place in a reactor, helps in interpreting the results, and serves as a guide for the design of pilot- and full-scale reactors. An established model can be verified using experimental results and can be improved if necessary. Therefore, experimenting and modeling are complementary to each other. Modeling also allows the rapid prediction and simulation of a process under different scenarios. It can be used to assess the sensitivity of predictions to variations in input parameters. This is particularly important in systems that combine adsorption and biological removal, since influent concentrations or flow rates often vary with respect to time or multiple substrates are encountered that have different adsorption and biodegradation characteristics.

This chapter describes the general approaches adopted in the modeling of GAC adsorbers with biological activity (BAC reactors) and activated sludge systems with PAC addition (PACT systems), which represent the combination of adsorption with attached- and suspended-growth systems, respectively. In the chapter, basic equations are provided for the delineation of fundamental mass transport, adsorption, and biological removal processes. Typical substrate and biomass balance equations are presented for the above-mentioned systems. A brief outline is given on the use of dimensionless numbers in fixed-bed GAC/BAC reactors. The chapter also discusses prevalent modeling efforts in BAC reactors and in PACT systems.

6.1
Modeling of GAC Adsorbers with Biological Activity

6.1.1
Introduction

GAC filtration is a common operation for the removal of organic and inorganic pollutants in both water and wastewater treatment. If a GAC filter is gradually

Activated Carbon for Water and Wastewater Treatment: Integration of Adsorption and Biological Treatment.
First Edition. Ferhan Çeçen and Özgür Aktaş
© 2011 WILEY-VCH Verlag GmbH & Co. KGaA, Weinheim.
Published 2011 by WILEY-VCH Verlag GmbH & Co. KGaA

converted into a BAC filter, by formation of a biofilm, biodegradation processes gain importance in addition to adsorption. Thus, a BAC filter is essentially a biofilm reactor involving the added complexity of adsorption. In the initial and intermediate stages of operation, when biological processes are taking place to a limited extent only, it is appropriate to consider this filter as a GAC adsorber. In later stages of operation, after exhaustion of the adsorptive capacity of the carbon bed, adsorption processes lose their importance and the filter can be regarded as a biofilm reactor. Since the differentiation between adsorptive GAC and BAC operations is difficult, the term B(GAC) is frequently employed, where 'B' stands for probable biological activity. In this chapter, however, the simpler term 'BAC' is adopted whenever there is biological activity in a GAC adsorber.

Several research groups have studied and modeled the interactions between biodegradation and adsorption in fixed-, expanded-, and fluidized-bed configurations. A number of models conceptualize various aspects of mass transfer, biodegradation, and adsorption, and aim to simulate substrate removal and biofilm kinetics under steady-state and non-steady-state conditions. Moreover, some of the models have been developed by assuming complete mixing of bulk liquid, whereas others are based on plug flow assumption. Some models consider the dispersed-flow case, which is in between the completely mixed and plug flow conditions.

6.1.2
Fundamental Processes around a Carbon Particle Surrounded by a Biofilm

Before considering the behavior of BAC reactors, it is essential to examine the main processes taking place around and inside a single carbon particle covered by a biofilm (Figure 6.1). This type of discussion will also facilitate the evaluation of the benefits when both adsorption and biological removal are present. The same type of considerations also apply in the modeling of BAC filters in drinking water treatment (Chapter 11).

The main processes occurring around and within an activated carbon particle on which microorganisms form a biofilm can be placed in two groups:

Processes related to substrate transport and removal
- Transport of substrate to the surface of the biofilm
- Biosorption on biofilm
- Diffusion of soluble substrate into the biofilm
- Biodegradation inside the biofilm
- Biodegradation by suspended biomass
- Diffusion into activated carbon pores
- Adsorption on activated carbon
- Desorption from activated carbon

Processes related to biomass and activated carbon
- Deposition and attachment of microorganisms on carbon surface
- Growth of biofilm

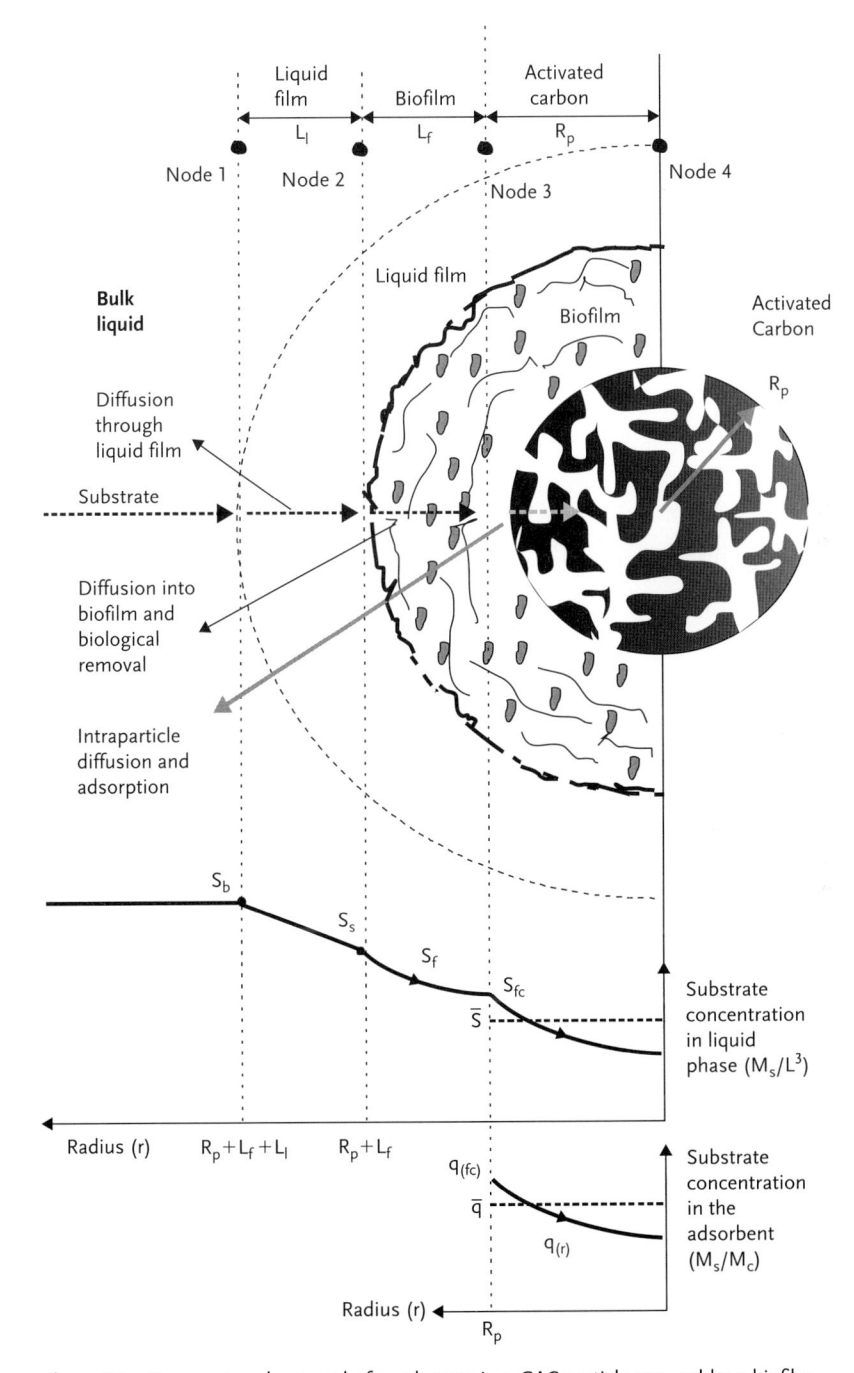

Figure 6.1 Transport and removal of a substrate in a GAC particle covered by a biofilm.

- Detachment and decay of biofilm
- Carbon aging

Many of these processes are described in the following sections.

6.1.2.1
Transport of Substrate to the Surface of Biofilm

In an activated carbon system with biological activity, the substrate has to be transported from the bulk liquid to the outer surface of the biofilm. As shown in Figure 6.1, it is often assumed that a stagnant liquid film, known as the diffusion layer or the external film, develops between the bulk liquid and the biofilm, providing resistance to the transport of substrate to biofilm surface. **Node 1** represents the bulk water–liquid film interface. The main substrate transport through the liquid film occurs by molecular diffusion, which is described by Fick's law:

$$J = -D_w \frac{\partial S}{\partial r} \qquad R_p + L_f \leq r \leq R_p + L_f + L_l \tag{6.1}$$

Assuming that the substrate gradient can be approximated by

$$\frac{\partial S}{\partial r} = \frac{(S_b - S_s)}{L_l} \tag{6.2}$$

As shown in Figure 6.1, the radial axis was selected outwards. Irrespective of the orientation of the radial axis, the magnitude of the flux in Eq. (6.1) can be shown as follows:

$$|J| = \frac{D_w}{L_l}(S_b - S_s) \tag{6.3}$$

The mass transfer coefficient is defined as the ratio of the molecular diffusion coefficient to the thickness of the diffusion layer:

$$k_{fc} = \frac{D_w}{L_l} \tag{6.4}$$

The flux into the liquid film is then expressed as follows:

$$|J| = k_{fc}(S_b - S_s) \tag{6.5}$$

where in these equations:

J = substrate flux into the liquid film ($M_s/L^2 \cdot T$),
D_W = molecular diffusivity of substrate in the liquid film (L^2/T),
r = radial coordinate (L),
S_b = substrate concentration in the bulk liquid (M_s/L^3),
S_s = substrate concentration at the liquid–biofilm interface (M_s/L^3),
L_1 = thickness of diffusion layer (L),
k_{fc} = liquid film or external mass transfer coefficient (L/T).

6.1.2.2
Diffusion and Removal of Substrate Within the Biofilm

The substrate mass reaching the biofilm surface (**Node 2**) undergoes diffusion into the biofilm. It is assumed that activated carbon particles are spherical with a radius of R_p and a homogeneous biofilm layer of thickness L_f is attached to the particle. The molecular diffusion of the substrate within the biofilm is expressed on the basis of Fick's law. Based on this law, the magnitude of flux in the biofilm, N, can be shown as in Eq. (6.6).

$$N = D_f \frac{\partial S_f}{\partial r} \qquad R_p \leq r \leq R_p + L_f \qquad (6.6)$$

where

N = magnitude of substrate flux into the biofilm ($M_s/L^2 \cdot T$),
S_f = substrate concentration inside the biofilm (M_s/L^3),
D_f = molecular diffusivity of substrate in the biofilm (L^2/T),
r = radial coordinate (L).

While substrate diffuses into the biofilm it is simultaneously removed by the microorganisms present. This removal creates a substrate concentration gradient which is the driving force for further diffusion (Figure 6.1).

If the substrate has a high molecular weight or is not soluble, it may be biosorbed onto the biofilm. After the hydrolysis step, it diffuses into the biofilm where it is consumed. Thus, also hydrolysis and/or biosorption of substrate can be regarded as important processes.

6.1.2.2.1 Mass Balance Inside the Biofilm A mass balance can be written for the substrate within a radial differential section of biofilm that is shown in Figure 6.2.

The accumulation of substrate in the slice Δr is expressed as follows:

$$\begin{bmatrix} \text{Accumulation rate of} \\ \text{substrate in the biofilm} \\ \text{slice} \end{bmatrix} = \begin{bmatrix} \text{Rate of substrate} \\ \text{entering the slice} \end{bmatrix}$$

$$- \begin{bmatrix} \text{Rate of substrate} \\ \text{leaving the slice} \end{bmatrix} - \begin{bmatrix} \text{Removal rate of} \\ \text{substrate in the biofilm slice} \end{bmatrix} \qquad (6.7)$$

$$A_b \cdot \Delta r \frac{\partial S_f}{\partial t} = [A_b N|_{r+\Delta r} - A_b N|_r] - r_{ut} A_b \Delta r \qquad (6.8)$$

where

A_b: biofilm surface area perpendicular to flux (L^2)

The first and second terms on the right-hand side represent the multiplication of the surface area with the magnitude of influx (N_{in}) and outflux (N_{out}), respectively.

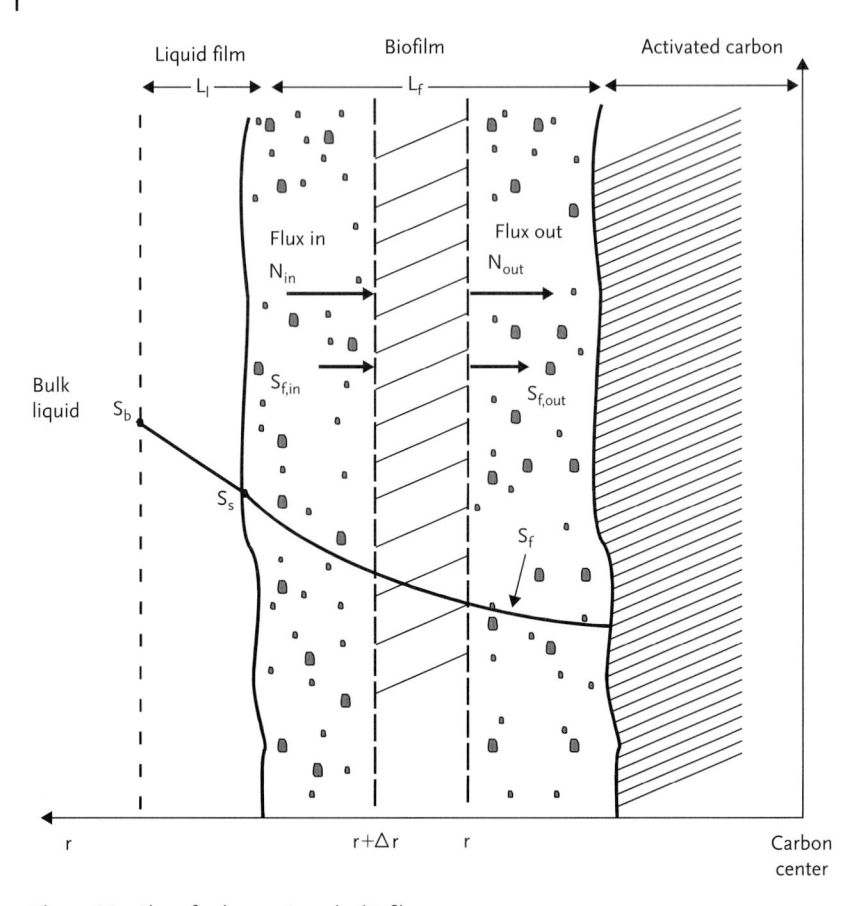

Figure 6.2 Flux of substrate into the biofilm.

Assuming that the biofilm area A_b at r and $r + \Delta r$ is constant, Eq. (6.8) simplifies to:

$$\frac{\partial S_f}{\partial t} = \frac{[N|_{r+\Delta r} - N|_r]}{\Delta r} - r_{ut} \qquad (6.9)$$

Letting $\Delta r \to 0$ leads to the following equation:

$$\frac{\partial S_f}{\partial t} = \left(\frac{\partial N}{\partial r}\right) - r_{ut} \qquad (6.10)$$

Substituting Eq. (6.6) into Eq. (6.10) leads to an expression which shows the change in substrate concentration in the biofilm with respect to time [1]:

$$\frac{\partial S_f}{\partial t} = D_f \frac{\partial^2 S_f}{\partial r^2} - r_{ut} \qquad (6.11)$$

Equation Eq. (6.11) shows that the accumulation of substrate is equal to the difference of diffusion and reaction. The rate of substrate utilization inside the biofilm, r_{ut}, sometimes denoted also as the intrinsic reaction rate, represents the amount of substrate removed per unit biofilm volume and time. It is often expressed according to the Monod model:

$$r_{ut} = \frac{1}{Y} \frac{\mu_{max} S_f}{K_s + S_f} X_f = \frac{k_{max} S_f}{K_s + S_f} X_f \qquad (6.12)$$

where

r_{ut} = intrinsic rate of substrate removal per unit volume of the biofilm $(M_s/L^3 \cdot T)$,
S_f = substrate concentration inside the biofilm (M_s/L^3),
D_f = molecular diffusivity of substrate in the biofilm (L^2/T),
μ_{max} = maximum specific growth rate $(1/T)$,
Y = yield coefficient (M_x/M_s),
k_{max} = maximum specific substrate utilization rate $\left(\frac{\mu_{max}}{Y}\right)$ $(M_s/M_x \cdot T)$,
X_f = density of biomass within biofilm (M_x/L^3),
K_s = half-velocity constant (M_s/L^3).

In the above units, the subscripts s and x denote the substrate and biomass, respectively.

In wastewater treatment, the metabolic reactions inside the biofilm are represented by r_{ut} and consist fundamentally of organics oxidation, nitrification, denitrification, and anaerobic degradation. Depending on the feed, operating conditions, and biomass properties, one or more of these processes may be relevant in a carbon system with biological activity.

In biofilm studies, several models have been proposed for expressing r_{ut}. It is hypothesized that diffusion resistance outside (external) and inside (internal) the biofilm affects the rate of removal of substrate and the order of reaction. The r_{ut} expression may take the form of a first- or zero-order reaction in S_f at low and high concentrations, respectively. Also, the substrate may fully diffuse into the biofilm or may be exhausted inside the biofilm, a situation that is referred to as partial penetration. For further information on biofilm kinetics the reader may refer to other sources [2–5].

If inside the biofilm diffusion and removal rates of substrate are equal to each other, the substrate concentration S_f reaches a steady state. Correspondingly, the variation of the substrate concentration along the thickness of the biofilm is given by the following equation:

$$D_f \frac{\partial^2 S_f}{\partial r^2} = X_f \frac{k_{max} S_f}{K_s + S_f} \qquad (6.13)$$

The solution of this equation for S_f requires knowledge of the parameters related to kinetics, mass transport, and biofilm properties.

The biofilm has two boundaries, as shown in Figures 6.1 and 6.2. The first boundary is the one shared with the liquid film (**Node 2**). In an integrated biofilm–carbon system, the second boundary is that shared with the surface of activated carbon (**Node 3**). The corresponding substrate concentrations at the liquid–biofilm (**Node 2**) and biofilm–carbon (**Node 3**) interfaces are denoted as S_s and S_{fc}, respectively [1].

At **Node 2**, conservation of mass requires that the flux from the liquid film is equal to the one at the liquid–biofilm interface:

$$k_{fc}(S_b - S_s) = D_f \left.\frac{\partial S_f}{\partial r}\right|_{r=R_p+L_f} \tag{6.14}$$

6.1.2.3
Diffusion into Activated Carbon Pores and Adsorption

If no biofilm layer is present, the transfer of substrate into the activated carbon particle involves only the diffusion from the bulk solution across the liquid film (external diffusion), diffusion within the pores of the carbon particle (pore diffusion), and diffusion along the internal surface of carbon particle (surface diffusion). The adsorption rate can be limited by any of these mass transfer processes.

However, if a biofilm has grown on the surface of activated carbon, the substrate has to pass additionally through the biofilm layer where it is partly consumed by biological reactions. Thus, the substrate having a concentration of S_{fc} at the biofilm–carbon interface (**Node 3** in Figure 6.1) diffuses further into the carbon particle, where it is removed by adsorption. Although biodegradation and adsorption mechanisms are acting concurrently, it is generally accepted that only adsorption is significant and biodegradation reactions can be neglected inside the pores of carbon particles. This assumption arises because bacteria have typically a diameter in the range of 200–2000 nm, much larger than the pores within the GAC. According to the IUPAC definition, the macropores, mesopores, and micropores of GAC have diameters greater than 50 nm, between 2 and 50 nm and less than 2 nm, respectively.

The internal substrate transport in a carbon particle or the intraparticle transport is described by two different models:

1. The homogeneous surface diffusion model (HSDM): This model assumes that the diffusion of substrate takes place due to a concentration gradient along the interior surface walls.
2. The pore diffusion model (PDM): According to this model, the adsorbate diffuses through the liquid found in the pores of activated carbon.

Actually, both surface and pore diffusion models can be combined for a more general description of intraparticle diffusion. However, studies show that in many cases the pore diffusion effects can be neglected. Therefore, it is generally sufficient to model the intraparticle diffusion using the HSDM which is given by the following partial differential equation [1]:

$$\frac{\partial q}{\partial t} = D_s \frac{1}{r^2}\frac{\partial}{\partial r}\left(r^2 \frac{\partial q}{\partial r}\right) = D_s \left(\frac{2}{r}\frac{\partial q}{\partial r} + \frac{\partial^2 q}{\partial r^2}\right) \tag{6.15}$$

where:

q = adsorbed substrate per amount of activated carbon (solid phase concentration) (M_s/M_c),
r = radial coordinate in the activated carbon, $0 \leq r \leq R_p$ (L),
D_s = intraparticle diffusivity of substrate (surface diffusion coefficient) in activated carbon (L^2/T).

In the above units, the subscript c stands for the activated carbon mass.

For the derivation of this equation in a carbon particle, which is based on Fick's law, the reader may refer to the literature [6]. This equation states that the substrate (adsorbate) uptake onto carbon is a function of time and the radial distance inside the carbon particle. The solution of Eq. (6.15) requires two boundary conditions:

At **Node 3** in Figure 6.1 (the biofilm–carbon interface), the mass flux of substrate from the biofilm into the carbon should be identical to the total mass adsorbed in the carbon [1]:

$$4\pi R_p^2 \cdot D_f \left. \frac{\partial S_f}{\partial r} \right|_{r=R_p} = \rho_p \frac{\partial}{\partial t} \left[\int_0^{R_p} 4\pi r^2 q \, dr \right] \qquad r = R_p, t \geq t_0 \qquad (6.16)$$

where

ρ_p: particle density of activated carbon (M_c/L^3),
R_p: radius of the carbon particle (L).

The main difference between the modeling of a biofilm system with adsorbing media and such modeling with nonadsorbing media can be deduced from Eq. (6.16). If the biofilm covers a nonadsorbing medium such as sand, the flux at **Node 3** will be equal to zero, as discussed in the literature [4].

At **Node 4** in Figure 6.1, the substrate flux at the center of the carbon particle is zero [1]:

$$\frac{\partial q}{\partial r} = 0 \quad \text{at } r = 0, t \geq t_0 \qquad (6.17)$$

The transport inside the carbon particle or the intraparticle transport can also be described in a more simplified way using the Linear Driving Force (LDF) model instead of Eq. (6.15). The LDF model shown in Eq. (6.18) assumes that the rate of substrate (adsorbate) uptake onto carbon is proportional to the difference between the substrate loading (solid phase concentration) at the external carbon surface (**Node 3** in Figure 6.1) and the mean substrate loading in the adsorbent [7]:

$$\frac{d\bar{q}}{dt} = k_p(q_{fc} - \bar{q}) \qquad (6.18)$$

where

k_p = particle phase mass transfer coefficient (1/T),

q_{fc} = substrate loading (solid phase concentration) at the external carbon surface (M_s/M_c),
$\bar{q}$ = mean substrate loading inside the carbon particle (the mean solid phase concentration) (M_s/M_c).

For spherical particles, the mean solid phase concentration is expressed as follows [6]:

$$\bar{q} = \frac{3}{R_p^3} \left[\int_0^{R_p} r^2 q \, dr \right] \tag{6.19}$$

Under normal conditions of operation, q_{fc} is greater than $\bar{q}$ since the substrate concentration at **Node 3** is higher than the average substrate concentration inside the carbon particle. In other words, the deficit in the adsorbed amount is the driving force for adsorption. However, as discussed extensively in Chapter 7, in some cases the solid phase substrate concentration inside the carbon particle may reach higher levels than outside. This concentration reversal is the underlying reason for bioregeneration of activated carbon.

6.1.2.3.1 Adsorption Isotherms

As shown in Figure 6.1, adsorption of substrate inside the carbon leads to a decrease of substrate concentration in the liquid phase along the particle radius. Therefore, to solve Eq. (6.16) or Eq. (6.19), it is essential to describe the solid phase concentration, q at any radius of the carbon particle (between **Node 3 and 4**), or alternatively a mean solid phase concentration, $\bar{q}$ inside the carbon particle.

Various types of isotherms (e.g., the Langmuir isotherm) can be used in describing the solid phase concentration at the surface and inside the carbon particle. However, the Freundlich isotherm shown in Eq. (6.20) is the most widely used one, since its applicability extends over a wide range of concentrations:

$$q_e = K_F S_e^{1/n} \tag{6.20}$$

where

q_e = substrate loading (solid phase concentration) on carbon (M_s/M_c),
S_e = equilibrium substrate concentration in the liquid phase (M_s/L^3),
K_F and $1/n$ = Freundlich isotherm constants.

At the biofilm–carbon boundary, the solid phase loading or substrate loading on the surface of carbon at equilibrium with the concentration S_{fc} is denoted as q_{fc}. Accordingly, the mean solid phase loading $\bar{q}$ inside the carbon particle is then related to the mean substrate concentration $\bar{S}$ inside the particle (Figure 6.1).

6.1.2.4
Biofilm Growth and Loss

The biofilm thickness shown in Figure 6.1 is actually not fixed, but undergoes changes due to biomass growth and loss. Since the biofilm is mostly responsible for biodegradation, equations describing the conservation of mass must also account for the biofilm thickness L_f, the biomass concentration X_f, or the total biofilm amount.

Detachment of biofilm greatly affects performance and stability of biofilm reactors. Detachment is highly dependent on the physical characteristics and the environmental conditions of the system. Detachment rate may increase with increasing biomass and fluid velocity [8]. The biofilm may detach as a result of the following mechanisms: (i) Erosion due to fluid shear, (ii) Sloughing, (iii) Grazing of the outer biofilm by protozoa, (iv) Abrasion of biofilm surface, and (v) Backwashing. A review emphasizes that the shear stress is the most researched factor in BAC reactors [9]. A summary of detachment rate expressions can be found in the literature [4].

Chang and Rittmann considered that the substrate diffuses through a boundary layer (as shown in Figure 6.1), the liquid–biofilm interface, which, however, can be moving with time [1]. According to this model, biofilm growth takes place due to substrate consumption, whereas biofilm decay and shear are regarded as separate factors leading to the loss of biofilm:

$$\frac{\partial (L_f X_f)}{\partial t} = \int_{R_p}^{R_p + L_f} \left(\frac{Y k_{max} S_f}{K_s + S_f} - b - b_s \right) X_f dr \tag{6.21}$$

where

Y = yield coefficient (M_x/M_s),
b = biofilm decay coefficient (1/T),
b_s = biofilm shear loss coefficient (1/T).

Biofilm density, X_f drops out when it is assumed constant.

Other researchers proposed another model (Eq. (6.22)) which considers the growth of the biofilm depending on substrate consumption and total decay of the biofilm [10]. Also, in this model the biofilm density can be assumed constant.

$$\frac{\partial L_f}{\partial t} = \frac{Y \left[\int_{R_p}^{R_p + L_f} \frac{k_{max} \cdot S_f}{K_s + S_f} X_f 4\pi r^2 dr \right]}{A_f X_f} - b_{tot} L_f \quad R_p \leq r \leq R_p + L_f \tag{6.22}$$

where

b_{tot} = total decay coefficient due to decay and fluid shear (1/T),
A_f = surface area of one BAC granule (L^2).

6.1.3
Benefits of Integrated Adsorption and Biological Removal:
Link to Fundamental Processes

The main benefits of integrated adsorption and biological processes are discussed in Chapter 3. Complementary to this, these benefits are better visualized when the fundamental diffusion, biodegradation, and adsorption phenomena in Section 6.1.2 are considered. In most cases, an integrated carbon–biological system is not simply a combination of adsorption and biological removal. In that context, the following issues are of interest:

1. Activated carbon adsorbs substrates directly when there is not yet any biological activity such as in the initial stages of GAC operation. Then, the presence of a biofilm surrounding activated carbon (Figure 6.1) would be neglected. The substrates adsorbed would be primarily those that are nonbiodegradable and inhibitory.
2. When activated carbon is placed into a suspended-growth biological system or filled into a reactor in the form of PAC and GAC, respectively, it may retain a number of inhibitory organic as well as inorganic substances (such as heavy metals). This adsorptive ability of activated carbon prevents both attached and suspended biomass from inhibition in PACT and BAC systems, respectively. This property of activated carbon reduces the effects of hazardous compounds under fluctuating flow rates or concentrations.
3. Activated carbon has the property of concentrating substrates, nutrients, dissolved oxygen, and so on, on its surface. The long retention time of substrates on carbon surface enables their contact with the suspended biomass or with the growing biofilm. This extended contact time is a factor increasing the possibility of biodegradation or biotransformation of various substrates.
4. If a biofilm develops on the surface of activated carbon as shown in Figure 6.1, the following consequences can be expected:

    a) The presence of a biofilm enables the removal of biodegradable or slowly biodegradable substrates, some of which may be nonadsorbable.
    b) Substrates are transported across the biofilm into the carbon until carbon saturation occurs (Figure 6.1). This mechanism is advantageous in the case of slowly biodegradable or nonbiodegradable substrates.
    c) The formation of a biofilm can, however, also hinder adsorption of substrates onto activated carbon (Figure 6.1). This is an unwanted outcome if these substrates are nonbiodegradable but adsorbable. On the other hand, the formation of biofilm on carbon surface may provide an advantage if substrates are biodegradable but slowly adsorbable or nonadsorbable.
    d) An aspect of all biofilms is their ability to respond to shock loadings of toxic and inhibitory chemicals. As shown in Figures 6.1 and 6.2, the resistance to the diffusion of substrate into the biofilm is the underlying reason for this observation. If a sudden loading of a toxic chemical occurs, the outer parts of the biofilm are affected, whereas the inner parts may still be active.
    e) If the thickness of biofilm increases, the biodegradable substrate concentration S_f, shown in Figures 6.1 and 6.2, may be decreased to zero at some depth of the biofilm. Also, the bulk liquid concentration (S_b) may drop to very low values. This low concentration may then favor the desorption of substrates from the carbon. Thus, activated carbon may temporarily adsorb substances which, depending on substrate concentrations, are then released into the surrounding biofilm or bulk liquid where they are biodegraded.

This mechanism may result in bioregeneration of activated carbon, leading to a more effective use of the activated carbon, as discussed in detail in Chapter 7. The desorption and biodegradation of substrate also has another beneficial effect, namely that of enabling the survival of biomass under low substrate loading rates. In that respect, biofilm systems with 'adsorbing' GAC media are significantly different from those which employ 'nonadsorbing' media such as sand or anthracite.

f) A GAC reactor may gradually be converted into a biofilm reactor where the sludge retention time (SRT), also called the sludge age, is approximated as follows:

$$\theta_x = \frac{\text{active biomass in the biofilm}}{\text{production rate of active biomass}} \quad (6.23)$$

Under steady-state conditions, the growth of biofilm is balanced by its loss, and the biofilm attains a steady thickness. When Eq. (6.23) is adapted to represent a steady-state biofilm it takes the following form [11]:

$$\theta_{x,\text{avg}} = \frac{X_f L_f A_b}{b_{\text{det}} X_f L_f A_b} = \frac{1}{b_{\text{det}}} \quad (6.24)$$

where

θ_x = sludge retention time (SRT) (T),
$\theta_{x,\text{avg}}$ = average θ_x for a steady-state biofilm (T),
b_{det} = detachment rate of biomass (1/T),
A_b = biofilm surface area (L^2).

Thus, the average SRT in a steady-state biofilm is equal to the reciprocal of the specific detachment rate as long as active biomass is not added to the biofilm by deposition [11]. Another SRT expression has also been proposed that incorporates the substrate flux into biofilm and biofilm detachment [4].

Since the amount of active biomass is high and the detachment rate is low in BAC reactors, the SRT is much higher than that in suspended-growth processes such as activated sludge. The high SRT is a factor favoring the establishment of slowly growing microorganisms such as nitrifiers or methanogens inside the biofilm. The high SRT further increases the exposure time of biomass to nonbiodegradable or slowly biodegradable organics and enables efficient biodegradation.

g) In the modified suspended-growth processes such as the aerobic PACT, anaerobic PACT or membrane processes with PAC addition (PAC-MBR), to some extent the surface of PAC serves as an attachment medium for microorganisms. This attachment generally increases at higher SRT, where the benefits of PAC addition become more obvious. However, in PACT systems the extent of biomass attachment is still much lower than that in BAC reactors.

h) Activated carbon may also adsorb intermediate or end products of biological reactions taking place in the biofilm or suspended phases. As discussed in

Chapter 3, the ability of activated carbon to adsorb Soluble Microbial Products (SMP) and Extracellular Polymeric Substances (EPS) generally leads to an improvement of effluent quality and sludge characteristics.

6.1.4
Modeling Approaches in GAC/BAC Reactors

Models of integrated adsorption and biodegradation mainly center around the synergism created between the biofilm and activated carbon. In addition to the mechanisms covered in Section 6.1.2, in integrated reactors the main factors to be considered include substrate type, reactor type, hydraulic mixing, mixing of the solid (adsorbent) phase, type of GAC, void volume, EBCT, hydraulic loading rate, growth and loss of biofilm, microbial species, type and frequency of backwashing, presence of multisolutes, competition between solutes, and so on. Moreover, fluctuations in flow rate and influent concentrations, or the dynamic conditions in a reactor, have to be taken into consideration.

Several models developed in the literature are based on simplifying assumptions for mass transport resistance, microbial kinetics, and biofilm thickness in order to gain an understanding of the BAC process [12].

All models involve mass balances of the substrate and biomass. In wastewater treatment systems, mass balances for substrate are usually based on sum parameters such as COD, BOD, or TOC if the target pollutant is the bulk organic substrate of undefined composition, which is normally the case in domestic and industrial wastewaters. The models may, however, be further elaborated to characterize the different fractions in substrate as biodegradable, slowly biodegradable or nonbiodegradable, or readily adsorbable, moderately adsorbable, nonadsorbable, and so on. The models can also be extended to involve oxygen consumption, nutrient consumption, and product formation. They can also be constructed to describe the change within the reactor in attached or suspended biomass or both.

6.1.4.1
Mass Balances in GAC/BAC Reactors

The discussion presented in this section addresses the simplest case, namely the mass balance made in a typical fixed-bed GAC filter. A continuous down- or up-flow fixed bed differs from a completely mixed reactor in that the liquid phase inside the reactor is usually not mixed, and a concentration gradient exists along the length of the flow. Therefore, in such reactors, a mass balance can only be made around the infinitesimal slice having a thickness Δz and a volume $\Delta V_B = A \Delta z$, as shown in Figure 6.3.

Mass balances can be constructed for the substrate considering the following phases in the reactor:

- liquid phase (bulk water)
- solid phase (adsorbent).

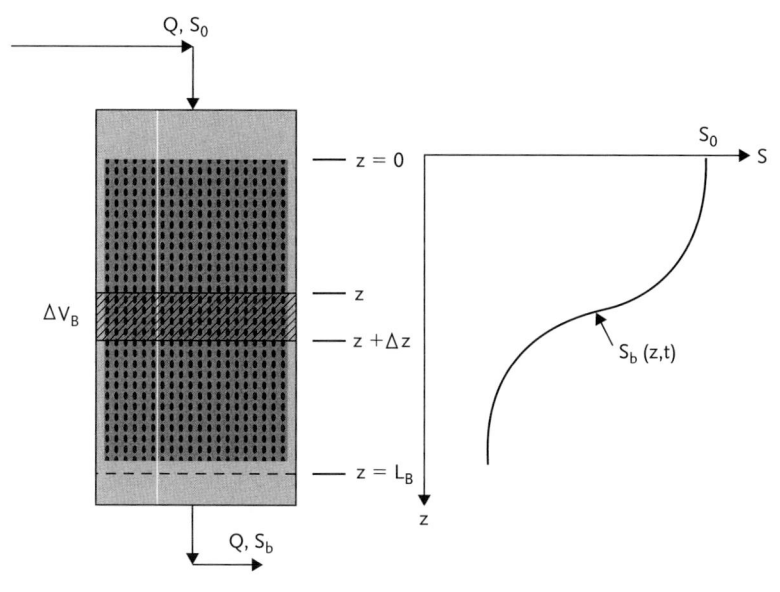

Figure 6.3 Fixed-bed GAC adsorber and variation of substrate concentration along the length of the adsorber.

6.1.4.1.1 Substrate Mass Balance in the Liquid Phase of the Reactor In the whole discussion in this section the term substrate is preferred over the term 'adsorbate' since substrate is not removed by adsorption only. The main reactions leading to the removal of substrate from the liquid phase are adsorption onto carbon and biological removal. The verbal description of these phenomena is as follows:

$$\begin{bmatrix} \text{Accumulation rate of} \\ \text{substrate} \end{bmatrix} = \begin{bmatrix} \text{Net rate of substrate} \\ \text{input by dispersion} \end{bmatrix}$$
$$+ \begin{bmatrix} \text{Net rate of substrate} \\ \text{input by advection} \end{bmatrix} - \begin{bmatrix} \text{Adsorption rate of} \\ \text{substrate} \end{bmatrix}$$
$$- \begin{bmatrix} \text{Biological removal rate of} \\ \text{substrate} \end{bmatrix} \quad (6.25)$$

In mathematical form, this mass balance in the liquid phase can be expressed as:

$$\varepsilon_B \Delta z A \frac{\partial S_b}{\partial t} = \varepsilon_B A \left(-D \frac{\partial S_b}{\partial z} + v_i S_b \right)_z - \varepsilon_B A \left(-D \frac{\partial S_b}{\partial z} + v_i S_b \right)_{z+\Delta z} \\ - r_{ads} \Delta z A - r_{biofilm} \Delta z A \quad (6.26)$$

where

S_b = concentration of substrate in the bulk liquid (M_s/L^3),
D = dispersion coefficient within the porous media at the interstitial velocity v_i (L^2/T),
ε_B = void ratio in the bed (unitless),
v_i = interstitial velocity: $Q/A\varepsilon_B$ (L/T),
Q = flow rate (L^3/T),
z = distance along the flow path (L),
A = cross-sectional area of the reactor (L^2),
r_{ads} = rate of substrate adsorption per unit volume of reactor ($M_s/L^3 \cdot T$),
$r_{biofilm}$ = rate of substrate removal in the biofilm per unit volume of reactor ($M_s/L^3 \cdot T$).

The substrate concentration S_b in the above equation varies with respect to operation time t and the distance z. The first two terms on the right-hand side account for the flux of substrate into and out of the slice Δz by dispersion and advection, respectively. Since the mass balance on the substrate is based on the liquid phase of the reactor, the void ratio (porosity) in the bed, ε_B, is included in these expressions.

Dividing Eq. (6.26) by $\varepsilon_B \Delta z\, A$ yields:

$$\frac{\partial S_b}{\partial t} = \frac{\left(D\frac{\partial S_b}{\partial z}\big|_{z+\Delta z} - D\frac{\partial S_b}{\partial z}\big|_z\right)}{\Delta z} - \frac{(v_i S_b|_{z+\Delta z} - v_i S_b|_z)}{\Delta z} - \frac{r_{ads}}{\varepsilon_B} - \frac{r_{biofilm}}{\varepsilon_B} \quad (6.27)$$

Letting $\Delta z \to 0$ leads to the following partial differential equation showing the net change in substrate concentration with respect to time:

$$\frac{\partial S_b}{\partial t} = D\frac{\partial^2 S_b}{\partial z^2} - v_i\frac{\partial S_b}{\partial z} - \frac{r_{ads}}{\varepsilon_B} - \frac{r_{biofilm}}{\varepsilon_B} \quad (6.28)$$

Under steady-state conditions, the net rate of change in substrate concentration will be equal to zero:

$$0 = D\frac{\partial^2 S_b}{\partial z^2} - v_i\frac{\partial S_b}{\partial z} - \frac{r_{ads}}{\varepsilon_B} - \frac{r_{biofilm}}{\varepsilon_B} \quad (6.29)$$

This equation states that the net substrate transferred to the slice Δz by dispersion and advection will be balanced by the uptake onto activated carbon and biodegradation in the biofilm.

Some researchers have proposed differentiating between two types of biological removal: (i) biological removal inside the biofilm and (ii) biological removal by suspended biomass in the interstices [13]. However, it is generally accepted that most biodegradation takes place in the biofilm phase at a rate of $r_{biofilm}$.

Biological Removal in the Biofilm Phase of the Reactor If no biological reaction is present, but only adsorption takes place, as is usually the case in the initial periods of operation, the biological reaction term can be neglected. However, in later periods of operation, once biofilm formation has taken place, a significant part of substrate is removed inside the biofilm.

Under such conditions, the spherical carbon particles contained in the control volume ΔV_B are surrounded by a biofilm. The r_{ut} is a function of the substrate concentration inside the biofilm and should therefore be integrated over the whole biofilm thickness L_f. The following relationship has been proposed for describing the removal of substrate in the infinitesimal volume ΔV_B [10]:

$$r_{\text{biofilm}} \Delta z \, A = N_p \left[\int_{R_p}^{R_p + L_f} \frac{k_{\max} \cdot S_f}{K_s + S_f} X_f \, 4\pi r^2 dr \right] \quad (6.30)$$

$$N_p = \frac{A \, \Delta z (1 - \varepsilon_B)}{V_p} \quad (6.31)$$

where

N_p = the number of GAC particles in ΔV_B,
V_p = volume of a carbon particle, (L^3).

Substituting Eq. (6.31) into Eq. (6.30) gives the rate of substrate removal in the biofilm phase:

$$r_{\text{biofilm}} = \frac{(1 - \varepsilon_B)}{V_p} \left[\int_{R_p}^{R_p + L_f} \frac{k_{\max} \cdot S_f}{K_s + S_f} X_f \, 4\pi r^2 dr \right] \quad (6.32)$$

The mass balances shown up to now assume that initially all of the substrate is soluble and ready to diffuse into the biofilm. In the case of an insoluble or macromolecular substrate, hydrolysis is a necessary mechanism leading to solubilization prior to biological consumption.

Biological Removal in the Suspended Phase of the Reactor If biological removal by suspended biomass cannot be neglected, another term has to be added to the mass balance as proposed in some studies [13, 14]. The following is an example of expressing this contribution of suspended biomass:

$$\varepsilon_B \Delta z \, A \frac{\partial S_b}{\partial t} = \varepsilon_B A \left(-D \frac{\partial S_b}{\partial z} + v_i \, S_b \right)_z - \varepsilon_B A \left(-D \frac{\partial S_b}{\partial z} + v_i \, S_b \right)_{z + \Delta z} \quad (6.33)$$
$$- r_{\text{ads}} \Delta z \, A - r_{\text{biofilm}} \Delta z \, A - r_{\text{susp}} \Delta z \, A$$

In this equation r_{susp} is represented as follows:

$$r_{\text{susp}} = \frac{k_b \, S_b}{K_s + S_b} X_{\text{susp}} \quad (6.34)$$

where

r_{susp} = removal rate of substrate by suspended biomass per reactor volume $(M_s/L^3.T)$,
k_b = maximum rate of specific substrate utilization per volume of reactor $(M_s/M_x.T)$,
X_{susp} = suspended biomass concentration (M_x/L^3).

Adsorption onto Activated Carbon in the Reactor The solid phase concentration q is actually a function of time t, reactor length z, and the radial coordinate r. However, as shown in Eq. (6.18), a common approximation is to express adsorption in terms of $\bar{q}$, the mean solid phase concentration along the carbon particle.

In Eq. (6.33), the term representing the rate of substrate uptake onto activated carbon, r_{ads}, can be found from a mass balance made on the slice Δz:

$$\rho_p \Delta z A (1 - \varepsilon_B) \frac{\partial \bar{q}}{\partial t} = r_{ads} \Delta z A \tag{6.35}$$

where

$\frac{\partial \bar{q}}{\partial t}$ = the mean substrate uptake rate onto carbon $(M_s/M_c.T)$,
ρ_p = apparent density of GAC granules (M_c/L^3).

Upon simplification the following expression is obtained for the adsorption rate:

$$r_{ads} = \rho_p (1 - \varepsilon_B) \frac{\partial \bar{q}}{\partial t} \tag{6.36}$$

In a filter packed with GAC, adsorption equilibrium and kinetics of adsorption are closely linked to each other, at all times and at all points in the filter [15].

Overall Substrate Balance Insertion of Eqs. (6.32), (6.34), and (6.36) into Eq. (6.28) leads to the following one which involves dispersion, advection, adsorption, and biodegradation in the slice Δz:

$$\frac{\partial S_b}{\partial t} = D \frac{\partial^2 S_b}{\partial z^2} - v_i \frac{\partial S_b}{\partial z} - \frac{(1-\varepsilon_B)}{\varepsilon_B} \rho_p \frac{\partial \bar{q}}{\partial t}$$

$$- \left[\int_{R_p}^{R_p + L_f} \frac{k_{max} \cdot S_f}{K_s + S_f} X_f 4\pi r^2 dr \right] \frac{(1-\varepsilon_B)}{V_p \varepsilon_B} - \frac{1}{\varepsilon_B} \frac{k_b S_b}{K_s + S_b} X_{susp} \tag{6.37}$$

A plot of the S_b concentration at $z = L_B$ with respect to operation time is possible. The ratio of this S_b to the influent concentration S_0 would yield a curve known as the 'breakthrough curve.' In the case of a GAC adsorber with biological activity the breakthrough curves take a different shape from that of the adsorber illustrated in Chapter 2.

To solve Eq. (6.37) for the substrate concentration, the parameters requiring determination are: D (dispersion coefficient), ρ_g, ε_B, the mean substrate uptake rate onto the carbon, and the constants in biological rate expressions. There are different methods of approximating the mean substrate uptake take onto carbon [6]. Normally, the last term in Eq. (6.37) representing the removal by suspended biomass is neglected in most modeling efforts.

A simpler approach in constructing the mass balance is to consider the mass flux across the liquid film. If liquid film diffusion is controlling, the mass transferred through this film will be removed by biodegradation and/or adsorption. When the flux from the liquid film is multiplied by the total biofilm surface area surrounding all carbon particles, the last term on the right-hand side of Eq. (6.38) is obtained [16]:

$$\frac{\partial S_b}{\partial t} = D \frac{\partial^2 S_b}{\partial z^2} - v_i \frac{\partial S_b}{\partial z} - \frac{3(1-\varepsilon_B)(R_p + L_f)^2}{\varepsilon_B R_p^3} k_{fc}(S_b - S_s) \quad (6.38)$$

Special Case: Complete Mixing in the Liquid Phase If completely mixed conditions exist in the liquid phase, then the substrate gradient along the length of the reactor, $\partial S_b/\partial z$ will be equal to zero. Then, the control volume in the mass balance becomes the total reactor volume V_B. If the removal by suspended biomass is neglected, the following equation can be written:

$$\varepsilon_B V_B \frac{dS_b}{dt} = Q(S_0 - S_b) - V_B \, r_{ads} - V_B \, r_{biofilm} \quad (6.39)$$

where

V_B = fixed-bed volume (L^3),
S_0 = influent substrate concentration (M_s/L^3),
S_b = bulk liquid substrate concentration throughout the depth of the reactor (M_s/L^3).

Inserting the former expressions for adsorption rate and biofilm utilization leads to the following relationship:

$$\frac{dS_b}{dt} = \frac{Q}{V_B \varepsilon_B}(S_0 - S_b) - \frac{(1-\varepsilon_B)}{\varepsilon_B} \rho_p \frac{\partial \bar{q}}{\partial t}$$
$$- \left[\int_{R_p}^{R_p + L_f} \frac{k_{max} S_f}{K_s + S_f} X_f \, 4\pi r^2 dr \right] \frac{(1-\varepsilon_B)}{V_p \varepsilon_B} \quad (6.40)$$

Similarly, this equation too can be extended to include the removal of substrate by suspended biomass:

$$\frac{dS_b}{dt} = \frac{Q}{V_B \, \varepsilon_B}(S_0 - S_b) - \frac{(1-\varepsilon_B)}{\varepsilon_B} \rho_p \frac{\partial \bar{q}}{\partial t}$$

$$- \left[\int_{R_p}^{R_p + L_f} \frac{k_{max} \, S_f}{K_s + S_f} X_f \, 4\pi r^2 dr \right] \frac{(1-\varepsilon)}{\varepsilon_B \, V_p} - \frac{1}{\varepsilon_B} \frac{k_b \, S_b}{K_s + S_b} X_{susp} \qquad (6.41)$$

It is also possible to adapt Eq. (6.38) to the case of complete mixing:

$$\frac{\partial S_b}{\partial t} = \frac{Q}{V_B \, \varepsilon_B}(S_0 - S_b) - \frac{3(1-\varepsilon_B)(R_p + L_f)^2}{\varepsilon_B \, R_p^3} k_{fc}(S_b - S_s) \qquad (6.42)$$

6.1.4.1.2 Biomass Balance in the Reactor
Bacterial growth with time in a GAC filter is inevitable if the influent contains dissolved organics that are biodegradable. The change in biomass amount inside an adsorber alters the void ratio in the carbon bed and thus the interstitial velocity. As in the case of substrate, a mass balance equation can be constructed for the biomass present in the infinitesimal slice Δz.

The biofilm growth and decay have been incorporated into a number of models [1, 16, 17]. The change in biofilm thickness may be expressed as a function of both time and the distance along the length of a reactor, as shown for the case of a fluidized-bed biofilm reactor [16]:

$$\frac{\partial L_f(z,t)}{\partial t} = \frac{k_{max} \, \overline{S_f}(z,t) \, L_f(z,t)}{K_s + \overline{S_f}(z,t)} - K_d \, L_f(z,t) \qquad (6.43)$$

where

$\overline{S_f}$: average substrate concentration inside the biofilm (M_s/L^3),
K_d: overall biofilm loss coefficient (1/T).

Only a few models also consider the change in suspended biomass [1]. In the following model, shown in Eq. (6.44), the first three terms on the right-hand side represent the variation of suspended biomass due to the substrate consumption, the decay, and the loss with the effluent stream, respectively. The last term indicates the increase in suspended biomass due to shear of attached biomass from the carbon surface [1]:

$$\frac{dX_{susp}}{dt} = \left(\frac{Y \, k_{max} \, S_b}{K_s + S_b} - b - \frac{Q}{V_B \, \varepsilon_B} \right) X_{susp} + \frac{3 \, X_w \, b_s}{V_B \, \varepsilon_B \, \rho_p \, R_p} L_f \, X_f \qquad (6.44)$$

where

X_w: total mass of carbon in the reactor, (M_c).

6.1.4.2
Characterization of BAC Reactors by Dimensionless Numbers
In GAC/BAC reactors, dimensionless analysis of the system helps in modeling and serves as a guide in experimental work. The relative importance of each mass

6.1 Modeling of GAC Adsorbers with Biological Activity

transport mechanism and the adsorption and biological processes can be determined by the use of dimensionless numbers [10, 17].

6.1.4.2.1 Definition of Dimensionless Parameters

$$\text{Dimensionless substrate}: S_b^* = \frac{S_b}{S_0} \quad S_s^* = \frac{S_s}{S_0} \quad S_f^* = \frac{S_f}{S_0} \tag{6.45}$$

$$\text{Dimensionless reactor length}: z^* = \frac{z}{L_B} \tag{6.46}$$

$$\text{Dimensionless time}: t^* = \frac{t}{L_B/v_i} \tag{6.47}$$

$$\text{Dimensionless radius}: r^* = r/R_p \tag{6.48}$$

$$\text{Dimensionless solid phase concentration}: \bar{q}^* = \bar{q}/q_0 \quad q_{fc}^* = q_{fc}/q_0 \tag{6.49}$$

6.1.4.2.2 Dimensionless Form of Mass Transfer Equations

The equations presented in Section 6.1.4.1 can be expressed in dimensionless form by substituting each variable with its dimensionless counterpart.

At **Node 2**, the liquid film–biofilm boundary, the dimensionless form of Eq. (6.14) is as follows:

$$\frac{k_{fc} R_p}{D_f}(S_b^* - S_s^*) = \frac{\partial S_s^*}{\partial r^*}\bigg|_{r^* = 1 + L_f/R_p} \tag{6.50}$$

The term on the left-hand side corresponds to the **Sherwood number**, which represents the ratio between the liquid film transport and the biofilm transport shown in Figure 6.1:

$$Sh = \frac{k_{fc} R_p}{D_f} \tag{6.51}$$

Then, Eq. (6.50) takes the following form:

$$Sh(S_b^* - S_s^*) = \frac{\partial S_s^*}{\partial r^*}\bigg|_{r^* = 1 + L_f/R_p} \tag{6.52}$$

At **Node 3** in Figure 6.1, Eq. (6.16) denotes the mass transport from the biofilm into activated carbon and can also be expressed in dimensionless form. Using the expression for the mean surface concentration, $\bar{q}$, as defined in Eq. (6.19), yields the following equation:

$$\frac{\partial S_f^*}{\partial r^*}\bigg|_{r^* = 1} = \frac{\rho_p q_0 R_p^2 v_i}{D_f S_0 \, 3 \, L_B}\left(\frac{d\bar{q}^*}{dt^*}\right) \tag{6.53}$$

This equation simply shows that the adsorption rate onto carbon depends on the flow of mass from the biofilm into carbon. If Eq. (6.18) is expressed in dimensionless form, the following equation is obtained:

$$\frac{d\bar{q}^*}{dt^*} = \frac{L_B}{v_i} k_p (q_{fc}^* - \bar{q}^*) \tag{6.54}$$

The particle phase mass transfer coefficient k_p in Eq. (6.18) is approximated as follows [6, 7]:

$$k_p = 15 D_s / R_p^2 \tag{6.55}$$

Substituting Eqs. (6.54) and (6.55) into 6.53 yields:

$$(q_{fc}^* - \bar{q}^*) = \left.\frac{\partial S_f^*}{\partial r^*}\right|_{r^*=1} \left(\frac{1}{5} \frac{D_f S_0}{D_s \rho_p q_0}\right) = \left.\frac{\partial S_f^*}{\partial r^*}\right|_{r^*=1} \cdot Bi_s' \tag{6.56}$$

Normally, in systems with adsorption only, the Biot number (Bi_s) represents the ratio between the liquid film diffusion and the intraparticle surface diffusion and is shown as follows [6]:

$$\text{Biot number } (Bi_s) = \frac{k_{fc} R_p S_0}{D_s \rho_p q_0} \tag{6.57}$$

On the other hand, the dimensionless term on the right-hand side of Eq. (6.56) can be interpreted as a modified form of the Biot number. When a biofilm layer exists next to the carbon surface instead of a liquid film, the dimensionless form of the mass balance equation can be written in terms of the modified Biot number (Bi') which compares the biofilm transfer and surface diffusion.

6.1.4.2.3 Dimensionless Form of Reactor Mass Balance
The substrate mass balance in a reactor, Eq. (6.28), can also be expressed in a dimensionless form:

$$\frac{S_0}{L_B/v_i} \frac{\partial S_b^*}{\partial t^*} = D \frac{S_0}{L_B^2} \frac{\partial^2 S_b^*}{\partial z^{*2}} - v_i \frac{S_0}{L_B} \frac{\partial S_b^*}{\partial z^*} - \frac{r_{ads}}{\varepsilon_B} - \frac{r_{biofilm}}{\varepsilon_B} \tag{6.58}$$

Dividing each side by $S_0 v_i / L_B$ gives:

$$\frac{\partial S_b^*}{\partial t^*} = \left(\frac{D}{L_B v_i}\right) \frac{\partial^2 S_b^*}{\partial z^{*2}} - \frac{\partial S_b^*}{\partial z^*} - \left(\frac{r_{ads}}{\varepsilon_B S_0 v_i / L_B}\right) - \left(\frac{r_{biofilm}}{\varepsilon_B S_0 v_i / L_B}\right) \tag{6.59}$$

The $\varepsilon_B S_0 v_i / L_B$ term in the denominator is equal to the substrate loading rate QS_0/V_B. Therefore, the third and fourth terms on the right-hand side of this equation stand for the dimensionless adsorption and biofilm removal rates, respectively:

$$\frac{\partial S_b^*}{\partial t^*} = \left(\frac{D}{L_B v_i}\right) \frac{\partial^2 S_b^*}{\partial z^{*2}} - \frac{\partial S_b^*}{\partial z^*} - r_{ads}^* - r_{biofilm}^* \tag{6.60}$$

The $D/L_B v_i$, termed the dispersion number, is often used in studies to characterize the flow regime in packed-bed biofilm reactors [18]. The reciprocal of this is the Peclet number which expresses the relative importance of substrate advection to dispersion along the column:

$$\text{Pe} = \frac{L_B v_i}{D} \quad (6.61)$$

6.1.4.2.4 Dimensionless Expression of the Adsorption Rate in Reactor
Using Eq. (6.36) showing r_{ads}, the dimensionless adsorption rate in Eq. (6.60) can be found as follows:

$$r^*_{ads} = \frac{\rho_p (1 - \varepsilon_B) \frac{\partial \bar{q}}{\partial t}}{\varepsilon_B S_0 v_i / L_B} \quad (6.62)$$

Rearranging the terms leads to the following equation:

$$r^*_{ads} = \frac{\rho_p (1 - \varepsilon_B) \frac{\partial \bar{q}}{\partial t}}{S_0} (\text{EBCT}) \quad (6.63)$$

where EBCT is the empty-bed contact time as defined in Chapter 2.

Or, alternatively, using Eq. (6.62) and the dimensionless expressions for t (Eq. (6.47)) and $\bar{q}$ (Eq. (6.49)), the following equation can be written:

$$r^*_{ads} = \left(\frac{\rho_p (1 - \varepsilon_B)}{\varepsilon_B} \frac{q_0}{S_0} \right) \frac{\partial \bar{q}^*}{\partial t^*} = D_g \frac{\partial \bar{q}^*}{\partial t^*} \quad (6.64)$$

D_g is denoted as the **solute distribution parameter** or the capacity factor [6]. It is independent of the bed size and indicates the ratio of substrate adsorbed onto activated carbon to that in the liquid phase.

6.1.4.2.5 Dimensionless expression of the biofilm removal rate in reactor
The biofilm removal rate expression in Eq. (6.32) has to be integrated if S_f and K_s are comparable to each other. In this section, the following simplified cases are considered:

Zero-Order Kinetics Inside the Biofilm If $S_f \gg K_s$, the removal rate term reduces to

$$\frac{k_{max} S_f}{K_s + S_f} X_f = k_{max} X_f \quad (6.65)$$

Assuming that X_f is a constant, Eq. (6.32) simplifies to:

$$r_{biofilm} = \frac{(1 - \varepsilon_B) k_{max} X_f}{V_p} \left[\int_{R_p}^{R_p + L_f} 4\pi r^2 dr \right] \quad (6.66)$$

The integral term in this equation is equal to the biofilm volume, $V_{biofilm}$. Then, the following equation is obtained:

$$r^*_{biofilm} = \left(\frac{r_{biofilm}}{\varepsilon_B S_0 v_i / L_B} \right) = \frac{(1 - \varepsilon_B)(V_{biofilm}/V_p) k_{max} X_f}{Q S_0 / V_B} \quad (6.67)$$

The volume of a carbon particle and the approximate biofilm volume around it are as follows:

$$V_p = \frac{4}{3} \pi R_p^3 \qquad V_{biofilm} = 4\pi R_p^2 L_f \qquad (6.68)$$

The expression in Eq. (6.67) takes the following form:

$$r_{biofilm}^* = \left(\frac{r_{biofilm}}{\varepsilon_B S_0 \, v_i / L_B}\right) = \frac{\frac{3(1-\varepsilon_B)}{R_p} k_{max} X_f L_f}{S_0} (\text{EBCT}) \qquad (6.69)$$

The $k_{max} X_f L_f$ term on the right-hand side of this equation is equal to the substrate removal rate per unit biofilm surface area. It represents the flux of substrate at the surface of biofilm ($J_{s,0}$) in the case of full penetration into the biofilm. $J_{s,0}$ is also referred to as the surface removal rate and is expressed in units of $M_s/L^2.T$ [4]:

$$J_{s,0} = k_{max} X_f L_f \qquad (6.70)$$

First-Order Kinetics Inside the Biofilm In the extreme case when $S \ll K_s$, the biofilm removal rate reduces to

$$\frac{k_{max} S_f}{K_s + S_f} X_f = k_{1,f} X_f S_f \qquad (6.71)$$

where

$k_{1,f}$ = first-order rate constant, $\frac{k_{max}}{K_s}$, $(L^3/M_x.T)$.

In this case, the rate expression has to be integrated since it varies with the substrate concentration inside the biofilm:

$$r_{biofilm} = \frac{(1-\varepsilon_B)}{V_p} X_f k_{1,f} \left[\int_{R_p}^{R_p + L_f} S_f 4\pi r^2 dr\right] \qquad (6.72)$$

It is also possible that substrate penetrates partially into the biofilm since it is totally consumed at some depth of the biofilm – a phenomenon usually encountered in the case of relatively thick biofilms in wastewater treatment. Detailed information exists in the literature about the modeling of biofilms with non-adsorbing media [4]. For nonadsorbing media, expressions can be found about the concentration profiles over the thickness of the biofilm (S_f variation with r).

6.1.4.2.6 Total dimensionless expression of the reactor mass balance

The final expression involving all dimensionless terms can be represented as follows assuming that zero-order biofilm kinetics (Eq. (6.65)) is valid. Using the dimensionless adsorption rates in Eqs. (6.63) or (6.64) leads to Eqs. (6.73) and (6.74), respectively:

$$\frac{\partial S_b^*}{\partial t^*} = \frac{1}{Pe}\frac{\partial^2 S_b^*}{\partial z^{*2}} - \frac{\partial S_b^*}{\partial z^*} - \frac{\rho_p(1-\varepsilon_B)\frac{\partial \overline{q}}{\partial t}}{S_0}\,(\text{EBCT})$$
$$- \frac{\frac{3(1-\varepsilon_B)}{R_p}k_{max}X_f L_f}{S_0}\,(\text{EBCT}) \tag{6.73}$$

$$\frac{\partial S_b^*}{\partial t^*} = \frac{1}{Pe}\frac{\partial^2 S_b^*}{\partial z^{*2}} - \frac{\partial S_b^*}{\partial z^*} - D_g\frac{\partial \overline{q}^*}{\partial t^*} - \frac{\frac{3(1-\varepsilon_B)}{R_p}k_{max}X_f L_f}{S_0}\,(\text{EBCT}) \tag{6.74}$$

The third term on the right-hand side of Eq. (6.74) stands for the dimensionless adsorption and is composed of terms related to adsorption equilibrium and adsorption kinetics. Similarly, the fourth term shows the dimensionless biofilm removal. Depending on the stage of operation, either adsorption or biodegradation or both will be of importance in the reactor. For example, when a biofilm has not developed yet ($L_f=0$), the last term will be equal to zero.

Alternatively, Eq. (6.38), which does not differentiate between adsorptive and biological removal, may be written in dimensionless terms:

$$\frac{\partial S_b^*}{\partial t^*} = \frac{1}{Pe}\frac{\partial^2 S_b^*}{\partial z^{*2}} - \frac{\partial S_b^*}{\partial z^*} - \frac{L_B}{v_i}\frac{3(1-\varepsilon_B)}{\varepsilon_B}\frac{(R_p+L_f)^2}{R_p^3}k_{fc}(S_b^*-S_s^*) \tag{6.75}$$

The last term on the right-hand side shows the liquid film mass transfer to the surface of biofilm. Assuming that the thickness of biofilm is negligibly small, this equation takes the following form:

$$\frac{\partial S_b^*}{\partial t^*} = \frac{1}{Pe}\frac{\partial^2 S_b^*}{\partial z^{*2}} - \frac{\partial S_b^*}{\partial z^*} - \frac{L_B}{v_i}\frac{3(1-\varepsilon_B)}{\varepsilon_B R_p}k_{fc}(S_b^*-S_s^*) \tag{6.76}$$

The last term on the right-hand side of this equation contains the **Stanton number**, which represents the ratio between the liquid film transfer and the advective (bulk) transport in the reactor [6, 15]:

$$\text{Stanton number (St)} = \frac{3(1-\varepsilon_B)k_{fc}L_B}{\varepsilon_B v_i R_p} \tag{6.77}$$

However, in modeling, often **the modified Stanton number** is used [15]:

$$\text{Modified Stanton number (St}^*) = \frac{(1-\varepsilon_B)k_{fc}L_B}{\varepsilon_B v_i R_p} \tag{6.78}$$

Accordingly, Eq. (6.76) can be written as follows:

$$\frac{\partial S_b^*}{\partial t^*} = \frac{1}{Pe}\frac{\partial^2 S_b^*}{\partial z^{*2}} - \frac{\partial S_b^*}{\partial z^*} - 3\,\text{St}^*(S_b^*-S_s^*) \tag{6.79}$$

6.1.4.2.7 Importance of Dimensionless Numbers in a Biofilm–Carbon System

As shown in Table 6.1, a number of dimensionless groups can be used in the characterization of GAC reactors with biological activity (BAC reactors). Complementary to

Table 6.1 Main dimensionless groups characterizing GAC adsorbers with biological activity (BAC reactors).

Dimensionless group	Physical meaning	Relevant point
Sherwood number, Sh	$\dfrac{\text{Rate of liquid film transport}}{\text{Rate of biofilm diffusive transport}}$	Node 2: Basic analysis at liquid film–biofilm boundary
Solute distribution parameter, D_g	$\dfrac{\text{Contaminant at the adsorbent surface}}{\text{Contaminant in the liquid phase}}$	Node 3: Basic analysis around a carbon particle in the absence of biofilm
Modified Biot number, Bi_s'	$\dfrac{\text{Rate of biofilm difusive transport}}{\text{Rate of surface diffusion in the particle}}$	Node 3: Basic analysis at the biofilm–carbon boundary
Peclet Number, Pe	$\dfrac{\text{Rate of advective transport}}{\text{Rate of transport by axial dispersion}}$	Reactor analysis
Stanton number (St) or modified Stanton number (St*)	$\dfrac{\text{Rate of liquid film transport}}{\text{Rate of advective transport}}$	Reactor analysis
Damköhler number Type I, Da_I	$\dfrac{\text{Rate of surface biological reaction}}{\text{Rate of advective transport}}$	Reactor analysis
Damköhler number Type II, Da_{II}	$\dfrac{\text{Rate of surface biological reaction}}{\text{Rate of liquid film transport}}$	Node 2: Basic analysis at liquid film-biofilm boundary
	or	
	$\dfrac{\text{Rate of surface biological reaction}}{\text{Rate of biofilm diffusive transport}}$	Biofilm analysis

this table, Table 6.2 presents typical values of key parameters that can be used for estimating dimensionless numbers for experimenting and modeling of BAC reactors.

The Peclet number (Pe) given by Eq. (6.61) shows the relative importance of advective transport with respect to axial dispersion. In reality, in all environmental systems such as the one shown in Figure 6.3 there is some degree of dispersion. In the limiting cases when Pe → 0 ($D → \infty$, very high dispersion) the reactor has a completely-mixed regime and when Pe → ∞ ($D →$ 0, no dispersion) it behaves as a plug flow reactor. If the $L_B v_i$ term is much larger than the dispersion coefficient D, as in the case of high filter velocities, the dispersion term can be neglected.

Knowing the maximum adsorption capacity q_0 which is in equilibrium with the influent concentration S_0, and the parameters related to carbon bed, one can estimate the solute parameter D_g in Eq. (6.64). A higher D_g shifts the adsorption equilibrium toward the solid phase and thus retards breakthrough.

Table 6.2 Typical values of parameters related to mass transport, biofilm degradation, and adsorption processes.

Parameters	Typical values	Units	Reference
Flow conditions			
Dispersion coefficient, D	0.04–0.45	cm²/s	[10,13]
	0.813	cm²/s	[17]
	0.00152	cm²/s	[19]
Interstitial velocity, v_i	30	m/h	[13]
EBCT	5–30	min	
Liquid film			
Mass transfer coefficient, k_{fc}	1.74×10^{-3}	cm/s	[16]
	7.21×10^{-3}	cm/s	[21]
	6.82×10^{-4}	cm/s	[19]
Liquid film thickness, L_l	20 (rapid sand filter)	μm	[4]
	20 (fluidized-bed)	μm	[4]
	100 (submerged biofilm reactor)	μm	[4]
Diffusivity of substrate in water, D_w	5.6×10^{-6}	cm²/s	[16]
	0.05×10^{-4}–0.24×10^{-4} (inorganic species)	cm²/s	[4]
	0.003×10^{-4}–0.011×10^{-4} (soluble BOD)	cm²/s	[4]
Biofilm			
Biofilm density, X_f	7000	mg L^{-1}	[14]
	6440	mg L^{-1}	[21]
	35 000 (water environment)	mg L^{-1}	[17]
	31 000	mg L^{-1}	[16]
	5×10^{13}	CFU/L	[13]
	10 000–60 000	mg VSS/L	[4]
	12 830–159 330	mg L^{-1}	[20]
Max. biofilm thickness, $L_{f,max}$	10	μm	[14]
	60	μm	[16], [19]
	4 (water environment)	μm	[17]
Biofilm diffusivity, D_f	4.48×10^{-6}	cm²/s	[16]
	8.05×10^{-6}	cm²/s	[13]
	3.5×10^{-6}	cm²/s	[17]
	6.75×10^{-6}	cm²/s	[21]
	5.07×10^{-6}	cm²/s	[19]
Maximum specific substrate utilization rate, k_{max}	9.6	mgS/mgX.d^a	[16]
	26 (water environment)	mgS/mgX.d^a	[17]
	13.7	mgS/mgX.d^a	[21]
	10.28	mgS/mgX.d^a	[19]
Half-velocity constant, K_s	0.24	mg L^{-1}	[21]
Activated carbon			
Intraparticle diffusivity of substrate, D_s	1×10^{-8}	cm²/s	[16]
	3.63×10^{-8}	cm²/s	[19]

(Continued)

Table 6.2 (Continued)

Parameters	Typical values	Units	Reference
	4.62×10^{-9}	cm^2/s	[21]
Particle density, ρ_p	0.8	g/cm^3	[14]
	0.88	g/cm^3	[13]
Particle radius, R_p	0.15	cm	[13]
	0.39	cm	[19]
Void ratio of the bed, ε_B	0.4	–	[13]

^amg substrate/mg biomass.day.

If the Sherwood number (Sh) (Eq. (6.51)) is high, the difference between the bulk concentration and the concentration at the surface of the biofilm will be small, indicating that diffusion through the liquid film is not limiting. If this number is low, this difference increases, indicating that liquid film resistance is important.

Equations (6.67) and (6.69) simply show the importance of biofilm removal rate with respect to substrate loading rate. Thus, they represent the dimensionless biofilm removal rates. In that sense, they are analogous to the Damköhler number of type I (Da_I), which represents the ratio of biological reaction rate to advection rate [22].

The Damköhler number of type II (Da_{II}) relates to surface reactions limited by external or internal mass transfer. Depending on the system, it can be expressed in many different ways. In general, it is taken as the ratio of the maximum biochemical surface reaction rate to the corresponding maximum mass transfer rate [17, 23, 24]. For example, if external (liquid film) diffusion is of interest, in the present biofilm–carbon system, the Damköhler number may be expressed as follows:

$$Da_{II,l} = \frac{J_{s,0}}{k_{fc} S_b} = \frac{k_{max} X_f L_f}{k_{fc} S_b} \qquad (6.80)$$

where $J_{s,0}$ is the substrate removal rate per biofilm surface area (Eq. (6.70)) and the denominator shows the maximum liquid film mass transfer. When $Da_{II,l}$ is much greater than unity, then the process is limited by the rate of liquid film diffusion.

Similarly, the diffusivity of a substrate in a biofilm may be of interest. In this case, the Damköhler number can be represented as the ratio of substrate removal rate per biofilm surface area to the rate of diffusion at biofilm surface:

$$Da_{II,f} = \frac{J_{s,0}}{(D_f S_s)/L_f} = \frac{k_{max} X_f L_f}{(D_f S_s)/L_f} = L_f^2 \frac{k_{max} X_f}{D_f S_s} \qquad (6.81)$$

When this Damköhler number is much greater than unity, then the process is limited by the rate of biofilm diffusion. According to Eq. (6.81), a lower $Da_{II,f}$ number would mean that the rate of biodegradation inside the biofilm is low compared to the diffusion into the biofilm. In this case, more substrate would reach the biofilm–carbon boundary for diffusion into the carbon.

In purely adsorbing systems, the Biot numbers represent either the ratio between the liquid film diffusion and the intraparticle surface diffusion (Bi_s) or the ratio between the liquid film diffusion and the intraparticle pore diffusion (Bi_p). However, in the present biofilm–carbon system shown in Figure 6.1, the modified Biot number, as derived in Eq. (6.56), represents the ratio between biofilm diffusion and intraparticle surface diffusion. In this respect, a low Biot number would mean that the diffusion through the biofilm is controlling rather than the surface diffusion in the carbon.

The influence of other dimensionless groups on breakthrough has been extensively discussed in the literature for the case of no biological activity [6, 15].

6.1.5
Prevalent Models in BAC Reactors Involving Adsorption and Biodegradation

6.1.5.1
Initial Steps in Modeling: Recognizing the Benefits of Integrated Systems

Weber and his co-workers were among the first to recognize the benefits of concurrent adsorption and biodegradation in a single reactor, which they referred to as the Integrated Biological–Physicochemical Treatment (IBPCT) scheme [25, 26]. In the early 1970s a numeric solution model was proposed by Weber and co-workers under the Michigan Adsorption Design and Applications Model (MADAM) program for adsorber beds, involving the dynamic aspects of fluidization and solids mixing, liquid dispersion and channeling, competitive adsorption of solutes, and biological activity [25]. The MADAM is specified as a set of alternative models which also consider the extent of solids (adsorbent) mixing. These models considered four different conditions in the liquid and solid phases (adsorbent) of biologically active carbon beds: plug flow and stationary, plug flow and completely mixed, dispersive and stationary, and dispersive and completely mixed. Model predictions and experimental data clearly demonstrated the benefit of a bioactive adsorber over a nonbioactive one. Studies also revealed the advantage of adsorbing media over a nonadsorbing one [26].

The initial modeling studies conducted by Weber and co-workers indicated an apparent synergism between biological activity and activated carbon adsorption. The researchers concluded that concurrent biodegradation and adsorption processes offered the following advantages: (i) adsorption of biologically resistant compounds that may be toxic and/or carcinogenic, (ii) biological removal of compounds that would be partially removed by activated carbon treatment alone, and (iii) extension of carbon service time, and reduction in regeneration costs and energy utilization [26].

6.1.5.2
Consideration of Substrate Removal and Biofilm Formation

Initial GAC/BAC modeling efforts focused on the dynamics of the biofilm layer surrounding the carbon particles. The effect of bacterial growth was taken into account first by Andrews and Tien, who analyzed the transport and transformation of substrate among the solution, biofilm, and adsorbent [27]. The predictive capacity of this model was verified using valeric acid as a substrate in fluidized-bed adsorbers.

Peel and Benedek developed a model for GAC adsorbers that were operated in the fixed-bed mode. The hydraulic regime was described by plug flow conditions

[28]. In this study, and later in many others, the assumption was made that the biofilm had a fixed thickness. However, it was seen that there was a need to consider the development of the biofilm as operation time progressed. Otherwise, if a constant biofilm thickness was assumed at the beginning of operation, models predicted a higher removal than was achieved in reality.

In reality, initially no bacteria are attached to carbon surface in a GAC adsorber, and biofilm grows only slowly after deposition and attachment of biomass. Ying and Weber accounted for this gradual development of biofilm as a function of time and depth in the GAC adsorber [26]. They assumed that the biofilm attained a steady thickness after a certain time. The model did not include the mass transfer resistance caused by the biofilm since a very thin layer was assumed. The model was applicable to both plug flow fixed-bed and completely mixed fluidized-bed adsorbers. It was considered adequate at concentration levels at which the biofilm grew in the form of patches on the carbon surface [29].

Later, a more elaborate approach was adopted by Andrews and Tien [27]. Their model considered a fluidized bed and complete mixing conditions for both the liquid and the solid phases (carbon particles) of the reactor. Two main interactions were taken into account in this model. First, the model allowed the occurrence of a thick biofilm. Consequently, the model considered the resistance caused by the biofilm to the transport of substrate to the carbon surface, a factor which was formerly not taken into account. Second, substrate concentrations reached very low values in the thick biofilm. Therefore, this situation could reverse the concentration gradient and lead to bioregeneration. Accordingly, these researchers were the first to model the bioregeneration of activated carbon when the solid phase became saturated with substrate, as discussed in detail in Chapter 7.

The model proposed by Andrews and Tien [27] was then generalized by Wang and Tien [30]. This improved model considered the biofilm to be composed of two regions, namely an aerobic region at the outer surface of biofilm followed by an anoxic region near the surface of the adsorbent. This model is a more accurate representation of reality in wastewater treatment, since oxygen becomes available only at outer parts of the biofilm as the biofilm thickness increases. The validity of this bilayer model was demonstrated further by experiments [31].

6.1.5.3
Integration of Adsorption into Models

The biofilm on activated carbon (BFAC) model constitutes one of the pioneering models [1]. The model is an extension of a former biofilm model [32]. The incorporation of the adsorption into this model does not, however, alter the equations of the biofilm model. The differential equations, boundary conditions, and initial conditions shown in Section 6.1.2 represent the fundamentals of this model, which include liquid film transport, biodegradation in the biofilm, intraparticle diffusion, and adsorption of substrate. The homogeneous surface diffusion model (HSDM), shown in Eq. (6.15), is included to describe the movement of the substrate in activated carbon. In the BFAC model, the researchers also took into account the growth of biofilm with respect to time, as shown in Eq. (6.21).

The BFAC model predicted steady-state conditions only, and was developed for a completely mixed reactor. A completely mixed biofilm reactor containing biofilm on activated carbon can be regarded as a stirred tank reactor or column reactor having a sufficiently high recycle flow rate. The model was verified with biofilms grown on spherical activated carbon packed in the column reactor [21]. The model simulated well the experimental results for the transient growth of biofilm on activated carbon and on glass beads and the bioregeneration of activated carbon. However, the model predicted lower effluent suspended biomass than was found in experimental studies. The authors concluded that further research was needed to understand the mechanisms of the shearing loss of biofilms [21].

Concurrently, a model was developed by Speitel and co-workers [12] that was somewhat similar to the one developed by Chang and Rittmann [1]. However, this model considered a plug flow fixed-bed configuration. The model assumed a single substrate, liquid film resistance, diffusional resistance inside the biofilm, biofilm growth according to Monod kinetics, and biofilm decay. Further, surface diffusion was assumed within the adsorbent, and the adsorption was described by the Freundlich isotherm. The major progress brought about by this model was that it also took into account the bioregeneration of activated carbon as explained in Chapter 7.

In subsequent years the aim was the refinement of models of simultaneous adsorption and biodegradation in GAC columns that were used for the elimination of low concentrations of synthetic organic compounds (SOCs) [33]. The sensitivity of models to individual parameters was investigated, as well as alternative formulations for mass transport resistances within the biofilm and at the biofilm–carbon interface. As the substrate concentrations were low, biofilm diffusion was considered negligible.

6.1.5.4
Modeling in the Case of High Substrate Concentrations

While the former models dealt mostly with low-strength and single-substrate conditions, the models developed by Pirbazari and co-workers have a predictive capacity for high-strength wastewaters [19]. For example, a predictive model was developed for a recycle fluidized-bed (RFB) adsorber configuration [16]. The model assumed liquid film transfer resistance, diffusion and biodegradation within the biofilm, biofilm growth, and adsorption within the activated carbon. The model was tested with two biodegradable compounds, sucrose and glucose, and with two different types of wastewaters, namely dairy wastewater and landfill leachate, consisting of a large range of substrates. Comparative experiments were carried out in bioactive and nonbioactive RFB adsorbers. Comparison of breakthrough data and model profiles in these two types of RFB adsorbers indicated better performance in the bioactive type.

Figure 6.4 shows typical breakthrough profiles in a GAC adsorber and one with biological activity (BAC reactor).

In the GAC adsorber shown in Figure 6.4 the breakthrough profile is determined by adsorption mechanisms only. The effluent concentration rises sharply with respect to operation time as adsorption capacity is diminished. The breakthrough

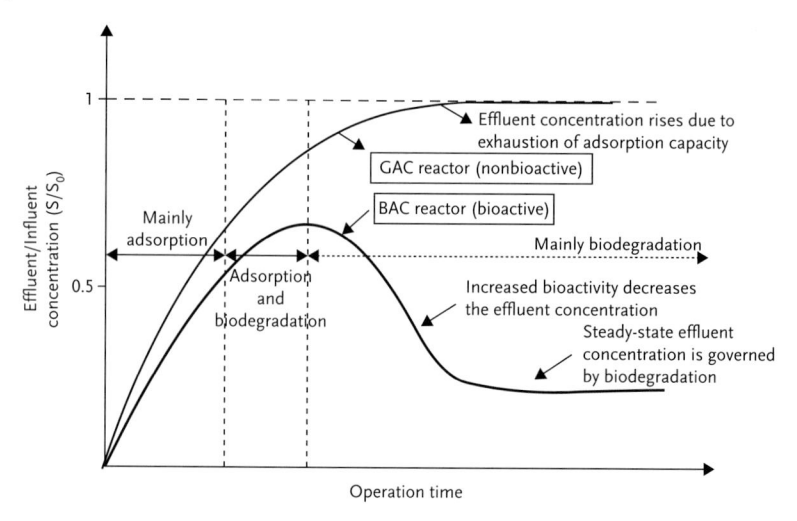

Figure 6.4 Breakthrough curves in the case of nonbioactive (GAC) and bioactive (BAC) adsorbers.

curve shown in this figure is based on the assumption that the influent does not contain any nonadsorbable matter. Similarly, in the BAC reactor, in the initial phases of operation, when biological activity has developed to a small extent, adsorption is the dominant mechanism. However, in this type of reactor, biological activity inside the reactor leads gradually to the establishment of a biofilm. Therefore, lower concentrations are achieved in the effluent compared to a GAC adsorber since some part of substrate is also removed by biodegradation. As time passes, the increase of biological activity inside the BAC reactor may reduce the effluent concentrations to very low levels. At later times of operation, biodegradation becomes the sole removal mechanism. Thus, under steady loading conditions, the steady-state effluent concentrations achieved in this period are governed by biodegradation. As shown in the modeling work, the whole shape of the breakthrough curve in a BAC reactor is very much influenced by the biological parameters such as k_{max}, K_s, and L_f [16, 19]. Also, the operational parameter EBCT has an effect on breakthrough. As seen from Eq. (6.69), an increase in EBCT would lead to a higher biofilm removal rate. Similarly, high k_{max} and L_f would favor biological degradation and shift the effluent concentrations to lower levels. In contrast to this, an increase in K_s would mean a resistance to substrate removal and would have an opposite effect [19].

6.1.5.5
Modeling the Case of Very Low Substrate (Fasting) Conditions

Hassan and co-workers developed a model applicable to packed-bed biosorbers that were operating under fasting conditions [34]. The model included liquid film resistance, internal diffusional resistance accompanied by a biological reaction according to Monod kinetics, and diffusion and adsorption onto activated carbon particles. The analyses were made under a number of conditions ranging from a continuous-flow stirred tank reactor (CSTR) to a plug flow reactor. The substrate concentration profiles in various regions of the reactor showed that the response to the fasting perturbation was very fast in the bulk liquid and the biofilm, while it was extremely slow inside the carbon particles. Thus, the benefits of integrating activated carbon adsorption with biofilm processes are highlighted in this study. During the fasting period, the substrate concentration never reached zero within carbon particles. Even though in the bulk liquid and inside the biofilm no substrate was available to microorganisms, they would survive, as substrate was desorbed from activated carbon. In this study, bioregeneration is not the main subject, however, the results presented point to bioregeneration of activated carbon.

6.1.5.6
Modeling the Step Input of Substrate in a Three-Phase Fluidized-Bed Reactor
A draft tube gas–liquid–solid fluidized-bed bioreactor (DTFB) consisting of activated carbon particles was subjected to step changes in influent phenol concentrations, and the dynamic behavior of phenol degradation was investigated experimentally [20]. The proposed mathematical model considered the transport of oxygen from the gas phase into the liquid phase, transport of phenol, oxygen, and other nutrients from the bulk liquid phase to the surface of the biofilm, diffusion and adsorption of phenol, and diffusion of oxygen and other nutrients inside the carbon particles. Oxygen was assumed not to be adsorbed inside activated carbon particles. A striking difference of this model from previous ones lies in the fact that biofilm properties such as thickness and density are allowed to vary. The growth kinetics expression was based on both phenol and oxygen concentration. Even when subjected to a 200% step increase in the influent phenol concentration, adsorption of phenol on activated carbon particles was a critical factor in achieving stability in the bioreactor.

6.1.5.7
Consideration of Carbon Properties in Modeling
Anaerobic GAC/BAC operation The treatment of toxic wastewater in an expanded-bed anaerobic GAC reactor was demonstrated using phenol and o-cresol as model substrates [35]. The steady-state model described the interactions between biodegradation and adsorption. The ideal adsorbed solution theory (IAST) discussed in Chapter 2 was used to describe the competitive adsorption between phenol and o-cresol. The adsorption model was combined with a biodegradation model that accounted for the possible inhibition of phenol by o-cresol. A very important aspect of this model is that it incorporates the impact of the mean GAC retention time in the anaerobic reactor. Replacement of GAC with virgin GAC is essential for long-term adsorption. However, it also has the disadvantage of leading to biomass loss. In this study, the concentration of adsorbable toxic compounds, the hydraulic retention time, and the GAC replacement rate were the three parameters that were strongly related and effectively controlled the process performance [35].

Consideration of the age and size of activated carbon Most models take into consideration the processes that are related to diffusion and removal of substrate in biofilm and adsorption of substrate in activated carbon particles. However, the properties of activated carbon are rarely integrated into these models. Traegner and co-workers developed a steady-state model to describe an activated carbon adsorber with continuous carbon replacement. Both age and size distributions of carbon particles were incorporated into the model. The effects of adsorption characteristics, the average age of carbon particles, surface diffusion coefficient, particle size, and carbon concentration on the effluent substrate concentration were investigated [36].

6.2
Modeling of the PACT Process

6.2.1
Mass Balances in the PACT Process

The fundamental characteristics of the PACT system are discussed in Chapter 3. In biological systems assisted by PAC, mass balance equations constructed in terms of substrate and biomass bear similarities to other suspended-growth biological systems. However, if total solids are considered in the mass balance, the dosage of PAC into the system has to be accounted for. Certainly, a balance can also be constructed in terms of PAC alone as shown in Section 6.2.2. In the case of substrate mass balance in a PACT system, the only difference is that an adsorption rate term is included into a classical mass balance, as shown in Section 6.2.3.

6.2.2
Mass Balance for PAC in the PACT Process

In activated sludge systems with PAC addition, the aeration tank can usually be described as a completely mixed reactor. In the mass balance involving PAC, the control section is shown in Figure 6.5. According to this process scheme, PAC is added to the influent stream to yield the PAC concentration $X_{PAC,inf}$.

$$\begin{bmatrix} \text{Accumulation} \\ \text{of} \\ \text{PAC} \end{bmatrix} = \begin{bmatrix} \text{Input of} \\ \text{PAC} \end{bmatrix} - \begin{bmatrix} \text{Output of} \\ \text{PAC} \end{bmatrix}$$
$$+ \begin{bmatrix} \text{Decay or generation} \\ \text{of PAC} \end{bmatrix} \tag{6.82}$$

In mathematical terms, the PAC mass balance equation simplifies to the following form since the third term on the right-hand side is zero:

$$V \frac{dX_{PAC}}{dt} = Q\, X_{PAC,inf} - P_{x,PAC} \tag{6.83}$$

where

V = volume of the aeration tank (L^3),
$X_{PAC,inf}$ = influent PAC concentration (M_c/L^3),

6.2 Modeling of the PACT Process

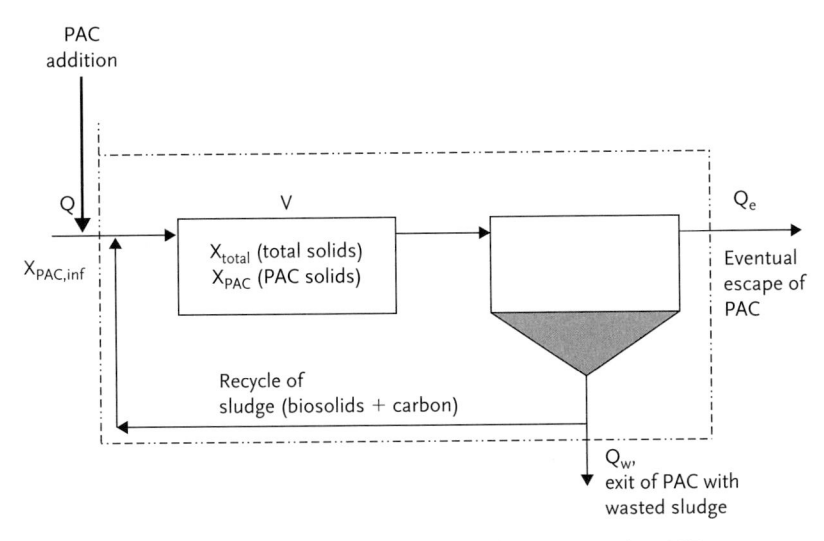

Figure 6.5 Mass Balance for Powdered Activated Carbon (PAC) in the PACT process.

X_{PAC} = PAC concentration inside the aeration tank (M_c/L^3),
$P_{x,PAC}$ = rate of PAC leaving the system with sludge wastage (M_c/T).

If the PAC is added to the influent (Figure 6.5), to the recycle line, or directly into the aeration tank at a mass of M_{PAC}, $X_{PAC,inf}$ can be expressed as M_{PAC}/Q. PAC leaves the whole system with sludge wastage (Q_w) and the effluent stream (Q_e). However, the amount of PAC in the effluent stream Q_e is usually negligibly small in the case of proper sludge settling.

θ_x is an operational parameter representing the mean solids retention time of the total suspended solids in the aeration tank, consisting of PAC and biological solids. This parameter is also called the sludge age or SRT. Thus, it is expressed as the amount of total suspended solids in the aeration tank per amount of suspended solids leaving the system daily:

$$\theta_x = \frac{V X_{total}}{P_x} \tag{6.84}$$

where

X_{total} = total suspended solids concentration in the aeration tank composed of biomass and PAC (M/L^3),
VX_{total} = total amount of sludge present in the aeration tank (M),
P_x = total rate of sludge wasted from the system (M/T).

It should be recognized that in this discussion the assumption is made that PAC, as a solid, behaves in the same way as the biological solids in the system and leaves the system with wasted sludge. Thus, the PAC amount leaving the system, $P_{x,PAC}$, constitutes a fraction of the total sludge P_x:

$$P_{x,PAC} = \frac{V X_{total}\, a}{\theta_x} = \frac{V X_{PAC}}{\theta_x} \quad (6.85)$$

where 'a' represents the PAC fraction in total sludge.

Substituting Eq. (6.85) into Eq. (6.83) gives the following relationship:

$$V \frac{dX_{PAC}}{dt} = Q X_{PAC,inf} - \frac{V X_{PAC}}{\theta_x} \quad (6.86)$$

Under steady-state conditions, the left-hand term in this equation becomes zero. Then, the steady-state concentration of PAC in the aeration tank is expressed as follows:

$$X_{PAC} = X_{PAC,inf} \frac{\theta_x}{\theta} \quad (6.87)$$

where

θ = hydraulic retention time (HRT=V/Q) (T).

This relationship shows that under steady-state conditions the PAC concentration in the mixed liquor is related to the PAC dosage and the ratio of sludge age to hydraulic retention time (θ_x/θ). Thus, the concentration of PAC will be increased in the aeration tank at high sludge ages. As shown in Section 6.2.3 on PAC modeling, the substrate removal depends on both the sludge age and the PAC concentration in the aeration tank. Since PAC is retained in the system for a period of time equal to the sludge age, that is, usually about 10 or more days, under steady-state conditions equilibrium between activated carbon and bulk substrate concentration is easily established.

6.2.2.1
Determination of Biomass or Carbon Concentration in a PACT Sludge

As shown in Section 6.2.2, addition of PAC to a biological tank, whether aerobic or anaerobic, leads to a mixed liquor which is composed of both biomass and PAC. Measurement of the mixed liquor volatile suspended solids (MLVSS), as is done in many cases in order to estimate biomass concentration, becomes complicated in the presence of activated carbon, which itself is a volatile solid.

Theoretically, the activated carbon concentration in the PACT sludge can be estimated from Eq. (6.87), which is derived for steady-state conditions. Knowing the concentration of PAC in the PAC sludge (X_{PAC}) would enable estimation of the biomass in the total MLVSS expression. However, the variations in the influent quality often make it necessary that carbon dosing ($X_{PAC,inf}$) to the biological tank is changed during operation. Additionally, the wastewater flow rate may exhibit fluctuations. Therefore, the steady-state assumption used in the determination of the PAC concentration (Eq. (6.87)) is often violated. To overcome this difficulty, a technique was proposed to calculate the concentration of biomass–MLVSS in a PACT sludge [37]. The method requires basically the measurement of (i) MLSS, (ii) MLVSS at 400 °C, and (iii) MLVSS at 550 °C on both the PACT sludge and the

PAC used. The following approximations are used to calculate the biomass-associated MLVSS:

$$\mathrm{CAR} = \frac{V_{400} - 0.9\, V_{550}}{X_{400} - 0.9\, X_{550}} \tag{6.88}$$

where

CAR = fraction of PAC in MLSS,
V_{400} = fraction of MLSS volatilized at 400 °C,
V_{550} = fraction of MLSS volatilized at 550 °C,
X_{400} = fraction of PAC volatilized at 400 °C,
X_{550} = fraction of PAC volatilized at 550 °C.

Then, the biomass-associated MLVSS ($\mathrm{MLVSS_b}$) can be calculated using the following equation:

$$\frac{\mathrm{MLVSS_b}}{\mathrm{MLSS}} = V_{550} - (\mathrm{CAR})\, X_{550} \tag{6.89}$$

Consequently, the carbon–MLVSS can be found from the difference between total MLVSS and $\mathrm{MLVSS_b}$.

6.2.3
Models Describing Substrate Removal in the PACT Process

Several research groups worked on the modeling of the PACT process. The early PACT model suggested by Benedek and co-workers accounts for biological processes, adsorption, and solids–liquid separation [38]. It is mainly based on the application of the model proposed by Dold and co-workers [39] considering the behavior of soluble and particulate substrate. The model could also be applied for the integrated adsorption–biological system. In the model, the initial organic substrate, S_0, expressed in terms of COD, was fractionated into four: biodegradable soluble substrate (S_{bs}), biodegradable particulate substrate (S_{bp}), nonbiodegradable soluble substrate (S_{us}), and nonbiodegradable particulate substrate (S_{up}).

The model assumed that the maximum specific substrate utilization rate for the soluble biodegradable substrate (S_{bs}) increases upon PAC addition. Also, the assumption was made that the nonbiodegradable soluble fraction is removed by adsorption onto activated carbon. Adsorption was generally considered only in connection with the soluble, nonbiodegradable organic matter, since the following assumptions were made: (i) adsorption of nonbiodegradable molecules on activated carbon is easier, and (ii) under steady-state conditions the ratio of biodegradable to nonbiodegradable molecules is low. However, under extreme conditions, for example, at the sudden, large PAC dosing at the start of a batch experiment, adsorption of soluble biodegradable molecules is also possible. The model also assumed that the settleability of activated sludge flocs is usually enhanced in the presence of PAC.

This model takes into account adsorption and biodegradation, but neglects the eventual volatilization of various organic compounds such as organic solvents. According to the model, the total mixed liquor concentration in the aeration tank is defined as X_l and consists mainly of four fractions: active fraction of the sludge (active biomass) (X_a), stored substrate (X_s), inert material (residue after volatilization) (X_i), and concentration of PAC in the sludge (X_{PAC}) [40]. Under batch conditions, the model describes the processes after the addition of PAC to activated sludge. The model gives the time variation of the four different fractions in substrate (S_{bs}, S_{bp}, S_{us}, S_{up}), X_s, X_a, and the oxygen requirement, while input parameters are S_0, T, $X_{a(t=0)}$ and $X_{PAC(t=0)}$.

Also, under continuous-flow operation the model is capable of simulating a PAC-fed activated sludge treatment plant under fluctuating flow rates and concentrations. The model describes the processes taking place in the aeration basin and final clarifier under particular conditions of recirculation and withdrawal of sludge.

The model was verified using the results of several batch and continuous-flow operations and proved to have a high reliability in both cases [38]. In the early 1980s, PAC addition was tested at full scale at the wastewater treatment plant of a large organic chemicals company in Hungary (Nitrochemical Works). The wastewater treatment plant was overloaded, and the wastewater contained organics such as substituted phenols, derivatives of benzene, nitrated aromatic substances, derivatives of aniline, chlorinated hydrocarbons, solvents, pesticides, and so on, which were not degraded in the case of unacclimated sludge [41].

Although initially fluctuations were seen in the performance upon PAC addition, later biological reactors exhibited a superior performance over control units. The PAC unit could be overloaded by 20% without any decrease in efficiency. An important result of the process was that a 20–50% decrease in the concentration of micropollutants was achieved compared to the control unit. Moreover, the mathematical model was successfully used in explanation of results obtained from full-scale operation in this plant [40].

Other researchers also focus on the removal of priority pollutants such as lindane, a widely used insecticide, in a PACT system [42]. The removal of lindane took place primarily by adsorption onto activated carbon. The equilibrium data was described by the Freundlich isotherm shown in Eq. (6.20).

The parameters evaluated in the PACT system included the carbon concentration, lindane concentration, sludge age, and hydraulic retention time [42]. Steady-state adsorption capacities for the PACs in continuous-flow reactors were lower than the equilibrium values obtained in batch reactors.

Weber and co-workers also studied the enhanced removal of nine specific organic compounds that were at $\mu g\ L^{-1}$ levels in a PACT system [43]. Mass balance equations were constructed in continuous-flow activated sludge reactors for the selected organic compounds: benzene, toluene, ethylbenzene, o-xylene, chlorobenzene, 1,2-dichlorobenzene, 1,2,4-trichlorobenzene, nitrobenzene, and lindane. Comparative continuous-flow studies were conducted without and with PAC that varied in the range of 0–200 mg L^{-1}. Prior to that, for each compound, completely mixed batch rate and equilibrium studies were carried out to quantify the mechanisms of volatilization, biodegradation, biosorption, and carbon adsorption.

In description of biodegradation, the Monod model was simplified to a first-order model since the pollutants concentrations were at μg L^{-1} level:

$$\frac{dS}{dt} = -k_b S \tag{6.90}$$

where

S = substrate concentration in the bulk liquid (M_s/L^3),
k_b = biodegradation rate constant (1/T).

Similarly, the rate of volatilization was modeled as an overall first-order process where k_v showed the volatilization rate constant. Biosorption was described by a linear model and was only significant in the case of lindane; benzene, toluene, ethylbenzene, o-xylene, nitrobenzene, chlorobenzene, and 1,2-dichlorobenzene were not biosorbed to sludge. Adsorption onto activated carbon was modeled using the Freundlich model shown in Eq. (6.20). The characteristics of some selected pollutants are shown in Table 6.3.

Predicted and measured effluent concentrations were compared under steady-state conditions of continuous-flow operation. Benzene, toluene, ethylbenzene, o-xylene, and nitrobenzene were effectively biodegraded under aerobic conditions. For those biodegradable compounds, carbon dosages of 100 mg L^{-1} did not have a significant effect on the steady-state effluent and off-gas concentrations. However, carbon dosing still enhanced their removal during the acclimation period. In activated sludge, 1,2-dichlorobenzene was poorly biodegraded whereas 1,2,4-trichlorobenzene and lindane were nonbiodegradable. However, PAC dosing led to removal in these

Table 6.3 Biodegradation, volatilization, and adsorption potential of some selected priority pollutants at μg L^{-1} levels (adapted from [43]).

Compound	Biodegradation rate constant, k_b (1/min)	log K_{ow}	Volatilization rate constant, k_v (1/min)	Freundlich parameters K_F(μg/mg) (L/μg)$^{1/n}$	1/n
Benzene	0.43–0.49^a	1.9	0.075–0.081^a	0.5	0.46
Chlorobenzene	0.26^a	2.84	0.050–0.052^a	1.8	0.40
1,2-Dichlorobenzene	0.017–0.018^a	3.38	0.025–0.026^a	3.2	0.41
1,2,4-Trichlorobenzene	nonbiodegradable	4.26	0.0236^b	6.2	0.44
Lindane	nonbiodegradable	3.72	nonvolatile	4.0	0.39

^{a} values are calculated from mass balances in completely mixed flow reactors (SRT:3–12 days).
^{b} values obtained from completely mixed batch reactors.

compounds. The addition of PAC reduced not only the effluent concentrations, but also the off-gas concentrations of volatile compounds. For example, without PAC addition 90% of influent 1,2,4-trichlorobenzene was found in the off-gas while the rest stayed in the liquid effluent. However, this compound also has a high sorption potential onto activated carbon. PAC dosages ranging from 25 to 200 mg L^{-1} removed this compound by 69–94%. The nonadsorbed fraction remained mostly in the off-gas. Similarly, in the absence of PAC, 93% of the nonvolatile lindane (Henry constant $H_c = 4.3 \times 10^{-7}$ atm.m^3/mol) stayed in the effluent while 7% of it was biosorbed. PAC dosages at 25–200 mg L^{-1} reduced lindane to nondetectable levels.

In another study, an empirical model was developed to define the removal of soluble substrates as a function of sludge age and carbon dosage in a PACT system [44]. The model was applied to the removal of TOC and the priority pollutant 4,6-dinitro-o-cresol in the absence and presence of activated carbon. The results clearly showed that at the same sludge age the effluent concentrations of TOC and the priority pollutant were decreased to lower values at increased PAC dosages.

Another attempt to model the PACT process was made by O'Brien [45]. This model takes into consideration the behavior of organic priority pollutants and includes biodegradation, adsorption, and stripping. The following mass balance is made for each priority pollutant:

$$\begin{bmatrix} \text{Accumulation} \\ \text{of compound i} \end{bmatrix} = \begin{bmatrix} \text{Input rate} \\ \text{of i} \end{bmatrix} - \begin{bmatrix} \text{Output rate} \\ \text{of i} \end{bmatrix} - \begin{bmatrix} \text{Stripping rate} \\ \text{of i} \end{bmatrix} - \begin{bmatrix} \text{Biodegradation rate} \\ \text{of i} \end{bmatrix} - \begin{bmatrix} \text{Adsorption} \\ \text{rate of i} \end{bmatrix} \quad (6.91)$$

In numerical terms, this mass balance is expressed as follows:

$$V \frac{dS_e}{dt} = Q \cdot S_0 - Q S_e - K_{st} S_e V - K_{bio} S_e X V - K_{ads} X_{PAC} S_e V \quad (6.92)$$

where

Q: flow rate (L^3/T),
S_0: influent compound concentration (M$_s$/L^3),
S_e: effluent compound concentration (M$_s$/L^3),
X: biomass concentration (M$_x$/L^3),
K_{st}: stripping rate coefficient (1/T),
K_{bio}: biodegradation rate coefficient (L^3/M$_x$.T),
K_{ads}: adsorption rate coefficient (L^3/M$_c$.T),
X_{PAC}: aeration tank PAC concentration (M$_c$/L^3),
V: volume of the aeration tank (L^3).

Under steady-state conditions the accumulation term is set to zero:

$$0 = Q S_0 - Q S_e - K_{st} S_e V - K_{bio} S_e X V - K_{ads} X_{PAC} S_e V \quad (6.93)$$

As seen in Eq. (6.93), the model assumed that biodegradation is first-order in both substrate and biomass concentration. Cometabolic removal and toxic effects of substrates were ignored. Adsorption varied linearly with the substrate and PAC concentration in the aeration tank. Stripping was assumed to be proportional to the concentration of the pollutant in the aeration tank. For various priority pollutants, where most of them were chlorinated aliphatics and aromatics, the stripping, biodegradation, and adsorption rate constants were determined.

The model results were compared with those obtained from pilot plant operation. The pilot plants were run with the primary effluent of the WWTP at DuPont's Chamber Works. As outlined in Chapter 3, the PACT process emerged as a result of these efforts to solve the problems in this plant. The presence of PAC had a great effect on suppressing the stripping of most compounds compared to sole activated sludge operation. For example, the model predicted that chlorobenzene stripping was 10% in PAC added cases whereas in the activated sludge it was 55%. The model results were in good agreement with real data except in the case of three multichlorinated volatile compounds (1,2-dichloroethane, chloroform, and methylene chloride) due to the high variability of these compounds in the influent. Thus, this steady-state model was effective in predicting the removal of priority pollutants at a 2520 $m^3 d^{-1}$ industrial WWTP that used the PACT process. The model showed that the competing effects of both biodegradation and carbon adsorption reduced the stripping of volatile organic compounds (VOCs). Further, the model was corrected for scale-up purposes and used to size a second-stage PACT system.

Since the steady-state model did not accurately predict the effluent concentrations of some priority pollutants when their influent concentrations varied with time, the model was extended to predict dynamic (non-steady-state) conditions in a WWTP [46]. The dynamic behavior of the system was evaluated basically using Eq. (6.92). The predicted and measured effluent concentrations of volatile, semi-volatile, and nonvolatile compounds were in agreement with each other. Particularly, the non-steady-state model was more accurate than the steady-state model in predicting the effluent concentrations from secondary clarifiers [46, 47].

More recently, a mathematical model was developed for a PACT system to represent the effect of PAC on microbial growth kinetics [48]. The model combined biochemical processes (biomass growth, substrate consumption, and product generation) with PAC adsorption. The substrate used in experiments was both biodegradable and adsorbable onto PAC. The model considered the removal of substrate by adsorption, desorption, and removal by microorganisms. The proposed model was used to simulate the effect of the initial substrate and PAC concentrations on soluble COD, biomass concentration, and Oxygen Uptake Rate (OUR) as a function of time. PAC was shown to first adsorb and then desorb the substrate. If microbial activity decreased the substrate concentration below the sorption equilibrium value, substrate was desorbed from surface. Also in this study the simulations showed that PAC acted as a buffer of substrate, in the same way as GAC, yielding slower responses to the change in substrate concentration, and increasing the stability of reactor operation.

References

1 Chang, H.T. and Rittmann, B.E. (1987) Mathematical modeling of biofilm on activated carbon. *Environmental Science and Technology*, **21**, 273–280.

2 Arvin, E. and Harremoes, P. (1990) Concepts and models for biofilm reactor performance. *Water Science and Technology*, **22**, 171–192.

3 Beg, S.A and Chaudhry, M.A.S. (1999) A review of mathematical modelling of biofilm processes: advances in modeling of selected biofilm processes. *International Journal of Environmental Studies*, **56**, 285–312.

4 Morgenroth, E. (2008) Modelling biofilms, in *Biological Wastewater Treatment, Principles, Modelling and Design* (eds M. Henze, M.C.M. van Loosdrecht, G. A. Ekama, and D. Brdjanovic), IWA Publishing, pp. 457–492.

5 Morgenroth, E. (2008) Biofilm reactors, in *Biological Wastewater Treatment, Principles, Modelling and Design* (eds M. Henze, M.C.M. van Loosdrecht, G. A. Ekama, and D. Brdjanovic), IWA Publishing, pp. 493–511.

6 Sontheimer, H., Crittenden, J., and Summers, R.S. (1988) *Activated Carbon for Water Treatment*, 2nd edn, Forschungstelle Engler–Bunte-Institute, Universität Karlsruhe, Germany.

7 Worch, E. (2008) Fixed-bed adsorption in drinking water treatment: a critical review on models and parameter estimation. *Journal of Water Supply: Research and Technology—AQUA*, **57** (3), 171–183.

8 Nakhla, G. and Suidan, M.T. (2002) Determination of biomass detachment rate coefficients in anaerobic fluidized-bed-GAC reactors. *Biotechnology and Bioengineering*, **80** (6), 660–669.

9 Chaudhary, D.S., Vigneswaran, S., Ngo, H.H., Shim, W.G., and Moon, H. (2003) Review: biofilter in water and wastewater treatment. *Korean Journal of Chemical Engineering*, **20** (6), 1054–1065.

10 Liang, C.H., Chiang, P.C., and Chang, E.E. (2007) Modeling the behaviors of adsorption and biodegradation in biological activated carbon filters. *Water Research*, **41**, 3241–3250.

11 Rittmann, B.E. and McCarty, P.L. (2001) *Environmental Biotechnology: Principles and Applications*, McGraw-Hill.

12 Speitel, G. E. Jr, Dovantzis, K. and Digiano, F.A. (1987) Mathematical modelling of bioregeneration in GAC columns. *Journal of Environmental Engineering*, **113**, 32–48.

13 Liang, C.H. and Chiang, P.C. (2007) Mathematical model of the non-steady-state adsorption andbiodegradation capacities of BAC filters. *Journal of Hazardous Materials*, **B139**, 316–322.

14 Tsai, H.H, Ravindran, V., and Pirbazari, M. (2005) Model for predicting the performance of membrane bioadsorber reactor process in water treatment applications. *Chemical Engineering Science*, **60**, 5620–5636.

15 Sontheimer, H., Frick, B.R., Fettig, J., Hörner, G., Hubele, C., and Zimmer, G., (1985) *Adsorptionsverhalten zur Abwasserreinigung*, DVGW-Forschungsstelle am Engler-Bunte Institut der Universität Karlsruhe (TH), G.Braun GmbH, Karlsruhe.

16 Kim, S.H. and Pirbazari, M. (1989) Bioactive adsorber model for industrial wastewater treatment. *Journal of Environmental Engineering*, **115** (6), 1235–1256.

17 Badriyha, B.N., Ravindran, V., Den, W., and Pirbazari, M. (2003) Bioadsorber efficiency, design, and performance forecasting for alachlor removal. *Water Research*, **37**, 4051–4072.

18 Çeçen, F. and Gönenç, I.E. (1995) Criteria for nitrification and denitrification of high-strength wastes in two upflow submerged filters. *Water Environment Research*, **67** (2), 132–142.

19 Ravindran, V., Kim, S.H., Badriyha, B. N., and Pirbazari, M. (1997) Predictive modeling for bioactive fluidized bed and stationary bed reactors: application to dairy wastewater. *Environmental Technology*, **18**, 861–881.

20 Tang, W.-T, Wisecarver, K., and Fan, L-S. (1987) Dynamics of a draft tube

gas–liquid–solid fluidized bed bioreactor for phenol degradation. *Chemical Engineering Science*, **42** (9), 2123–2134.
21 Chang, H.T. and Rittmann, B.E. (1987) Verification of the model of biofilm on activated carbon. *Environmental Science and Technology*, **21**, 280–288.
22 Weber, W.J. Jr and DiGiano, F.A. (1996) *Process Dynamics in Environmental Systems*, John Wiley & Sons, New York, NY.
23 Bailey J. E. and Ollis D. F. (1986) *Biochemical Engineering Fundamentals*, McGraw-Hill, Inc., New York, NY.
24 Gagnon, G.A. and Huck, P.M. (2001) Removal of easily biodegradable organic compounds by drinking water biofilms: analysis of kinetics and mass transfer. *Water Research*, **35**, 2254–2264.
25 Weber, W.J. and Crittenden, J.C. (1975) MADAM I- A numeric method for design of adsorption systems. *Journal of Water Pollution Control Federation*, **47** (5), 924–940.
26 Ying, W.C. and Weber, W. J. Jr (1979) Bio-physicochemical adsorption model systems for wastewater treatment. *Journal of Water Pollution Control Federation*, **55** (11), 2661–2677.
27 Andrews, G. F. and Tien, C. (1981) Bacterial film growth in adsorbent surfaces. *American Institute of Chemical Engineers (AIChE) Journal*, **27** (3), 396–403.
28 Peel, R.G. and Benedek, A. (1977) The modeling of activated carbon adsorbers in the presence of bio-oxidation. American Institute of Chemical Engineers (AIChE). *Symposium Series*, **73** (166), 25–35.
29 Olmstead, K.P. and Weber, W.J. (1991) Interactions between microorganisms and activated carbon in water and wastewater treatment operations. *Chemical Engineering Communications*, **108**, 113–125.
30 Wang, S.C.P. and Tien, C. (1984) Bilayer film model for the interaction between adsorption and bacterial activity in granular activated carbon columns, Part I: Formulation of equations and their numerical solutions. *American Institute of Chemical Engineers (AIChE) Journal*, **30** (5), 786–794.
31 Wang, S.C.P. and Tien, C. (1984) Bilayer film model for the interaction between adsorption and bacterial activity in granular activated carbon columns, Part II: Experiment. *American Institute of Chemical Engineers (AIChE) Journal*, **30** (5), 794–801.
32 Rittmann, B. E. and McCarty, P. L. (1980) Model of steady-state-biofilm kinetics, *Biotechnology and Bioengineering*. **22**, 2343–2357.
33 Speitel, G. E. Jr and Zhu, X.J. (1990) Sensitivity analyses of biodegradation/ adsorption models. *Journal of Environmental Engineering*, **116** (1), 32–48.
34 Hassan, M. M., Beg, S. A., and Biswas, M. E. (1998) Analysis of dynamic responses of a biosorber under fasting conditions. *International Journal of Environmental Studies*, **54** (1), 57–72.
35 Nakhla, G.F. and Suidan, M.T. (1992) Modeling of toxic waste-water treatment by expanded-bed anaerobic GAC-reactors. *Journal of Environmental Engineering*, **118**, 495–512.
36 Traegner, U. K., Suidan, M. T., and Kim, B. R. (1996) Considering age and size distributions of activated-carbon particles in a completely-mixed adsorber at steady state. *Water Research*, **30** (6), 1495–1501.
37 Arbuckle, W. and Griggs, A.A., (1982) Determination of biomass MLVSS in PACT sludges. *Journal of Water Pollution and Control Federation*, **54**, 1553–1557.
38 Benedek, P., Major, V., and Takács, I. (1985) Mathematical model suggested for a carbon-activated sludge system. *Water Research*, **19** (4), 407–413.
39 Dold, P.L., Ekama G.A., and Marais, G.v.R. (1980) A general model for the activated sludge process. *Progress in Water Technology*, **12** (6), 47–77.
40 Takács, I. (1986) Application of a mathematical model for activated sludge treatment. *Water Science and Technology*, **18**, 163–174.
41 Benedek, P., Takács, I., and Vallai, I. (1986) Upgrading of a wastewater factory. *Water Science and Technology*, **18**, 75–82.

42 Weber, W. J. Jr, Corfis, N.H., and Jones, B.E. (1983) Removal of priority pollutants in integrated activated sludge-activated carbon treatment systems. *Journal of Water Pollution Control Federation*, **55** (4), 369–376.

43 Weber, W. J. Jr, Jones, B.E., and Katz, L.E. (1987) Fate of toxic compounds in activated sludge and integrated PAC systems. *Water Science and Technology*, **19**, 471–782.

44 Garcia-Orozco, J.H., Fuentes, H.R., and Eckenfelder, W. W. Jr (1986) Modeling and performance of the activated sludge-powdered activated carbon process in the presence of 4,6 dinitro-o-cresol. *Journal of Water Pollution Control Federation*, **58** (4), 320–325.

45 O'Brien, G.J. (1992) Estimation of the removal of organic priority pollutants by the powdered activated carbon treatment process. *Water Environment Research*, **64** (7), 877–883.

46 O'Brien G.J. (1993) Comparison of a steady-state and dynamic model for predicting priority pollutant removal. *Environmental Progress*, **12** (1), 76–80.

47 O'Brien G.J. and Teather, E.W. (1995) A dynamic model for predicting effluent concentrations of organic priority pollutants from an industrial wastewater treatment plant. *Water Environment Research*, **67** (6), 935–942.

48 Orozco, A.M.F., Contreras, E.M., and Zaritzky, N.E. (2010) Dynamic response of combined activated sludge-powdered activated carbon batch systems. *Chemical Engineering Journal*, **157** (2–3), 331–338.

7
Bioregeneration of Activated Carbon in Biological Treatment

Özgür Aktaş and Ferhan Çeçen

Bioregeneration is the renewal of carbon's adsorptive capacity by the action of microorganisms so that further adsorption can take place in activated carbon applications combined with biological processes [1]. It is a very important phenomenon, providing an efficient and economic integration of adsorption and biodegradation. This chapter gives an overview of the theory of bioregeneration, the relationship between adsorption reversibility and bioregeneration, the factors affecting bioregeneration, the methods for determination and quantification of bioregeneration, and the mathematical modeling approaches adopted in bioregeneration. For a better understanding of bioregeneration, the reader is advised first to refer to Chapter 6, which describes the models that represent carbon adsorption with biological activity.Bioregeneration can be achieved either by mixing bacteria with saturated activated carbon in offline systems [2–6] or in simultaneous (concurrent) treatment as in PACT or BAC systems [7–11]. Figure 7.1 illustrates the bioregeneration process. As shown in this figure, the regeneration of activated carbon is usually incomplete since some sites on carbon are not available for bioregeneration.

7.1
Mechanisms of Bioregeneration

In the literature, two main theories of bioregeneration of activated carbon are mentioned.

7.1.1
Bioregeneration Due to Concentration Gradient

One of the bioregeneration theories postulates the biodegradation of target compounds following desorption due to a concentration gradient between the activated carbon surface and bulk liquid. According to this theory, on the surface of carbon the adsorption concentration rises more than the equilibrium adsorption concentration through the decrease of the bulk concentration by biological activity.

Activated Carbon for Water and Wastewater Treatment: Integration of Adsorption and Biological Treatment.
First Edition. Ferhan Çeçen and Özgür Aktaş
© 2011 WILEY-VCH Verlag GmbH & Co. KGaA, Weinheim.
Published 2011 by WILEY-VCH Verlag GmbH & Co. KGaA

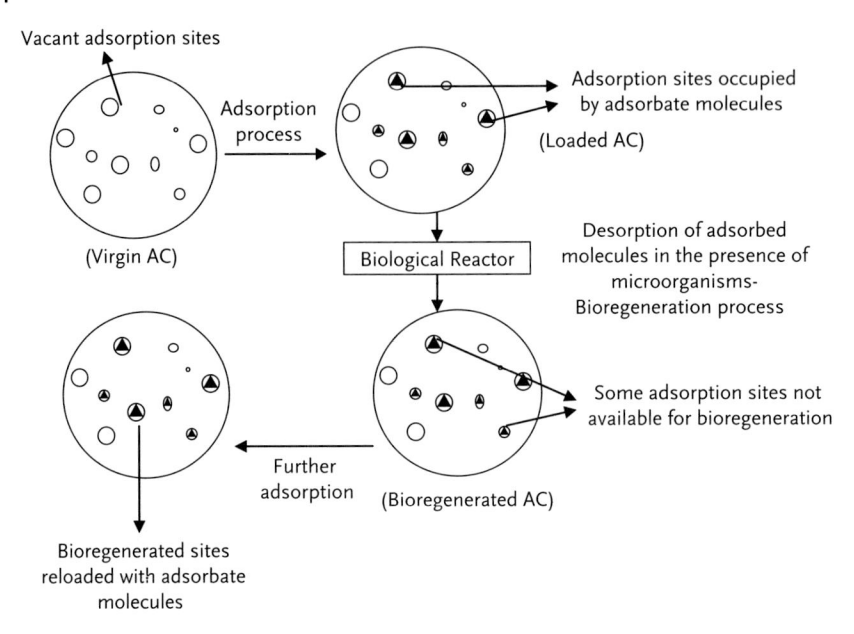

Figure 7.1 Schematic representation of bioregeneration.

Consequently, adsorbed organics are desorbed due to a concentration gradient between activated carbon surface and bulk liquid [9, 12, 13]. At equilibrium, an adsorbable substance is distributed between bulk solution and activated carbon. Once microorganisms degrade adsorbable substances, the equilibrium is disturbed, resulting in desorption, which partially replenishes the solution concentration. This regeneration process may continue until all of the reversibly adsorbed substances are degraded. This biodegradation depends on a number of factors such as the nature of the microbial population, adsorbability, and reversibility of adsorbate [11]. Besides the concentration gradient, the difference in Gibbs free energy between the adsorbate molecules in solution ($-\Delta G°_{ads}$) and the adsorbate molecules inside the porous structure ($-\Delta G°_{mod}$) was suggested to be the driving force for bioregeneration [14].

The mechanism of bioregeneration is illustrated in Figure 7.2. Similarly to Figure 6.1 in Chapter 6, this figure illustrates the decrease in substrate concentration through the liquid film and biofilm toward the activated carbon surface. However, contrary to Figure 6.1, a hypothetical plane of zero gradient (PZG) exists in this case. PZG is defined as the line in the biofilm where the concentration gradient ($\partial S_f/\partial r$) is zero. At this hypothetical line inside the biofilm, the concentration of substrate reaches a minimum value. PZG separates the region in which substrate in the bulk liquid is biodegraded from the region in which previously adsorbed substrate on the activated carbon surface is biodegraded. As the bulk liquid concentration decreases, the PZG moves toward the liquid film. Thereby, the flux of the substrate is reversed; this means that the substrate flux is

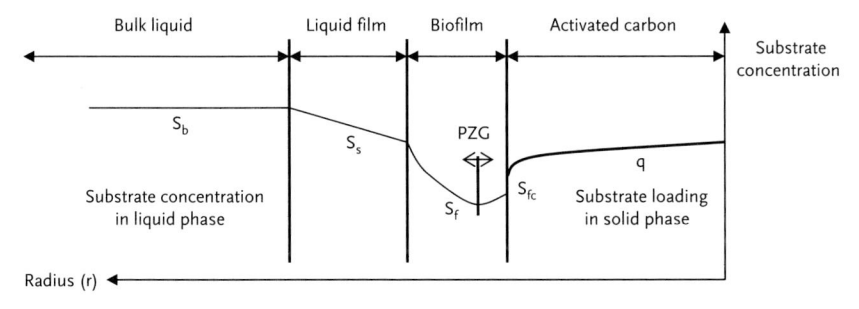

Figure 7.2 Concentration profile of a contaminant during bioregeneration.

in the direction from the solid phase (carbon) to the bulk liquid. The consequence is the bioregeneration of activated carbon. Whenever the bulk liquid concentration increases, the PZG moves toward the activated carbon surface, and bioregeneration does not occur [15].

An example of bioregeneration is shown in Figure 7.3. The figure shows that phenol loaded on two types of activated carbon (SA4 and CA1) was removed much better in the presence of active microorganisms compared to abiotic desorption alone [16]. Abiotic desorption alone led to a slight decrease in the amount of phenol on carbon since equilibrium was quickly achieved between the solid and liquid phases. However, in the presence of microorganisms in the bulk liquid, continuous desorption was maintained in order to reach equilibrium. Figure 7.3 shows that, depending on the type of activated carbon, the adsorbate remaining on the activated carbon may be decreased to very low values by bioregeneration. Comparison of Figure 7.3a and b shows that an activated carbon type (e.g., CA1 in this case) may be much more bioregenerable compared to another one (e.g., SA4). The underlying reasons for the dependence of bioregeneration on activated carbon grade are discussed later in this chapter.

7.1.2
Bioregeneration Due to Exoenzymatic Reactions

The second bioregeneration theory includes desorption of organic matter due to reactions catalyzed by exoenzymes (extracellular enzymes). According to this theory, the bioregeneration mechanism involves exoenzymes [13, 17, 18], although some researchers have suggested that it involves only desorption due to a concentration gradient and state that nondesorbable compounds could not be bioregenerated [9, 12, 19, 20]. According to the theory including exoenzymes, bacteria, with sizes at a level of micrometers, are too large to diffuse into activated carbon pores that are at nanometer levels. However, some enzymes excreted by bacteria could easily diffuse through the bulk liquid into activated carbon pores and react with adsorbate. These exoenzymes may also be adsorbed before they could react with the adsorbate [21]. Consequently, hydrolytic decay of adsorbate may occur. In addition, the adsorbability of organic matter may be weakened after reacting with the enzyme. Therefore, further desorption of adsorbate may also take place.

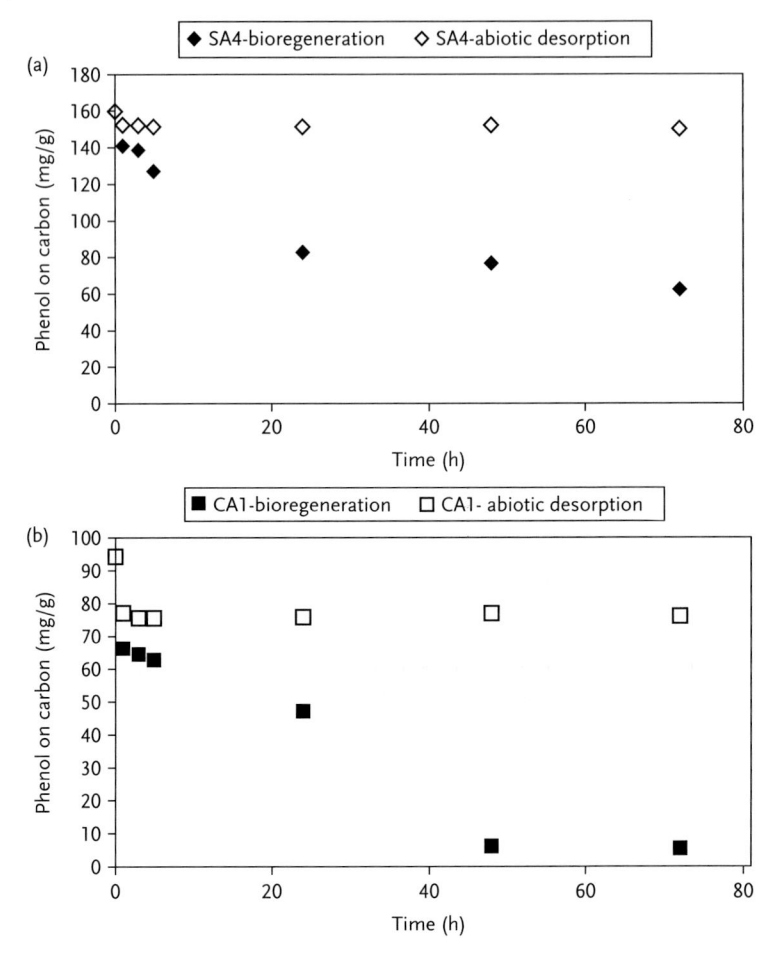

Figure 7.3 Profiles of phenol loadings on powdered activated carbons (a) Norit SA4 (thermally activated) and (b) Norit CA1 (chemically activated) in bioregeneration reactors integrated with activated sludge and during desorption in the absence of activated sludge (redrawn after [16]).

On the other hand, it is also suggested that enzyme molecules are larger than micropores, so that bioregeneration cannot occur due to enzymatic reactions [22]. According to this view, for an enzyme to actively act as a catalyst in a pore, the pore diameter should be at least three times greater than the enzyme size to allow the enzyme to approach and induce a fit [22]. Considering that the average molecular diameter of a monomeric enzyme (molecular weight between 13 000 and 35 000) is above 31–44 Å, the diameter of the pore in which the exoenzyme can catalyze should be larger than 10 nm (100 Å). Thus, this excludes micropores ($\phi < 2$ nm) and some of mesopores ($\phi = 2$–50 nm) [22]. Since low-molecular-weight aromatic compounds (e.g. phenol) are mainly adsorbed in pores with a diameter under 0.7 nm [23], the effect of hypothetical exoenzymatic reactions is expected to be limited in such cases. Other researchers also state that micropores of activated carbon are occupied by adsorbed molecules and are not subject to bioregeneration [14, 24]. However, there are also studies indicating that a biofilm can be established on BAC, so that bacteria occupy macropores whereas their exoenzymes are found in micropores [4].

Some researchers support the idea that the inaccessibility of micropores for exoenzymes does not contradict the probable exoenzymatic activity in meso- and macropores [18, 25]. However, it was also stated that no evidence exists that exoenzymes are involved in ring hydroxylation and ring fission reactions of aromatic compounds. Thereby, it was proposed that the bioregeneration hypothesis including exoenzymatic reactions is not valid for these compounds [22]. Previously, it was also argued that exoenzymes were probably incapable of breaking down molecules and that degradation in pore volumes was insignificant [26]. Others suggested that bioregeneration mechanisms including enzymes would be extremely slow due to low diffusivity of large hydrolytic enzymes within the pores [27].

The bioregeneration efficiencies of thermally activated carbons that were loaded with phenol were found to be much higher than their total desorbabilities [24]. This indicated that some exoenzymatic reactions might have occurred so that bioregeneration exceeded the expectation. Considering that phenol was adsorbed mainly on micro- and mesopores and exoenzymes were not expected to reach micropores, exoenzymatic reactions were expected to take place in mesopores. In the case of chemically activated carbons, however, exoenzymes were not expected to be effective on bioregeneration since the dominant adsorption mechanism was physisorption and the nondesorbed phenol probably resided in the micropores where enzymes did not reach [28]. The results of the same researchers indicated that exoenzymatic bioregeneration was possible, but was very much related to the activation type, adsorption characteristics, and porosity of carbon [24]. Also, in another study, bioregeneration efficiencies were shown to exceed the total desorbabilities [10]. It was also suggested that desorption from micropores took place due to a reverse concentration gradient. On the other hand, desorption from mesopores was attributed to the activity of exoenzymes. Therefore, the process of bioregeneration was featured by the two noncontradictory hypotheses [18].

7.1.3
Bioregeneration Due to Acclimation of Biomass

Adsorption-induced acclimation of microorganisms is another mechanism proposed for bioregeneration [29–31]. As shown in Figure 3.6 in Chapter 3, the adsorption of refractory organics onto activated carbon that is in intimate contact with microorganisms induces the acclimation of attached microorganisms to these compounds. This idea differs from the previous ones in that in this case the nature of bacterial culture changes such that refractory organics can be used as substrates.

Adsorption-induced acclimation does not seem to be valid for PAC-added activated sludge systems operating at low sludge ages since there may not be enough time for acclimation. In accordance with this, the hypothesis of bioregeneration is more likely to be valid in nitrification systems operating at higher sludge ages rather than those aiming primarily at carbonaceous BOD removal. Also, in BAC filtration of potable water, acclimation of bacteria to adsorbed substrates was not considered as a significant mechanism [29].

7.2
Offline Bioregeneration

Offline bioregeneration employs desorption and biological activity in a closed-loop recirculating batch system. It includes taking exhausted activated carbon out of service and regenerating it by recycling a mixture of acclimated bacteria, nutrients, and dissolved oxygen (DO) through the activated carbon column in an environment favorable to enhancement of biodegradation of adsorbed organic matter [2]. Offline bioregeneration could be more easily applied compared to bioregeneration in simultaneous (concurrent) BAC treatment [2]. This is based on the limitations of BAC operation such as availability of nutrients and dissolved oxygen, persistence of organic compounds, and operational difficulties including hydraulic short-circuiting and excessive head loss. It was also reported that an offline BAC bioregeneration system led to higher degradation rates than concurrent BAC operation [32]. For example, an offline system metabolized $70\,\text{mg}\,\text{L}^{-1}$ of 2-chlorophenol (2-CP), whereas in a concurrent system only $30\,\text{mg}\,\text{L}^{-1}$ could be degraded. Also, the bioregeneration efficiency for a loaded activated carbon was 95% after a 1-day cycle in an offline system, whereas this was limited to 30% after 30 days of operation in a concurrent system [33].

Offline GAC bioregeneration is also achieved by the recirculation of microorganism-free filtered permeate through a GAC column and biodegradation in a separate bioreactor [34, 35] (Figure 7.4). In this case, bioregeneration was improved compared with recirculation of unfiltered microorganisms through the column, because microbial fouling was prevented in the GAC reactor [35]. A patented work combined GAC filter with a sequencing batch biofilm reactor (SBBR) such that biodegradable pollutants were desorbed from the GAC filter

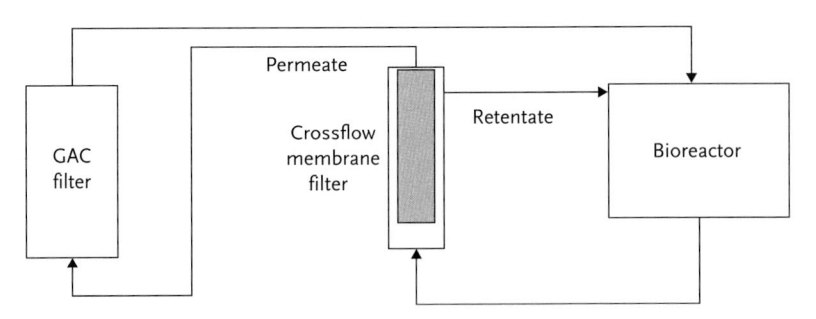

Figure 7.4 Closed-loop offline bioregeneration system with pre-filtration of microorganisms (drawn based on [34]).

through recirculation of wastewater and biodegraded in the SBBR using specialized microbes [36]. This type of offline bioregeneration can also be achieved by the recirculation of a solvent through the exhausted column. Desorption efficiency is increased in such cases since the bulk concentration of substrate is lowered. In another study, highly explosive compounds RDX (Royal Demolition eXplosive) and HMX (High Melting eXplosive) were leached from the exhausted GAC filter by the use of an ethanol–water mixture. These compounds were further degraded by acclimated bacteria in a fixed-bed bioreactor. Thereby, GAC was regenerated through combined chemical–biological regeneration [37]. The same procedure was also applied in batch studies for regeneration of GAC that was loaded with RDX and HMX. As solvents, methanol, ethanol, anionic and nonionic synthetic surfactants and cyclodextrins were used [38].

7.3
Concurrent (Simultaneous) Bioregeneration in PACT and BAC Systems

Several researchers state that activated carbon is also bioregenerated in PACT processes [7, 12, 13, 39, 40]. For example, in the treatment of a synthetic wastewater containing high concentrations of phenol and p-methylphenol, bioregeneration of PAC took place during concurrent adsorption and biodegradation in sequencing batch reactor (SBR) operation [41]. However, there are also studies showing that the bioregeneration hypothesis is not always valid. In some cases the PACT process is a simple combination of adsorption and biodegradation [22, 42–44].

The regeneration of activated carbon by a bacterial biofilm in a BAC system has also been reported by several authors [10, 11, 31, 45]. BAC systems involve GAC particles which are covered by a biofilm. The processes in BAC filtration could be separated into three phases [18]. The initial phase consists of the preferred adsorption of contaminants from wastewater when the adsorption rate considerably surpasses the biodegradation rate. The second phase is the equilibrium when the rate of adsorption and biodegradation are comparable. The third phase consists of dynamic conditions when the biodegradation rate is higher than the adsorption rate, and desorption from the pores occurs resulting in regeneration of carbon.

However, in BAC systems biofouling can occur due to excessive growth of microorganisms. Consequently, biofouling and bioregeneration may be two mechanisms opposing each other [4, 46].

7.4
Dependence of Bioregeneration on the Reversibility of Adsorption

Desorption of an adsorbate from activated carbon occurs in response to changes in operating conditions such as the decrease in liquid phase concentration, displacement of adsorbates by competitive adsorption (refer to Section 2.2.5.2), and some other changes (such as pH) in the liquid phase decreasing adsorbability [47]. As discussed in Section 7.1.1, desorption is a prerequisite step for bioregeneration. Thus, any condition that affects desorption of an adsorbate eventually affects the extent of bioregeneration.

Bioregeneration is not only restricted to readily desorbable compounds [48]. However, bioregeneration of activated carbon seems to be limited in the case of some complex substrates such as humic substances because these are likely to be irreversibly adsorbed. Hence, bioregeneration is controlled by the reversibility of adsorption [9, 12]. When activated carbon was loaded with nondesorbable compounds it could not be bioregenerated [12, 19, 20]. Such apparent irreversibility is commonly referred to as hysteresis or nonsingularity [10].

There are two possible mechanisms leading to irreversible adsorption. One of them is the high-energy bonding of adsorbate molecules to specific functional groups on the active sites of the carbon surface resulting in covalent bonding [10, 49, 50]. Depending on the type of surface functional group and sorbate, a sufficiently strong bond can form resisting desorption. Therefore, chemisorption appears to be the most logical explanation for irreversible adsorption, the degree of which is directly related to the number of high-energy (chemisorptive) bonds [49]. On the other hand, reversible adsorption was attributed to adsorption as a result of van der Waals forces and/or weaker charge-transfer complexes that occur at adsorption sites [49]. Since the adsorption energy is weaker in physisorption, adsorption is more reversible.

The second possibility is the oxidative polymerization of phenolic compounds onto activated carbon in the presence of oxygen [11, 50]. Phenol molecules (PhOH) undergoing an oxidative coupling reaction may be irreversibly adsorbed on activated carbon, which in turn may result in less bioregeneration [44, 51]. Molecular oxygen can act as an initiator in oxidative coupling reactions by reacting with phenol molecules to produce phenoxy radicals (PhO*) as shown in Eq. (7.1).

$$PhOH + O_2 \rightarrow PhO^* + HO_2^* \tag{7.1}$$

The phenolate ion (PhO$^-$) can also react with oxygen to produce phenoxy radicals:

$$PhO^- + O_2 \rightarrow PhO^* + O_2^* \tag{7.2}$$

The HO$_2$* formed through the reaction shown in Eq. (7.1) can also react with another phenol molecule:

$$PhOH + HO_2^* \rightarrow PhO^* + H_2O_2 \tag{7.3}$$

Hydrogen peroxide reacts with another phenol molecule to produce phenoxy radicals:

$$PhOH + H_2O_2 \rightarrow PhO^* + H_2O + HO^* \tag{7.4}$$

$$PhOH + HO^* \rightarrow PhO^* + H_2O$$

Phenoxy radicals (PhO*) formed by the removal of a hydrogen atom from each phenolic molecule can participate in direct coupling with other phenoxy radicals at room temperature while the activated carbon surface serves as a catalyst [51].

In general, oxidative coupling of phenolic compounds, except nitrophenols, was found to decrease the reversibility of adsorption [51]. Functional groups influence the probability of the adsorbate being oxidized [50]. In general, unsaturated groups (e.g., carboxyl and nitro groups) decrease the susceptibility to oxidation, whereas saturated groups (e.g., methyl groups) increase the probability of oxidation. Also, in another study it was hypothesized that phenol and 2,4-dichlorophenol undergo oxidative coupling on the GAC surface and the availability of oxygen promotes irreversible adsorption [11]. Adsorption of phenolic compounds was not fully of the physical type. However, in the literature, oxidative polymerization is not reported for nonphenolic compounds. For example, a study demonstrated that adsorption of a phenolic compound, o-cresol, was increased in the presence of molecular oxygen due to oxidative polymerization, whereas adsorption of a non-phenolic compound, 3-chlorobenzoic acid, was not affected by the presence of oxygen [50].

Oxidative coupling of phenolic compounds is also dependent on the type of carbon activation. Various studies report that thermally activated carbons exhibit a higher degree of irreversible adsorption (Figure 7.5) and lower bioregeneration [9, 24, 28, 50]. This was ascribed to the affinity of thermally activated carbons toward oxygen and changes in surface chemistry upon contact with oxygen [50]. Thermal activation of carbons is originally carried out in the absence of oxygen, leading to a more reactive surface towards oxygen. Contrary to this, chemically

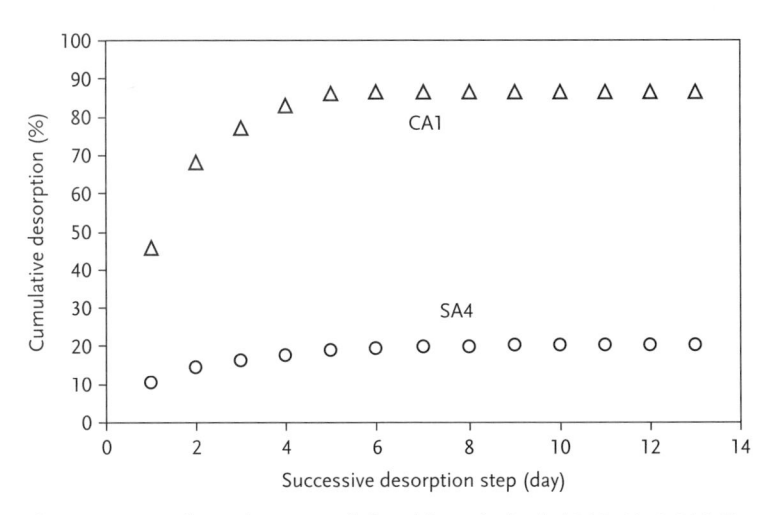

Figure 7.5 Cumulative desorption of phenol from the loaded PACs Norit SA4 (thermally activated) and Norit CA1 (chemically activated) at the end of each successive desorption step (redrawn after [28]).

activated carbons have a surface with fully oxidized active sites such that interaction with oxygen does not affect the surface.

The energy of adsorption is particularly important from the point of view of adsorption reversibility and bioregeneration. The higher the change in the Gibbs free energy of adsorption ($-\Delta G°_{ads}$) of an adsorbate, the lower is the bioregeneration [14]. This can be attributed to the high energy of adsorptive binding with the adsorbent surface that results in a lower reversible adsorption. For example, aniline was not available for biodegradation in the bulk liquid since it was adsorbed on PAC with a higher energy than that of phenol [52]. As such, the energy of adsorption might provide information about which organic compound could be conveniently removed by simultaneous adsorption and biodegradation.

When ozonation is applied before the bioregeneration process, the carbon surface may get into contact with the ozone remaining in solution. This contact eventually results in the formation of carboxyl and phenol functional groups that increase the hydrophilicity of carbon surface and decrease the Gibbs free energy in adsorption of organic matter. Consequently, this reduction in Gibbs free energy increases bioregeneration efficiency [53].

From the point of view of water pollution control, for nonbiodegradable compounds adsorption should ideally be irreversible. Otherwise, desorption of such compounds from carbon would lead to the deterioration of effluent quality unless other measures are taken. Therefore, special attention should be paid to the possible desorption of an adsorbed organic compound that is hardly biodegradable such as 4-nitrophenol (4-NP) [54]. In contrast to this, desorption of biodegradable compounds will lead to bioregeneration, in other words to the renewal of the adsorptive capacity of activated carbon. On the other hand, irreversible adsorption of biodegradable compounds will unnecessarily deteriorate the adsorptive potential and shorten the service life of activated carbon [50].

7.5
Other Factors Affecting Bioregeneration

Bioregeneration can be optimized by varying the nature of microorganisms, environmental conditions, and loading on activated carbon [11]. For an effective bioregeneration to take place, certain microbiological and technological prerequisites are required such as the presence of microbiological agents capable of utilizing the adsorbate, presence of mineral components (nitrogen, phosphorus, sulfur, and so on), creation of optimum conditions for the vitality of microorganisms (temperature, dissolved oxygen concentration, and so on) and optimization of the concentration ratio between microorganisms and adsorbate [25]. The adsorption–desorption balance, residence time, and spatial distribution of molecules in carbon pores were suggested as the other factors determining the efficiency of bioregeneration [55]. The loading on carbon, the adsorption characteristics (i.e., Freundlich K_F and $1/n$), and the location within the GAC column were reported as the additional factors affecting the rate and extent of bioregeneration [56].

7.5.1
Biodegradability

The bioregeneration hypothesis is generally valid for biodegradable, slowly biodegradable, and adsorbable organic compounds. Slowly biodegradable compounds can be biodegraded if a sufficient contact time with biomass is provided. In PACT systems, these compounds are in contact with the biomass for a length of time equal to the sludge age if they are adsorbed by PAC and thus enter the solid phase [57]. On the other hand, in a conventional biological treatment system without activated carbon, slowly biodegradable compounds may leave with the effluent stream since they may not be biodegraded. For a further insight into these issues, the reader may refer to Chapter 6.

Various compounds that are considered nonbiodegradable (such as chloroform and chlorinated benzenes) can be adsorbed on activated carbon and subsequently metabolized by attached microorganisms [58]. Increased retention of biomass in the system favors the metabolization of organics. Therefore, theoretically, bioregeneration should increase with sludge age [7]. Slowly biodegradable organics are first adsorbed onto activated carbon and retained in the system for a long period of time. Upon desorption from the carbon surface, they may be biodegraded by attached and suspended biomass. The consequence is the bioregeneration of activated carbon. Hence, adsorption acts as a prerequisite step for biodegradation.

In the whole bioregeneration process, the rates of both desorption and biodegradation are important. Compared to desorption, biodegradation is considered to be the rate-limiting step in bioregeneration, particularly for slowly biodegradable substances. In the region surrounding activated carbon, an increase in microbial uptake rate would lower the bulk concentration of a substance. This would increase the driving force for diffusion of substrate from carbon, which consequently would increase the rate of bioregeneration. In that respect, acclimation of biomass to the compounds in question is expected to increase both the rate and efficiency of bioregeneration. For example, compared to nonacclimated biomass, the presence of acclimated biomass [59] resulted in much faster and higher bioregeneration of activated carbons that were loaded with the biologically resistant compound 2-chlorophenol (2-CP) [60].

On the other hand, readily biodegradable organic matter is generally nonadsorbable and can be removed by biological activity alone, whereas the removal of nonbiodegradable and adsorbable compounds takes place through simple adsorption. It was stated that bioregeneration was due to simple desorption of sorbed compounds followed by biodegradation. Hence, bioregeneration could only occur in the case of compounds that are both biodegradable and adsorbable [20]. For example, activated carbons loaded with 2-CP could not be bioregenerated since this compound is nonbiodegradable [60]. However, carbons loaded with the biodegradable phenol were successfully bioregenerated by the same activated sludge [24]. In another study, 4-NP could be successfully removed through desorption–biodegradation, although total bioregeneration of activated carbon was not achieved [61]. In other studies, the bioregeneration concept was extended to include readily biodegradable and less biodegradable compounds as well as toxic compounds [10, 58]. However, there are also other studies denying the hypothesis of bioregeneration. For example, in a BAC process, bioregeneration was shown not to be valid in the case of biodegradable compounds such as phenol [22].

In the case of complex organics such as the natural organic matter (NOM) in natural waters, ozonation of water increased the extent of bioregeneration of activated carbon [53]. As discussed in Chapter 8, ozonation mostly increases the biodegradability of NOM components. Eventually, this increases the bioregenerability of activated carbon. However, bioregeneration of activated carbon in drinking water treatment is still a debated issue, as discussed in Section 9.1.3 of Chapter 9.

7.5.2
Chemical Properties of Substrate

The extent of bioregeneration is also dependent on the compound adsorbed. For example, the nonphenolic compound, 3-chlorobenzoic acid, was more available for bioregeneration of PAC compared to the phenolic compound, o-cresol [9]. In another study, phenol with a smaller Gibbs free energy of adsorption ($-\Delta G°_{ads}$) was found to be more suitable for bioregeneration than nonionic and anionic surfactants [14]. Another study showed that the extent of bioregeneration was greater for phenol-loaded carbon compared to that loaded with p-methylphenol-, p-ethylphenol-, and p-isopropylphenol [41]. Carbons loaded with alkyl-substituted phenols exhibited less bioregenerability, and bioregenerability decreased with an increase in the length of the alkyl chain [41]. In another study, aromatic compounds with electron-donating substituent groups (e.g., phenol with a hydroxyl group) exhibited higher irreversible adsorption compared to aromatic compounds with electron-attracting substituent groups [62]. The extent of GAC bioregeneration was higher with phenol compared to the halogen-substituted 2,4-dichlorophenol (2-DCP) in other studies [11, 63]. The lower desorbability and biodegradability of 2-CP resulted in much less bioregenerability compared to phenol [60]. In another study, desorption and bioregeneration took place more rapidly for p-nitrophenol (PNP) compared to phenol [64]. Although not stated by the authors, this finding may be due to irreversible adsorption of phenol which is caused by oxidative coupling. On the other hand, oxidative coupling does not occur with nitrophenols [51]. Other studies also showed that PACs loaded with 2-nitrophenol (2-NP) [65] and PNP [66] were bioregenerated to a higher extent compared with phenol. This finding was also explained by the same reasoning that oxidative coupling did not take place in the case of nitrophenols.

7.5.3
Carbon Particle Size

The bioregeneration hypothesis seems to be more valid in BAC filters than in PAC-added activated sludge processes such as the aerobic PACT. The first reason

Table 7.1 Extent of reversible adsorption and bioregeneration for various activated carbon types loaded with phenol (adapted from [24]).

Carbon type	Physical form	Activation type	Reversibility of adsorption (%)	Bioregeneration efficiency (%)
SA4	powdered	thermal	20.3	58.1
CA1	powdered	chemical	86.6	93.6
PKDA	granular	thermal	25.8	66.6
CAgran	granular	chemical	87.5	84.8

is that the sludge age is much higher in attached-growth processes compared with activated sludge. Thus, microorganisms often have enough time to degrade adsorbed refractory organics. The second reason, which is also related to the first one, is that GAC provides an intimate contact between activated carbon pores and microorganisms by providing suitable attachment sites for the latter [31].

The size of activated carbon might be another factor influencing bioregeneration. However, the evaluation of PAC and GAC countertypes with similar physical characteristics revealed that adsorption reversibility [28, 60] and bioregenerability [24] were comparable in each case. Obviously, the carbon size was not an effective factor for the reversibility of adsorption and bioregenerability of activated carbon (Table 7.1). For thermally or chemically activated PACs the degrees of hysteresis were only slightly higher compared to their granular countertypes [28]. This showed that desorbability from PAC was slightly less compared to a similar GAC with a more macroporous structure. On the other hand, desorption was faster from PACs, probably due to the higher intraparticle diffusivity (diffusivity inside carbon) compared to GACs that have larger diameters. Other researchers also stated that the diffusive transport inside carbon was slower, and the opportunities for exchange between the carbon and liquid phases diminished in the case of GAC particles, which are much larger than PAC particles [8]. In a GAC filter, the resistance to diffusive transport was the rate-limiting step in bioregeneration. However, in another study conducted in a batch system it was shown that the diffusivity controlled only the desorption rate from carbon, but not the total desorbability [28]. The importance of intraparticle diffusivity is addressed in Chapter 6.

7.5.4
Carbon Porosity

The porous structure of activated carbon is a factor that determines to a great extent both the rate and degree of bioregeneration [14]. Particularly, considering the adsorption of large molecules, mesopores are the most useful part of this porous structure in activated carbon. A mesoporous activated carbon was found to be more bioregenerable than a microporous one, particularly when the bulk substrate concentrations decreased rapidly under the dynamic conditions of a GAC filter [25]. Another study suggested that at the end of bioregeneration most of the toluene remaining on activated carbon was likely to reside in micropores [56].

7.5.5
Carbon Activation Type

The method used in the activation of carbon is of crucial importance with respect to the extent of irreversible adsorption and/or bioregeneration. Thermally activated PACs, produced from peat or coal, exhibited a higher extent of irreversible adsorption compared to chemically activated wood-based GACs [49]. Reversibility of adsorption and bioregeneration were considerably higher for chemically activated wood-based PAC than thermally activated peat-based PAC in other studies [9, 50]. Chemically activated PAC (CA1 from Norit) exhibited a higher extent of bioregeneration than the thermally activated PAC (SA4 from Norit) loaded with o-cresol and 3-chlorobenzoic acid, although SA4 exhibited a higher adsorptive capacity than CA1 for the target compounds [9].

Other studies with the same PACs as above (SA4 and CA1) and their granular (GAC) counterparts (thermally activated PKDA and chemically activated CAgran from Norit) with similar physical characteristics, revealed that for the chemically activated carbons CA1 and CAgran adsorption reversibility was higher [28, 60], and a higher bioregeneration was achieved [24] (Table 7.1). Although carbon characteristics such as the carbon surface, molasses number, iodine adsorption, and pore volume, indicated better adsorption capacity for CA1 and CAgran, in reality the carbons SA4 and PKDA surprisingly exhibited a higher adsorption capacity [28, 60]. CA1 is a PAC grade which is chemically activated with phosphoric acid. Chemical activation creates fully oxidized active sites on the surface. Therefore, the interaction with oxygen is unlikely to have an effect on such a surface. However, thermal activation is performed in the absence of free oxygen. The resulting PAC surface then becomes more reactive toward oxygen. Therefore, the surface chemistry may change upon contact with oxygen. These differences in surface characteristics could have an effect on the extent of irreversible adsorption [50]. Another study also showed that, in comparison to thermal treatment of activated carbon, nitric acid treatment resulted in higher bioregeneration of activated carbon loaded with molinate [67]. All these results showed that, the activation method and/or raw material of carbon have a greater influence on the reversibility of adsorption than the physical characteristics of carbon. For long-term operation, an activated carbon grade that is more amenable to bioregeneration should be chosen since this increases the service time of the carbon.

7.5.6
Physical Surface Properties of Carbon

The physical characteristics of activated carbons (porosity, raw material) do not play an important role in the development of biological activity since microorganisms are mainly fixed on the external surface of activated carbon [68]. The effect of bacterial adhesion on bacterial activity was investigated for different types of steam-activated PACs [69]. The authors concluded that surface characteristics like BET-specific surface areas, total surface acidity, functional groups, total surface basicity, surface charge, pH_{pzc}, iodine number, metal concentrations, electrical resistance, free radical concentration, and formate adsorption capacity were not related to the stimulation of biological activity. However, porosity of activated carbon affects the bioregeneration capacity since desorption from meso- and macropores may be easier than that from micropores. Faster and complete biodegradation occurred particularly at the outer macroporous surface of activated carbon [33]. The presence of larger pores, mainly meso-, and macropores, favored the bioregeneration of activated carbons that were loaded with molinate [67].

7.5.7
Desorption Kinetics

A compound which is slowly desorbed may be regarded as irreversibly adsorbed. In other words, a compound may not be desorbed although its adsorption is reversible [50]. Hence, the kinetics of desorption may also play an important role in the extent of adsorption irreversibility. Some bioregeneration studies have revealed slow rates of bioregeneration for organic compounds that also desorb slowly [2, 9, 19, 50]. The rate of desorption depends on culture conditions, fluid dynamics, metabolic activity of microorganisms, and the type and density of carbon particles [70]. A peak bioregeneration rate was achieved upon establishment of microbial activity in a GAC bed because the substrate in the outer portion of GAC was readily available to microorganisms [8]. However, bioregeneration rate declined over time because the majority of sorption sites were limited by the diffusive transport resistance in GAC. Hence, as the surface diffusivity for substrate increased, the potential for bioregeneration also increased [8].

The pore size distribution of activated carbon also affects desorption kinetics [50]. Desorption from the PAC grade SA4 was more gradual than that from CA1, and consisted of two phases. A fast initial phase of desorption, during which most of the desorption took place, was followed by a slow desorption phase. This dual rate suggested that the PAC grade SA4 contained two regions with different diffusion modes. A network of relatively wide macropores allowed rapid diffusion, whereas a region of narrower micropores resulted in relatively slower diffusion. Contrary to this, CA1 had a much more open structure and a much higher total pore volume and mesopore volume than SA4. This caused better accessibility and faster diffusion for CA1.

7.5.8
Substrate–Carbon Contact Time

In offline bioregeneration studies, adsorption and biodegradation processes are applied sequentially. Activated carbon is initially contacted with the adsorbate.

Following the adsorption process, bioregeneration is achieved in a separate biological reactor [2]. The initial contact time of activated carbon with the adsorbate, either in batch or continuous-flow contactors, is important for the extent of succeeding bioregeneration. The bioregenerable fraction of an adsorbed compound may decrease with the increase in this contact time [9]. Another study showed that the reversibility of adsorption varied with contact time [50]. The desorbable fraction was found to decrease with an increase in the contact time between PAC and adsorbate. In offline bioregeneration, depending on MLVSS, increasing the initial contact time between activated carbon and adsorbate from 24 to 96 h resulted in an increase of bioregeneration from about 10–60% to 60–75% in succeeding biological reactors [2].

As the duration of bioregeneration increased, the breakthrough performance of phenol-loaded GAC improved when the same carbon was used in the subsequent adsorption period [34]. In another successive (offline) bioregeneration study performed with preloaded GAC columns, bioregeneration increased at high empty-bed contact times (EBCTs) [56]. This was observed because a higher EBCT implied a lower bulk concentration inside the filter due to more efficient removal. The lowering in bulk concentration resulted in a larger driving force for desorption and subsequent biodegradation of adsorbed compound.

During concurrent bioregeneration (see Section 7.3), the contact time between adsorbate (substrate) and activated carbon is also important. At a sludge age of 3 days in a GAC-SBR system, in the case of phenol and 2,4-DCP the bioregeneration efficiencies were 39% and 38%, respectively. At a sludge age of 8 days, these values increased to 48% and 43% for phenol and 2,4-DCP, respectively [71]. Also other researchers stated that bioregeneration increased with increase in the EBCT and/or sludge age [55, 57]. However, in a denitrification study, bioregeneration rate decreased with an increase in EBCT, although the concentration gradient between the carbon surface and the bulk medium was greater at higher EBCT [13]. This was attributed to the formation of a thicker biofilm on GAC surface at lower EBCT since biomass loss was less in that case. Therefore, bioregeneration is not a process simply controlled by a concentration gradient; sometimes it may also be affected by biological factors such as biofilm thickness [13].

7.5.9
Concentration Gradient and Carbon Saturation

Bioregeneration of activated carbon may also depend on the concentration of the target compound in bulk solution and its loading on activated carbon. The extent of bioregeneration was found to be smaller when a lower equilibrium concentration was reached in a BAC column [56]. The dependence of bioregeneration on concentration is related to the low loading on GAC surface at low bulk concentrations. This is attributed to the following factor: at low concentrations compounds are adsorbed on high-energy adsorption sites of activated carbon, eventually resulting in a highly irreversible adsorption [56]. As discussed in Section 9.1.3 of Chapter 9, in the BAC treatment of drinking water, bioregeneration of activated carbon is usually questionable since pollutants (natural or synthetic) are often present at very low concentrations. Therefore, the same reasoning as above can also be applied to this case. The low loading of activated carbon is likely to result in irreversible adsorption.

In batch adsorption studies, prior to contact with a bacterial consortium, the 4-chlorophenol (4-CP) loading on activated carbon was raised by increasing the supply of this compound. Then, upon contact with microorganisms, higher bioregeneration efficiencies were obtained in batch bioregeneration flasks [72]. Although only 21% bioregeneration was achieved at a loading of 10.8 mg 4-CP/g GAC, bioregeneration efficiencies increased to 24% at 54 mg 4-CP/g GAC, to 56% at 108 mg 4-CP/g GAC and to 75% at 216 mg 4-CP/g GAC. These findings can also be attributed to the adsorption of adsorbate on high-energy sites at lower loadings [72].

Another phenomenon is that at low bulk concentrations of substrate large concentration gradients are not developed inside GAC [56]. Putz and co-workers suggested that the process tends to be diffusion-limited beyond the initial period of bioregeneration, so that the smaller concentration gradients associated with lower concentrations yielded slower diffusion and thus slower bioregeneration rates [56]. Speitel and DiGiano also reported that the liquid phase concentration surrounding GAC was extremely important; it was a determinant factor for the rate and extent of bioregeneration in simultaneous bioregeneration. However, a lower bulk concentration resulted in a higher total bioregeneration [8]. Actually, these findings do not contradict each other, because the former [56] deals with the rate of bioregeneration at later periods of the process, whereas the latter [8] addresses the extent of total bioregeneration.

It is also claimed that bioregeneration might vary temporally and spatially within a GAC column [56]. As shown in Chapter 2, a GAC column contains three zones of varying length: exhausted GAC (saturated zone), partially exhausted GAC (the mass transfer zone: MTZ), and virgin GAC. Some authors suggested that the biological activity in the exhausted GAC zone held by far the most promise for bioregeneration since this exhausted zone had the most potential for desorption and subsequent biodegradation of adsorbed biodegradable organic matter [56]. Others also showed that, after about 11 days of operation, cumulative bioregeneration was highest at the influent of the GAC column (about 16%), decreased at the mid-column region (about 15%), and was least at the effluent end of the column (about 5.5%). However, in all of these three zones bioregeneration efficiencies were less than 1% at the end of the first 3 days of operation. However, the differences in bioregeneration efficiency between the zones appeared after 3 days and increased to the values given above at the eleventh day [8].

7.5.10
Biomass Concentration

Another determinant for the extent and rate of bioregeneration may be the concentration of biomass. In one study, increasing the average MLVSS concentration from 126 to 963 mg L^{-1} resulted in an increase in bioregeneration from about 10% to 60% at the end of 24 h [2]. It was also suggested that bioregeneration was dependent on the ratio of biological solids to activated carbon solids. More carbon particles are expected to be surrounded by bacteria at high MLSS [57]. However, others stated that increasing the initial MLVSS shortened the time to reach equilibrium in batch bioregeneration systems, but had little effect on the magnitude of bioregeneration [11].

Immobilization of microorganisms on the activated carbon surface is a factor that increases bioregeneration efficiency. As discussed in Section 3.2.2 of Chapter 3, activated carbon serves as a supporting material for microorganisms. Further, its high adsorption capacity provides substrates to microorganisms. Immobilized *Pseudomonas* and *Candida* cells could achieve 90% bioregeneration of GAC and tolerated phenol concentrations up to 15 000 mg L^{-1}, whereas free cells could not tolerate >1500 mg L^{-1} [73]. Co-immobilization of biomass and exoenzymes together with activated carbon may be another way of creating a concentrated environment for the adsorbent and the microorganisms and of increasing the contact between them. A system consisting of PAC and a special fungus culture *P. Chrysosporium* co-immobilized with Ca-alginate within a hydrogel matrix degraded pentachlorophenol (PCP) on PAC more efficiently than nonimmobilized systems [74].

7.5.11
Dissolved Oxygen Concentration

Considering that in aerobic systems the availability of DO is important for biodegradation, one can easily conclude that the DO level will also determine the degree of bioregeneration. A phenol-loaded GAC column was successfully bioregenerated at an initial DO concentration of 9 mg L^{-1}, whereas no bioregeneration occurred at 4 mg L^{-1}. The low DO level was sufficient for biodegradation of phenol in the influent stream only, but not for biodegradation of phenol that was adsorbed on GAC [39]. The DO level in GAC columns requires attention in a bioregeneration process, because in bioregeneration a large amount of adsorbed substrate becomes very rapidly available to microorganisms, and the need for DO is raised [56].

7.5.12
Microorganism Type

Another important determinant for bioregeneration is the nature of the microbial population. In particular, in the case of slowly biodegradable compounds or organics that are classified as nonbiodegradable in conventional biological treatment works, specific or acclimated microorganisms are required for degradation of target compounds. For example, GAC loaded with the hardly biodegradable insecticide atrazine was inoculated with an atrazine-degrading bacterium *Rhodococcus rhodochrous* [75]. The bacteria attached to the GAC significantly extended the bed life through bioregeneration of activated carbon. In another study, a bacterial consortium was isolated from the rhizosphere of *Phragmitis communis* in order to colonize GAC and degrade the hardly biodegradable 4-chlorophenol (4-CP) that was loaded on activated carbon [72]. A consortium of *Pseudomonas* strains was also successfully employed for removing bioresistant surface-active substances from activated carbon [25]. In another study, consortia of bacteria including *Flavobacterium* sp. were able to degrade azo dyes by breaking azo bonds [76]. Acclimation of a mixed activated sludge to xenobiotics was also a very effective method for the bioregeneration of activated carbons that were loaded with compounds such as 2-CP and 2-NP [59, 65].

7.5.13
Substrate and Biomass Associated Products of Biodegradation

The quality of the BAC surface deteriorates during bioregeneration. This is attributed to the adsorption of lyzed cells, slowly biodegradable substances, and metabolites. Therefore, bioregeneration of activated carbon can hardly be a way of avoiding the deterioriation of adsorbent quality, but bioregeneration can still prolong the usage time of the adsorbent [54]. In any case, bioregeneration alone will not be sufficient for the complete recovery of adsorption capacity [55].

In biological systems assisted by activated carbon, often a slime matrix including decay products covers microorganisms and activated carbon. Figure 7.6 shows the Scanning Electron Microscopy SEM images of activated carbon samples taken from activated sludge reactors [16]. The slime is recognized as a bright white gelatinous structure on the carbon surface.

Correspondingly, at high biomass concentrations (MLSS >2500 mg L^{-1}) in a PACT process, the carbon particles become trapped within the floc matrix, and the carbon pores are closed [77]. The difficulty of desorption of microbial products and the filling of pores with decay products of microbial cells are also mentioned in other studies [5, 18, 55, 78].

In an offline bioregeneration study, in a GAC column the breakthrough performance deteriorated gradually after each successive bioregeneration step, even in the

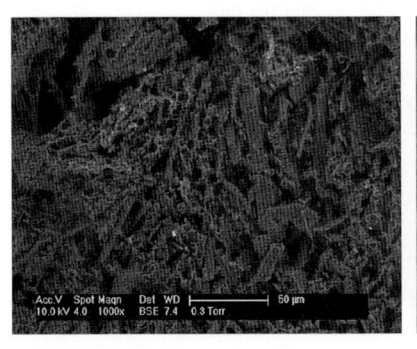

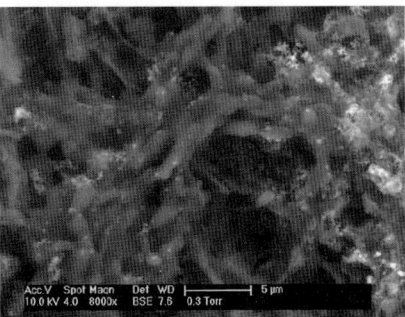

Figure 7.6 Scanning Electron Micrographs of slime matrix covering microorganisms and carbon surface forming a bright white gelatinous structure [16].

case of recirculation of filtered permeate through the column [35]. Although the authors attributed this finding to accumulation of adsorbates, irreversible adsorption of metabolic end products might also have contributed to this phenomenon.

In another study, 82.5% of TCE adsorbed on activated carbon was bioregenerated in successive treatments. However, bioregenerated activated carbon had a much lower TCE adsorption capacity compared to virgin GAC, probably due to adsorption of soluble microbial products (SMPs) [79]. Other researchers observed that nearly 50% of SMPs were adsorbed onto PAC in a PACT system, but only 4% of these adsorbed SMPs were biodegraded by microorganisms [12]. In another study, in a GAC-FBR system receiving toluene-contaminated water, the loss of adsorption capacity was attributed to the irreversible adsorption of SMP rather than to the adsorption of toluene, whose adsorption is highly reversible [80]. The importance of EPS and SMP in integrated adsorption and biological removal is discussed in Chapter 3.

7.5.14
Presence of Multiple Substrates

Most of the research on bioregeneration addresses the case of a single compound. However, wastewaters usually contain multiple compounds that might influence desorption and biodegradation of each other. Putz and co-workers stated that mixtures of organic compounds are difficult to treat because each compound can vary in its ability to be biodegraded or adsorbed [56].

As discussed in Chapter 3, cometabolism is the biological transformation of a nongrowth substrate by nonspecific enzymes of bacteria which can only be induced by a growth substrate. Most xenobiotic organic compounds cannot be normally metabolized by microorganisms. However, some of them can be removed as nongrowth (cometabolic) substrates by cometabolism. In integrated adsorption and biodegradation, adsorption of xenobiotic substances onto activated carbon also increases the chance of cometabolic removal. Consequently, activated carbon is bioregenerated as a result of cometabolic processes.

Bioregeneration of activated carbon was studied in the case of two competing compounds, namely phenol and 2,4-DCP, that were simultaneously present in a BAC system [10, 71]. In such systems, the percentage of bioregeneration of each compound is related to its biodegradability and degree of hysteresis [10]. The bioregeneration of 2,4-DCP in a bisolute system was higher than that in a single solute system, probably due to the possibility of cometabolic removal in the presence of phenol as a growth substrate. However, bioregeneration of phenol was suppressed in the bisolute system, probably due to preferential adsorption of 2,4-DCP.

Various activated carbons loaded with 2-CP were bioregenerated through cometabolism in the presence phenol as the growth substrate [59]. Bisolute bioregeneration efficiencies reached up to 55% for phenol and 67% for 2-CP. Another bisolute study also indicated cometabolic bioregeneration, where 65% of 2-NP loaded on activated carbon could be bioregenerated in the presence of phenol [65].

Cometabolic transformation of 4-CP was also evident in a hybrid GAC-MBR in which bioregeneration increased the reactor performance [81]. In another study, the presence of peanut oil accelerated the biodegradation of polycyclic aromatic compounds (PAHs). This resulted in partial bioregeneration of activated carbon that was loaded with anthracene [82].

Bioregeneration was investigated in GAC columns using mixtures of biodegradable (benzene or toluene) and nonbiodegradable (perchloroethylene or carbon tetrachloride) synthetic organic compounds (SOCs) [56]. The loading on the GAC before and after bioregeneration showed a marked decrease in biodegradable SOC as well as an increase in nonbiodegradable SOC. Bioregeneration of GAC took place in the case of the biodegradable SOC. Then, regenerated pores of GAC were loaded with the nonbiodegradable SOC [56]. In another study, in the coexistence of biodegradable and nonbiodegradable compounds having similar adsorbabilities, in the case of the nonbiodegradable compound, the service life of GAC increased up to 1.5 times compared to adsorption alone [83]. Biodegradation of a weakly adsorbed compound improved the adsorption capacity for the nonbiodegradable compound only slightly. This is observed because weakly adsorbed chemicals do not compete for GAC adsorption sites as well as strongly adsorbed ones. Therefore, their biodegradation would not reopen many adsorption sites [56]. In bisolute systems consisting of a biodegradable and a nonbiodegradable compound, bioregeneration increased with the adsorbability of the biodegradable compound [56].

7.6
Determination of Bioregeneration

The extent of bioregeneration is usually indefinite in concurrent treatment systems such as PACT and BAC. Often, studies reporting an increase in the bed life of BAC or an increase in the removal efficiency of PACT systems do not consider bioregeneration alone, but also include biodegradation. However, a qualitative and quantitative study of bioregeneration is important in order to optimize the elimination of organic pollutants [11]. It is recognized that the literature does not contain much quantitative data on bioregeneration. The experimental measurement of bioregeneration is difficult, since biodegradation, adsorption, and desorption occur simultaneously within concurrent systems [13].

On the other hand, the extent of bioregeneration is much easier to assess in offline systems where preloaded activated carbon is consecutively biologically treated. In this respect, batch studies can provide useful data that can be extrapolated to the design of continuous systems.

Bioregeneration of activated carbon can be quantified either by the direct measurement of substrate loading on activated carbon or by the indirect estimation of substrate consumption through the measurement of CO_2 production. For these purposes, instrumental respiration methods or radiochemical analysis are employed, respectively. Examples of quantification of bioregeneration are presented in Table 7.2.

Table 7.2 Examples of bioregeneration efficiencies reported in the literature.

Compound	Biomass	Reactor Type	Max. Bioregen. (%)	Reference
Phenol	Acclimated	Batch	100	[12]
Phenol	*Pseudomonas putida*	Batch PAC+GAC	47	[84]
Phenol	*Pseudomonas putida*		87	[85]
Phenol	*Pseudomonas, Candida*	Batch GAC	90	[73]
Phenol	Acclimated	Batch	30	[86]
Phenol	Activated sludge	BAC column	17–21	[87]
Phenol	Activated sludge	BAC column	100	[54]
Phenol	Acclimated	GAC column	75.2	[2]
Phenol	Activated sludge	Batch	93.6	[24]
Phenol	Acclimated	PAC	56±6	[66]
p-Nitrophenol	Acclimated	PAC	64±8	[66]
Phenol (trace)	Acclimated	BAC column	15	[8]
p-Nitrophenol (trace)	Acclimated	BAC column	22	[8]
Phenol	Acclimated	Batch	83.8	[63]
2,4-Dichlorophenol	Acclimated	Batch	64.3	[63]
Phenol	Acclimated	Batch	76.3	[10]
2,4-Dichlorophenol	Acclimated	Batch	60.2	[10]
Phenol	Acclimated	Batch	31.4	[11]
2,4-Dichlorophenol	Acclimated	Batch	14.3	[11]
Phenol	Acclimated	BAC-SBR	48	[71]
2,4-Dichlorophenol	Acclimated	BAC-SBR	43	[71]
Phenol	Acclimated	SBR	77±4	[41]
p-methylphenol	Acclimated	SBR	69±4	[41]
p-ethylphenol	Acclimated	SBR	68±4	[41]
p-isopropylphenol	Acclimated	SBR	58±6	[41]
4-Chlorophenol	*Phragmitis communis*	Batch	75	[72]
2-Chlorophenol	Activated sludge	Batch PAC+GAC	12.9	[60]
2-Chlorophenol	Acclimated	Batch PAC+GAC	66.6	[59]
2-Nitrophenol	Acclimated	Batch PAC	64.9	[65]
o-Cresol	Activated sludge	Batch	15	[9]
3-Chlorobenzoic acid	*Pseudomonas sp.*	Batch	85	[9]
Tetrachloroethylene	*Pseudomonas sp.*	BAC column	39.4	[56]
Toluene	*Pseudomonas sp.*	BAC column	45.5	[56]
Benzene	*Pseudomonas sp.*	BAC column	38.2	[56]
Carbon tetrachloride	*Pseudomonas sp.*	BAC column	33.2	[56]
Fluorobenzene	Pure culture, *F11*	Batch GAC	58–80	[88]
Trichloroethylene	Phenol-utilizers	GAC column	82.5	[79]
Surfactants mixture	*Pseudomonas sp.*	Batch	35	[25]
Surfactants mixture	*Pseudomonas sp.*	GAC column	69	[25]
Sulfonol	*Pseudomonas sp.*	GAC column	22	[25]
Polyoxyethylene	Acclimated	GAC filter	53	[55]
Molinate	Mixed culture	Batch PAC	67	[67]
Azo dye	*Pseudomonas luteola*	Batch BAC	89	[89]
Azo dye	Mixed culture	Batch	52	[76]

7.6.1
Investigation of the Extent of Reversible Adsorption

Knowledge about the extent of reversible and irreversible adsorption is the key in the selection of both the optimum activated carbon type and refreshment rate [50]. The reversibility of adsorption is expressed as the degree of hysteresis (w) by using Eq. (7.5).

$$w(\%) = \left(\frac{1/n_{ads}}{1/n_{des}} - 1\right) \times 100 \tag{7.5}$$

where $1/n_{des}$ and $1/n_{ads}$ represent the values of desorption and adsorption intensity obtained from Freundlich isotherms, respectively.

The $1/n_{des}$ value obtained from the desorption isotherm is lower than the $1/n_{ads}$ value obtained from the adsorption isotherm. This implies that the rate of desorption is lower than the rate of adsorption. In a literature study, higher activated carbon dosages resulted in a lower degree of hysteresis (irreversibility) compared to lower dosages [28]. In the reversibility of adsorption, the adsorption energies onto different pores seem to play an important role. The initial substrate/carbon ratio affects the distribution of adsorption energies among carbon pores; it is therefore decisive for adsorption reversibility [28].

In another study, the bioregenerable fraction of adsorbed substrate was lower than the theoretically desorbable fraction. This showed that bacteria were unable to reach nondesorbable compounds or influence their desorbabilities. Hence, bioregeneration was controlled by the desorbability of the compounds. However, this result is in contradiction to the bioregeneration theory stating that organic matter is desorbed according to exoenzymatic reactions. Bearing this in mind, the term 'bioregeneration' should be used with caution. Regeneration of loaded activated carbon can also be achieved through abiotic desorption, for example, by leaching of a compound from loaded activated carbon. Some authors prefer to use the 'bioregeneration' term only for cases when a direct interaction has been shown between microorganisms and adsorbed compound [22]. A study has shown that the bioregenerable fraction may exceed the desorbable fraction in some cases, probably due to exoenzymatic reactions [24]. The researchers concluded that this direct interaction should certainly be called 'bioregeneration.' However, most authors use the term 'bioregeneration' as long as microorganisms are responsible for the removal of the dissolved compound in the bulk fluid, leading to desorption of the sorbed compound [12, 19].

7.6.2
Use of Adsorption Isotherms

A bioregeneration study in a batch system involved loaded GAC, acclimated microorganisms, mineral salt solution, and the adsorbate [11]. In this system, bioregeneration was quantified using Freundlich adsorption isotherm constants and equilibrium concentrations. GAC was initially equilibrated with a known concentration of adsorbate. After saturation, the concentration in the supernatant was measured to determine the amount of adsorbate before bioregeneration. For the investigation of bioregeneration, this mixture was then inoculated with acclimated microorganisms and mineral salts solution. It was aerated and the residual adsorbate concentration was measured at different time intervals. At the end of bioregeneration, the supernatant was decanted, and GAC was rinsed and put into an autoclave to terminate the activity of the microorganisms. Then, the sterilized GAC was reloaded and equilibrated with a known concentration of adsorbate. Finally, the residual adsorbate concentration was measured. Using the Freundlich parameters representing adsorption capacity and intensity respectively, K_F and $1/n$, previously obtained from adsorption isotherms, the amount of substance adsorbed before the bioregeneration step was calculated using the following equation:

$$q = (S_0 - S_e) \cdot V/W = K_F S_e^{1/n} \qquad (7.6)$$

where,

q = amount of a compound adsorbed per unit weight of carbon (M_s/M_c),
S_0 = initial concentration (M_s/L^3),
S_e = equilibrium concentration (M_s/L^3),
V = total volume of sample (L^3),
W = weight of activated carbon (M_c).

In the units, s and c denote the substrate and carbon, respectively.

After bioregeneration, the amount of adsorbed substrate left on activated carbon (q') was determined by Eq. (7.7):

$$q' = K_F S_e'^{1/n} \qquad (7.7)$$

where,

S_e' = adsorbate concentration in the bulk liquid, (M_s/L^3).

After addition of substrate and equilibration, the additional amount of substrate adsorbed (Δq_2) was calculated by Eq. (7.8):

$$\Delta q_2 = (S_2 - S_{2e}) \cdot V/W \qquad (7.8)$$

where,

S_2 = concentration at the beginning of equilibration (M_s/L^3),
S_{2e} = equilibrium concentration (M_s/L^3).

Hence, the total adsorbability (q_2) was calculated as follows:

$$q_2 = q' + \Delta q_2 = K_F S_{2e}^{1/n} \qquad (7.9)$$

Then, the quantity of bioregenerated phenol was calculated as follows:

$$\text{Quantity bioregenerated } (M_s/M_c) = q_1 - q' = q_1 - (q_2 - \Delta q_2) \qquad (7.10)$$

$$\text{Percentage of bioregeneration } (\%) = 100 * (q_1 - q')/q_1 \qquad (7.11)$$

7.6.3
Direct Measurement by Using Adsorption Capacities

As seen in Eq. (7.12), bioregeneration can also be quantified by considering the relative equilibrium adsorption capacities of fresh and bioregenerated activated carbons [25].

$$\% \text{ bioregeneration} = 100 * \alpha_{reg}/\alpha \tag{7.12}$$

where $\alpha_{reg.}$ and α stand for equilibrium adsorption capacities of regenerated and fresh activated carbon, respectively.

Goeddertz and co-workers used a similar calculation method for a GAC column [2]. Jaar and Wilderer reported that a fixed-bed GAC reactor loaded with 3-chlorobenzoate and/or thioglycolic acid and operated in sequencing batch mode lost only 10% of its adsorption capacity after about 14 months of operation, indicating 90% bioregeneration [90]. Hutchinson and Robinson determined the extent of bioregeneration using breakthrough curves for both fresh and bioregenerated GAC columns [34, 35].

7.6.4
Direct Measurement by Solvent Extraction

Using a direct measurement method in batch reactors, Ha and Vinitnantharat studied the bioregeneration of GAC loaded with a single or bisolute [10]. GAC was initially equilibrated with a known concentration of adsorbate before bioregeneration. After saturation of carbon, the supernatant was taken out and the concentration in the supernatant was measured. The GAC was then contacted with a known weight of acclimated microorganisms. The amount of adsorbate remaining on the GAC was monitored over a time period. For analysis, the adsorbate remaining on activated carbon was extracted with methylene chloride. The percentage of bioregeneration was calculated from the following equation:

$$\% \text{ bioregeneration} = 100 * (q_i - q_f)/q_i \tag{7.13}$$

where,

q_i = initial amount of adsorbate adsorbed on GAC (M_s/M_c),
q_f = amount of adsorbate remaining after contact with biomass (M_s/M_c).

7.6.5
Quantification of Bioregeneration in Simultaneous Adsorption–Biodegradation

Unlike previous methods, bioregeneration can also be quantified in batch systems where adsorption and biodegradation take place simultaneously [91]. For this purpose, batch systems were used with (i) bacteria immobilized on GAC (BAC system), (ii) bacteria immobilized on sand, (iii) GAC with no biological activity, and (iv) free bacterial cells. The bacterial cells immobilized on sand were used to investigate the advantage of immobilization over free cells. Higher biomass growth rates were found in the case of immobilized cells because substrate became concentrated on the surface. As discussed in Section 3.2.2 of Chapter 3, immobilization of cells on activated carbon can increase the biodegradation potential of target compounds. Further, adsorption of toxic substances from the bulk liquid onto activated carbon has a positive influence on biodegradation of target compounds. As adsorbates the researchers used nonbiodegradable azo dyes and biodegradable anthraquinone dyes [91]. They determined the improvement in removal as the difference between the removal obtained by bacteria immobilized on GAC (i) and the total removal obtained by two other systems, bacteria immobilized on sand (ii), and GAC with no biological activity (iii). The difference (i−(ii+iii)) may arise due to bioregeneration and stimulation of biological activity by adsorption of toxic substances onto activated carbon. However, if the compound used in the bioregeneration study is not toxic to microorganisms, this difference can be the result of bioregeneration only. In another study, the same authors used this quantification method together with the Monod equation to describe the biodegradation kinetics in BAC modeling [48]. No desorption took place in BAC beds in the case of acid dyes found in textile industry wastewater which had azo and di-azo structures; consequently, no bioregeneration was recorded.

7.6.6
Measurement of Biodegradation Products

The extent of bioregeneration can be determined by the measurement of biodegradation products such as CO_2, chloride, and methane. In one study, bioregeneration was determined in a fast and accurate way by measuring CO_2 production in a batch culture [9]. The batch culture used in bioregeneration studies consisted of acclimated biomass, a mineral salts medium, and a known amount of preloaded PAC. The PAC saturated with the adsorbate was separated by centrifugation and the PAC pellet was then used in bioregeneration experiments. Batches with the following compositions were used as blanks: unloaded PAC with biomass, biomass without PAC, and loaded PAC without biomass. At the end of the experiments, when the CO_2 curves had reached a stable level indicating that growth had ceased, all batches were acidified in order to purge all CO_2 into the headspace. The percentage of CO_2 in the headspace was determined using a gas chromatograph. The authors state that the direct measurement of bioregeneration was unsuitable due to the contact time-dependent recoveries in the extraction of loaded PAC.

Another method in the quantitative determination of bioregeneration is to use a substrate that is radiolabeled with ^{14}C. In the effluent stream of the biological reactor, the amount of CO_2 containing ^{14}C and the radiolabeled substrate shows that preadsorbed compounds have been desorbed; thus the amount of bioregeneration can be calculated [8, 12, 56]. The $^{14}CO_2$ production rate is of principal interest because it allows the calculation of the biodegradation rate of adsorbed substrate. The $^{14}CO_2$ originates from the radiolabeled carbon in the substrate at the start of the experiment. Production of $^{14}CO_2$ occurs mainly due to biodegradation of substrate. However, another potential source of $^{14}CO_2$ is the endogenous decay of ^{14}C-containing biomass that is attached to GAC. A mass balance on ^{14}C biomass in the GAC column allows the $^{14}CO_2$ production rate to be corrected for endogenous decay [8]. It was also possible to quantify bioregeneration by comparing the profiles of radiolabeled phenol inside the activated carbon before and after biological treatment with the assumption that labeled and unlabeled phenol were adsorbed and biodegraded to the same extent [86].

Biodegradation of chlorinated compounds results in dechlorination, which is observed as the release of chloride ions. Therefore, an indirect way of determining the biodegradation of adsorbed organics is measurement of chloride ion in the bulk liquid. This is often done in the case of chlorinated organic compounds such as trichloroethylene (TCE) [79] and 4-chlorophenol (4-CP) [72]. This method is also applied for organics that contain other halogens than chlorine. For example, bioregeneration of GAC loaded with fluorobenzene was quantified by the measurement of fluoride in solution [88].

In anaerobic biological systems involving activated carbon, bioregeneration was determined and/or quantified by the measurement of biogas production [58, 92]. In another study, in an anaerobic BAC system, bioregeneration was determined by radiolabeling phenol in the feed and measuring the radiolabeled methane and carbon dioxide in the gaseous effluent [15].

7.6.7
Use of Respirometry in Aerobic Systems

Respirometric methods are evaluated for their potential in the monitoring of bioregeneration. In one study, the activity of microorganisms present on carbon was estimated using Warburg's apparatus [54]. The specific rate of oxygen consumption was monitored during bioregeneration. For this purpose, a definite amount of activated carbon was taken from the column several times during bioregeneration, and its oxygen consumption was monitored in a BOD bottle.

Bioregeneration was also quantified by oxygen uptake measurements with a manometric respirometer placed in a constant temperature cabinet [41]. The oxygen consumption in biodegradation alone was compared with that in simultaneous adsorption and biodegradation of phenol and alkyl-substituted phenols. The substrate removal by biodegradation was determined by oxygen uptake. Substraction of this amount from the initial substrate concentration yielded the amount of substrate adsorbed by activated carbon. The difference between the initial and final loading on carbon equaled the bioregenerated amount. Ng and co-workers quantified bioregeneration of activated carbon loaded with phenolic compounds by oxygen uptake measurements in both concurrent and successive adsorption–biodegradation [66]. The sequential adsorption–biodegradation approach provided a good estimate of the upper limit of bioregeneration that could be achieved in simultaneous adsorption–biodegradation.

7.6.8
Investigation by Scanning Electron Microscopy (SEM)

SEM investigations help to determine whether microorganisms attach on the outer surface of carbon or at inner pores. Microorganisms tend to attach to sites where substrate is concentrated for an efficient uptake. It can be expected that microorganisms attaching to inner sides can significantly contribute to the assimilation of adsorbed compounds. The microbial activity at inner sides enhances the bioregeneration of activated carbon by inducing a progressive desorption of adsorbate to bulk solution [71] or by excreting extracellular enzymes through activated carbon pores [24]. The Environmental SEM (ESEM) investigation of PACs showed that microorganisms attached themselves on both the external surface and the interval cavities of PAC particles [24]. The external surface contained mainly protozoa and filamentous, long rod-, and spiral-shaped bacteria. Groups of short rod- and cocci-shaped bacteria attached only in the internal cavities of carbon particles, probably due to the turbulent fluid dynamics on the external surface. In the study of Vuoriranta and Remo, SEM investigations showed that bacteria (filamentous and rod-shaped) were present only inside the holes and pores, but not on the surface of GAC particles in an FBR with a turbulent fluid dynamics [46]. In another study, SEM investigations showed that bacteria were attached on both the interval cavities and the outer surface of PAC particles [70]. In a bioregeneration study with azo dyes, SEM investigations have shown that bacteria successfully colonized the macropores of GAC [76].

SEM analysis also showed that at high sludge ages, the micropores in activated carbon were more densely covered with microorganisms [71]. This higher attachment of microorganisms probably increased the bioregeneration efficiency through assimilation of easily accessible organic compounds that were adsorbed on activated carbon.

In another study, ESEM analyses showed that groups of cocci-shaped bacteria with diameters of 1–2 μm dominated the microflora [16]. Considering that the activated sludge used in bioregeneration studies had been acclimated to phenol and 2-CP for more than a year, these dominant bacteria can be considered as phenol-oxidizers. These cocci-shaped bacteria attached to the outer surface or internal cavities of activated carbon, usually in groups resembling a bunch of grapes (Figure 7.7). On the other hand, larger protozoan-like microorganisms and filamentous bacteria were less often encountered.

7.7
Bioregeneration in Anaerobic/Anoxic Systems

Although most studies on bioregeneration are performed under aerobic conditions, bioregeneration is also reported in anaerobic biological processes involving activated carbon [58, 92]. It was noted in a study that the carbon equivalent of gaseous products (methane and carbon dioxide) exceeded the removal achieved in

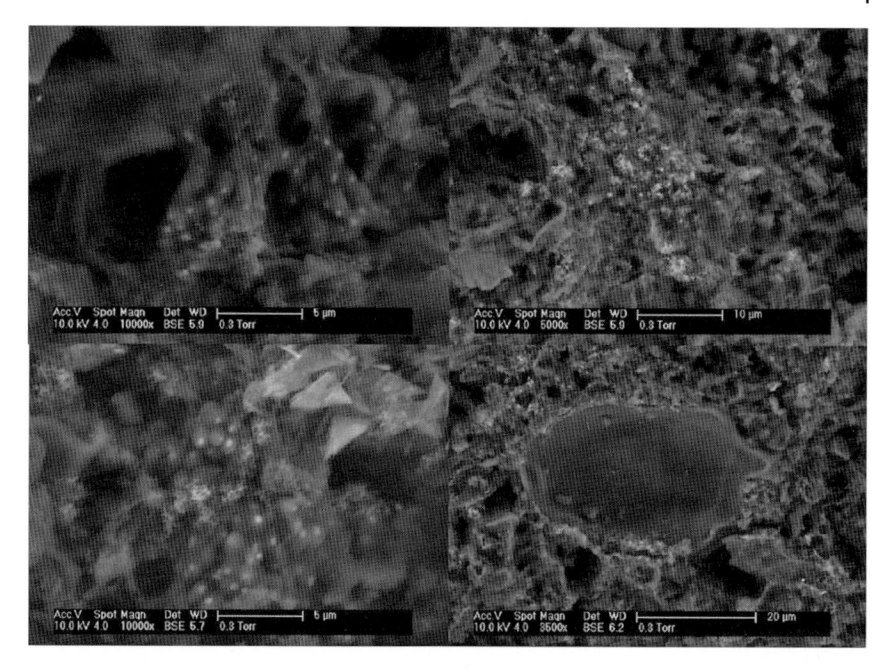

Figure 7.7 Scanning Electron Micrographs of microorganisms on activated carbon [16].

organic carbon. The extra gas production was due to bioregeneration in an anaerobic BAC system treating catechol [92]. In another study that was conducted in an expanded-bed anaerobic BAC reactor, it was shown by methane formation that when the influent concentration of a compound (phenol, chloroacetaldehyde (CAA), pentachlorophenol (PCP), and tetrachloroethylene (PCE)) exceeded the biological degradation capacity, the compound was initially adsorbed on GAC. It was then gradually biodegraded, resulting in bioregeneration of GAC [93]. Bioregeneration under anaerobic conditions was also reported in the case of bleach plant effluent, where microbial activity was evidenced from steadily increasing alkalinity and biogas production [94].

Bioregeneration was also investigated by others under the anoxic conditions of a denitrifying BAC filter that was supplied with sucrose as the organic carbon source [13, 95]. In this case, BAC served as a reservoir and supplied sucrose to denitrifying bacteria when organic carbon was deficient in the bulk liquid. This process resulted in bioregeneration of loaded activated carbon. Bioregeneration was dependent on the ratio of carbon to nitrogen in the environment surrounding BAC [13].

7.8
Models Involving Bioregeneration of Activated Carbon

As discussed in Chapter 6, over the years many models have been developed to explain and predict the behavior of biologically active fixed- and fluidized-bed GAC

Table 7.3 Summary of bioregeneration models.

Substrate	Diffusion	Adsorption Equilibrium	Microbial Kinetics	Reference
Valeric acid	No external diffusion, Biofilm, Intraparticle	Langmuir isotherm	First-order	[96]
Single (phenol or p-nitrophenol)	External, Biofilm, Intraparticle	Freundlich Isotherm	Monod	[64]
Single (phenol)	External	Freundlich Isotherm	Haldane	[2]
Multi-solute MDBA model	External, Biofilm, Intraparticle	IAST (Ideal Adsorbed Solution Theory)	Monod	[97]
Single (phenol), Bisolute (phenol and p-cresol)	Biofilm, Intraparticle	Multi-solute Fritz-Schluender isotherm	Haldane	[34]
Single and Bisolute (phenol and 2,4-dichlorophenol)	–	Freundlich Isotherm	–	[63]
Single (phenol, p-nitrophenol or toluene)	External, Biofilm, Intraparticle	Freundlich Isotherm	Monod	[98]
Single (phenol or p-nitrophenol)	External	–	First-order	[99]

adsorbers. On the other hand, only a few of them address bioregeneration of activated carbon. However, in order to have further insight into the observed performance, the models must also satisfactorily predict biodegradation of adsorbed substrate and bioregeneration of activated carbon. In modeling of bioregeneration, the same basic equations are valid as those presented in Section 6.1.2 in Chapter 6. Recognizing the features influencing bioregeneration as well as developing models that include bioregeneration would also help in understanding these phenomena in full-scale systems. Prevalent bioregeneration models to date are summarized in Table 7.3. These models are described in the following subsections.

7.8.1
Modeling of Bioregeneration in Concurrent Adsorption and Biodegradation

In the early 1970s Weber and co-workers interpreted *in-situ* bioregeneration by postulating a multilayer around the activated carbon particle. According to their concept, the first layer around a carbon particle consisted of an anaerobic biofilm, while the next layer, in contact with bulk liquid, was aerobic. However, the researchers emphasized that further studies are required to define the mechanism of bioregeneration [100].

An attempt to model bioregeneration of activated carbon was made by Andrews and Tien [96]. According to this model, in a thin biofilm layer grown on activated carbon, the substrate concentration at the biofilm–carbon interface is at S_{fc}, as

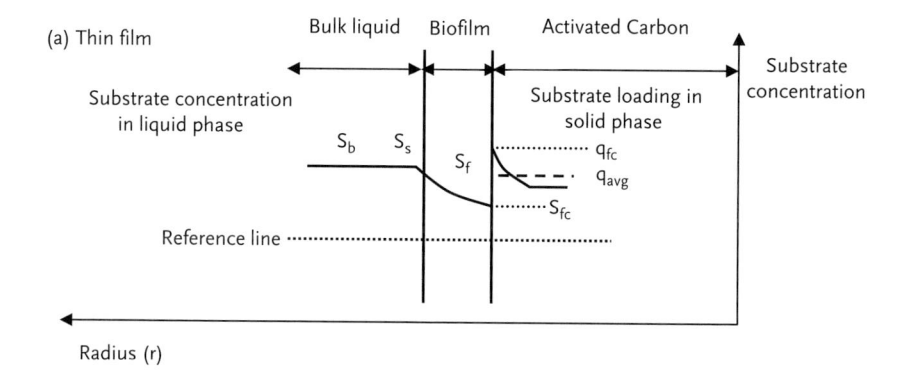

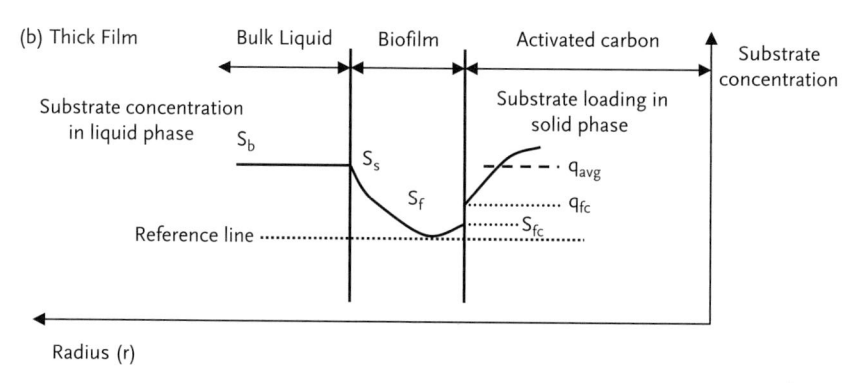

Figure 7.8 Growth of (a) thin and (b) thick film on the surface of activated carbon and substrate profiles in liquid and solid phases (redrawn after [96]).

shown in Figure 7.8a [96]. The corresponding solid phase concentration is shown as q_{fc}. The driving force for adsorption is the difference between the equilibrium adsorbate concentration q_{fc} and the average adsorbate concentration, q_{avg}, inside the carbon, as shown in Figure 7.8a. However, in the case of a thick biofilm, due to consumption of substrate by biological reactions, inside the biofilm the substrate concentration may decrease to very low levels, as shown in Figure 7.8b. In such a case, the solid-phase concentration q_{fc} at the biofilm–carbon interface is also reduced, while that inside the carbon particle, q_{avg}, is at an elevated level due to adsorption of substrate. This reversed concentration gradient then leads to desorption of substrate from the carbon into the biofilm, which is the basic requirement of bioregeneration [96].

Using valeric acid as a model substrate, Andrews and Tien showed that the growth of bacteria to a thick biofilm on activated carbon led to this phenomenon known as bioregeneration [96]. Their work is significant in the sense that it showed the interactions between the two processes, namely adsorption and biofilm growth.

7.8.2
Modeling of Bioregeneration in Single Solute Systems

The main BAC models that incorporate bioregeneration of activated carbon were developed by Speitel and co-workers. The first model developed by this group was used in prediction of the sorbed as well as the bulk liquid concentration of the substrate at specific locations of a GAC column [64]. The model considered a single substrate, phenol or p-nitrophenol at trace concentrations. The model incorporated the same basic processes as those discussed in Chapter 6, namely the liquid film resistance, diffusion through the biofilm, and biodegradation according to Monod kinetics. The model considered growth and decay of the biofilm over time, but excluded the transport resistance at the biofilm–carbon interface. A thin, flat-plate biofilm was assumed that consisted of a homogeneous matrix of microorganisms and EPS. According to the model, the biomass growth in suspended phase was negligible. The model assumed homogeneous spherical GAC particles and reversible adsorption. The intraparticle transport of substrate was expressed by the homogeneous surface diffusion model, whereas adsorption was described by the Freundlich isotherm. In the reactor mass balance, the dispersion term present in Eq. (6.38) in Chapter 6 was neglected, and the time-dependent variation of substrate concentration in the liquid phase of the reactor was described as follows:

$$\frac{\partial S_b}{\partial t} = -v_i \frac{\partial S_b}{\partial z} - \frac{3(1-\varepsilon_B)}{\varepsilon_B} \frac{(R_p + L_f)^2}{R_p^3} k_{fc}(S_b - S_s) \qquad (7.14)$$

where

S_b = substrate concentration in the bulk liquid (M_s/L^3),
S_s = substrate concentration at the liquid-biofilm interface (M_s/L^3),
ε_B = void ratio in the bed (unitless),
R_p = radius of the carbon particle (L),
L_f = biofilm thickness (L),
k_{fc} = liquid film or external mass transfer coefficient (L/T).
v_i = interstitial velocity, (L/T)
z = distance along the flow path, (L)
t = time (T)

In the model of Speitel and co-workers, the substrate mass balance in the biofilm phase involves the same equation as Eq. (6.11) in Chapter 6. On the other hand, the variation in biofilm thickness L_f is modeled as follows [64]:

$$\frac{\partial L_f}{\partial t} = k_{max} Y \int_{R_p}^{R_p + L_f} \left(\frac{S_f}{K_s + S_f} dr \right) - k_d L_f \qquad (7.15)$$

where

k_{max} = maximum specific substrate utilization rate $\left(\frac{\mu_{max}}{Y}\right)$ ($M_s/M_x \cdot T$),
k_d = biomass loss coefficient accounting for both decay and shearing (1/T),

Y = yield constant (M_x/M_s),
K_s = half-velocity constant (M_s/L^3),
S_f = substrate concentration inside the biofilm (M_s/L^3),
r = radial coordinate extending from the center of activated carbon particle (L),
R_p = radius of activated carbon particles (L).

The performance of the model was compared with experimental data obtained by the measurement of radiolabeled phenol or p-nitrophenol and CO_2. The model was initialized after an acclimation period during which sufficient biomass attachment was achieved. The model was able to adequately predict bioregeneration rates only after reaching a peak bioregeneration rate and could not predict well the rates during the initial phase.

However, in this model the predicted liquid phase concentrations were higher than measured values. Therefore, the model was prone to underestimate the total extent of bioregeneration in the long term. This model was not largely predictive because of an inadequate understanding of biofilms, including density, diffusional transport resistance, and loss rate through shearing at the time of the study.

Speitel and Zhu performed sensitivity analyses for parameter values such as the half-velocity constant K_s, surface diffusion coefficient (D_s), and the amount of biomass initially attached to the biomass [101]. The meaning of these parameters for a biofilm–carbon system is discussed in Chapter 6. The researchers tested three biodegradation–adsorption models to examine alternative formulations of mass transport resistances within the biofilm and at the biofilm–GAC interface. For each model, the hypothetical concentration profiles during bioregeneration are shown in Figure 7.9. The 'biofilm' model adopted was that initially developed by Speitel and co-workers [64], which assumed no mass transfer resistance at the biofilm–GAC interface (Figure 7.9a). Another model is the 'one-liquid film' model proposed by Ying and Weber, which included only one-liquid film transport resistance between the bulk liquid and the biofilm (Figure 7.9b) [102]. However, in this model the mass transfer resistance within the biofilm and at the biofilm–GAC interface are neglected. This model was mainly applicable to GAC columns having very thin biofilms where desorbed substrate is readily available to microorganisms.

The 'two-liquid film' model was developed by Speitel and Zhu [101] using the previous model of Speitel and co-workers [64], which could not simultaneously

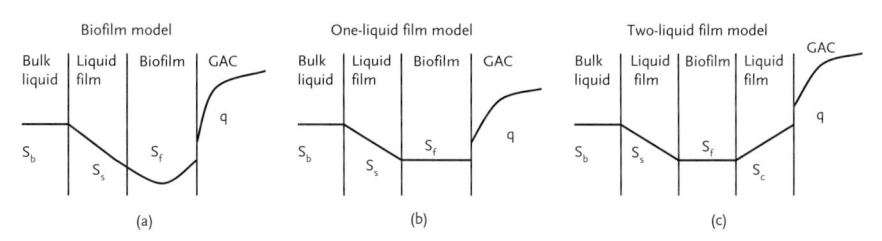

Figure 7.9 Concentration profiles in three different models during bioregeneration (redrawn after [101]).

provide a satisfactory prediction of bioregeneration rate and effluent concentration. The previous 'biofilm' model gave unsatisfactory results because of inaccurate parameter values and an inaccurate description of mass transport resistance. It assumed that microorganisms were present as a thick, constant biofilm [64]. However, in reality, at low substrate concentrations a true biofilm may not exist, and microbial growth may occur in the form of scattered colonies on GAC surface. This difference was important in accounting for mass transport resistances and affected the predicted bioregeneration rates [101].

The 'two-liquid film' model aimed to account for scattered surface growth of microorganisms (Figure 7.9c) [101]. Scattered surface growth means that substrate desorbing from GAC moves first into the liquid phase and travels some distance before reaching the biofilm. Mass transfer resistance occurs within this hypothetical liquid film; this factor was previously not considered in the first model [64]. The mathematical formulation of scattered growth involved the assumption of a very thin biofilm around each GAC particle. The diffusive transport resistance within the thin biofilm was assumed to be negligible. Therefore, the mathematical expression for substrate concentration in the biofilm was changed to exclude the diffusive transport resistance within the biofilm Eq. (7.16) as in the case of the 'one-liquid film' model developed by Ying and Weber [102]. In addition, all biomass at any axial position in the GAC column was assumed to be exposed to the same substrate concentration. The following equation was used in describing the change in substrate concentration inside the biofilm:

$$\frac{\partial S_f}{\partial t} = \frac{k_{fc}}{L_f}(S_b + S_c - 2S_f) - \frac{k_{max} X_f S_f}{K_s + S_f} \tag{7.16}$$

where

S_c = substrate concentration at the liquid GAC interface (M_s/L^3),
X_f = density of biomass within biofilm (M_x/L^3).

Simulations of the 'biofilm' (Figure 7.9a) and 'one-liquid film' models (Figure 7.9b) resulted in good agreement with experimental data with regard to bioregeneration rate and effluent concentrations. The 'one-liquid film' model resulted in a better fit compared with the 'biofilm' model, because scattered surface growth of microorganisms prevailed at low substrate concentrations, and diffusive transport resistance was negligible within the biofilm. Model results did not fit the experimental bioregeneration rates when an additional liquid film mass transfer resistance was considered between the GAC and the biofilm in the two-liquid film model (Figure 7.9c). The 'two-liquid film' model included an excessive mass transport resistance at the GAC–biofilm interface although it was actually negligible under conditions that favored scattered surface growth of microorganisms. The simplest model, the 'one-liquid film' model, provided the best fits [101]. However, the drawbacks of the models were the difficulty of determining the amount of initial biomass and neglecting irreversible adsorption. Irreversible adsorption leads to overestimation of bioregeneration. This is a problem commonly encountered in the modeling of bioregeneration.

The Biofilm on Activated Carbon (BFAC) model, previously mentioned in Section 6.1.5.2 in Chapter 6, incorporates the mechanisms of liquid film transfer, biodegradation, and adsorption of a substrate, as well as biofilm growth [103]. The solution of the model quantitatively illustrated the mechanism of bioregeneration [104]. Fluxes across the biofilm–carbon interface showed a complete reversal, with phenol diffusing out of the activated carbon during the period of bioregeneration, when the phenol concentration was reduced at the biofilm–carbon interface as the biofilm grew thicker. The BFAC model accurately described the substrate concentration and the sequence of bioregeneration.

As discussed in Section 7.6.6, in biological systems the CO_2 production can be related to microbial substrate consumption. Chang and Rittmann used substrate and CO_2 measurements when they studied bioregeneration using BAC media and glass beads. Glass beads were used as supporting media for biofilm growth and served as a control of the BAC reactor [104]. Tests with glass beads were representing 'biodegradation' only since glass beads did not adsorb phenol. On the other hand, in BAC media both adsorption and biodegradation were taking place. The conceptual drawing in Figure 7.10 illustrates that in both systems, namely BAC and glass beads with biofilm growth, the decrease in substrate concentration is accompanied by an increase in CO_2 production. The substrate concentration in the case of glass beads exhibited a profile which is typical for a nonadsorbing biofilm reactor operating at non-steady-state conditions in the startup period (Figure 7.10a). On the other hand, the substrate concentration in the effluent of the BAC reactor was low at the startup because of adsorption. Due to the exhaustion of GAC adsorption capacity in the course of time, the effluent substrate concentration increased steadily and reached a peak indicating the breakthrough point of GAC. Later, the substrate concentration decreased steadily since biodegradation prevailed. The substrate concentration then leveled off when equilibrium was established between adsorption, biodegradation, and bioregeneration (Figure 7.10a). However, as seen in Figure 7.10b the production of CO_2 from phenol mineralization was at its peak even after the substrate concentration leveled off, indicating bioregeneration of activated carbon. Following the peak, CO_2 production declined and reached nearly a constant level equal to the one in control (glass bead) reactors. The peak in BAC indicated that in this reactor more phenol was available to the biofilm than in the case of the nonadsorbing glass beads. This extra CO_2 production was explained by desorption of previously adsorbed phenol which was subsequently mineralized by the biofilm. After reaching the peak value, bioregeneration started to decline due to the decrease in adsorbed phenol (Figure 7.10b).

Biodegradation and bioregeneration models are also developed for biological systems that are assisted by PAC. For example, similarly to the case of GAC, a model was developed for completely mixed membrane bioadsorbers in which PAC served as an attachment medium [98]. The following processes were taken into consideration: (i) biological reaction in the bulk liquid, (ii) film transfer from bulk liquid phase to the biofilm, (iii) diffusion with biological reaction inside biofilm, (iv) adsorption equilibrium at the biofilm-adsorbent interface, and (v) surface

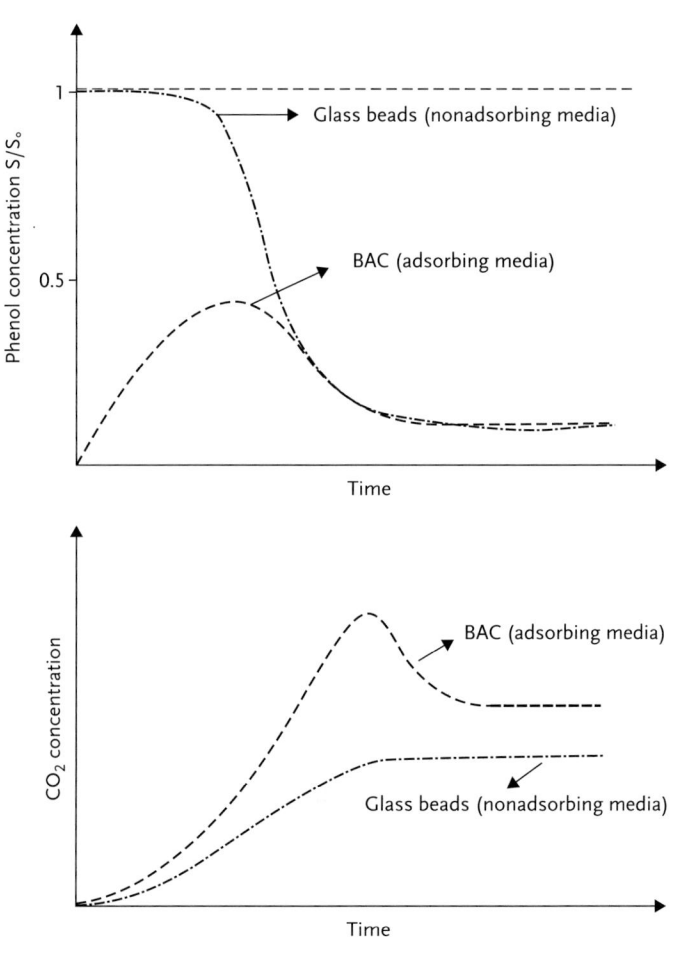

Figure 7.10 Non-steady-state (a) substrate (upper figure) and (b) CO_2 concentrations in the case of biological activated carbon (BAC) and glass beads with biofilm growth (lower figure) (conceptual drawing based on [104]).

diffusion within PAC particles. The model simulated the process dynamics for the model compounds phenol, p-nitrophenol (PNP), and toluene. The model sensitivity was tested with respect to adsorption mass transfer coefficients (surface and pore diffusion coefficients D_s and D_p, respectively), adsorption equilibrium parameters (K_F and $1/n$), biological kinetic parameters (Y, K_s, k_{max}), biofilm diffusion coefficient (D_f), the initial and maximum biofilm thickness (L_f), the influent substrate concentration (S_0), and the hydraulic retention time (HRT). The process efficiency was relatively insensitive to adsorption and biofilm transport parameters.

The relative contributions of adsorption and biodegradation to contaminant removal were investigated to gain an insight into bioregeneration of carbon.

Therefore, simulation studies were conducted for three different cases: (i) adsorption and biodegradation inside the biofilm and by suspended biomass, (ii) absence of adsorption, but biodegradation inside the biofilm and by suspended biomass, and (iii) adsorption alone. The effluent substrate concentrations under three different scenarios are illustrated in Figure 7.11 for each substrate.

During the initial phase of operation, adsorption prevailed over biological activity for each compound. As concluded from the respective Freundlich isotherm parameters, PNP and phenol adsorb better than toluene. Correspondingly, Figure 7.11b and c demonstrates that during the initial phase, better removals are predicted for PNP and phenol than toluene (Figure 7.11a). As time progressed, biological activity became more significant as the microorganisms began utilizing the available substrate for biofilm growth and eventually dominated over adsorption even before the biofilm was fully developed. After biodegradation becomes predominant, the steady-state removal is governed by biodegradation rates. Under steady-state conditions, removals exceeding 99.9% are predicted for the three model contaminants.

In this system, activated carbon stored the substrate when biological activity was low, and released it when biodegradation became dominant. Comparisons of the adsorbent capacity utilization for the 'adsorption' and 'adsorption and biodegradation' cases provided estimates of adsorbent bioregeneration for PNP, phenol, and toluene as 75%, 60%, and 60%, respectively. However, in this MBR system biodegradation by attached or suspended biomass could be not distinguished from each other.

In another study, a modeling approach for BAC also showed that during the initial period of operation, a steady mass flux of TCE was transported into the GAC particle due to diffusion and adsorption. The aqueous substrate concentration in the vicinity of the biofilm/particle interface gradually became lower than that of the solid concentration as the system gradually became bioactive. This caused the substrate to diffuse out of the carbon for biofilm degradation [105].

7.8.3
Modeling of Bioregeneration in Multicomponent Systems

In reality, single solute systems hardly exist, either in wastewater or in water treatment. Therefore, mathematical models have been developed to predict the bioregeneration in BAC columns treating mixtures of biodegradable and nonbiodegradable organic compounds [56]. Erlanson and co-workers described a two-component equilibrium-based model referred to as the biodegradation/adsorption-screening model (BASM) [83]. The model considered only one biodegradable and one nonbiodegradable chemical. When the nonbiodegradable chemical controlled the service life of the activated carbon column, the only significant gain in service life occurred when both chemicals had similar adsorbabilities. On the other hand, if the biodegradable chemical controlled the service life, the service life was 1.2- to 7-fold higher than in the case of adsorption alone, depending on the relative adsorbability of the two chemicals. Thus, the service life

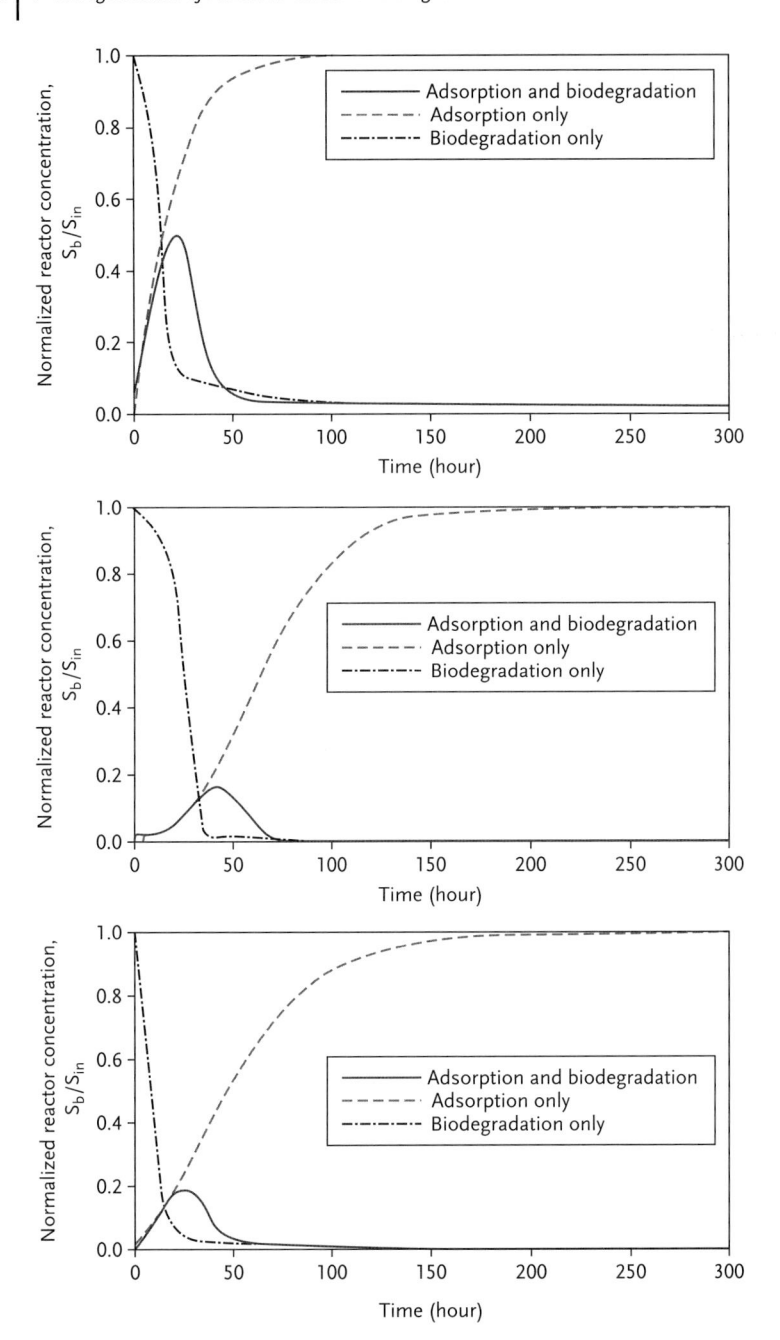

Figure 7.11 Normalized effluent profiles in PAC-assisted MBR for (a) toluene (upper figure) (b) PNP (middle figure) and (c) phenol (lower figure) under three different scenarios [98]. (permission received).

could be maximized, ensuring that biodegradation began as soon as possible after startup. Putz and co-workers suggested that modeling becomes much simpler by considering only equilibrium situations in BAC columns compared to kinetic models [56]. Equilibrium conditions reflect the steady-state situation, whereas kinetic models also consider non-steady conditions, which makes modeling more complex. The BASM model helped to determine in which cases it was possible to benefit from employing simultaneous biodegradation and adsorption rather than solely adsorption by applying several hundreds of hypothetical scenarios.

Speitel and co-workers also developed a kinetic model called the Multiple-Component Biofilm Diffusion Biodegradation and Adsorption model (MDBA), which described both adsorption and biodegradation in multicomponent GAC columns [97]. The MDBA model combined the single component adsorption and biodegradation model developed by the same group [64] with the Ideal Adsorbed Solution Theory (IAST). Details of the IAST are presented in Chapter 2. The MDBA model adjusted the IAST equation by using a correction factor to account for differences between predicted and measured equilibrium concentrations, assumed homogeneous surface diffusion inside activated carbon considering that pore diffusion was insignificant, and assumed that biodegradation of multicomponents occurred simultaneously in order to simplify the model. In another study, a good correlation was found between the MDBA model fits and the measured effluent concentrations and between the simulated and measured loadings on activated carbon [56]. The authors used the model to predict the cumulative bioregeneration over time [56]. In that study, bioregeneration was measured to take place more rapidly than in the simulated model, but the same value was reached at the end.

The Freundlich adsorption concept as used by Vinitnantharat and co-workers [11] for quantification of bioregeneration was employed by the same group in another study [63] to develop a predictive isotherm model. The model aimed to evaluate the extent of bioregeneration, as described in Section 7.6.2. In this study, the modeling of bioregeneration was conducted with phenol and 2,4-dichlorophenol in both single- and bisolute systems and also in the presence and absence of biodegradation by-products. The loadings on activated carbon in batch and BAC-Sequencing Batch Reactors (BAC-SBR) were estimated using Freundlich isotherm constants. The results were compared with experimental loadings that were obtained by the direct measurement method, in which the solute was extracted with methylene chloride, as described in Section 7.6.4. When metabolic by-products were excluded from isotherms, the loadings on carbon were overestimated. However, when metabolic by-products were included in isotherms, the model predicted bioregeneration with reasonable consistency, in the case of both single and bisolute systems and in batch and BAC-SBR reactors. The study indicated that metabolic intermediates and by-products should be taken into account in the modeling of bioregeneration [63]. This necessity arises because by-products forming as a result of incomplete biodegradation may decrease the adsorption capacity of activated carbon for target compounds.

7.8.4
Modeling of Offline Bioregeneration

The properties of an offline bioregeneration system are outlined in Section 7.2. Goeddertz and co-workers presented a predictive model in an offline GAC bioregeneration system using phenol as the model compound [2]. The Freundlich adsorption isotherm constants, the external mass transfer through the biofilm, and the Haldane type biodegradation kinetics were incorporated into the model. The model successfully predicted the bulk liquid substrate concentrations as well as the extent of bioregeneration. The authors stated that significant bioregeneration occurred when the desorption step was rate-limiting rather than biodegradation, as shown with experimental data and the predicted model.

Another model was developed to predict the offline bioregeneration for single (phenol) and bisolute (phenol and p-cresol) systems in a fixed-bed GAC adsorber [34]. The model included liquid film and intraparticle mass transfer. It described the adsorption equilibrium by the multisolute Fritz–Schluender isotherm. Further, it assumed plug flow conditions in the GAC column and CSTR conditions in the separate bioregeneration reactor. Microbial growth was described by Haldane kinetics for the single substrate, whereas a simplified growth model was used that assumed a constant specific growth rate for the two components in the bisolute system. The researchers suggested that the model could be used to predict the breakthrough behavior of both fresh and bioregenerated activated carbon. However, in multisolute systems, the model was only applicable when growth parameters belonging to each substrate, such as phenol and p-cresol, were similar. The model would not be useful in the case of two compounds with very different molecular structures.

In a subsequent study, this model could not predict the breakthrough of a bioregenerated GAC when a microbial culture was recirculated through the GAC column [35]. The reason was microbial fouling of carbon pores, which significantly deteriorated the carbon capacity. However, in the same study, the model agreed well with the findings of GAC bioregeneration when microorganism-free filtered permeate was recirculated through the GAC column. In that case, the presence of a separate bioreactor prevented microbial fouling. In the bisolute system, the model tended to overpredict effluent phenol concentrations and slightly underpredict effluent p-cresol levels.

7.8.5
Modeling the Kinetics of Bioregeneration

Vinitnantharat and co-workers calculated the bioregeneration rate constant assuming that biodegradation followed first-order kinetics [11]. The first-order kinetics was also used in other studies to define bioregeneration with respect to adsorbed quantity [13, 24, 54]. According to first-order bioregeneration, the rate of bioregeneration at time t is proportional to the amount of adsorbate left over on the surface of GAC. Hence, the renewal of active sites occurres very fast at the beginning and declines gradually. This can be shown by the following equations:

$$dq'/dt = -k.q' \qquad (7.17)$$

$$\ln q'/q_1 = -kt \qquad (7.18)$$

where

q_1 = amount of adsorbed substrate at the beginning of bioregeneration (M_s/M_c),
q' = amount of adsorbed substrate left at time t during bioregeneration (M_s/M_c),
k = first-order bioregeneration rate constant (1/T).

A first-order kinetic model was used to describe the bioregeneration of PAC that was loaded either with phenol or p-nitrophenol [99]. The kinetic model used the bulk phenol or p-nitrophenol concentrations during the bioregeneration process. The aim was to determine the first-order rate constants of desorption and biodegradation. The model considered only desorption and biodegradation of loaded PAC, but not the initial adsorption step. The results showed that the kinetic model fitted to experimental data. The biodegradation rate constant was much higher than the desorption rate constant. Hence, desorption came out as the rate-determining step in bioregeneration. Therefore, the desorption rate could be introduced into the model to express the bioregeneration rate. The change of substrate concentration in the bulk liquid during bioregeneration was expressed in Eq. (7.19).

$$\frac{dS_b}{dt} = mk_{des}q' - k_{bio}S_b \qquad (7.19)$$

where,

S_b = bulk liquid substrate concentration (M_s/L^3),
m = PAC dose (M_c/L^3),
k_{des} = first-order desorption rate constant during bioregeneration (1/T),
k_{bio} = first-order biodegradation rate constant (1/T).

In another study, the rate of bioregeneration was found to be dependent on substrate type [54]. Bioregeneration was slower when the substrate consisted of a mixture of substituted phenols than when it consisted of phenol alone. In the study of Vinitnantharat and co-workers, the first-order bioregeneration rate was found to be lower for 2,4-dichlorophenol than for phenol, which is more biodegradable [11].

Literature studies have pointed out that mathematical modeling of bioregeneration is a challenging task that requires special attention. Particularly, predictive modeling approaches provide useful design data for optimization of bioregeneration processes. Several studies, outlined above, have led to advances in the modeling of bioregeneration. However, further studies on modeling are still needed, particularly for multicomponent systems with more than two adsorbates.

References

1 Aktaş, Ö. and Çeçen, F. (2007) Bioregeneration of activated carbon: a review. *International Biodeterioration & Biodegradation*, **59**, 257–272.
2 Goeddertz, J.G., Matsumoto, M.R., and Weber, A.S. (1988) Offline bioregeneration of granular activated carbon. *Journal of Environmental Engineering*, **114**, 1063–1076.
3 Holst, J., Martens, B., Gulyas, H., Greiser, N., and Sekoulov, I. (1991) Aerobic biological regeneration of dichloromethane-loaded activated carbon. *Journal of Environmental Engineering*, **117**, 194–208.
4 Scholz, M. and Martin, R.J. (1997) Ecological equilibrium on biological activated carbon. *Water Research*, **31**, 2959–2968.
5 Roy, D., Maillacheruvu, K., and Mouthon, J. (1999) Bioregeneration of granular activated carbon loaded with 2,4-D. *Journal of Environmental Science and Health*, **34**, 769–791.
6 Silva, M., Fernandes, A., Mendes, A., Manaia, C.M., and Nunes, O.C. (2004) Preliminary feasibility study for the use of an adsorption/bioregeneration system for molinate removal from effluents. *Water Research*, **38**, 2677–2684.
7 Sublette, K.L., Snider, E.H., and Sylvester, N.D. (1982) A review of the mechanism of powdered activated carbon enhancement of activated sludge treatment. *Water Research*, **16**, 1075–1082.
8 Speitel, G.E. and Digiano, F.A. (1987) The bioregeneration of GAC used to treat micropollutants. *Journal of American Water Works Association*, **79**, 64–73.
9 de Jonge, R.J., Breure, A.M., and van Andel, J.G. (1996) Bioregeneration of powdered activated carbon (PAC) loaded with aromatic compounds. *Water Research*, **30** (4), 875–882.
10 Ha, S.R. and Vinitnantharat, S. (2000) Competitive removal of phenol and 2,4-dichlorophenol in biological activated carbon system. *Environmental Technology*, **21**, 387–396.
11 Vinitnantharat, S., Baral, A., Ishibashi, Y., and Ha, S.R. (2001) Quantitative bioregeneration of granular activated carbon loaded with phenol and 2,4-dichlorophenol. *Environmental Technology*, **22**, 339–344.
12 Schultz, J.R. and Keinath, T.M. (1984) Powdered activated carbon treatment process mechanisms. *Journal of Water Pollution Control Federation*, **56** (2), 143–151.
13 Kim, D., Miyahara, T., and Noike, T. (1997) Effect of C/N ratio on the bioregeneration of biological activated carbon. *Water Science and Technology*, **36**, 239–249.
14 Klimenko, N., Winther-Nielsen, M., Smolin, S., Nevynna, L., and Sydorenko, J. (2002) Role of the physico-chemical factors in the purification process of water from surface-active matter by biosorption. *Water Research*, **36**, 5132–5140.
15 de Sa, F.A.C. and Malina, J.F. Jr (1992) Bioregeneration of granular activated carbon. *Water Science and Technology*, **26**, 2293–2295.
16 Aktaş, Ö. (2006) Bioregeneration of activated carbon in the treatment of phenolic compounds. Ph.D. Thesis, Bogazici University, Institute of Environmental Sciences, Istanbul, Turkey.
17 Perrotti, A.E. and Rodman, C.A. (1974) Factors involved with biological regeneration of activated carbon. *American Institute of Chemical Engineering Symposium Series*, **70**, 316–325.
18 Sirotkin, A.S., Koshkina, L.Y., and Ippolitov, K.G. (2001) The BAC process for treatment of waste water containing non-iogenic synthetic surfactants. *Water Research*, **35** (13), 3265–3271.
19 Speitel, G.E., Lu, C.J., Turakhia, M., and Zhu, X.J. (1989) Biodegradation of trace concentrations of substituted phenols in granular activated carbon

columns. *Environmental Science and Technology*, **23**, 66–74.
20 Olmstead, K.P. and Weber, W.J. (1991) Interactions between microorganisms and activated carbon in water and waste treatment operations. *Chemical Engineering Communications*, **108**, 113–125.
21 Li, A.Y.L. and DiGiano, F.A. (1983) Availability of sorbed substrate for microbial degradation on granular activated carbon. *Journal of Water Pollution Control Federation*, **554**, 392–399.
22 Xiaojian, Z., Zhansheng, W., and Xiasheng, G. (1991) Simple combination of biodegradation and carbon adsorption. Mechanism of the biological activated carbon process. *Water Research*, **25** (2), 165–172.
23 Martin, M.J., Artola, A., Balaguer, M.D., and Rigola, M. (2002) Enhancement of the activated sludge process by activated carbon produced from surplus biological sludge. *Biotechnology Letters*, **24**, 163–168.
24 Aktaş, Ö. and Çeçen, F. (2006) Effect of activation type on bioregeneration of various activated carbons loaded with phenol. *Journal of Chemical Technology and Biotechnology*, **81**, 1081–1092.
25 Klimenko, N., Smolin, S., Grechanyk, S., Kofanov, V., Nevynna, L., and Samoylenko, L. (2003) Bioregeneration of activated carbons by bacterial degraders after adsorption of surfactants from aqueous solutions. *Colloids and Surfaces, A*, **230**, 141–158.
26 Benedek, A. (1980) Simultaneous biodegradation and activated carbon adsorption – a mechanistic look. In Activated Carbon Adsorption of Organics from the Aqueous Phase, vol. 2 (eds I.H. Suffet and M.J. McGuire), Ann Arbor Science Publishers, Inc., Michigan, USA.
27 Andrews, G.F. and Tien, C. (1980) The inter-reaction of bacterial growth, adsorption and filtration in carbon columns treating liquid waste. *American Institute of Chemical Engineering Symposium Series*, **152**, 164–175.

28 Aktaş, Ö. and Çeçen, F. (2006) Effect of type of carbon activation on adsorption and its reversibility. *Journal of Chemical Technology and Biotechnology*, **81**, 94–101.
29 Maloney, S.W., Bancroft, K., and Suffet, I.H. (1983) Comparison of adsorptive and biological total organic carbon removal by granular activated carbon in potable water treatment, in *Treatment of Water by Granular Activated Carbon* (eds M.J. Mc Guire and I.H. Suffet), American Chemical Society, Washington, DC, USA, pp. 279–302.
30 Ng, A. and Stenstrom, M.K. (1987) Nitrification in powdered activated carbon-activated sludge process. *Journal of Environmental Engineering*, **113**, 1285–1301.
31 Imai, A., Onuma, K., Inamory, Y., and Sudo, R. (1995) Biodegradation and adsorption in refractory leachate treatment by the biological activated carbon fluidized bed process. *Water Research*, **29**, 687–694.
32 Kolb, F.R. and Wilderer, P.A. (1997) Activated carbon sequencing batch biofilm reactor to treat industrial wastewater. *Water Science and Technology*, **35**, 169–176.
33 Klimenko, N., Marutovsky, R.M., Pidlisnyuk, V.V., Nevinnaya, L.V., Smolin S.K., Kohlmann, J., and Radeke, K.H. (2002) Biosorption processes for natural and wastewater treatment-Part I: Literature review. *Engineering and Life Sciences*, **2**, 10, 317–323.
34 Hutchinson, D.H. and Robinson, C.W. (1990) A microbial regeneration process for granular activated carbon-I. Process modelling. *Water Research*, **24**, 1209–1215.
35 Hutchinson, D.H. and Robinson, C.W. (1990) A microbial regeneration process for granular activated carbon-II. Regeneration studies. *Water Research*, **24**, 1217–1223.
36 Irvine, R.L., Ketchum, L.H., Wilderer, P.A., and Montemagno, C.D. (1992) Granular Activated Carbon-Sequencing Batch Biofilm Reactor. U.S. Patent No: 5,126,050.

37 Kzenovich, J.P., Daniels, J.I., Stenstrom, M.K., and Heilmann, H.M. (1994) Chemical and biological systems for regenerating activated carbon contaminated with high explosives. Proceedings Demil'94, November 14–16, 1994, Luxembourg.

38 Morley, M.C. and Speitel, G.E. (1999) Biodegradation of High Explosives on Granular Activated Carbon: Enhanced Desorption of High Explosives from GAC – Batch Studies. Report No: ANRCP-1999-11, Amarillo National Resource Center for Plutonium, Texas, USA.

39 Chudyk, W.A. and Snoeyink, V.I. (1984) Bioregeneration of activated carbon saturated with phenol. *Environmental Science and Technology*, **18** (1), 1–5.

40 Seo, G.T., Ohgaki, S., and Suzuki, Y. (1997) Sorption characteristics of biological powdered activated carbon in BPAC-MF (biological powdered activated carbon-microfiltration) system for refractory organic removal. *Water Science and Technology*, **35**, 163–170.

41 Lee, K.M. and Lim, P.E. (2005) Bioregeneration of powdered activated carbon in the treatment of alkyl-substituted phenolic compounds in simultaneous adsorption and biodegradation process. *Chemosphere*, **58**, 407–416.

42 Çeçen, F. (1994) Activated carbon addition to activated sludge in the treatment of kraft pulp bleaching wastes. *Water Science and Technology*, **30** (3), 183–192.

43 Bornhardt, C., Drewes, J.E., and Jekel, M. (1997) Removal of organic halogens (AOX) from municipal wastewater by powdered activated carbon (PAC)/activated sludge (AS) treatment. *Water Science and Technology*, **35**, 147–153.

44 Garner, I.A., Watson-Craik, I.A., Kirkwood, R., and Senior, E. (2001) Dual solute adsorption of 2,4,6-trichlorophenol and N-(2-(2,4,6-trichlorophenoxy) propyl amine on to activated carbon. *Journal of Chemical Technology and Biotechnology*, **76**, 932–940.

45 Badriyha, B.N., Ravindran, V., Den, W., and Pirbazari, M. (2003) Bioadsorber efficiency, design and performance forecasting for alachor removal. *Water Research*, **37**, 4051–4072.

46 Vuoriranta, P. and Remo, S. (1994) Bioregeneration of activated carbon in a fluidized GAC bed treating bleached kraft mill secondary effluent. *Water Science and Technology*, **29**, 239–246.

47 Thacker, W.E., Snoeyink, V.I., and Crittenden, J.C. (1983) Desorption of compounds during operation of GAC adsorption systems. *Journal of American Water Works Association*, **75**, 144–149.

48 Walker, G.M. and Weatherley, L.R. (1997) A simplified predictive model for biologically activated carbon fixed beds. *Process Biochemistry*, **32**, 327–335.

49 Yonge, D.R., Keinath, T.M., Poznanska, K., and Jiang, Z.P. (1985) Single-solute irreversible adsorption on granular activated carbon. *Environmental Science and Technology*, **19**, 690–694.

50 de Jonge, R.J., Breure, A.M., and van Andel, J.G. (1996) Reversibility of adsorption of aromatic compounds onto powdered activated carbon. *Water Research*, **30** (4), 883–892.

51 Vidic, R., Suidan, M.T., and Brenner, R.C. (1993) Oxidative coupling of phenols on activated carbon: impact on adsorption equilibrium. *Environmental Science and Technology*, **27**, 2079–2085.

52 Orshansky F. and Narkis, N. (1997) Characteristics of organics removal by PACT simultaneous adsorption and biodegradation. *Water Research*, **31**, 391–398.

53 Klimenko, N., Savchina, L.A., Kozyatnik, I.P., Goncharuk, V.V., and Samsoni-Todorov, A.O. (2009) The effect of preliminary ozonization on the bioregeneration of activated carbon during its long term service. *Journal of Water Chemistry and Technology*, **31** (4), 220–226.

54 Ivancev-Tumbas, I., Dalmacija, B., Tamas, Z., and Karlovic, E. (1998) Reuse of biologically regenerated activated carbon for phenol removal. *Water Research*, **32**, 1085–1094.

55 Sirotkin, A.S., Ippolitov, K.G., and Koshkina, L.Y. (2002) Bioregeneration of activated carbon in BAC filtration. Proceedings of Biological Activated Carbon Filtration IWA Workshop, 29–31 May 2002, Delft University of Technology, Delft, the Netherlands.

56 Putz, A.R.H., Losh, D.E., and Speitel, G.E. Jr (2005) Removal of nonbiodegradable chemicals from mixtures during granular activated carbon bioregeneration. *Journal of Environmental Engineering*, **131** (2), 196–205.

57 DeWalle, F.B. and Chian, E.S. (1977) Biological regeneration of powdered activated carbon added to activated sludge units. *Water Research*, **11**, 439–446.

58 Kim, B.R., Chian, E.S.K., Cross, W.H., and Cheng, S. (1986) Adsorption, desorption and bioregeneration in an anaerobic granular activated carbon reactor for the removal of phenol. *Journal of Water Pollution Control Federation*, **58**, 35–40.

59 Aktaş, Ö. and Çeçen, F. (2009) Cometabolic bioregeneration of activated carbons loaded with 2-chlorophenol. *Bioresource Technology*, **100**, 4604–4610.

60 Aktaş, Ö. and Çeçen, F. (2007) Adsorption, desorption and bioregeneration in the treatment of 2-chlorophenol with activated carbon. *Journal of Hazardous Materials*, **141**, 769–777.

61 Li, P.H.Y., Roddick, F.A., and Hobday, M.D. (1998) Bioregeneration involving a coal based adsorbent used for removing nitrophenol from water. *Journal of Chemical Technology and Biotechnology*, **73**, 405–413.

62 Tanthapanichakoon, W., Ariyadejwanich, P., Japthonh, P., Nakagawa, K., Mukai, S.R., and Tamon, H. (2005) Adsorption–desorption characteristics of phenol and reactive dyes from aqueous solution on mesoporous activated carbon prepared from waste tires. *Water Research*, **39**, 1347–1353.

63 Ha, S.R., Vinitnantharat, S., and Ishibashi, Y. (2001) A modelling approach to bioregeneration of granular activated carbon loaded with phenol and 2,4-dichlorophenol. *Journal of Environmental Science and Health*, **36**, 275–292.

64 Speitel, G.E. Jr, Dovantzis, K., and Digiano, F.A. (1987) Mathematical modelling of bioregeneration in GAC columns. *Journal of Environmental Engineering*, **113**, 32–48.

65 Aktaş, Ö. and Çeçen, F. (2010) Adsorption and cometabolic bioregeneration in activated carbon treatment of 2-nitrophenol. *Journal of Hazardous Materials*, **177**, 956–961.

66 Ng, S.L., Seng, C.E., and Lim, P.E. (2009) Quantification of bioregeneration of activated carbon and activated rice husk loaded with phenolic compounds. *Chemosphere*, **75** (10), 1392–1400.

67 Coelho, C., Oliveira, A.S., Pereira, M.F.R., and Nunes, O.C. (2006) The influence of activated carbon surface properties on the adsorption of the herbicide molinate and the bio-regeneration of the adsorbent. *Journal of Hazardous Materials*, **B138**, 343–349.

68 Labouyrie, L., Bec, R.L., Mandon, F., Sorrento, L.J., and Merlet, N. (1997) Comparison of biological activity of different types of granular activated carbons. *Environmental Technology*, **18**, 151–159.

69 Morinaga, H., Nishijima, W., and Okada, M. (2003) Stimulation of biological activity by the addition of different PACs. *Environmental Technology*, **24**, 179–186.

70 Abu-Salah, K., Shelef, G., Levanon, D., Armon, R., and Dosoretz, C.G. (1996) Microbial degradation of aromatic and polyaromatic toxic compounds adsorbed on powdered activated carbon. *Journal of Biotechnology*, **51**, 265–272.

71 Ha, S.R., Vinitnantharat, S., and Ozaki H. (2000) Bioregeneration by mixed organisms of granular activated carbon with a mixture of phenols. *Biotechnology Letters*, **22**, 1093–1096.

72 Caldeira, M., Heald, S.C., Carvalho, M. F., Vasconcelos, I., Bull, A.T., and Castro, P.M.L. (1999) 4-Chlorophenol degradation by a bacterial consortium: development of a granular activated carbon biofilm reactor. *Applied Microbiology and Biotechnology*, **52**, 722–729.

73 Ehrhardt, H.M. and Rehm, H.J. (1985) Phenol degradation by microorganisms adsorbed on activated carbon. *Applied Microbiology and Biotechnology*, **21**, 32–36.

74 Lin, J.E., Wang, H.Y., and Hickey, R.F. (1991) Use of coimmobilized biological systems to degrade toxic organic compounds. *Biotechnology and Bioengineering*, **38**, 273–279.

75 Jones, L.R., Owen, S.A., Horrell, P., and Burns, G. (1998) Bacterial inoculation of granular activated carbon filters for the removal of atrazine from surface water. *Water Research*, **32**, 2542–2549.

76 Walker, G.M. and Weatherley, L.R. (1998) Bacterial regeneration in biological activated carbon systems. *Process Safety and Environmental Protection*, **76**, 177–182.

77 Marquez, M.C. and Costa, C. (1996) Biomass concentration in PACT process. *Water Research*, **30**, 2079–2085.

78 Martin, M.J., Serra, E., Ros, A., Balaguer, M.D., and Rigola, M. (2004) Carbonaceous adsorbents from sewage sludge and their application in a combined activated sludge-powdered activated carbon (AS-PAC) treatment. *Carbon*, **42**, 1389–1394.

79 Nakano, Y., Hua, L.Q., Nishijima, W., Shoto, E., and Okada, M. (2000) Biodegradation of trichloroethylene (TCE) adsorbed on granular activated carbon (GAC). *Water Research*, **34**, 4139–4142.

80 Zhao, X., Hickey, R.F., and Voice, T.C. (1999) Long-term evaluation of adsorption capacity in a biological activated carbon fluidized bed reactor system. *Water Research*, **33**, 2983–2991.

81 Li., Y. and Loh, K.C. (2007) Hybrid hollow fiber membrane bioreactor for cometabolic transformation of 4-chlorophenol in the presence of phenol. *Journal of Environmental Engineering*, **133**, 404–410.

82 Pannu, J.K., Singh, A., and Ward, O.P. (2003) Influence of peanut oil on microbial degradation of polycyclic aromatic hydrocarbons. *Canadian Journal of Microbiology*, **49**, 508–513.

83 Erlanson, B.C., Dvorak, B.I., Lawler, D.F., and Speitel, G.E. Jr (1997) Equilibrium model for biodegradation and adsorption of mixtures in GAC columns. *Journal of Environmental Engineering*, **123**, 469–478.

84 Sigurdson, S.P. and Robinson C.W. (1978) Substrate inhibited microbiological regeneration of granular activated carbon. *Canadian Journal of Chemical Engineering*, **56**, 330–339.

85 Wallis, D.A. and Bolton, E.E. (1982) Biological regeneration of activated carbon. *AIChE Journal*, **78** (219), 64–69.

86 Le Cloirec, P., Martin, G., and Bernard, T. (1986) Bioregeneration of granular activated carbon: an investigation by radiochemical compound and microbreakdown. *Water SA*, **12** (4), 169–172.

87 Lu, C.J. (1992) Bioregeneration of phenol saturated GAC columns. *Journal of the Chinese Institute of Environmental Engineering*, **2** (2), 81–89.

88 Carvalho, M.F., Duque, A.F., Gonç alvez, I.C., and Castro, P.M.L. (2007) Adsorption of fluorobenzene onto granular activated carbon: isotherm and bioavailability studies. *Bioresource Technology*, **98**, 3424–3430.

89 Lin, Y.-H. and Leu, J.Y. (2008) Kinetics of reactive azo-dye decolorization by Pseudomonas luteola in a biological activated carbon process. *Biochemical Engineering Journal*, **39**, 457–467.

90 Jaar, M.A.A. and Wilderer, P.A. (1992) Granular activated carbon sequencing batch biofilm reactor to treat problematic wastewaters. *Water Science and Technology*, **26** (5–6), 1195–1203.

91 Walker, G.M. and Weatherley, L.R. (1999) Biological activated carbon treatment of industrial wastewater in

stirred tank reactors. *Chemical Engineering Journal*, **75**, 201–206.
92 Suidan, M.T., Cross, T., and Fong, M. (1980) Continuous bioregeneration of granular activated carbon during the anaerobic degradation of catechol. *Progress in Water Technology*, **12**, 203–214.
93 Tsuno, H., Kawamura, W., and Oya, T. (2006) Application of biological activated carbon anaerobic reactor for treatment of hazardous chemicals. *Water Science and Technology*, **53** (11), 251–260.
94 Jackson-Moss, C.A., Maree, J.P., and Wotton, S.C. (1992) Treatment of bleach plant effluent with the biological granular activated carbon process. *Water Science and Technology*, **26** (1–2), 427–434.
95 Sison, N.F., Hanaki, K., and Matsuo, T. (1996) Denitrification with external carbon source utilizing adsorption and desorption capability of activated carbon. *Water Research*, **30**, 217–227.
96 Andrews, G.F. and Tien, C. (1981) Bacterial film growth in adsorbent surfaces. *American Institute of Chemical Engineers (AIChE) Journal*, **27** (3), 396–403.
97 Speitel, G.E., Lu, C.J., Zhu, X.J., and Turakhia, M. (1989b) Biodegradation and adsorption of a bisolute mixture in GAC columns. *Journal of Water Pollution Control Federation*, **61**, 221–229.
98 Tsai, H.H., Ravindran, V., and Pirbazari, M. (2005) Model for predicting the performance of membrane bioadsorber reactor process in water treatment applications. *Chemical Engineering Science*, **60**, 5620–5636.
99 Ng, S.L., Seng, C.E., and Lim, P.E. (2010) Bioregeneration of activated carbon and activated rice husk loaded with phenolic compounds: kinetic modeling. *Chemosphere*, **78**, 510–516.
100 Weber, W.J., Jr, Friedman, L.D., and Bloom, R., Jr (1973) Biologically-extended physicochemical treatment. Proceedings of Sixth Conference of the International Association on Water Pollution Research, Pergamon Press, Oxford, England.
101 Speitel, G.E. Jr and Zhu, X.J. (1990) Sensitivity analyses of biodegradation/adsorption models. *Journal of Environmental Engineering*, **116** (1), 32–48.
102 Ying, W.C. and Weber, W.J. Jr (1979) Bio-physicochemical adsorption model systems for wastewater treatment. *Journal of Water Pollution Control Federation*, **55** (11), 2661–2677.
103 Chang, H.T. and Rittmann, B.E. (1987) Mathematical modeling of biofilm on activated carbon. *Environmental Science and Technology*, **21**, 273–280.
104 Chang, H.T. and Rittmann, B.E. (1987) Verification of the model of biofilm on activated carbon. *Environmental Science and Technology*, **21**, 280–288.
105 Den, W. and Pirbazari, M. (2002) Modeling and design of vapor-phase biofiltration for chlorinated volatile organic compounds. *AIChE Journal*, **48** (9), 2084–2103.

8
Combination of Activated Carbon Adsorption and Biological Processes in Drinking Water Treatment

Ferhan Çeçen

8.1
Introduction

Water used for drinking purposes is abstracted from both surface and groundwater supplies. Traditionally, in surface water treatment plants the main aim is the removal of color, taste and odor, turbidity, and pathogens from the water. The decomposition of plant, animal, and microbial material in both water and soil environments leads to a variety of complex organics in water bodies that are collectively denoted as Natural Organic Matter (NOM) [1]. However, as a result of industrialization, in addition to background NOM, a large number of anthropogenic organic compounds have been introduced into surface waters collectively termed as Synthetic Organic Compounds (SOCs). The conventional type of water treatment consisting of screening, coagulation, flocculation, rapid sand filtration, and disinfection steps proves to be insufficient in the removal of NOM and SOCs [2].

Similarly, in many cases, groundwaters are extensively contaminated with anthropogenic substances, and traditional treatment methods aimed at removal of gases and dissolved substances prove to be insufficient. Therefore, in the last decades many treatment alternatives have been proposed for the treatment of surface and groundwaters.

This chapter discusses the fundamentals of a process used in drinking water treatment which integrates the merits of adsorption and biological removal in one unit, namely the biological activated carbon (BAC) filtration. BAC filtration is an advanced form of GAC filtration in which microorganisms are active in the removal of organic (NOM and organic micropollutants) as well as some inorganic pollutants from drinking water. In some cases, the whole process, consisting of preozonation and bioactive GAC filtration, may take this designation. Sometimes, the process is also named 'biological granular activated carbon' and is abbreviated as BGAC or B(GAC).

Activated Carbon for Water and Wastewater Treatment: Integration of Adsorption and Biological Treatment.
First Edition. Ferhan Çeçen and Özgür Aktaş
© 2011 WILEY-VCH Verlag GmbH & Co. KGaA, Weinheim.
Published 2011 by WILEY-VCH Verlag GmbH & Co. KGaA

8.2
Rationale for Introduction of Biological Processes in Water Treatment

Many of the unit processes and operations that are applied today in water treatment date back to the beginning of the twentieth century. On the other hand, the engineered application of biological processes is relatively new. Even professionals in environmental sciences and engineering are surprised when they learn about such a concept, since the common belief is that drinking water treatment utilizes physicochemical processes only. Nevertheless, biological processes occur unavoidably in all water treatment units, in particular in slow sand filters that have been used for filtration of coagulated and settled water.

In drinking water treatment, the finished water has to fulfill criteria such as microbiological safety, chemical stability, and toxicity. Moreover, since the 1970s, a new concept referred to as 'biological stability' has emerged. Water is designated as 'biologically stable' if it does not contain substances that are likely to be used by bacteria within the water treatment units and/or the distribution system. In causing biological instability, the biodegradable fraction in NOM, namely the Biodegradable Organic Matter (BOM), plays the major role. However, inorganic substances such as iron, manganese, sulfide, and ammonium may also lead to biological instability as a result of oxidation under aerobic conditions. Further, nitrate, a common pollutant in groundwater supplies, may also cause instability because of the possibility of denitrification under anoxic conditions, either in water treatment units or in the water distribution system.

If BOM, ammonia, and others are not sufficiently eliminated during water treatment, even if they are present at very low concentrations this may lower the biological stability of the water. Bacterial re-growth will then be the main concern in the water distribution system. Bacterial re-growth or aftergrowth is usually defined as the increase in heterotrophic plate count (HPC) during the distribution of water in the network [3]. Re-growth of bacteria has undesired consequences such as re-growth of pathogens, increased turbidity, formation of taste and odor, consumption of dissolved oxygen (DO), and accelerated corrosion in the water distribution system [4]. In order to avoid bacterial re-growth, the common measure in plant operation is to maintain a high chlorine residual in the water distribution system. However, bacterial re-growth may not always be effectively suppressed by this procedure. Then, for stabilization of water, it becomes inevitable for plant authorities to seek for biological processes such as BAC filtration.

8.3
Significance of Organic Matter in Water Treatment

A large proportion of organic matter found in surface waters such as rivers, lakes, and ponds consists of NOM. In surface waters, the level and composition of NOM fluctuates over time depending on physicochemical and biological processes taking place in the environment. This in turn affects the characteristics of the raw water entering a treatment plant.

8.3.1
Expression and Fractionation of Organic Matter

NOM cannot be measured as a unique parameter. Therefore, the common approach is to use surrogate parameters.

8.3.1.1
Expression of Organic Matter in Terms of Organic Carbon

Dissolved Organic Carbon (DOC) The dissolved fraction of organic matter in a water body is often of interest and is indicated by the term Dissolved Organic Matter (DOM). DOM is a better expression than NOM because in surface or groundwaters it is difficult to distinguish between organic matter of natural and anthropogenic origin. However, DOM is an acronym and not a water quality parameter. On the other hand, a parameter frequently measured in water treatment works is the Dissolved Organic Carbon (DOC), which indicates the concentration of carbon in the organic molecules present in a sample after filtration through a 0.45-μm pore size filter. In general, the DOC concentrations in surface waters lie in the range of 1–10 mg L^{-1}. The main factor leading to elevated DOC levels may be excessive algae growth [5].

Total Organic Carbon (TOC) NOM consists of dissolved and particulate fractions. The TOC parameter indicates all of the organic carbon in a sample and consists of Dissolved Organic Carbon (DOC) and Particulate Organic Carbon (POC). The measurement of TOC gains importance if the water contains much particulate matter such as in a eutrophic lake.

Expression of Organic Matter by Spectral Measurements Sophisticated analyses used in characterization of NOM prove to have no practicality for routine monitoring in treatment plants. Therefore, a more preferred procedure is to use nonspecific simple parameters. In general, spectrophotometric measurements in the UV range of 254–280 nm reflect the presence of compounds with unsaturated double bonds or π–π electron interactions. In this regard, UV_{254} is the commonly used parameter for expression of organic matter in water treatment. The UV_{254} parameter represents the existence of unsaturated carbon bonds including aromatic compounds, which are generally inert to biodegradation.

The UV absorbance at a fixed wavelength is often normalized by DOC to give the Specific Ultraviolet Absorbance (SUVA), commonly expressed in units of L/mgC.m. This normalization allows the comparison of various water bodies. A high SUVA value in water generally indicates that the extent of humification, aromaticity, and hydrophobicity of organic matter is high, featuring low biodegradability. Another parameter that correlates to some extent with DOC is the color of water. Similarly, the color may be expressed in a normalized way by dividing by DOC.

8.3.1.2
Measurement of the Biodegradable Fraction in NOM

NOM consists of both biodegradable and nonbiodegradable fractions. The biodegradable fraction (BOM) is generally considered to consist of carbohydrates and low-molecular-weight (LMW) compounds. BOM is of significance in water treatment works since it serves as an energy and carbon source for bacteria, thereby promoting bacterial re-growth. Some LMW biodegradable compounds can be directly used for metabolism, whereas high-molecular-weight (HMW) ones have first to undergo enzymatic attack.

Humic substances found in water account for about 50–75% of the DOC [6]. Although humic substances are generally regarded as resistant to biodegradation, some studies reveal the contrary. For example, it was strikingly noted that humic substances accounted for a large proportion of the BOM in a stream [7]. In relatively recent studies, a small fraction of humic substances found in rivers and lakes was also shown to contribute to BOM [8]. Obviously, the timescale has to be considered in order to have an accurate idea about biodegradability of humic substances [6].

Bearing in mind the fact that BOM is an acronym and not a parameter, two different parameters came into use, namely:

- Biodegradable Dissolved Organic Carbon (BDOC)
- Assimilable Organic Carbon (AOC).

Biodegradable Dissolved Organic Carbon (BDOC) As the name implies, the BDOC test aims at the determination of the DOC fraction that is bioavailable for bacteria. BDOC constitutes the total amount of DOC that is removed from water by heterotrophic microorganisms due to both cell synthesis and mineralization. According to the original BDOC procedure, the water sample is filter sterilized and then reinoculated with the same sample that was filtered through a 2-μm pore size filter for removal of particles and protozoa [9]. Then, the sample is incubated in the dark at 20 °C. In the BDOC test, the DOC remaining in solution is measured with respect to time. The incubation period lasts up to 30 days until the achievement of a plateau in DOC. Then, the BDOC value is calculated from the difference between the initial and final DOC.

The procedure in BDOC determination is somewhat similar to the BOD procedure, but the BDOC test takes much longer than the BOD test. However, the BOD test cannot be conducted at the very low organic carbon levels found in drinking water supplies, and the BDOC tests provides other advantages. In the BOD test oxygen-consuming compounds such as ammonia may interfere with the oxygen demand determination. On the other hand, the BDOC test truly reflects the biodegradability of organic matter, since it is based on the direct measurement of organic carbon.

Depending on the rate of biodegradation, the DOC in a water sample may be divided into fractions, namely fast-BDOC, slow-BDOC and non-BDOC or NBDOC, representing the truly nonbiodegradable portion [10]. The fast-BDOC or the readily biodegradable fraction in BDOC can be directly used by bacteria. The slowly biodegradable fractions in BDOC undergo hydrolysis at variable rates [11]. Usually, extending the duration of the BDOC test to 30 days allows adequate estimation of different fractions.

Assimilable Organic Carbon (AOC) This parameter expresses the fraction of DOC that can easily be assimilated into biomass during bacterial growth. Thus, unlike the BDOC, which represents the organic carbon that is removed in energy reactions and cell synthesis, AOC merely reflects the organic carbon that is taken up for cell synthesis. Therefore, AOC constitutes a small fraction of BDOC, as shown in Figure 8.1. AOC is typically less than 10% of the BDOC in a water sample [12].

The original AOC method developed in the 1980s by van der Kooij is based on the growth measurement with a mixture of two pure cultures in a sample of pasteurized water [3]. The microorganisms used in the test are the *Pseudomonas fluorescens* P17, which is capable of utilizing a wide range of LMW compounds at very low concentrations, and the *Spirillum* sp. strain NOX, which utilizes carboxylic acids only. As shown in Eq (8.1), the AOC concentration is calculated from the maximum colony counts of these strains, using their yield coefficients (Y) for acetate. The numerical value of Y has to be determined experimentally [13]. Then, the AOC value is expressed in units of µg acetate-C equivalents per L.

$$\text{AOC} = \frac{N_{P17}}{Y_{P17}} + \frac{N_{NOX}}{Y_{NOX}} \quad (8.1)$$

where N_{P17} and N_{NOX} are the bacterial concentrations in units of CFU L^{-1} and Y_{P17} and Y_{NOX} are the yield coefficients for P_{17} and NOX, respectively, in units of CFU µg^{-1} acetate-C.

AOC constitutes about 0.1–9% of TOC and comprises a wide range of easily biodegradable organics such as sugars, organic acids, nucleic acids, aldehydes, ketones, and alcohols [14].

AOC and BDOC are useful surrogates for BOM. However, both rely on biological activity and may therefore be susceptible to limitations and errors. In fact, AOC is an expression of the biological response to the assimilable carbon in water. Thus, in contrast to the BDOC parameter, it is not a direct measure of the carbon

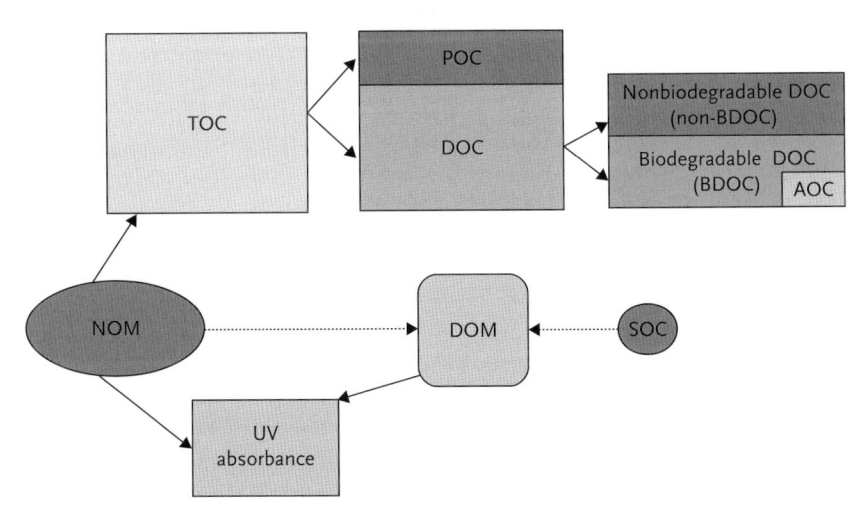

Figure 8.1 Fractionation and measurement of organic matter in water.

concentration itself. Both tests, BDOC and AOC, require a relatively long period of time for completion.

Modifications in the BDOC method The original BDOC method requires relatively long incubation times. Several modifications have been made to the original BDOC method that was proposed by Servais and co-workers [9], mainly to increase the rate of organic carbon uptake and shorten the duration of the test. The BDOC may not reflect the true amount of biodegradable organic carbon in every case. Therefore, another aim is the accurate determination of BDOC by the increase of biomass concentration and diversity. Some of the modifications to the original test procedure are as follows:

- *Increase in biomass concentration in the test:* The concentration of biomass can be increased by addition of sand from a sand filter containing biomass [15].
- *Increase in the speed of BDOC uptake*: Continuous circulation of water over a biofilm attached to sintered porous glass may accelerate the BDOC uptake [16, 17]. In addition, the use of attached bacteria [18, 19] or a plug flow reactor [7] may increase the speed of uptake.
- *Increase the domain of BDOC:* In rapid tests the determination of relatively labile DOC is possible only whereas 42-day incubations have been used with measurement of headspace CO_2 for determination of labile and refractory DOC components [20].
- *Increase the biomass concentration and diversity:* The classical procedure may lead to underestimation of actual BDOC. A recent study revealed that the underlying reason was that most of the time either bacteria were in insufficient numbers in the original sample or their diversity was not high enough to reflect the total biodegradable fraction. Application of this newly proposed BDOC procedure based on the use of acclimated bacteria led to more accurate results [21, 22].

Modifications in the AOC method The AOC parameter is very widely used as an indication of the re-growth potential in water distribution. The modified AOC methods are based on the common principle of measuring the growth of bacteria. The significant difference in these tests is the utilization of different types of inoculums, such as natural microbial communities or pure cultures [23]. Additionally, in some modifications the measurement of growth is based on different analytical methods such as plating, ATP analysis, turbidity, flow cytometry, and luminescence. Also, factors used in conversion of cell or biomass measurements into the AOC expression may differ [23]. Accordingly, all these factors affect the accurate determination of AOC.

In comparison to the AOC test, which merely reflects the easily assimilable part in DOC, the BDOC test encompasses almost all of biodegradable organics. Interestingly, no definite statements can be made about the relationship between these two parameters in raw and treated waters. Therefore, it is recommended that both tests are conducted in water treatment works, since measuring one alone can potentially under- or overestimate the bacterial re-growth potential of water [24, 25].

Other methods of assessing the re-growth potential Since the measurement of BDOC and AOC is time consuming, more practical substitutes have been sought for the assessment of re-growth potential in a water distribution system. For example, a fiber optical sensor was employed to detect changes in biofilm growth, and these were then related to the BDOC of water [26]. Additionally, the Biofilm Formation Rate (BFR) may be taken as an indicator of biological re-growth potential. BFR can be correlated to the AOC concentration; these two parameters can then be simultaneously interpreted [3]. Besides this method there are also others measuring the re-growth potential or biostability of drinking water such as the biofilm annular reactor (BAR) [27].

8.3.1.3
Nonbiodegradable Dissolved Organic Carbon (non-BDOC or NBDOC)

This fraction in DOC, shown in Figure 8.1, is of special interest in water treatment because it acts as a precursor in the formation of Disinfection By-Products (DBPs). On the other hand, in the water distribution system it has little effect on bacterial re-growth because of inherent lack of biodegradability. This fraction can either be removed by destructive pretreatment techniques such as ozonation or by nondestructive ones such as activated carbon adsorption or the like.

8.3.1.4
Fractionation of NOM in Terms of Molecular Size and Chemical Behavior

Using the current advanced analytical techniques, NOM can be analyzed in a very detailed way. These techniques include elemental composition, Nuclear Magnetic Resonance (NMR) spectroscopy, infrared spectroscopy, fluorescence spectroscopy, and pyrolysis/GC-MS for determination of molecular structure of NOM, functional groups on NOM, and individual NOM components.

The molecular size distribution of NOM is commonly measured using membrane filtration with different molecular weight cutoffs. However, more detailed information is obtained about the fractions in NOM by the use of Size Exclusion Liquid Chromatography coupled to Organic Carbon Detection (SEC-OCD) and Organic Nitrogen Detection (OND).

As shown in Figure 8.2, these techniques allow the separation of NOM into fractions according to size and chemical behavior. The hydrophilic part in DOC is denoted

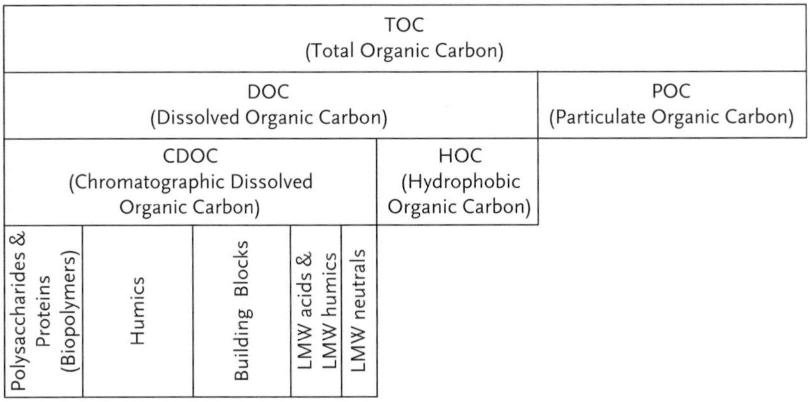

TOC (Total Organic Carbon)					
DOC (Dissolved Organic Carbon)					POC (Particulate Organic Carbon)
CDOC (Chromatographic Dissolved Organic Carbon)				HOC (Hydrophobic Organic Carbon)	
Polysaccharides & Proteins (Biopolymers)	Humics	Building Blocks	LMW acids & LMW humics	LMW neutrals	

Figure 8.2 Breakdown of TOC in fractions which can be analyzed by SEC-OCD (redrawn after Ref. [1]).

as the chromatographic DOC (CDOC). The hydrophobic part (HOC) remaining on the column is calculated from the difference between DOC and CDOC [1].

The CDOC has a molecular range from 20 000 to 100 Da and can be further separated by the SEC column into five fractions; 1 Da represents the mass of a hydrogen atom. The biopolymer fraction is often designated as the polysaccharide fraction and actually contains all polymeric (>20 kDa) and colloidal material, and includes mainly polysaccharides and proteins. Humics having a molecular weight of around 1 kDa correspond to 50–60% of the CDOC and are more hydrophobic than other fractions. Also, early characterizations showed that the humic fraction of NOM was more hydrophobic than nonhumic fractions [28]. Nowadays, ^{13}C NMR spectroscopy is often used to examine the aromaticity/aliphaticity of humic substances. Humic substances constitute about 50% of NOM, whereas in highly colored waters this fraction may rise to 90%. The nonhumic fraction includes biochemically well-defined compounds such as hydrophilic acids, proteins, lipids, amino acids, and carbohydrates. Building blocks are mainly degradation products of humics and have a molecular weight of 300–500 Da. The LMW acids and the LMW humics are together called LMW organics. Their molecular weight is less than 350 Da. Neutrals are defined as a mixture of neutral and amphiphilic compounds, and their molecular weight is also <350 Da [1].

The importance of this NOM fractionation for biological activated carbon (BAC) filtration is further discussed in Chapter 9.

8.4
Removal of NOM in Conventional Water Treatment

8.4.1
Rationale for NOM Removal

The presence of NOM poses mainly the following problems in drinking water treatment and distribution:

- *NOM serves as a precursor in formation of Disinfection By-Products (DBPs)*: NOM compounds are not necessarily harmful themselves, but they may lead to undesired consequences. In the disinfection of water, the most widely used chemical is chlorine gas or NaOCl. In the presence of NOM, chlorine-containing compounds act either as an oxidizing or a substitution agent and produce a number of halogenated DBPs. Also, prechlorination of water after water intake can lead to DBP formation.

  Many DBPs are proven or potential carcinogens, and their presence is increasingly regulated. Removal of NOM gained particular importance in the 1970s when the presence of organic chlorination by-products was discovered. Although initially the focus was on trihalomethanes (THMs) and haloacetic acids (HAAs), other halogenated DBPs are also known today. For example, brominated organohalogen compounds can be formed in water if both bromide and NOM are present.

- *NOM acts as a substrate for biological re-growth in distribution systems:* The presence of BOM increases the re-growth potential of bacteria, while this potential is expressed by the AOC and BDOC parameters.
- *NOM binds metals and hydrophobic organic chemicals:* Binding of NOM to such substances leads to complexes that are not removed in water treatment plants and therefore passed on to the distribution system.
- *NOM causes taste, odor, and color in drinking water:* Raw waters contain many naturally occurring compounds causing taste and odor such as the algal metabolites 2-methylisoborneol (MIB) and 1,10-*trans*-dimethyl-*trans*-9-decalol (geosmin), and color-imparting substances such as humics.
- *NOM accelerates membrane fouling:* Membrane reactors are widely employed in water treatment, as strict regulations are set for specific contaminants. The presence of BOM in water may lead to unwanted bacterial growth on membrane surfaces and may accelerate biofouling.
- *NOM exerts increased coagulant and disinfectant/oxidant demand:* A limited fraction of NOM can be removed by coagulation. At high NOM levels the required coagulant dose increases. Further, in the disinfection step the presence of NOM would increase the disinfectant dose needed to overcome biological re-growth.

8.4.2
Extent of NOM Removal

In conventional water treatment, the efficiency of removal of NOM depends on the characteristics of organic matter (structure, molecular size, functionality), the inorganic composition in water, and the operation and design of treatment plant [29]. The traditional water treatment consisting of coagulation, flocculation, sedimentation, and filtration steps proves to be ineffective in NOM removal. In such units, NOM removal occurs mainly via precipitation and adsorption to flocs. However, these operations remove primarily the hydrophobic organic fraction in NOM consisting mainly of humic substances and other large molecules, leaving the hydrophilic compounds, constituting mainly the biodegradable fraction, largely unaffected [30]. Therefore, in many cases in conventional treatment the efficiency of removal of NOM may be just around 30% [31]. In particular, coagulation has no effect on the AOC of water due to the low molecular size of compounds contributing to this parameter [32, 33]. The AOC fraction in total DOC is usually rather low, less than 5%. Still, it is this part of NOM that causes the instability in the water distribution system.

The SUVA value indirectly indicates the potential for NOM removal. Low SUVA values (<4 L/mgC.m) denote that the DOC present is composed of humic material that is difficult to remove by coagulation [34]. Also, coagulation of raw water is not very effective at low DOC concentrations [29]. Enhanced coagulation is a procedure improving DOC and BDOC removal; however, this type of coagulation leads primarily to the removal of HMW and hydrophobic fractions in NOM [8]. Enhanced coagulation was shown to decrease the initial concentration of NOM while simultaneously increasing the strongly adsorbable fraction [35]. Factors such

8.5
Use of Activated Carbon in Water Treatment

8.5.1
Powdered Activated Carbon (PAC) Addition

Among the alternatives for removal of NOM, adsorption onto activated carbon comes top of the list. In water treatment plants, PAC is usually dosed temporarily or intermittently into water in the case of deterioration of raw water quality or for alleviation of short-term operational problems. As shown in Figure 8.3, under typical circumstances, PAC is added either into rapid mixing (coagulation), flocculation, or filtration units. The PAC, added at any point, is separated from water in the filtration step.

Depending on the dose of PAC and the turbidity of the water, algal metabolites imparting taste and odor, for example, MIB and geosmin, can be effectively eliminated from the water [36]. It is recommended to consider the addition of PAC at the water intake if there is objectionable taste and odor and the water is suspected to contain microcystin toxins [37].

Addition of PAC can also bring about the elimination of several micropollutants within the scope of SOC. The compounds within this definition are present at ppb or even lower levels in natural waters and are suspected to be harmful to health. PAC may also be dosed into membrane systems in order to increase the removal of specific organics.

In conventional treatment units such as coagulation tanks, elimination of micropollutants is limited, even in PAC-added cases. This is attributed to the relatively low hydraulic residence times in these units. At low hydraulic residence times the carbon does not reach equilibrium conditions. The limited contact time between bulk water and PAC particles hinders an efficient removal of natural as well as anthropogenic compounds. As a consequence, large carbon doses have to be added, which then results in additional sludge production and reduction of filter efficiency.

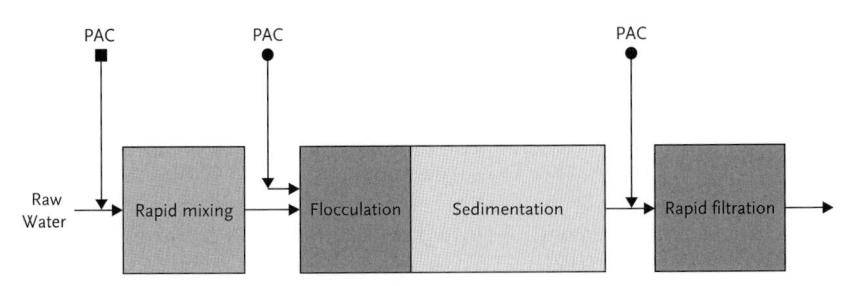

Figure 8.3 Addition of PAC to different points of the water treatment train.

8.5.2
Granular Activated Carbon (GAC) Filtration

GAC filtration is introduced as a separate and long-term operation for alleviation of problems encountered in permanently loaded raw waters. In a typical water treatment plant, GAC filtration is often placed after sand filtration, although other configurations are possible. As mentioned in Chapter 2, GAC filtration was applied mainly for the removal of taste and odor at an early date [38]. Since the 1960s, GAC filtration has received attention as a process aimed at the removal of SOCs.

In modern water treatment works, GAC is more widely utilized than PAC, mainly due to the possibility of regeneration and easy handling. The main advantage of a GAC adsorber is that it provides a high contact time between water and carbon, a factor improving the possibility of adsorption and effective use of activated carbon. In general, GAC is used in fixed- or expanded-bed filters or integrated into already existing sand or sand/anthracite filters. The major characteristics of fixed-bed GAC filters are discussed in Chapter 2.

8.6
Biological Activated Carbon (BAC) Filtration

8.6.1
History of BAC Filtration in Water Treatment

The concept of BAC filtration in water treatment can be traced back to GAC filtration. Ozonation and GAC filtration were initially introduced for the removal of taste and odor [38]. In the 1970s it was recognized that bacteria proliferated in conventional GAC filters and were effective in the removal of NOM. This type of removal was even more enhanced when GAC filters received ozonated water. The enhancing effect of ozonation was attributed to the increase in the biodegradable fraction, which consequently accelerated biodegradation in GAC filters. The end result was that in GAC filters the removal of organics exceeded expectations.

The observation of biodegradation in GAC filters paved the way for a groundbreaking development in water treatment, the engineered application of biological processes in Europe, mainly in Germany, Holland, and France [8]. It was recognized that integration of adsorption and biological degradation in the same unit would help in solving the occurrence of biological instability in finished water. As commonly accepted, the removal of contaminants by biological means is preferred to other alternatives, since generally no harmful products are generated and costs are lower. Thus, if a significant part of NOM could be made amenable to biodegradation in GAC filters, this would decrease the need for costly physicochemical processes.

The pioneering biofiltration process for the removal of NOM is the Mülheim process that was developed in 1976 in Germany by the Rheinisch–Westfälische Wasserwerksgesellschaft (RWW) GmbH [39]. The process was named after a large-scale water treatment plant in Mülheim, Germany. In the 1960s and 1970s the Ruhr river was highly polluted due to wastewater discharges. The classical surface water treatment before 1976 consisted of flocculation/sedimentation with chlorine and limestone powder addition, rapid sand filtration, GAC filtration, artificial groundwater recharge, and final disinfection with chlorine. However, this operation scheme led to insufficient purification of water. Reaction of DOC with chlorine resulted in the formation of AOX. Moreover, extremely high ammonia concentrations were recorded in the raw water, and high doses of chlorine had to be used to oxidize ammonia by breakpoint chlorination. Therefore, RWW looked for a new process in water treatment [39]. The Mülheim process was tested at pilot- and full-scale in Mülheim. Instead of the former use of chlorine, the oxidation of water was carried out by ozonation in combination with BAC filtration.

The main target of the Mülheim process is to decrease the DOC concentration in order to reduce the need for disinfectants and to avoid re-growth of microorganisms in the water distribution system. In this process, refractory organics constituting the HMW fraction in NOM are partially oxidized by ozonation, leading to an increase in biodegradability. This is then reflected as an increase in the BDOC and AOC of water. In later filtration steps, organic matter is biodegraded by biofilm bacteria attached to activated carbon. At the time of development, the Mülheim process attracted considerable interest, since ammonia was also oxidized biologically in spite of low temperatures. The process permitted for the first time the production of high quality water out of polluted Ruhr water without the use of much chlorine or long residence times during infiltration [38]. The scientific supporter of the project was Prof. Sontheimer. The first practical experience of the Mülheim process is discussed in his book about adsorption [40].

The Mülheim process at the Dohne water works in Germany has operated since that time in the following way: raw water catchment directly from the runoff of the Ruhr river, preozonation and flocculation, main ozonation, biological double layer filtration, BAC filtration, infiltration for artificial groundwater recharge, re-catchment by siphon well systems after a soil passage of 1–2 days, a final disinfection by chlorine (0.1–0.15 ppm), deacidification by sodium hydroxide, and finally water distribution [39, 41]. Some operating results are given in Chapter 10.

Compared to applications in Europe, in the USA the water industry was initially reluctant to implement biological processes in water treatment because of concerns about the probable introduction of microorganisms into the water distribution system as a result of carryover of GAC particles containing biomass [2]. This topic is further discussed in Section 9.9.5. Due to this hesitation, in North America such processes have gained special attention only since the late 1980s [41]. However, latest reports in the USA reveal that BAC filters are added to the treatment train and some existing filters are converted to BAC filters in water treatment works [42]. Apparently, the use of activated carbon will become more common in water treatment due to the increased use of ozone as a preoxidant and the concern over biological re-growth in the water network.

In fact, prior to BAC filtration, natural treatment systems such as river bank filtration (RBF) and soil aquifer treatment (SAT), and engineered/mechanical treatment systems such as slow sand filtration (SSF) and biological rapid sand filtration (BSF) have been in use for a long time, particularly in Western Europe. These treatment systems employing microbial activity to varying extents can also be considered as biological filtration processes [43]. However, they have been excluded from the scope of this book in view of the fact that they do not employ activated carbon. However, in some places in this book activated carbon filtration is compared with sand and anthracite/sand filtration, a very common operation in water treatment.

8.6.2
Combination of Ozonation and BAC Filtration

Ozone finds wide application in the pretreatment of water, in order to fulfill different requirements such as color, odor, and taste removal, oxidation of inorganics such as iron and manganese, and as an aid in coagulation. Ozonation can also lead to the removal of many micropollutants. Since ozone is a powerful oxidant it acts against many microorganisms including viruses.

Ozonation usually precedes BAC filtration. As outlined in detail in Section 8.7.2, ozonation converts slowly degradable or nonbiodegradable compounds into biodegradable ones, thus enabling their biological removal. Otherwise, these compounds would create nuisances in treatment steps such as disinfection.

Depending on water characteristics and the aim of treatment, ozonation units may be placed at the head of water treatment only (preozonation) or at intermediate stages (Figure 8.4). As discussed in Section 8.7.2, pre- and post-ozonation applications change the adsorbability and biodegradability of organic compounds in water; therefore, they also affect BAC filtration.

In Europe, multiple applications of small ozone doses at several points of treatment is more common than single-stage preozonation [44]. Experiences show the superiority of multistage ozonation over single-stage ozonation in the removal of DOC.

Ozone has been used for a long time as a last-stage disinfectant and was not regarded as leading to the formation of DBPs. However, nowadays it is recognized that ozonation also leads to by-products such as bromate that are of health concern. Other preoxidation alternatives are also available in water treatment. For example, oxidation with UV/H_2O_2 does not lead to bromate formation. However, this application has the disadvantage of nitrite formation [45].

8.6.3
Current Use of BAC Filtration in Water Treatment

GAC and BAC filtration may be applied in the case of surface and groundwater treatment, mainly for the purpose of removing NOM and specific organics. Over

Pre-ozonation → Rapid mixing → Flocculation → Sedimentation → Rapid filtration → Ozonation → BAC filtration → Disinfection →

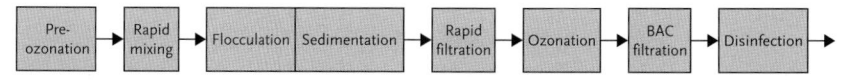

Figure 8.4 Positioning of ozonation and BAC filtration in drinking water treatment.

the entire course of their operation, GAC filters are gradually converted into microbially active BAC filters that combine the merits of adsorption and biodegradation. In modern water treatment works, the scope of BAC filtration is extended to serve the following purposes:

- Reduction of biodegradable matter for increasing the biostability of water
- Biological nitrogen removal (ammonium and nitrate removal)
- Reduction of chlorine demand at the disinfection stage
- Reduction of DBP precursors
- Reduction of DBPs
- Removal of organic micropollutants
- Removal of inorganic micropollutants
- Pretreatment for Membrane Filtration
- Assistment of Membrane Filtration
- Groundwater Bioremediation.

8.7
Adsorption and Biodegradation Characteristics of Water

8.7.1
Raw Water NOM

Raw water should be initially well defined in order to predict the removal of organic matter by adsorption and/or biodegradation in BAC filters. Organic matter in water is usually characterized by the DOC parameter, which arises from adsorbable and biodegradable, adsorbable and nonbiodegradable, nonadsorbable and biodegradable, and nonadsorbable and nonbiodegradable compounds [46].

NOM consists of different fractions of variable adsorptivity. Another important criterion in removing organic matter in BAC filtration is the biodegradability of water. The hydrophilic fraction of NOM consists of high polarity and LMW organic compounds and is more readily biodegraded. Molecular size distribution in NOM affects the biodegradability also. Smaller molecules are generally more amenable to biodegradation than larger ones.

An indirect indicator of biodegradability of raw and ozonated waters is the SUVA value. A lower SUVA value indicates a higher biodegradation potential, as shown in Figure 8.5. A SUVA value less than 2 L/mgC.m generally indicates that water contains biodegradable organics which can be removed by biofiltration. On the other hand, high SUVA and color values mainly indicate the presence of humic substances that are resistant to biofiltration unless preozonation is applied [8].

8.7.2
Impact of Ozonation on NOM Characteristics

It is possible to express the ozone dose in ppm or $mg\,L^{-1}$ units. However, the ozone dose is commonly normalized by the initial DOC of water and is expressed

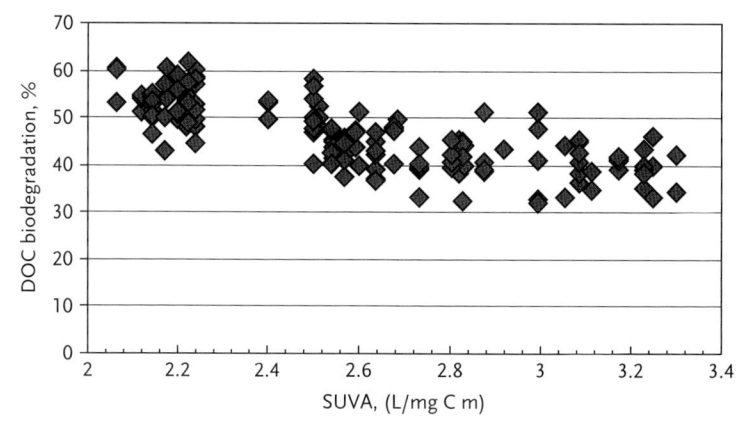

Figure 8.5 Dependence of DOC biodegradation on the SUVA value of water (redrawn after Ref. [47]).

in terms of $mg\,O_3/mg\,DOC$, referred to as the specific ozone dose. In preozonation of water, specific ozone doses range usually from 0.5 to $3\,mg\,O_3/mg\,DOC$. Generally, the ozone doses employed for preozonation of water are much lower than those for post-ozonation.

In a typical water treatment system, both direct reactions of molecular ozone and indirect reactions involving hydroxyl radicals are possible. The extent of each reaction depends on the conditions prevalent in ozonation as well as the water characteristics. Recent studies show that the direct action of ozone is mainly responsible for the formation of small organic molecules [48].

The introduction of pre- or post-ozonation prior to BAC filtration brings about a decrease in the the parameters TOC, DOC, and UV absorbances.

Pre- or post-ozonation of water as shown in Figure 8.6 has two main consequences:

- increase in the biodegradability of original water,
- change in the adsorbability of original water.

8.7.2.1
Increase in the Biodegradability of NOM

Changes in NOM structure and composition Ozonation breaks down the structure of the NOM and enhances the transformation of HMW compounds into LMW ones. Substantial structural changes occur in humic acids, which mainly constitute the nonbiodegradable fraction. A strong and rapid decrease is seen in color and UV_{254} due to loss of aromaticity and depolymerization of molecules [49].

In addition, ozonation solubilizes particulate organic carbon (POC), as shown in the case of a eutrophic lake water [50]. Ozonation of algal cells leads to the release of organic matter and increases the AOC and DOC of water [48].

Increase in BDOC and AOC upon ozonation The increase observed in BDOC and AOC parameters upon ozonation is an indication of biodegradability improvement. The common by-products of ozonation are aldehydes (formaldehyde,

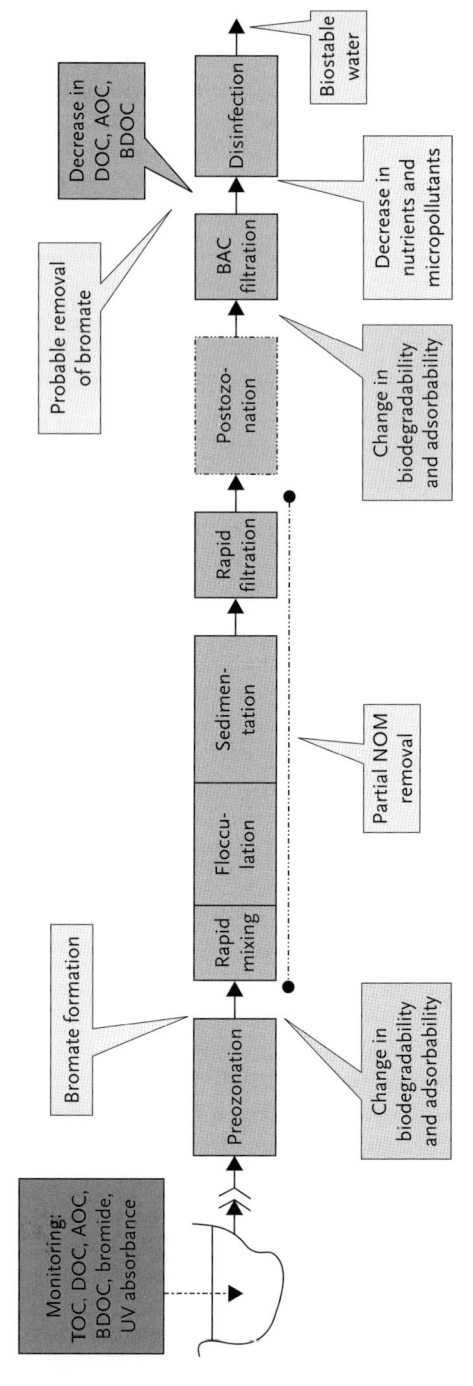

Figure 8.6 Effects of ozonation and BAC filtration in drinking water treatment train.

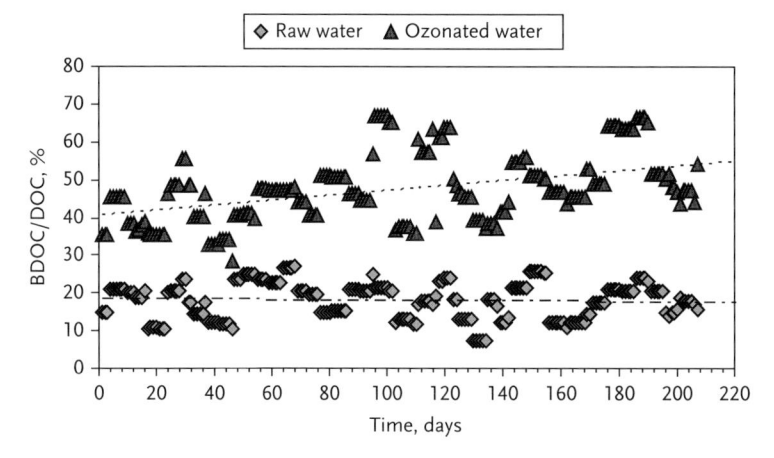

Figure 8.7 The BDOC/DOC ratios in raw and ozonated waters in the sampling period [21].

acetaldehyde, glyoxal, methylglyoxal) and carboxylic acids (formic, acetic, glyoxylic, pyruvic, and ketomalonic acids), which fall into the category of LMW oxygen-containing compounds having generally a high biodegradability [49, 51]. The AOC in water is usually attributed to the presence of such LMW organics. Contrary to former investigations, in newer studies the AOC was shown to be mainly composed of specific acids and not aldehydes and ketones [52].

No definite predictions of the increase in AOC and BDOC of water upon ozonation can be given since the characteristics of the water play a major role. For example, in one case at an ozone dose of 1.5 mg O_3/mg DOC, BDOC was shown to increase by 31% [53]. Figure 8.7 shows that depending on the seasonal characteristics of a raw surface water, the BDOC/DOC ratio could be raised from about 0.2 up to >0.5 at an ozone dose of 2 mg O_3/mg DOC. At this dose, the BDOC of the water was increased by 163%, from about 0.68 up to 1.76 mg L^{-1}.

When water is ozonated, a higher reduction in UV_{254} compared to that of DOC is generally observed, since ozone effectively breaks conjugated bonds, and this is reflected by the former parameter. Accordingly, the $SUVA_{254}$ (UV_{254}/DOC) value also decreases upon ozonation. In the raw surface water shown in Figure 8.7, the average $SUVA_{254}$ value of about 2.5 was decreased to about 1.3 by ozonation.

The increase in biodegradability can also be monitored by the AOC parameter. For example, 2.9- and 11.7-fold increases have been recorded in AOC at 1 and 4 mg L^{-1} ozone concentrations, respectively [54]. In a water treatment plant that switched from chlorination to a combination of chlorination and ozonation, AOC and BDOC increased by 127% and 49%, respectively [25].

Preozonation can usually increase the AOC concentration in water as much as threefold. However, depending on the composition of the raw water, only in extreme cases can ozonation lead to a two- to threefold increase in BDOC concentration. Generally, a higher increase is achieved in AOC than in BDOC.

Investigations in recent years give an idea of the origin of AOC. Ozonation of filtered and algal cell enriched water at laboratory- and pilot-scale revealed that only a fraction of the AOC formation arose due to oxidation of DOM [52]. On the other hand, most of AOC formation was attributed to oxidation of algal cells [48]. The AOC formed from DOM could be well described by organic acids, whereas the algal-derived AOC was very different in composition. It is recommended that these findings should be considered when ozonation is planned at the front of a treatment train. In such a case, ozone would oxidize algal matter and lead to high AOC formation [1]. Whether this situation is desirable or not depends on the introduction of subsequent treatment steps.

Advanced analytical techniques showed that AOC formation was linked to the increase in LMW acid concentration upon ozonation of water [55].

Importance of the ozone dose The ozone dose is also crucial in controlling the formation of the different fractions in BDOC. Biodegradation becomes more efficient when the fraction of readily biodegradable and hydrolyzable matter increases upon ozonation. However, no definite values can be given for the formation of BDOC fractions, since this depends on ozonation conditions and water characteristics.

For example, Carlson and Amy observed that the rapidly degradable BDOC increased up to an ozone dose of about 1.0 mg O_3/mg DOC, but leveled out at higher doses [56]. At this optimal dose, the readily degradable BDOC fraction in DOC was about 14%. By employing the optimal ozone dose, the rapidly degradable BDOC can be maximized, while the formation of slowly biodegradable BDOC can be minimized [57]. For a lake water in which initially no BDOC was present, ozone doses as high as 3 mg O_3/mg DOC had to be applied to lead to an insignificant BDOC formation. On the other hand, for a river water in which the BDOC was 1.15 mg L^{-1}, an ozone dose of 0.5 mg/mg DOC increased the BDOC to 2.31 mg L^{-1}, while higher doses up to 3 mg O_3/mg DOC made only a slight contribution. Furthermore, most of the BDOC formed upon ozonation consisted of readily biodegradable BDOC. The same response was also observed in the case of a lake water which initially had a high BDOC of about 5.04 mg L^{-1} [10].

Mineralization of DOC and the optimum ozone dose An important issue in ozonation is to find the optimum dose leading to maximum BDOC and/or AOC formation for subsequent BAC filtration. Simultaneously, the ozone dose should be kept at a level to avoid mineralization of TOC or DOC into CO_2. Complete mineralization of organic matter would require large doses of ozone and make further biological removal redundant. The optimum dose is found experimentally and depends on raw water characteristics. For example, for a surface water the optimum dose leading to minimal mineralization of organic carbon (about 5%), but maximum increase in BDOC was found as 2 mg O_3/mg DOC [21].

Consequences of biodegradability increase Since mostly the organic carbon rather than nitrogen or phosphorus in water serves as an electron donor and limits bacterial re-growth, the increase in biodegradability, shown by an increase in AOC and/or BDOC, has unwanted consequences unless biofiltration is applied. Therefore, BOM removal in BAC filtration is an effective option in the control of biological re-growth.

8.7.2.2
Change in the Adsorbability of NOM

In most cases, ozonation has a negative effect on subsequent GAC/BAC operation because it increases the polarity and hydrophilicity of NOM, leading to a reduction of adsorbability [53]. In addition, ozonation leads to the formation of biodegradable compounds that adsorb at a lower rate than nonbiodegradable ones [58].

8.7.3
Determination of Adsorption and Biodegradation Characteristics of Water

Prior to a BAC operation, it is essential to distinguish between the biodegradable, nonbiodegradable, adsorbable, and nonadsorbable fractions in NOM and to assess the changes brought about by ozonation. The importance of this NOM subdivision with regard to adsorbability and biodegradability was addressed in the early 1980s [38, 59]. This approach has also been adopted in the modeling of BAC filtration (Chapter 11).

The fractionation of NOM with respect to adsorbability is usually carried out in batch GAC adsorption tests. In that respect, the chemistry of the carbon surface (surface charge and polarity of the carbon), pore size distribution of the carbon, and NOM properties such as molecular size distribution, polarity, acidity, and aromaticity are of interest [60]. Similarly, NOM can be fractionated with regard to biodegradability using batch tests.

Batch tests may hint at the adsorbability, desorbability, and biodegradability potential of raw and ozonated waters and facilitate the choice of the right GAC grade. However, such results cannot be directly extrapolated to continuous-flow BAC filtration, since in the latter process operational factors also play a role.

In full-scale systems, nonbiodegradable organics should ideally be irreversibly adsorbed, otherwise desorption would deteriorate the effluent quality. On the other hand, desorption of biodegradable organics from GAC would result in the enlargement of the adsorption capacity of the GAC for nonbiodegradable compounds. This arises because biological activity eliminates substances that would otherwise compete for adsorption sites. Therefore, as exemplified in the following study (Example 8.1), it is important to examine the adsorption and desorption properties of the biodegradable and nonbiodegradable fractions in raw and ozonated waters [61].

Example 8.1: Adsorbability and Desorbability of Organic Matter in Raw and Ozonated Waters In raw and treated waters, NOM is composed of a number of organic compounds that have different adsorbabilities and desorbabilities. In this study, the main aim was to examine the adsorption and desorption behavior of the biodegradable and nonbiodegradable fractions in a surface water [61]. The raw water was taken from the reservoir Ömerli in Istanbul, Turkey, and had the following characteristics: DOC: 4.23 mg L^{-1}, pH: 7.6, alkalinity: 58 mg L^{-1} CaCO$_3$, NH$_4^+$–N: 0.22 mg L^{-1}, Fe: 0.025 mg L^{-1}, Mn: 0.087 mg L^{-1}, and UV$_{254}$: 0.056 cm^{-1}.

Table 8.1 Types of water used in adsorption and desorption studies.

Water Type	Characteristics
Raw water	Filtered water
Ozonated raw water	Water subjected to ozonation at a dose of 2 mg O_3/mg DOC
Biologically treated water	Residual water after the BDOC test
Ozonated and biologically treated water	Residual water after the BDOC test on ozonated water

In order to study the adsorption and desorption, four different types of water were used, as shown in Table 8.1:

Raw and ozonated waters comprise biodegradable as well as nonbiodegradable organics. The biodegradable fraction is expected to increase upon ozonation. On the other hand, 'biologically treated water' contains mainly the nonbiodegradable organics in raw water. Similarly, the 'ozonated and biologically treated water' contains the nonbiodegradable organics after elimination of biodegradable compounds from ozonated water. The nonbiodegradable fraction in water represents the substances that would also not be biodegraded in BAC filtration. This fraction should be sufficiently adsorbed by activated carbon, since otherwise it would leave with the effluent.

In the study, the importance of GAC characteristics was also evaluated by using different GAC grades [61]. Adsorption and desorption experiments were carried out with each water type using three thermally activated (Row Supra, Norit 1240 and HD 4000 supplied from NORIT) GAC grades and one chemically activated (CAgran supplied from NORIT) GAC grade.

Adsorbability of NOM onto activated carbon The average initial DOC concentrations in raw, ozonated, biologically treated, and ozonated–biologically treated waters were found as 4.23, 4.12, 3.88, and 3.43 mg L^{-1}, respectively.

Adsorption isotherms constructed with 'raw water' or 'ozonated water' showed that the DOC in water could be divided into fictive components with different adsorbabilities. As shown in Figure 8.8, three main different adsorption regions could be distinguished. In **Region III** only those organics are removed by adsorption that have a high affinity to activated carbon at low carbon doses. Most of the NOM components are characterized by **Region II**, corresponding to medium carbon doses. The Freundlich isotherm constants belonging to this region are shown in Table 8.2. On the other hand, **Region I** corresponds to high GAC doses and indicates the concentration of the weakly adsorbable/nonadsorbable DOC, which was estimated as 0.7–0.9 mg L^{-1} for thermally activated carbons. Similarly, other researchers report that the nonadsorbable DOC in raw as well as ozonated waters ranged from 0.50 to 0.88 mg C L^{-1} for the initial DOC range of 5–6.2 mg C L^{-1} [62].

Moreover, ozonation of water influenced the adsorbability of NOM. However, the influence of ozonation was dependent on the carbon grade. Isotherms showed

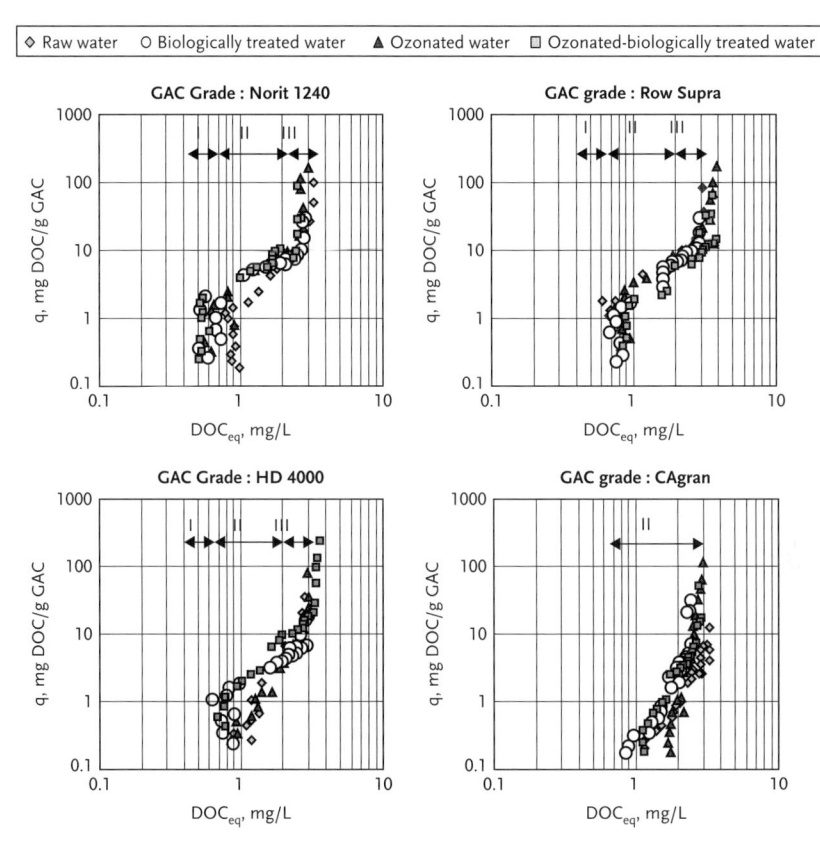

Figure 8.8 Adsorption isotherms showing the uptake of DOC from water onto various GAC grades (redrawn after Ref. [61]).

that the Freundlich adsorption capacity K_F and intensity $1/n$ were not significantly changed upon ozonation when the three thermally activated carbons were used (Figure 8.8, Table 8.2). However, adsorption onto the chemically activated carbon CAgran was obviously lower than that onto other carbon grades. Compared to others, CAgran had a larger pore volume and larger macropores. Ozonation even worsened the adsorption onto this carbon grade. Probably, upon ozonation, smaller NOM molecules were formed that could not be held by this carbon grade.

For effective adsorption, the surface charges of NOM and GAC are also relevant. As seen in Figure 8.9a, in all water types the surface charge of NOM was negative. Ozonation increased the negative charge on NOM compared to raw water, indicating that more polar compounds were formed that were hydrophilic.

Figure 8.9b shows the surface charges of GAC grades. The chemically activated carbon, CAgran had a pH_{PZC} of about 4. Thus, under test conditions (pH: 7.6), the surface charge of this carbon was negative, which caused repulsion of negatively charged NOM. The thermally activated carbons having higher pH_{PZC} exhibited

Table 8.2 Freundlich isotherm constants belonging to Region II (medium carbon doses) in Figure 8.8.

Water Type and GAC Grade	K_F (mg/g) (L/mg)$^{1/n}$	$1/n$	R^2	Remarks
Raw Water (Filtered water)				
Row Supra	2.96±0.09	1.51±0.13	0.94	Adsorption behavior of biodegradable
CAgran	0.13±0.004	3.44±0.41	0.93	and nonbiodegradable components in
Norit 1240	1.45±0.10	2.32±0.11	0.96	NOM
HD 4000	0.29±0.01	4.32±0.39	0.93	
Ozonated Water (ozonation of raw water at 2 mg O$_3$/mg DOC)				
Row Supra	2.96±0.25	1.33±0.17	0.96	Adsorption behavior of NOM upon
CAgran	0.001±0.002	10.34±0.82	0.94	ozonation (the biodegradable fraction is
Norit 1240	2.98±0.17	1.53±0.31	0.95	increased compared to raw water)
HD 4000	0.36±0.10	3.56±0.34	0.95	
Biologically Treated Water (Residual water after the BDOC test)				
Row Supra	1.79±0.10	1.92±0.29	0.97	Adsorption behavior of the
CAgran	0.26±0.03	3.59±0.15	0.94	nonbiodegradable components in raw
Norit 1240	3.73±0.26	0.87±0.03	0.95	water
HD 4000	1.70±0.12	1.33±0.12	0.96	
Ozonated and Biologically Treated Water (Residual water after the BDOC test on ozonated water)				
Row Supra	1.40±0.21	1.77±0.28	0.93	Adsorption behavior of the
CAgran	0.19±0.04	4.12±0.51	0.97	nonbiodegradable components in
Norit 1240	3.85±0.43	1.17±0.17	0.92	ozonated water
HD 4000	1.76±0.18	2.08±0.22	0.98	

higher NOM adsorption capacities for all four different water types. In samples denoted as 'biologically treated' and 'ozonated-biologically treated' the negative charge on NOM was reduced upon the elimination of BDOC (Figure 8.9a). This means that the BDOC had a more negative charge than the remaining nonbiodegradable organics. As a result of the decrease in the net negative charge the repulsion forces between GAC and NOM were reduced, and adsorption was slightly improved. This indicated that the nonbiodegradable fraction in NOM had a higher adsorptivity than BDOC. This is a desired result, since adsorption of nonbiodegradable organics is one of the main aims in GAC adsorption.

Desorbability of NOM from activated carbon It is also essential to assess the reversibility of NOM adsorption. Therefore, desorption isotherms were constructed for each type of water, as shown in Figure 8.10. These isotherms show the DOC loading still remaining on the GAC surface with respect to equilibrium DOC concentration. In contrast to adsorption, in desorption experiments a low surface loading (low q) would indicate favorable desorption. In the case of raw water, the maximum percent DOC desorption from CAgran, HD 4000, Row Supra, and

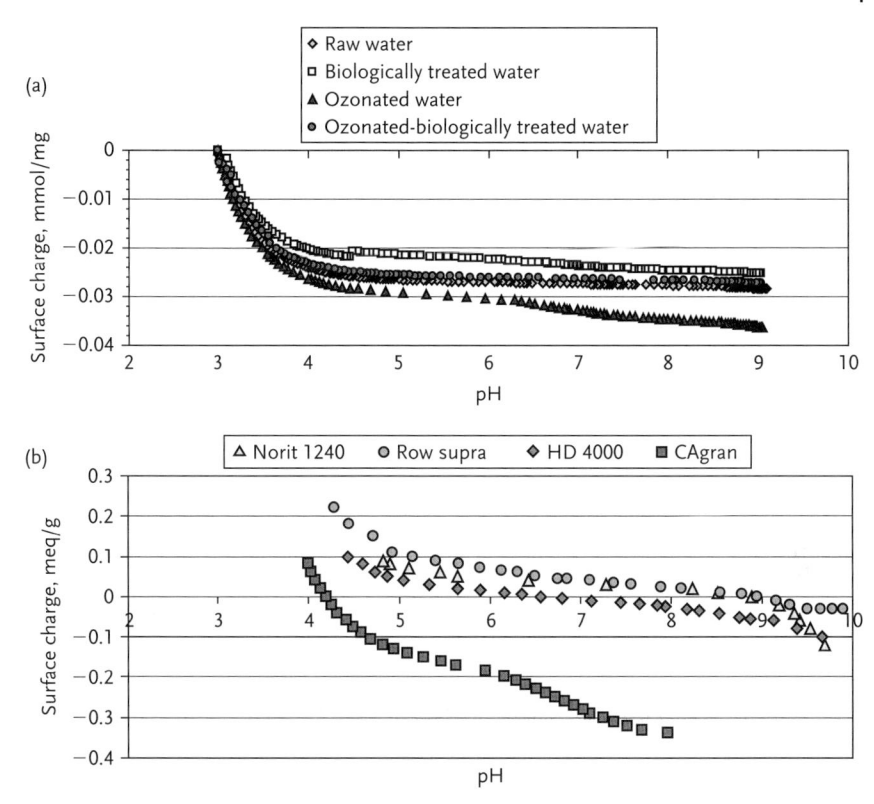

Figure 8.9 Surface charges on (a) NOM in raw and treated waters, (b) various GAC grades (redrawn after Ref. [61]).

Norit 1240 was 51%, 38%, 24%, and 4%, respectively. Compared to raw water, in ozonated waters, desorption slightly increased for all carbon grades. In the case of 'biologically treated water' and 'ozonated-biologically treated' waters, mainly containing nonbiodegradable organics, desorption from GAC was similar. In any case, the chemically activated carbon CAgran had the lowest surface loading (q), and thus it exhibited the highest desorption. Also, HD 4000 desorbed to some extent, whereas Norit 1240 and Row Supra exhibited relatively irreversible adsorption (Figure 8.10). In particular, in the case of 'biologically treated' and 'ozonated-biologically treated' waters it is important that desorption from activated carbon should be low, since these waters contain mainly nonbiodegradable matter that can only be removed by adsorption.

In general, adsorption results indicated that the surface chemistry of NOM and GAC grades was more influential on adsorption than the pore structure of GAC. Although the GAC grade Cagran has the highest total pore volume (especially high meso- and macropore volumes), it had a poorer adsorption capacity than others. Moreover, the type of GAC outweighed the importance of ozonation with respect to both adsorption and desorption of DOC [61]. The significance of adsorbability,

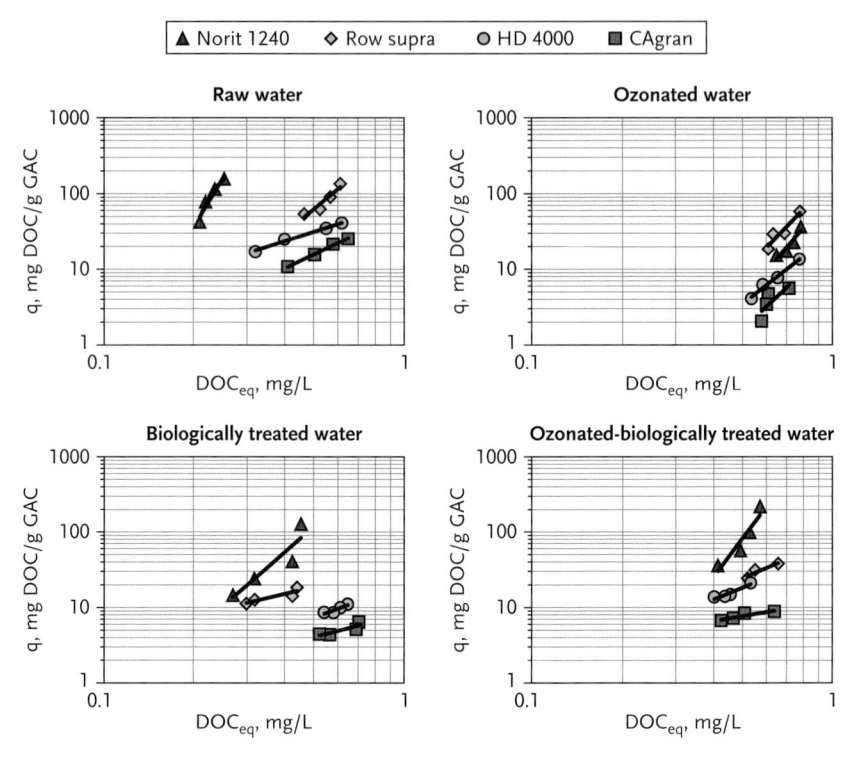

Figure 8.10 Desorption isotherms with various GAC grades and water samples (redrawn after Ref. [61]).

desorbability and biodegradability in raw or ozonated waters in BAC filtration is further discussed in Chapter 9.

References

1 EAWAG (Publ.) (2009) Wave21 final report – Drinking water for the 21st Century. Schriftenreihe Nr. 20.
2 Pontius, F.W. (2003) *Drinking Water Regulation and Health*, John Wiley & Sons, Inc.
3 van der Kooij, D. (2003) Managing regrowth in drinking water distribution systems, in *Heterotrophic Plate Counts and Drinking-water Safety* (eds J. Bartram, J. Cotruvo, M. Exner, C. Fricker, and A. Glasmacher), World Health Organization (WHO), IWA Publishing, London, UK, pp. 199–232.
4 Rittmann, B.E. and McCarty, P.L. (2001) *Environmental Biotechnology: Principles and Applications*, McGraw-Hill.
5 Schreiber, B., Schmalz, V., and Eckhard Worch, E. (2006) *Exportorientierte F&E auf dem Gebiet der Wasserver- und -entsorgung Teil I: Trinkwasser Band 2: Leitfaden (R&D in the field of water supply and waste water treatment under regional conditions Part I: Drinking water, Vol. 2: Recommendations)*, ISBN 3-00-015478-7, p. 200.
6 McDonald, S., Bishop, A.G., Prenzler, P. D., and Robards, K. (2004) Review:

analytical chemistry of freshwater humic substances. *Analytica Chimica Acta*, **527**, 105–124.

7 Volk, C.J.,Volk, C.M., and Kaplan, L.A. (1997) Chemical composition of biodegradable dissolved organic matter in stream water. *Limnology and Oceanography*, **42**, 39–44.

8 Juhna, T. and Melin E. (2006) Ozonation and Biofiltration in Water Treatment – Operational Status and Optimization Issues, Deliverable number D.5.3.1 B, *TECHNEAU Integrated Project funded by the European Commission under the Sixth Framework Programme* (contract number 018320).

9 Servais, P., Billen, G., and Hascoet, M.C. (1987) Determination of the biodegradable fraction of dissolved organic matter. *Water Research*, **21**, 445–450.

10 Yavich, A.A., Lee, K.H., Cehn, K.C., Pape, L., and Maten, S.J. (2004) Evaluation of biodegradability of NOM after ozonation. *Water Research*, **38**, 2839–2846.

11 Laurent, P., Prevost, M., Cigana, J., Niquette, P., and Servais, P. (1999) Biodegradable organic matter removal in biological filters: evaluation of the CHABROL model. *Water Research*, **33** (6), 1387–1398.

12 van der Aa, L.T.J., Rietveld, L.C., and van Dijk, J.C. (2010) Effects of ozonation and temperature on biodegradation of natural organic matter in biological granular activated carbon filters. *Drinking Water Engineering and Science Discussion*, **3**, 107–132.

13 Uhl, W. (2000) Simultane biotische und abiotische Prozesse in Aktivkohlefiltern der Trinkwasseraufbereitung in: 'Sorptionsverfahren der Wasser-und Abwasserreinigung – Schriftenreihe des SFB 193 (Biologische Behandlung industrieller und gewerblicher Abwässer) der TU Berlin', 13, 151–179, TU-Berlin Universitätsbibliothek, Berlin, ISBN3-7983-1827-1.

14 Hammes, F. and Egli, T.(2005) New method for assimilable organic carbon determination using flow-cytometric enumeration and a natural microbial consortium as inoculum. *Environmental Science and Technology*, **39**, 3289–3294.

15 Joret, J.C., Levi, Y., Dupin, T., and Gilbert, M. (1988) Rapid method for estimating bioeliminable organic carbon in water. Proceedings of the American Water Works Association Annual Conference, 1715–1725.

16 Frias, J., Ribas, F., and Lucena, F.(1992) A method for the measurement of biodegradable organic carbon in waters. *Water Research*, **26**, 255–258.

17 Ribas, F., Frias, J., Ventura, F., Mohedano, L., and Lucena, F.(1995) Assessment of biological activity and fate of organic compounds in a reactor for the measurement of biodegradable organic carbon in water. *Journal of Applied Bacteriology*, **79**, 556–568.

18 Trulleyová, S. and Rulík, M.(2004) Determination of biodegradable dissolved organic carbon in waters: comparison of batch methods. *Science of the Total Environment*, **332**, 253–260.

19 Park, S.K., Pak, K.R., Choi, S.C., and Kim, Y.K. (2004) Evaluation of bioassays for analyzing biodegradable dissolved organic carbon in drinking water. *Journal of Environmental Science and Health Part A—Toxic/Hazardous Substances and Environmental Engineering*, **A39**, 103–112.

20 McDowell, W.H., Zsolnay, A., Aitkenhead-Peterson, J.A., Gregorich, E.G., Jones, D.L., Jödemann, D., Kalbitz, K., Marschner, B., and Schwesig, D., (2006) A comparison of methods to determine the biodegradable dissolved organic carbon from different terrestrial sources. *Soil Biology and Biochemistry*, **38**, 1933–1942.

21 Yapsakli, K. (2008) Application of biological activated carbon (BAC) in drinking water treatment. Ph.D. Thesis, Bogazici University, Institute of Environmental Sciences, Istanbul, Turkey.

22 Yapsakli, K. and Çeçen, F. (2009) Use of enriched inoculum for determination of biodegradable dissolved organic carbon

(BDOC) in drinking water. *Water Science and Technology: Water Suply*, **9** (2), 149–157.

23 Hammes, F. (2008) A comparison of AOC methods used by the different TECHNEAU partners, TECHNEAU Integrated Project, Deliverable number 3.3.10, *TECHNEAU Integrated Project* funded by the European Commission under the Sixth Framework Programme (contract number 018320).

24 Volk, C. and LeChevallier, M.W. (2000) Assessing biodegradable organic matter. *Journal of American Water Works Association*, **92** (5), 64–76.

25 Escobar, I.C. and Randall, A.A. (2001) Assimilable organic carbon (AOC) and biodegradable dissolved organic carbon (BDOC): complementary measurements. *Water Research*, **35** (18), 4444–4454.

26 Schaule, G., Moschnitschka, D., Schulte, S., Tamachkiarow, A., and Flemming, H.C. (2007) Biofilm growth in response to various concentrations of biodegradable material in drinking water. *Water Science and Technology*, **55** (8–9), 191–195.

27 Sharp, R.R., Camper, A.K., Crippen, J.J., Schneider, O.D., and Leggiero, S. (2001) Evaluation of drinking water biostability using biofilm methods. *Journal of Environmental Engineering*, **127** (5), 403–410.

28 Liao, W., Christman, R.F., Johnson, J.D., and Millington, D.S. (1982) Structural characterization of aquatic humic material. *Environmental Science and Technology*, **16** (7), 403–410.

29 Volk, C. and LeChevallier, M.W. (2002) Effects of conventional treatment on AOC and BDOC levels. *Journal of American Water Works Association*, **94** (6), 112–123.

30 Jacangelo, J.G. (1995) Selected processes for removing NOM: an overview. *Journal of American Water Works Association*, **87** (1), 64–78.

31 Randtke, S.J. (1988) Organic contaminant removal by coagulation and related process configurations. *Journal of American Water Works Association*, **80** (5), 40–56.

32 Volk, C., Bell, K., Ibrahim, E., Verges, D., Amy, G., and LeChevallier, M. (2000) Impact of enhanced and optimized coagulation on removal of organic matter and its biodegradable fraction in drinking water. *Water Research*, **34** (12), 3247–3257.

33 Shi-hu, S., Min, Y., Nai-yun, G., and Wen-Jie, H. (2008) Molecular weight distribution variation of assimilable organic carbon during ozonation/BAC process. *Journal of Water Supply: Research and Technology-AQUA*, **57** (4), 253–258.

34 Edzwald, J.K. (1994) Coagulation concepts for removal of TOC. American Water Works Association WQTC Conference, San Fransisco, 6–10 November 1994.

35 Hong, S., Kim, S., and Bae, C. (2009) Efficiency of enhanced coagulation for removal of NOM and for adsorbability of NOM on GAC. *Desalination and Water Treatment*, **2**, 89–94.

36 Cook, D., Newcombe, G., and Sztajnbok, P. (2001) The application of powdered activated carbon for MIB and geosmin removal: predicting PAC doses in four raw waters. *Water Research*, **35** (59), 1325–1333.

37 Knappe, D.R.U., Belk, R.C., Bariley, D., Gandy, S., and Rastogi, N. (2004) Algae detection and removal strategies for drinking water treatment plants, Research Report/AWWA Research Foundation.

38 Sontheimer, H., Crittenden, J., and Summers, R.S. (1988) *Activated Carbon for Water Treatment*, 2nd edn, Forschungstelle am Engler–Bunte-Institute, Universität Karlsruhe, Germany.

39 Bundermann, G. (2006) 30 years of RWW's practical experience with an advanced microbiological water treatment system for Ruhr river water – 'the Mülheim treatment process 1976–2006', in *Recent Progress in Slow Sand and Alernative Biofiltration Processes* (eds R. Gimbel, N.J.G. Graham, and M.R. Collins), IWA Publishing, London, UK. pp. 30–38.

40 Sontheimer, H., Frick, B.R., Fettig, J., Hörner, G., Hubele, C., and Zimmer, G.

(1985) *Adsorptionsverfahren zur Wasserreinigung*, DVGW-Forschungsstelle am Engler–Bunte-Institute, Universität Karlsruhe, Germany.

41 Uhl, W. (2006) Biofiltration processes for organic matter removal, in *Biotechnology*, 2nd revised edn (eds Rehm H.J. and Reed G.), 11c, Wiley VCH, Weinheim, Germany.

42 Marda, S., Kim, D., Kim, M.J., Gordy, J., Pierpont, S., Gianatasio, J., Amirtharajah, A., and Kim, J.H. (2008) Plant conversion experience: ozone BAC process installation and disinfectant residual control. *Journal AWWA*, **100** (4).

43 Amy, G., Carlson, K., Collins M.R., Drewes, J., Gruenheid, S., and Jekel, M. (2006) Integrated comparison of biofiltration in engineered versus natural systems, in *Recent Progress in Slow Sand and Alernative Biofiltration Processes* (eds R. Gimbel, N.J.G. Graham, and M.R. Collins), IWA Publishing, London, UK. pp. 3–11.

44 Rice, R.C. and Netzer, A. (eds) (1982) *Handbook of Ozone Technology and Applications*, vol. **1**, Ann Arbor Science.

45 Ijpelaar, G.F., Harmsen, D.J.H., and Heringa, M. (2007) UV disinfection and UV/H_2O_2 oxidation: by-product formation and control. Deliverable number D 2.4.1.1, *TECHNEAU Integrated Project funded by the European Commission under the Sixth Framework Programme* (contract number 018320).

46 Nishijima, W., Kim, W.H., Shoto, E., and Okada, M. (1998) The performance of an ozonation-biological activated carbon process under long term operation. *Water Science and Technology*, **38** (6), 163–169.

47 Yapsakli, K. and Çeçen, F. (2010) Effect of type of granular activated carbon on DOC biodegradation in biological activated carbon filters. *Process Biochemistry*, **45**, 355–362.

48 Hammes, F., Meylan, S., Salhi, E., Köster, O., Egli, T., and von Gunten, U. (2007) Formation of assimilable organic carbon (AOC) and specific natural organic matter (NOM) fractions during ozonation of phytoplankton. *Water Research*, **41**, 1447–1454.

49 Camel, V. and Bermond, A. (1998) The use of ozone and associated oxidation processes in drinking water treatment. *Water Research*, **32**, 3208–3222.

50 Nishijima, W. and Okada, M. (1998) Particle separation as a pretreatment of an advanced drinking water process by ozonation and biological activated carbon. *Water Science and Technology*, **37** (10), 117–124.

51 Win, Y.Y., Kumke, M.U., Specht, C.H., Schindelin, A.J., Kolliopulos, G., Ohlenbusch, G., Kleiser, G., Hesse, S., and Friml, F.H. (2000) Influence of oxidation of dissolved organic carbon (DOM) on subsequent water treatment processes. *Water Research*, **34** (7), 2098–2104.

52 Hammes, F., Salhi, E., Köster, O., Kaiser, H.P., Egli, T., and von Gunten, U. (2006) Mechanistic and kinetic evaluation of organic disinfection by-product and assimilable organic carbon (AOC) formation during the ozonation of drinking water. *Water Research*, **40**, 2275–2286.

53 Kim, W.H., Nishijima, W., Shoto, E., and Okada, M. (1997) Competitive removal of dissolved organic carbon by adsorption and biodegradation on biological activated carbon. *Water Science and Technology*, **35** (7), 147–153.

54 Polanska, M., Huysman, K., and van Keer, C. (2005) Investigation of assimilable organic carbon (AOC) in Flemish drinking water. *Water Research*, **39**, 2259–2266.

55 van der Helm, A.W.C., Grefte, A., Baars, E.T., Rietveld, L.C., van Dijk, J.C., and Amy, G.L. (2009) Effects of natural organic matter (NOM) character and removal on ozonation for maximizing disinfection with minimum bromate and AOC formation. *Journal of Water Supply: Research and Technology-AQUA*, **56** (6), 373–385.

56 Carlson, K.H. and Amy, G. (1997) The formation of filter removable biodegradable organic matter during ozonation. *Ozone Science*

and Engineering, **19** (2), 179–199.

57 Carlson, K.H. and Amy, G. (2001) Ozone and biofiltration optimization for multiple objectives. *Journal of American Water Works Association*, **93** (1), 88–98.

58 Nishijima, W. and Speitel, G.E. Jr (2004) Fate of biodegradable dissolved organic carbon produced by ozonation on biological activated carbon. *Chemosphere*, **56**, 113–119.

59 Hubele, C. and Sontheimer, H. (1984) Adsorption and biodegradation in activated carbon filters treating pre-ozonated humic acid. Proceedings of the 1984 National Conference on Environmental Engineering of the American Society of Civil Engineers, held in Los Angeles, California , June 25–27, 1984, 376–381 (eds M. Pirbazari and J.S. Devinny).

60 Karanfil, T., Kitis, M., Kilduff, J.E., and Wigton, A. (1999) The role of GAC surface chemistry on the adsorption of organic compounds: Natural organic matter. *Environmental Science and Technology*, **33** (18), 3225–3233.

61 Yapsakli, K., Çeçen, F., Aktaş, Ö., and Can, Z.S. (2009) Impact of surface properties of granular activated carbon and preozonation on adsorption and desorption of natural organic matter. *Environmental Engineering Science*, **26**, 489–500.

62 van der Aa, L.T.J., Achari, V.S., Rietveld, L.C., Siegers, W.G., van Dijk, J.C. (2004) Modelling biological activated carbon filtration: determination adsorption isotherms of organic compounds, in Proceedings of the WISA Conference, Cape Town, 2–6 May, pp. 1143–1153.

9
Removal of NOM, Nutrients, and Micropollutants in BAC Filtration

Ferhan Çeçen

9.1
Removal of Organic Matter

9.1.1
Main Mechanisms

Figure 9.1 illustrates the main processes taking place inside an activated carbon filter used in water treatment. The relative importance of adsorption and biodegradation changes in the course of transition of a Granular Activated Carbon (GAC) filter into a Biological Activated Carbon (BAC) filter.

9.1.2
Breakthrough Curves

In drinking water biofiltration, the breakthrough curves indicate the fraction of nonbiodegradable and nonadsorbable matter as well as the fraction eliminated by biodegradation (Figure 9.2). The most easily removed fraction in dissolved organic carbon (DOC) is the one that is amenable to both adsorption and biodegradation.

9.1.2.1
Initial Stage of Operation: Removal by Adsorption

As shown in Figure 9.2, in the initial stages of operation (**Period A**), organic matter is primarily removed by adsorption onto GAC since a microorganism layer has not yet developed. In water treatment, this period of adsorption may last for several months, depending on the operation conditions and characteristics of the water. Both biodegradable and nonbiodegradable organic matter, represented by BDOC and non-BDOC (or NBDOC), respectively, may be removed by adsorption. However, the rate of adsorption of biodegradable compounds was shown to be lower than that of nonbiodegradable ones [1]. In **Period A**, the effluent consists mainly of nonadsorbable compounds.

Activated Carbon for Water and Wastewater Treatment: Integration of Adsorption and Biological Treatment.
First Edition. Ferhan Çeçen and Özgür Aktaş
© 2011 WILEY-VCH Verlag GmbH & Co. KGaA, Weinheim.
Published 2011 by WILEY-VCH Verlag GmbH & Co. KGaA

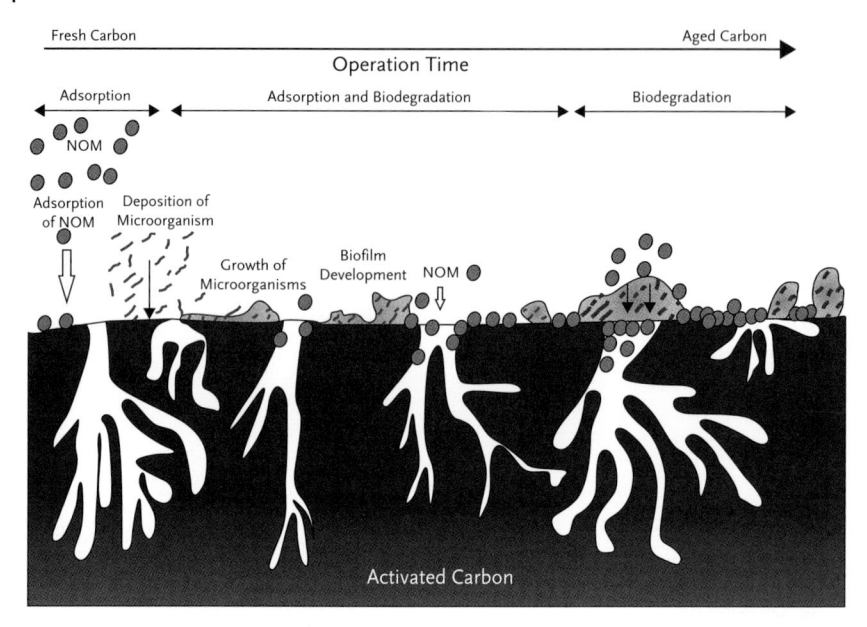

Figure 9.1 Adsorption, biodegradation, and biofilm formation in drinking water BAC filtration.

9.1.2.2
Intermediate and Later Stages of Operation

The formation of a biofilm on the GAC surface is a crucial requirement for biodegradation. The effluent of sand filtration, which often precedes GAC filtration, contains microorganisms, biodegradable organics, suspended cells, particulates, dissolved oxygen (DO), and dissolved nutrients. Microorganisms present in this water first deposit and attach onto the GAC surface and then begin to grow.

At intermediate stages of operation (**Period A–B in** Figure 9.2), organics are both adsorbed and biodegraded by microorganisms. Thus, most of the adsorbable and biodegradable fraction will be removed inside the filter. In any case, removal of NOM by biological means is preferred rather than adsorption. Biological removal of organics would lower the loading on the GAC surface and provide additional adsorption capacity for nonbiodegradable compounds. Consequently, this would increase the service life of the activated carbon. Further, biological removal of NOM would favor the adsorption of organic micropollutants that usually compete with NOM.

Figure 9.1 illustrates the fact that NOM in water can also be biosorbed onto the biofilm. In this, the surface charge of the biofilm as well as the molecular sizes and surface charges of organic molecules in NOM play a significant role. Many studies reveal that NOM carries a negative charge. An example of the charge of the NOM in various types of water is presented in Chapter 8 [2]. The net negative surface charge of biofilms usually hinders sorption and transport of NOM molecules [3].

In one study, in a fixed bed covered by a living biofilm, most of the NOM removal was shown to take place by biosorption of organics having a molecular weight smaller than 3 kDa. Ozonation at a dose of 1 mg O_3/mg DOC decreased the

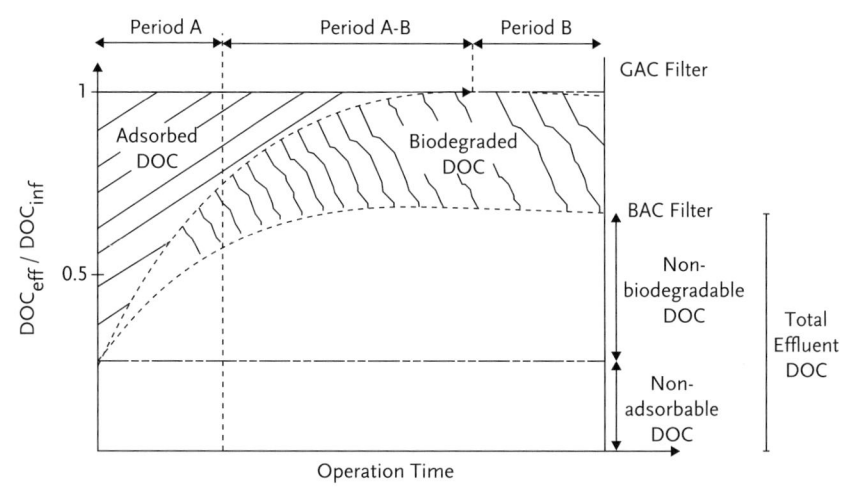

Figure 9.2 Breakthrough in DOC in operation of a BAC filter.

molecular size of organics, but also increased the acidity of the NOM. However, biosorption did not change upon ozonation. Apparently, these positive and negative effects counterbalanced each other [4].

In the intermediate stages of operation (**Period A–B**), adsorption gradually decreases whereas biodegradation gains importance. At later stages of operation (**Period B**), biological processes prevail, since the adsorption capacity of GAC is completely exhausted in most cases. When the surface of GAC is loaded with organic matter in wastewater treatment it may be renewed by bioregeneration. However, bioregeneration is a very debated issue in drinking water filtration, as discussed in Section 9.1.3.

Table 9.1 Removal of different NOM fractions during operation of GAC/BAC filters (based on [5]).

GAC Filtration	**Transition from GAC to BAC**	**BAC filtration**
	Mechanism of removal	
Adsorption	Adsorption and Biodegradation	Biodegradation
	Removal of NOM fractions	
• Removal of humics (up to 90%) • Removal of HMW and LMW organics depending on GAC properties • No removal of biopolymers	• Pore blockage by NOM<5 kDa: Decrease in adsorption capacity • Limited removal of humics • Onset of biological activity- Removal of LMW fractions • No removal of biopolymers	• Removal of biodegradable humics and LMW fractions • No removal of biopolymers

Table 9.2 Data on biological activated carbon (BAC) columns.

Type of Water	Raw and ozonated
Filters/GAC Grades	2 filled with CAgran (Norit Company): very open (meso and macro) pore structure, wood-based, chemically activated with phosphoric acid.
	2 filled with Norit 1240 (Norit Company): microporous structure, coal-based, produced by thermal activation
Column Height	1 m
Column Diameter	2 cm
Hydraulic Loading Rate	1.67 m h^{-1}
GAC depth	50 cm

The performance of BAC filtration is commonly monitored by the parameters DOC, BDOC, and AOC. Additionally, in the last decades sophisticated analytical techniques have come into use for the determination of the various fractions in NOM [5]. Table 9.1 represents the removal of NOM fractions with respect to operation time in a filter. In general, biopolymers representing the high molecular weight (HMW) organics are not removed in BAC filtration. These biopolymers, together with some fraction of humics, constitute the major part of non-biodegradable organics in the effluent.

In order to differentiate between the removal of organic matter by biodegradation and that by adsorption, laboratory- or pilot-scale tests have to be conducted. The following study is an example of this.

EXAMPLE 9.1: Removal of Organic Matter from Raw and Ozonated Water in GAC/BAC Filtration Studies to date show that the DOC fractionation in water is of great importance with respect to the breakthrough in BAC filtration. However, it is recognized that less attention has been paid to GAC properties, which could also have an influence on breakthrough profiles.

This example is related to Example 8.1 in Chapter 8. While the former example highlights the adsorption and desorption of organics using two GAC grades, the current example compares the performance when BAC filters that are filled with these GAC grades are fed with raw or ozonated water. Table 9.2 presents basic data on the experimental setup that is shown in Figure 9.3.

Differentiation between Adsorption and Biological Removal GAC filters fed with sterilized water ('sterile' or 'nonbioactive' GAC filters) were run in parallel with BAC filters. The main aim was to differentiate adsorptive removal from biodegradation. In sterile GAC filters only adsorption would be expected, whereas in BAC filters both adsorption and biodegradation mechanisms would be active before breakthrough. The breakthrough profiles in sterile GAC and corresponding BAC filters are illustrated in Figure 9.4a and b for the two different GAC grades, CAgran and Norit 1240, respectively [7].

Sterile GAC filters – Removal by Adsorption As seen in Figure 9.4a, after approximately 1000 bed volumes, the adsorption capacity of the sterile CAgran filter was

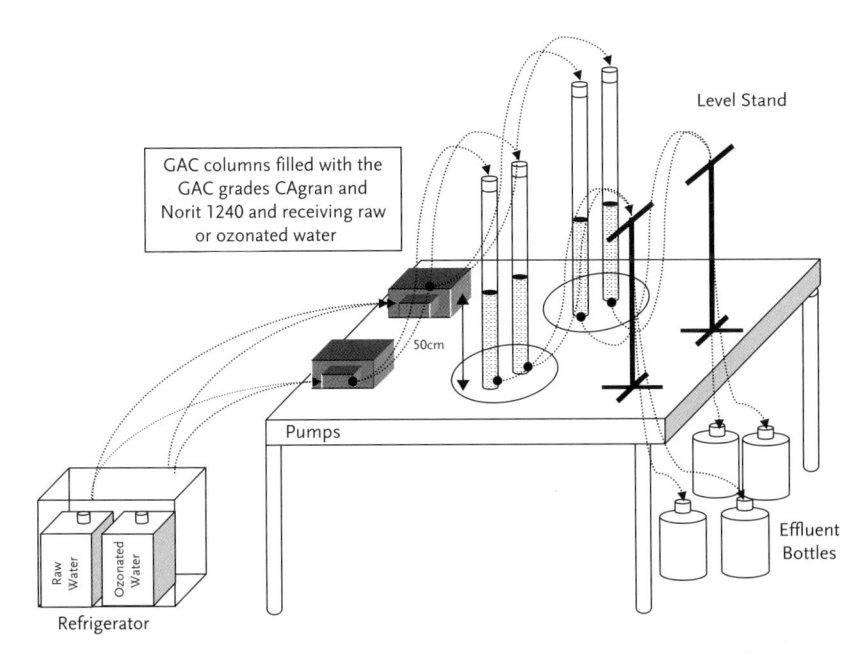

Figure 9.3 Laboratory-scale Biological Activated Carbon (BAC) setup (redrawn after [6]).

exhausted. On the other hand, Figure 9.4b shows that in the sterile Norit 1240 filter breakthrough commenced much later, at about 4000 bed volumes. In both filters, the nonadsorbable DOC fraction was about the same, 22% and 25%, respectively.

BAC filters: Removal by Adsorption and Biodegradation Breakthrough profiles in BAC filters (Figure 9.4) were compared with corresponding sterile filters. For a filter in which oxygen is not chemisorbed by the activated carbon surface, oxygen consumption measurements would provide important information on the extent of biodegradation. CAgran is a chemically activated carbon grade which has a very low affinity for oxygen. The DO measurements along the depth of the BAC filter filled with this carbon grade revealed the direct correlation of oxygen consumption with substrate removal [6]. Therefore, a great part of DOC removal taking place after the exhaustion of adsorptive capacity was attributed to biodegradation. BAC filters filled with CAgran reached breakthrough within 2000 bed volumes (Figure 9.4a). The biodegradable fraction in the initial DOC was about 47%.

On the other hand, as seen in Figure 9.4b, breakthrough did not commence in the Norit 1240 filter even after 18 000 bed volumes. Comparison of this operation with sterile filtration led to the conclusion that the effluent of this BAC filter was composed of nonbiodegradable and nonadsorbable matter. This indicated that biodegradation was still continuing in this filter. However, in this type of filter, no direct correlation can be made between oxygen consumption and biodegradation. Since Norit 1240 is a carbon grade which has an affinity for oxygen, in addition to biodegradation, some of the oxygen consumption recorded may be attributed to chemisorption of oxygen.

The most problematic part of NOM is the nonadsorbable and nonbiodegradable fraction. It is this fraction in DOC that would leave the filter and act as a precursor

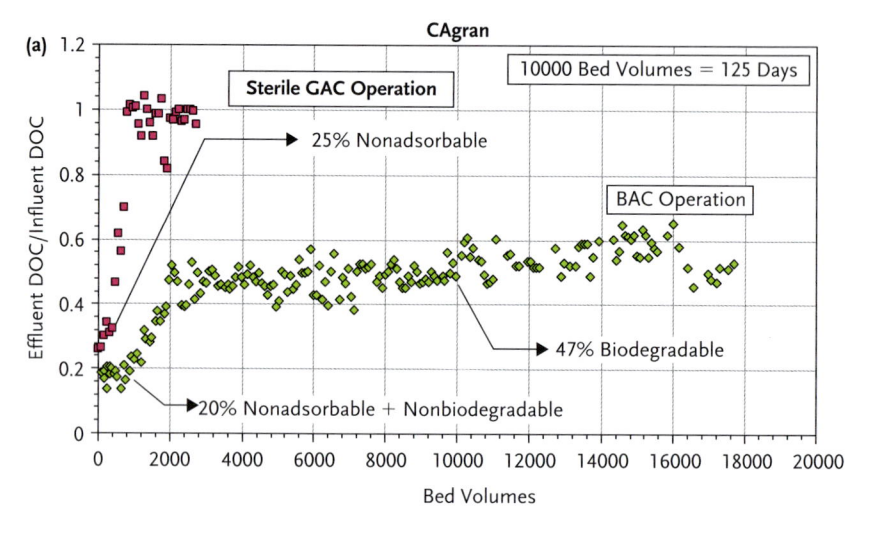

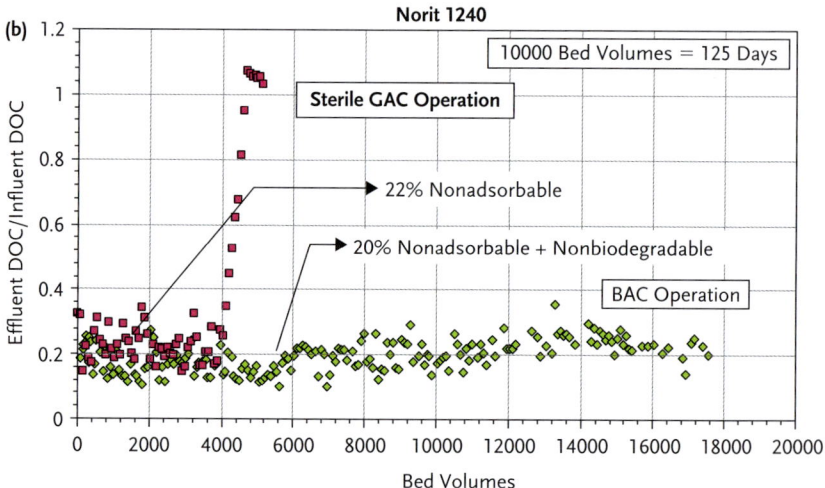

Figure 9.4 Breakthrough of DOC in sterile GAC and BAC operation with (a) CAgran and (b) Norit 1240 (redrawn after [7]).

for Disinfection By-Product (DBP) formation. As shown in Figure 9.4, this fraction was about 20% in this specific surface water [7].

Comparison of DOC and BDOC Removal   As shown in Figure 9.5, the performance of the Norit 1240 filter in DOC removal was better than that of the CAgran filter. The major advantage of Norit 1240 lies in the considerable adsorption of readily and slowly biodegradable organics, as concluded from the isotherm data presented in the example in Chapter 8. Another explanation for enhanced removal may be that catalytic processes are taking place on the surface of this carbon that

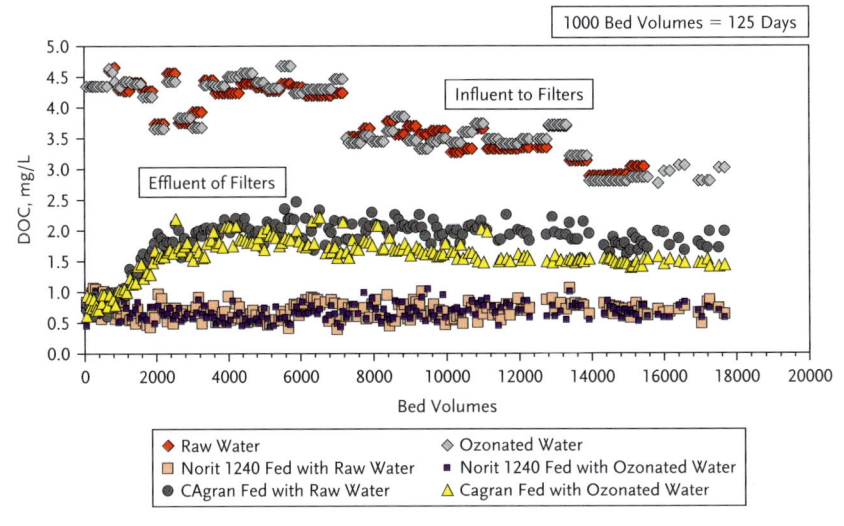

Figure 9.5 Influent DOC in raw water and ozonated water and the corresponding effluent DOC in Norit and CAgran filters fed with these waters (redrawn after [7]).

convert nonbiodegradable substances into biodegradable ones. Such reactions are probable since Norit 1240 chemisorbs large amounts of DO on its surface, a factor favoring biodegradation of slowly biodegradable organics, as outlined in Chapter 3.

Ozonation obviously increased the BDOC levels in the water compared to raw water (Figure 9.6). The maximum percent increase in BDOC upon ozonation was 245% with an average value of 166%. However, from the point of BAC filtration, for both carbon grades, ozonation did not have a noticeable impact on effluent

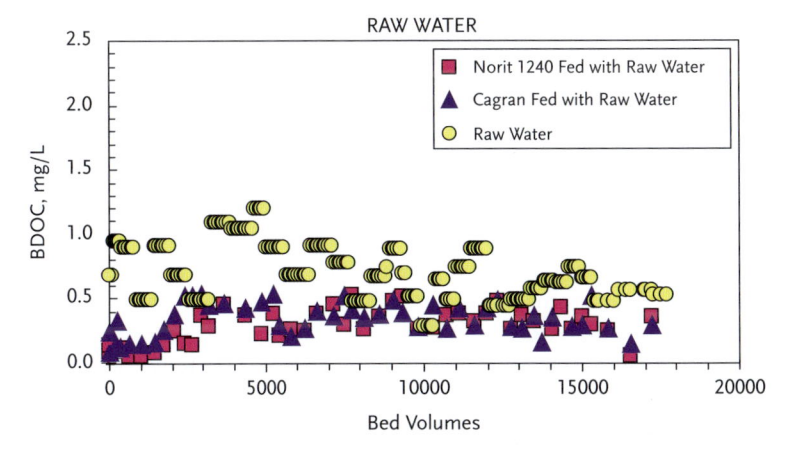

Figure 9.6 Influent and effluent BDOC in BAC filters receiving raw water (upper figure) and ozonated water (lower figure) (adapted from [6]).

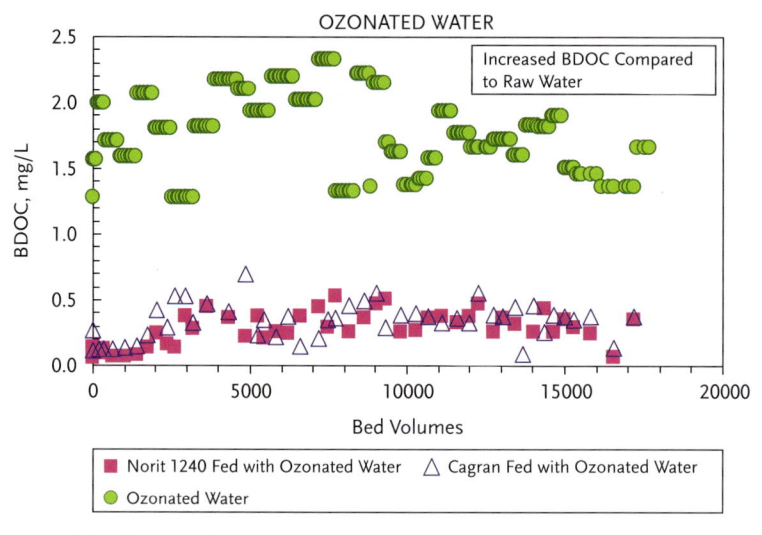

Figure 9.6 (Continued)

DOC, as seen in Figure 9.5. This was attributed to the fact that the surface water had initially a relatively high biodegradable fraction, as indicated by low SUVA (2.54 ± 0.56 L/mgC.m). Therefore, for this specific case ozonation proved to be redundant.

9.1.3
Bioregeneration of BAC Filters

Bioregeneration of activated carbon and modeling of bioregeneration are extensively discussed in Chapter 7, mainly under conditions pertaining to wastewater treatment. However, in drinking water treatment both the type of organic matter and the concentration differ from those in wastewater treatment.

In BAC filtration of water, contradictory arguments often exist about the bioregeneration of activated carbon. In the evaluation of bioregeneration, a common approach is to construct a mass balance for oxygen consumption and CO_2 production. Early studies on BAC filtration point to bioregeneration of activated carbon, whereas some contrary findings are also reported [8]. In new studies, some authors hold the opinion that bioregeneration of activated carbon surface by bacteria is possible [9]. However, many researchers state that adsorption and biological removal mechanisms act separately and deny the possibility of bioregeneration in drinking water BAC filtration.

As outlined in detail in Chapter 7, for bioregeneration to take place, one of the requirements is that a concentration gradient develops between the GAC surface and the bulk medium, and that this finally leads to desorption of substrate. In some cases of BAC filtration, such desorption was shown to take place when the concentration of micropollutants in the influent went down [10]. Presumably, bioregeneration of activated carbon took place, although it was not documented very clearly. In some cases, as in groundwater bioremediation, bioregeneration of activated carbon barriers was also demonstrated (Section 9.5). During periods of low biological activity, organic micropollutants were removed by adsorption in the GAC barriers placed in the groundwater. These presorbed pollutants then desorbed and were available for subsequent biodegradation, resulting in bioregeneration of activated carbon [11].

In summary, the issue of whether bioregeneration plays a significant role in drinking water BAC filtration is not yet resolved. In fact, bioregeneration is documented in only a few cases for specific micropollutants of small size, but no precise documentation exists on bioregeneration of activated carbon with respect to compounds contributing to NOM.

9.2
Factors Affecting the Performance of BAC Filtration

9.2.1
Comparison of GAC with Other Media

9.2.1.1
Biomass Attachment

A sand filter provides a higher specific surface area (surface area per unit volume of packed filter) for biomass attachment than GAC because the effective size of sand is usually smaller. However, the amount of biomass in a colonized GAC filter (BAC filter) is usually much higher than that in anthracite or sand media. For example, in the case of preozonated drinking water, the fixed bacterial biomass per gram of GAC was found to be about two orders of magnitude higher than that in a sand or anthracite filter [12, 13]. In the GAC Sandwich^{TM} filter, in which GAC is placed between the sand layers of a slow filter, the additional dissolved oxygen demand was thought to result from rapid colonization and microbiological activity of the GAC surface [14].

9.2.1.2
Removal Performance

In older treatment works, GAC has usually been integrated into already existing sand or anthracite filters. When the performance of biologically active GAC–sand and anthracite–sand filters were compared after preozonation, the higher TOC and AOC removal in the former was attributed to adsorption [15]. Further, a column test showed that GAC was a more appropriate material than anthracite in the removal of AOC [16].

A pilot study showed that GAC–sand filters achieved aldehyde removal sooner than anthracite–sand filters and were more resistant to temporary perturbations such as intermittent chlorination and out-of-service periods [17]. Additionally, the GAC–sand combination outperformed the anthracite–sand combination in the removal of glyoxal, a less readily biodegradable aldehyde. As discussed in

Chapter 3, the long retention of organics on the GAC surface, allowing biodegradation of slowly biodegradable substrates, is the main underlying reason.

The sorptive property of GAC is also of importance in the removal of ionic pollutants. For example, sorption of dissolved oxygen onto GAC enables efficient removal of perchlorate, since this reduces the competition of oxygen with perchlorate, as discussed in Section 9.6.3 [18].

9.2.2
Importance of GAC Grade

9.2.2.1
Coal- and Wood-Based GAC

Most GAC grades used in water treatment plants today are made from various kinds of natural coal, which are both cheap and readily available. However, wood-based GAC grades have also been used as successful alternatives [19]. In general, GAC grades made from wood have a softer texture than those made from coal. The relative hardness of the filter grain is considered to be essential for long-term durability of media, and this affects the cost of filtration [20].

Presumably, in water treatment, bioactivity in a filter does not depend on the GAC grade used. For example, coal- and wood-based GACs were shown to yield similar results in the biological removal of formaldehyde and glyoxal [17]. However, contrary results are also reported. For example, nitrification was improved at cold temperatures when wood-based GAC grades were employed. The underlying reason is believed to be the release of phosphate from this type of GAC [21].

9.2.2.2
GAC Activation

To date, the type and activation of GAC have received relatively less attention in BAC filtration than have other factors. As discussed in the example in Section 9.1, a recent study revealed the importance of the GAC type in BAC filtration. DOC biodegradation obviously continued longer in a filter packed with a thermally activated carbon than it did in a filter packed with a chemically activated one [7].

9.2.3
Empty Bed Contact Time (EBCT) and Hydraulic Loading Rate (HLR)

As to the design of GAC reactors, a few criteria emerged in the 1960s such as the hydraulic loading rate (HLR) and the empty-bed contact time (EBCT) [22]. A wide range is reported for the EBCTs and HLRs in practical applications. Table 9.3 shows the typical ranges in full-scale operation. Complementary to that, Table 9.4 provides examples of the ranges in laboratory-scale studies. HLRs, also termed superficial velocity (Chapter 2), range in full-scale systems from 5 to 15 m h^{-1}, with a typical value of about 12 m h^{-1} [22, 23].

The EBCT is an important parameter that significantly influences the removal of biodegradable organic matter (BOM) within biological filters [29–31]. Therefore, it

Table 9.3 Typical operation criteria in drinking water BAC filtration.

Criteria	Range
Empty-Bed Contact Time (EBCT)	5–30 min
Packed bed height (L_B)	1–3 m
Hydraulic loading rate (HLR)	5–15 m h^{-1}
Carbon particle size	0.5–3 mm
Regeneration frequency	3–24 months
Preozonation dose	0–3 mg O_3/mg DOC
Development of biological activity	Several weeks or months

Table 9.4 Typical operating conditions in laboratory-scale BAC filtration studies.

Study	Diameter (cm)	Column depth (cm)	EBCT (min)	Hydraulic loading rate (m h^{-1})	Biofilter development time	Ozone dose (mg O_3/mg DOC)
[24]	6	100	15			2.5
[25]	5.1	91	6	9		
[26]	10	70	15	2.44	2 months	
[27]	12	30	15	1.2		
[28]	25	40	15–20	0.01–0.015	4 months	<1
[1]	1.4	25	15	9.4	10 days (with inoculum)	1–2
[7]	2	50	18	1.67	Immediate upon inoculation	2

is a frequently used design and operation parameter in sand, GAC, and BAC filtration. In fact, the effective contact time in a filter (Chapter 2, Eq. (2.13)) is different from the EBCT. The effective contact time is not easy to determine since the interstitial volume changes during the course of filter operation as a result of biomass growth and detachment.

As shown in Chapter 2, the EBCT parameter is related to the HLR. A literature review of water treatment shows that the EBCT and not the HLR is the key parameter for biological BOM removal [12]. In general, beyond a minimum value, the use of very long EBCTs does not yield benefits. In nonadsorptive media such as sand, no significant differences were observed in TOC removal between sand filters that were fed with ozonated water (2–3.6 mg O_3/mg TOC) and were operated at 4, 10, and 20 minutes of EBCT [32]. Studies indicate that the minimum EBCT depends mainly on the type of water and presence of biodegradable matter. Very low EBCT values or very high HLRs led to inferior DOC removal since biomass formation was limited in such cases [33].

For removal of readily biodegradable substances, such as ozonation by-products and those within the scope of the AOC parameter, a lower EBCT is required than that for the removal of DBP precursors, which are less amenable to biodegradation [12].

In general, increasing the EBCT has the positive effect of delaying the breakthrough. For example, the results of a pilot-scale study at the Weesperkarspel Plant in Amsterdam indicate that DOC breakthrough was rapidly achieved at an EBCT of 7 minutes, whereas increasing the EBCT to 23 or 40 minutes retarded the breakthrough of DOC [34].

9.2.4
Filter Backwashing

In BAC filtration the presence of microorganisms and higher life forms leads to a rapid pressure buildup and clogging. Particularly at high temperatures, frequent and efficient backwashing procedures are needed. Backwashing is an important operational parameter in that it controls biomass losses. The efficiency of removal of BOM may be significantly reduced after backwashing in cold water. However, in winter months backwashing is carried out less frequently.

The detachment of biological and nonbiological particles during backwashing will influence the optimization of backwash strategies [35]. Bacteria are hydrophobic and are therefore more difficult to detach from surfaces than hydrophilic particles such as clay [36].

It is also relevant whether the backwashing is conducted with air, water, or chlorinated water [37]. Some treatment plants employ nonchlorinated backwash water, whereas others use constantly or occasionally chlorinated backwash water. The concentration of chlorine in the backwash water can affect the amount of biomass in filters and consequently the removal of BOM. Yet, most of the investigations about backwashing have been conducted on sand filters rather than on BAC filters. For example, a pilot plant study has shown that in anthracite–sand biofilters backwashing with chlorinated water reduced the biomass concentrations by approximately one order of magnitude compared to nonchlorinated water [38]. In general, chlorine in the backwash water seems to have adverse effects on the performance of anthracite–sand and GAC–sand filters when the concentration is above $1\,\text{mg}\,\text{L}^{-1}$. GAC filters tolerate chlorinated backwash water more than anthracite filters, both at low and high temperatures. Chloramine in the backwash water at $0.25\,\text{mg}\,\text{L}^{-1}$ did not affect BOM removal, either at low or high temperatures [39]. With regard to carbon attrition and pressure loss over 500 cycles of filter backwashing, wood- and coal-based GACs were shown to have similar durabilities [19].

9.2.5
Effect of Temperature

Many drinking water sources experience seasonal temperature variations. Theoretically, BOM removal would be expected to increase at higher temperatures as a result of enhanced mass transfer and microbial kinetics.

BOM removal within biofilters is reported to be related to temperature [39]. This effect became more pronounced under unfavorable conditions, namely in the case of chlorine in backwash water, anthracite media, or refractory organic matter.

Comparative bench-scale sand biofiltration at 5°C, 20°C, and 35°C revealed that at 5°C significantly lower NOM removal was achieved than at the other temperatures [40].

9.2.6
Effect of Oxidant Residuals

In water treatment, preoxidation steps leave some residuals such as ozone, hydrogen peroxide, chlorine, chlorine dioxide, and monochloramine, which then enter biofilters and lead to a reduction in biological removal [13]. For example, ozone residuals of 0.1–0.2 mg L^{-1} were shown to inhibit bacterial development in pilot-scale anthracite–sand filters and to reduce their performance [41].

In this regard, GAC serves another important function in that it decomposes chlorine and other oxidants through redox reactions [42]. Therefore, establishment of biological activity would still be possible in such filters. However, in the long term the presence of oxidants in influent water can lead to structural deterioration of GAC. To avoid this, fluidized-bed GAC filtration was proposed as a process for the safe destruction of ozone residuals [43].

9.3
Performance of BAC Filters: Organics Removal

The total removal of organic matter across a BAC filter is commonly monitored by the parameters TOC, DOC, and UV absorbance. Table 9.5 exemplifies the typical concentration ranges and the extent of removal at each treatment stage.

Of all the parameters, DOC is the most widely employed one. In units preceding BAC filtration, such as coagulation, flocculation, and sedimentation, the reduction in DOC is due to physicochemical mechanisms. In contrast to this, the reduction observed in BAC filtration is attributed to biodegradation as well as adsorption.

It is also important to know which fractions in NOM are removed in BAC filtration. Depending on water source and treatment conditions, the extent of removal achieved in DOC, UV$_{254}$, BDOC, and AOC may vary. In that respect, fractionation of these parameters with regard to molecular size is also important. In examining the fate of the biodegradable fraction in BAC filtration the best procedure is to monitor the change in BDOC and/or AOC parameters. The concentration ranges of BDOC and AOC vary appreciably from one source to another as shown in Table 9.5.

The organics contributing to the AOC parameter consist of small molecules having a MW < 1 kDa. Therefore, the AOC concentration in the influent water is only slightly changed in conventional water treatment consisting of flocculation, sedimentation, and filtration units. In contrast to this, in most cases, biological

Table 9.5 Typical concentrations of organic matter and ammonia in water treatment systems incorporating BAC filtration.

Water type	TOC (mg C L^{-1})	DOC (mg C L^{-1})	BDOC (mgC L^{-1})	AOC (µgC L^{-1})	UV_{254} (cm^{-1})	$SUVA_{254}$ (L/mgC.m)	Ammonia (mg ($NH_4^+ - N L^{-1}$))	Remarks	Reference
Raw water	2.95–5.12 (avg. 4.06)	2.85–4.89 (avg. 3.65)	0.25–1.16 (avg.0.68)	—	0.067–0.211 (avg.0.093)	2.02–3.11 (avg. 2.50)	0.15–0.98 (avg.0.34)	Laboratory-scale	[6, 7]
Ozonated water	2.89–4.88 (avg. 3.99)	2.78–4.67 (avg: 3.56)	1.25–2.30 (avg. 1.76)	—	0.026–0.137 (avg. 0.051)	1.03–1.73 (avg. 1.31)	0.14–0.96 (avg 0.34)	Ozone dose: 2 mg O_3/mg DOC GAC Grade/Water Type:	
Effluent of BAC filtration		0.39–1.05 (avg. 0.69)	0.0–0.5 (avg. 0.29)		0.0–0.027 (avg. 0.005)	0.0–0.74 (avg. 0.12)	0.0–0.17 (avg. 0.02)	Norit 1240 fed with raw water	
		0.42–1.03 (avg. 0.66)	0.0–0.72 (avg. 0.27)		0.0–0.09	0.0–0.24	0.0–0.15	Norit 1240 fed with ozonated water	
		1.53–2.47 (avg. 1.94)	0.0–0.49 (avg. 0.30)		(avg. 0.004) 0.018–0.039 (avg. 0.028)	(avg. 0.11) 0.22–7.12 (avg. 3.73)	(avg. 0.02) 0.0–0.48 (avg. 0.04)	Cagran fed with raw water	
		1.40–2.21 (avg. 1.62)	0.0–0.67 (avg. 0.33)		0.002–0.022 (avg. 0.015)	0.14–4.83 (avg. 2.27)	0.0–0.48 (avg. 0.05)	CAgran fed with ozonated water	
Raw water		2.795		103	0.040			AOC>1 kDa:15 µg L^{-1}, AOC<1 kDa:88 µg L^{-1}	[45]
Sedimentation		2.380		116	0.023			AOC>1 kDa:12 µg L^{-1}, AOC<1 kDa:104 µg L^{-1}	
Sand filtration		2.230		99	0.022			AOC>1 kDa:10 µg L^{-1}, AOC<1 kDa:89 µg L^{-1}	
Ozonated water		1.924		142	0.013			AOC>1 kDa:22 µg L^{-1}, AOC<1 kDa:120 µg L^{-1}	
Effluent of BAC filtration		1.812		92	0.012			AOC>1 kDa:12 µg L^{-1}, AOC<1 kDa:80 µg L^{-1}	

Chlorination effluent	–	144	–	AOC>1 kDa:22 µg L^{-1}, AOC<1 kDa:122 µg L^{-1} (BAC filtration-EBCT: 8.6 min)	[46]
Raw water	5.2–7.7 (tested cond. 6)		0.099–0.192 (tested cond. 0.13)	0.02–2.5	River water/Pilot-scale study
				MW fractionation / % in DOC: >30 kDa / 8.4%, 30–10 kDa/ 11.9%, 10–3 kDa/ 19.1%, 3–1 kDa/ 22.8%, <1 kDa/37.8%	
Ozonated water	5.75		~0.11	Ozone dose: 1–1.5 mg L^{-1}	
Effluent of coagulation, sedimentation, sand filtration	3.86		0.08		
Post-ozonation	~3.2		~0.075	Ozone dose: 2–2.5 mg L^{-1}	
Effluent of BAC filtration	2.65		0.03		
Raw water	4.32–7.04 (avg.6.2) 6.6 (studied water)		0.134–0.164 (avg.0.152)		[47]
Preozonation					

(*Continued*)

Table 9.5 (Continued)

Water type	TOC (mg C L^{-1})	DOC (mg C L^{-1})	BDOC (mgC L^{-1})	AOC (µgC L^{-1})	UV$_{254}$ (cm^{-1})	SUVA$_{254}$ (L/mgC.m)	Ammonia (mg ($NH_4^+ - NL^{-1}$))	Remarks	Reference
Sedimentation	4.97								
Filtration	4.66								
Post-ozonation	4.75							Removal capacity improved	
BAC filtration	2.84								
Raw water				122				Full-scale study, EBCT=6.6.min	[16]
								AOC-NOX : 104 µg L^{-1}	
				Range: 50–189				AOC-P17 : 18 µg L^{-1}	
Preozonation effluent				175					
				(Range: 117–491)				AOC-NOX : 170 µg L^{-1}	
								AOC-P17 : 5 µg L^{-1}	
Ozonation effluent				62					
				(Range: 33–116)				AOC-NOX : 52 µg L^{-1}	
								AOC-P17 : 10 µg L^{-1}	
BAC effluent				25				AOC-NOX : 23 µg L^{-1}	
				(Range: 13–38)				AOC-P17 : 2 µg L^{-1}	

9.3 Performance of BAC Filters: Organics Removal

Sample							Notes	Ref.
Chlorination effluent						30		
Raw water	2.99–4.71 (min-max.)					(Range: 17–42)		
				0.134–0.204 (min-max.)			AOC-NOX : 27 µg L^{-1} AOC-P17 : 3 µg L^{-1} Full-scale study, T: 2.8–10 °C Ozone dose = 1–2 mg O$_3$/mg TOC EBCT=26–30 min (media: GAC-phonolith-CaCO$_3$)	[52]
Ozonated water		0.75–1.22						
BAC effluent		0.30–0.48						
Finished water	2.21–2.85 (min-max.)			0.025–0.073 (min-max.)				
Raw water	4.29 ± 0.31	1.06 ± 0.25	44 ± 14	0.111–0.171 (min-max.)			Pilot-scale study/Direct biofiltration as a pretreatment for membrane operation EBCT: 31 min.	[53]
BAC effluent	3.79 ± 0.18	0.70 ± 0.16	35 ± 16					
Raw water		0.16 ± 0.10		0.058	3.12			[54]
Settled water (coagulation-flocculation-settling)		0.12 ± 0.05		0.045	2.96			
Ozonated water		0.37 ± 0.10		0.033	2.03		Saturated GAC filters:	
Filtered water (sand filtration)		0.35 ± 0.10		0.021	1.31			
BAC filtration								

(Continued)

Table 9.5 (Continued)

Water type	TOC (mg C L^{-1})	DOC (mg C L^{-1})	BDOC (mgC L^{-1})	AOC (μgC L^{-1})	UV$_{254}$ (cm^{-1})	SUVA$_{254}$ (L/mgC.m)	Ammonia (mg (NH$_4^+$ − NL^{-1}))	Remarks	Reference
			~0.28					EBCT: 1.8 min	
			~0.21					EBCT: 9.0 min	
Influent to BAC filtration:								Pilot-scale study with pretreated water (pretreatment in the full-scale plant as shown in Figure 9.8)	[34]
Influent to BAC-WPK1		4.5–6		50–90	0.08–0.11			Preozonated water, dose: 1.5–2 mg O$_3$ L^{-1}	
Influent to BAC-WPK5		5.5–6.5		20–50	0.12–0.15			Preozonated water, dose: 0.5 mgO$_3$ L^{-1}	
Influent to BAC-WPK6		5.5–6.5		40–150	0.10–0.12			Preozonated water, dose: 1.5 mg O$_3$ L^{-1}	
Influent to BAC-WPK7		5.5–6.5		70–170	0.08–0.10			Preozonated water, dose: 2.5 mg O$_3$ L^{-1}	—
–1Influent to BAC-WPK8		5.5–6.5		5–15	0.12–0.17			No preozonation	

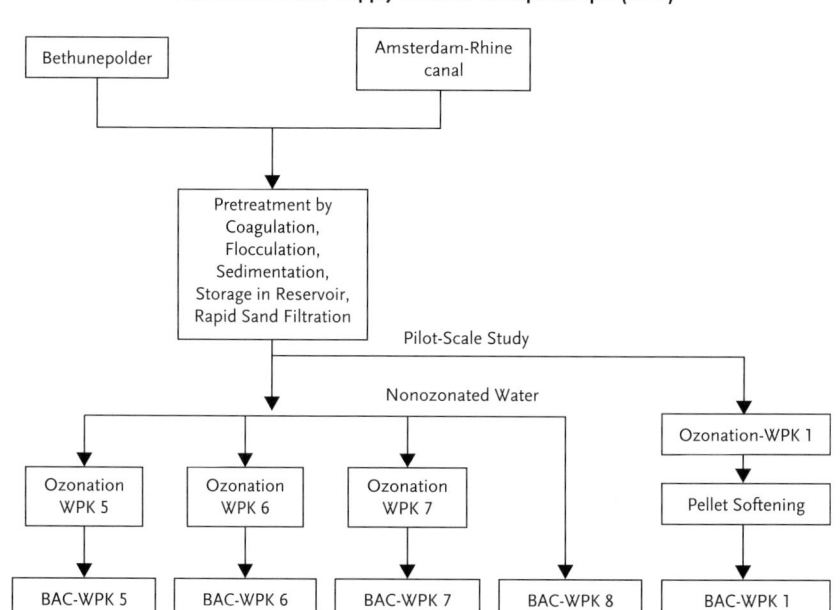

9.3 Performance of BAC Filters: Organics Removal | 283

Amsterdam Water Supply Location Weesperkarspel (WPK)

Figure 9.7 Flow scheme of the pilot-scale plant at Weesperkarspel/Netherlands (drawn based on [34]).

processes taking place inside BAC filters effectively reduce AOC, as shown in Table 9.5. For example, Chien and co-workers showed that preozonation increased the total AOC concentration, which, however, was reduced in biofiltration at pilot-scale [16, 44]. In one river water, the fraction of NOM with MW <1 kDa corresponded to 53–67% of DOC. The AOC constituted only 2.7–5.9% of DOC. Most of the AOC had a MW <1 kDa (Table 9.5). The ozonation/BAC process removed a large part of DOC, but did not necessarily result in a great reduction in AOC [45]. Another study showed that a large part of DOC had a molecular weight less than 1 kDa [46]. In another case, in combination with post-ozonation, BAC filtration could eliminate a large part of DOC in the <1 kDa MW fraction [47]. Further, enhanced biodegradation of TOC was observed for NOM sources with a lower $SUVA_{254}$ value, indicating that low aromaticity of organic matter favored effective removal [48].

Figure 9.7 shows the flow scheme of a pilot-scale study at the Weesperkarspel Plant in Amsterdam. The characteristics of waters fed to each BAC filter are presented in Table 9.5. The NOM in the source raw water consisted of large humic acid molecules. Preozonated water was fed to the filters denoted as BAC-WPK5, BAC-WPK6, and BAC-WPK7. In all filters, the EBCT was 40 min, while the HLR was about 3.15–3.18 m h^{-1}. Preozonation resulted in the formation of AOC, whereas without preozonation the AOC concentration in the influent of filter BAC-WPK8 was very small [34]. BAC filters receiving preozonated water removed

30–70% (up to 150 µg L^{-1} acetate-C) of the produced AOC. During the first 5000 bed volumes (BV), DOC removal is supposed to depend mainly on adsorption. After 5000 BV, DOC removal is supposed to depend mainly on biodegradation. All filters, except BAC-WPK8, continued to remove AOC, indicating biological activity. However, in this study, in none of the BAC filters were the AOC concentrations reduced to the level before preozonation.

In a complementary study the researchers aimed at quantifying the effects of ozonation and water temperature on the biodegradation of NOM. Removal of DOC, AOC, and DO and the production of CO_2 were taken as indicators of NOM biodegradation. The main effect of ozonation was to increase the AOC at 35 µg L^{-1} acetate-C/mg O_3. The removal of DOC and AOC correlated with each other and increased at higher ozone doses. However, their removal was not significantly influenced by the water temperature [49].

In BAC filtration, BDOC is usually less efficiently removed than AOC. In understanding this, it is necessary to compare the timescale in BAC filtration to that in the BDOC test. The BDOC test is typically conducted in a period of about 30 days and reflects all of the biodegradable fraction in DOC. However, in BAC filtration the EBCT is in the range of 5–30 min, and the effective retention time of the filter is even lower. Thus, only the readily biodegradable fraction in influent BDOC would be directly consumed by bacteria. Therefore, at low EBCTs DOC removal can be lower than predicted from the initial BDOC concentration. As a result, slowly degradable BDOC would leave the filter and enter the water distribution system unless other measures are taken [50].

The extent and mechanism of removal in BAC filters depends largely on the characteristics of influent organic matter. If the major part of compounds contributing to the DOC parameter are adsorbable or biodegradable, the main removal mechanisms would be adsorption and biodegradation, respectively.

A pilot-scale study comparing the efficiencies of the three processes, BAC filtration alone, preozonation and BAC filtration (O_3 + BAC), and chlorination and BAC filtration (Cl_2 + BAC) revealed that the efficiencies were about the same until adsorption saturation was reached. However, after saturation of GAC, biodegradation gained importance and the effluent DOC was lower in the O_3 + BAC process than others, probably due to enhanced biological activity in the filter [51].

Although most studies point to the negative effect of a temperature decrease, contradictory results are also reported. Operation of a small full-scale ozonation/biofiltration plant revealed that significant biological activity can be achieved and maintained in biofilters treating ozonated water even at low temperatures and phosphorus limited conditions [52].

9.3.1.1.1 Relevance of Preozonation

If preozonation is applied before BAC filtration, along with the increase in BDOC, the fraction of the readily biodegradable BDOC is increased, which is a factor favoring the removal of DOC. Many readily biodegradable by-products of preozonation are effectively removed in BAC filters. Accordingly, the ozone dose can be regarded as an important factor for DOC removal since it controls the formation of BDOC.

In most cases preozonation is regarded as an indispensable procedure in achieving biodegradation in BAC filters. However, depending on the level and type of biodegradable matter in raw water, biological activity may also be established without preozonation. Contrary to expectations, preozonation may not always significantly enhance DOC biodegradation in BAC filters. For example, a study has shown that in BAC filters receiving preozonated water, DOC biodegradation efficiencies did not differ from those for raw water despite the large increase in BDOC upon preozonation (first rows in Table 9.5). Thus, for this specific surface water, preozonation proved to be nonessential. This was attributed to the relatively high biodegradability of NOM in raw water [7]. The same type of finding is also reported by others [53].

9.4
Performance of BAC Filters: Nutrient Removal

9.4.1
Nitrification in BAC Filters

9.4.1.1
Importance of Ammonia Removal

Elevated ammonia levels indicate the pollution of the source with domestic and industrial wastewaters or the contribution of agricultural runoff. For example, in polluted lakes, where algal blooms are abundant, the ammonia nitrogen is measured in the range of 0.18–2.0 mg NH_3-N L^{-1} [56]. The presence of ammonia significantly increases the demand for chlorine at the disinfection stage. Fortunately, in BAC filters, ammonia can also be microbially oxidized by nitrifiers that are embedded in the biofilm or found in the bulk water. This removal lowers the overall chlorine demand at the disinfection stage. Consequently, it indirectly lowers the DBP formation potential.

Nitrifiers are abundant in many water sources and are passed on to water distribution systems where they grow readily if ammonia and oxygen are present. If ammonia is not removed from water in water distribution systems, these microorganisms produce metabolic by-products reducing the quality of the water and favoring the growth of other microorganisms, such as heterotrophic organisms.

Typical ammonia concentrations found in raw and treated waters are given in Table 9.5.

9.4.1.2
Factors Affecting Nitrification in BAC Filters

Typically, in cold winter months ammonia concentrations in a raw water are higher and the ammonia load to BAC filters is increased. Due to the low growth rate of nitrifiers and the additional ammonia load, nitrification may seriously be disturbed in such cases.

Andersson and co-workers investigated the impact of temperature on nitrification in BAC filters using two different GAC media, an open-superstructure wood-based activated carbon and a closed-superstructure activated carbon based on bituminous coal [57]. The study was conducted at pilot-scale (first-stage filters) and full-scale (second-stage filters). In the treatment plant, ammonia concentration ranged from 0.020 to 0.160 mg NH_4^+–N L^{-1}. In pilot-scale tests the influent water was enriched with ammonia at 0.4 mg NH_4^+–N L^{-1}. In full-scale operation ammonia concentration varied from 0.020 to 0.120 mg NH_4^+–N L^{-1}. Ammonia removal ranged from 40% to 90% in pilot filters, at temperatures above 10 °C, while for the same temperature range ammonia removal exceeded 90% in full-scale filters. At moderate temperatures (4–10 °C), pilot filters packed with open- and closed-superstructure GAC removed 10–40% of incoming ammonia. In full-scale filters, with open-superstructure GAC, ammonia removal exceeded 90%, while with closed-superstructure GAC it was limited to 45%. At temperatures below 4 °C in both media, removal efficiencies were below 30% at pilot and full scale. Attachment and detachment of nitrifiers plays an important role in nitrification. Attachment onto GAC was shown to be strong even after several weeks of running with cold water. The decrease in nitrification performance was attributed to the decline of bacterial activity rather than attachment of biomass.

For efficient attachment and colonization of nitrifying organisms, the type of GAC is important, particularly at low temperatures. The impact of backwashing on nitrifying activity was shown to be less in the case of an open-structure GAC than with a closed-structure one [55]. Nitrification studies in sand filters revealed that both the decay rate of biomass and backwashing procedures may have an impact on the amount of biomass in filters. Backwashing in sand filtration had the advantage that it controlled grazing and reduced the decay of biomass, but it could also lead to detachment and loss of nitrifiers [58].

Incomplete nitrification in BAC filtration leads to increased nitrite levels. Nitrite is an unwanted ion in water owing to its health significance. Although initially no nitrite was present in the influent, data from full-scale operation indicate that nitrite concentration could be increased in BAC filtration up to 0.06 mg L^{-1} as a consequence of incomplete nitrification [59].

9.4.2
Denitrification in BAC Filters

Denitrification of nitrate would depend on the presence of anoxic conditions and the level of BOM in a BAC filter. As shown in the following example, under normal operating conditions there is little evidence that significant denitrification occurs in BAC filtration. The reason can be attributed to the relatively high DO and low BDOC levels in BAC filters.

EXAMPLE 9.2: Nitrogen Removal in BAC Filters Receiving Raw and Ozonated Water This example shows the extent of nitrification along with the removal of organic carbon in BAC filters operated at 25 °C. The details of organic carbon removal have already been presented in the example in Section 9.1. The nitrification performance was evaluated in the case of two different GAC grades

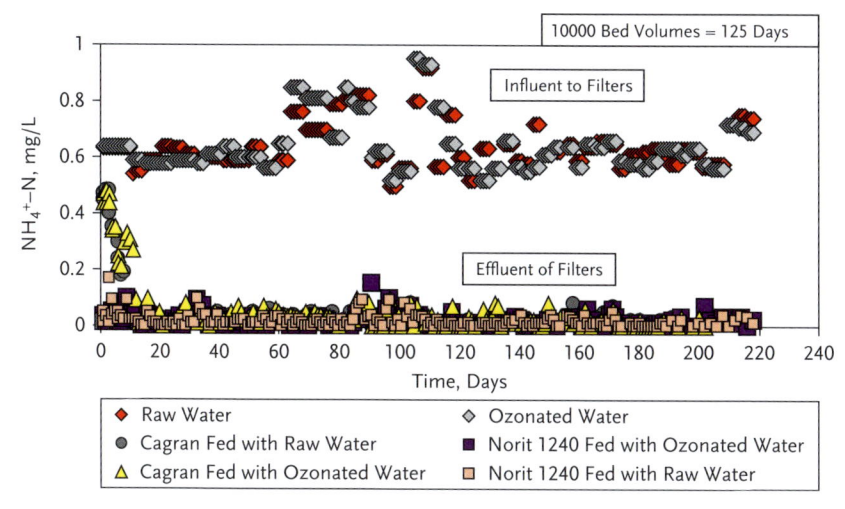

9.4 Performance of BAC Filters: Nutrient Removal | 287

Figure 9.8 Influent and effluent ammonia nitrogen concentrations in BAC columns fed with raw and preozonated water (adapted from [6]).

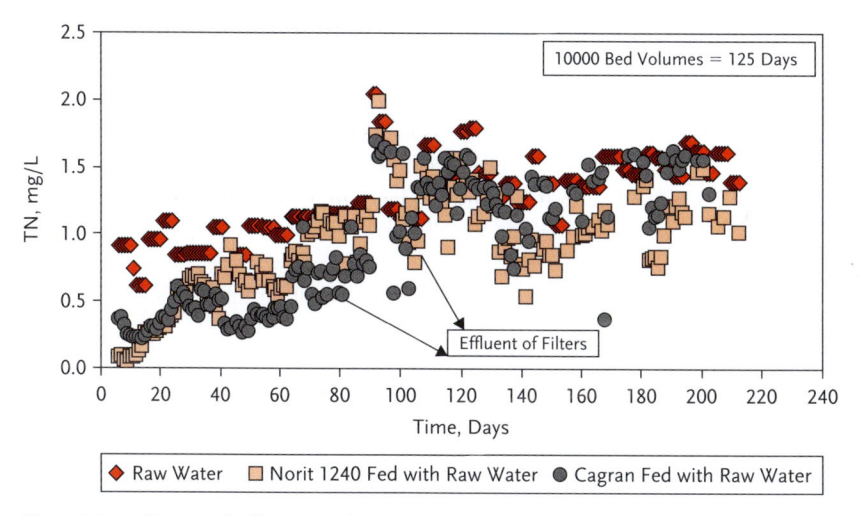

Figure 9.9 Influent and effluent Total Nitrogen (TN) concentrations in BAC columns fed with raw water (adapted from [6]).

(CAgran and Norit 1240) without and with preozonation at a dose of 2 mg O_3/mg DOC [60]. The NH_4^+−N concentration in filter influent fluctuated around 0.6 mg L^{-1} (Figure 9.8). In BAC filters, ammonia removal started from the first day of operation. This was in contrast to sand filters, in which ammonia removal took place much later after inoculation with nitrifiers [58]. As seen in Figure 9.8, no significant differences were detected between the two carbon grades in terms of NH_4^+−N removal, and nearly complete removal was achieved. Most of the incoming ammonia was removed in the upper parts of the columns. Preozonation of water had no beneficial effect on ammonia removal.

A mass balance was made in BAC filters with respect to the total nitrogen (TN), which consists of the fractions organic-N, $NH_4^+-N, NO_3^--N,$ and NO_2^--N. The measurements showed that organic nitrogen in influent water was negligibly small in this specific surface water. The negligible differences between influent and effluent TN were attributed to bacterial assimilation (Figure 9.9). Overall, the extent of denitrification seemed to be insignificant [6].

Moreover, nitrifying species in these filters were investigated by the use of molecular biology tools. The findings are reported in the example in Section 9.9.

9.5
Removal of Micropollutants from Drinking Water in BAC Systems

9.5.1
Occurrence of Organic Micropollutants in Water

In the last decades, a variety of organic micropollutants have been detected in raw and finished waters. These organic micropollutants fall mostly into the category of xenobiotic organic compounds (XOCs), which are considered to be even more problematic than NOM. Representatives of such organics include pesticides and biocides, halogenated hydrocarbons, tensides, aliphatic and aromatic hydrocarbons, phenols, aniline, amines, pharmaceuticals and personal care products (PPCPs), endocrine disrupting compounds (EDCs), synthetic organic complex builders, diagnostics, and household chemicals [60].

Since the 1990s, there has been increased concern about the presence of pharmaceutically active compounds, EDCs, and personal care products (PCPs) in raw and finished drinking waters because of the potential adverse effects on human health and the environment. In surface waters the concentrations of EDCs amount to a few µg L^{-1}. On the other hand, pharmaceuticals are found at levels as low as a few ng L^{-1}. Other pollutants such as methyl-*tert*-butylether (MTBE) are also found at very low levels [61]. An important aspect is that also some pollutants, such as N-nitrosodimethylamine (NDMA), can be generated in the course of treatment.

The use of ozonation combined with BAC filtration brings about a reduction in most micropollutants, among them the PPCPs, EDCs, pesticides, toxic algal metabolites known as microcystins, and taste and odor compounds. The presence of GAC has the main effect of concentrating the micropollutants on the surface from a 'low-substrate' environment, as discussed in Chapter 3. In order to understand their fate and removal in BAC filtration, it is first essential to have an insight into the adsorptive properties of these compounds.

9.5.2
Competition Between Background NOM and Organic Micropollutants

In GAC filtration, NOM usually acts as a background pollutant shielding the effective removal of micropollutants. However, BAC filtration, providing the

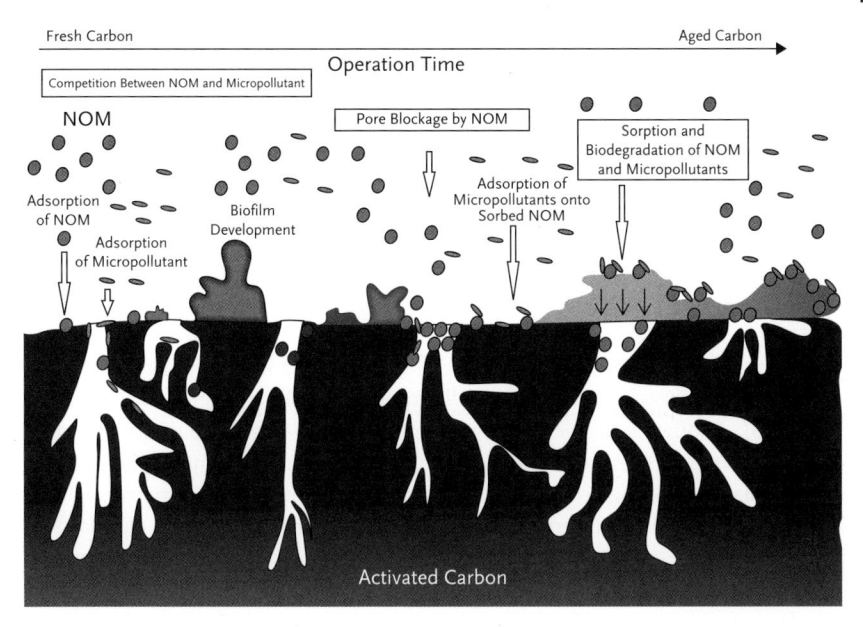

Figure 9.10 Competition between NOM and organic micropollutants in BAC filtration.

advantage of biodegradation, results in more effective NOM removal than that achieved by GAC filtration.

As shown in Figure 9.10, NOM hinders adsorption of organic micropollutants via two different mechanisms [62]:

- direct competition for adsorption sites on activated carbon,
- blockage of pores with NOM (preloading).

From one perspective it is desirable in water treatment that NOM components that impart color, odor, and taste to water and function as potential DBP precursors should be removed. From a different perspective, loading of the carbon surface with NOM components is undesirable if the primary concern is the removal of micropollutants. In essence, NOM components are not as effectively adsorbed as most target micropollutants. However, micropollutants have a concentration disadvantage since they are present at $\mu g\,L^{-1}$ to $ng\,L^{-1}$ level compared to NOM, which is at $mg\,L^{-1}$ level [63]. Therefore, adsorption of non- or slowly biodegradable target compounds, such as PCBs, toluene, benzene, and atrazine, commonly encountered in surface and groundwaters, is hindered in the presence of NOM [64, 65].

Most of the research on adsorption of organic micropollutants centers on pesticides. In particular, the interaction of NOM with atrazine, a commonly used herbicide, is investigated in detail. It was shown that the competitive effects in the adsorption of atrazine were stronger at a high DOC concentration corresponding to a high NOM level [65, 66]. Therefore, the capacity of carbon for atrazine adsorption becomes exhausted more rapidly at high background DOC. However, the fractionation of DOC is also decisive for the adsorption of micropollutants [67]. NOM is a multicomponent mixture which can be divided into fictive fractions of different adsorbability [68]. In general, the competitive effect of NOM depends on the character and concentration of the NOM and the type of activated carbon [69].

9.5.3
Adsorption of Organic Micropollutants onto Preloaded GAC

The negative effects of NOM preloading are exemplified in the hindered diffusion of trichloroethylene (TCE), a trace pollutant commonly encountered in groundwater [63]. Another example is the adsorption of the pharmaceuticals naproxen and carbamazepine and the endocrine disrupting compound nonylphenol, substances that are among emerging pollutants at the present time. Preloading with NOM has changed the Freundlich adsorption parameters for these compounds. The reduction in adsorption capacity was most severe for the acidic naproxen, followed by the neutral carbamazepine and the more hydrophobic nonylphenol [70, 71].

Competitive adsorption was also observed when the solution containing the pharmaceuticals (clofibric acid, diclofenac, fenoprofen, gemfibrozil, ibuprofen, indomethacin, ketoprofen, naproxen, and propyphenazone) was subjected to batch adsorption tests with PAC. Relatively less hydrophobic pharmaceuticals such as clofibric acid and ibuprofen were not efficiently adsorbed onto PAC [72].

9.5.4
Effect of GAC Characteristics on Adsorption

The raw material, activation conditions, and surface treatment of GAC also affect adsorption of NOM onto the carbon surface [73]. Such characteristics also play a role in the removal of micropollutants. Isotherm studies with surface-treated coal- and wood-based activated carbons indicated that carbon surface acidity was an important factor in adsorption of trichloroethylene (TCE) and trichlorobenzene (TCB). High surface acidity increased the polarity of surface and reduced adsorption [74].

Hydrophilicity of activated carbons is an important criterion for adsorption. For example, hydrophobic carbons were shown to adsorb TCE and MTBE more effectively than hydrophilic ones because enhanced water adsorption on the latter interfered with adsorption of micropollutants from solutions containing NOM. Also important was the pore size distribution of the adsorbent. It was concluded that an effective adsorbent should possess a micropore size distribution that extends to widths that are approximately twice the kinetic diameter of the target adsorbate. In this way the negative effect of NOM due to pore blockage/constriction would be prevented [75].

9.5.5
Adsorption and Biological Removal of Organic Micropollutant Groups in BAC Filtration

The difficulties in biological removal of micropollutants can mainly be ascribed to their very low concentrations in raw water, at about ng L^{-1} or low µg L^{-1}. As outlined in Chapter 3, very low concentrations would generally hinder the removal of a pollutant as a growth substrate in biological processes. Still, biological removal can take place through the use of the micropollutant as a secondary and/or as cometabolic substrate in the presence of a growth substrate (primary substrate). In integrated adsorption and biological removal, uptake of micropollutants from the bulk solution and their concentration on carbon surface makes their biodegradation more likely. For some micropollutants the pathway of biological removal is sufficiently explained, whereas for many others this still remains unclear.

9.5.5.1
Pesticides

In a BAC filter, which received preoxidized water, atrazine was removed more effectively than in GAC filtration since less NOM was adsorbed, and competition, pre-loading and pore blocking effects were reduced [76]. Similarly, another study points to concurrent biodegradation and desorption of atrazine under anoxic conditions of a BAC filter, whereby the effluent concentration was reduced to 0.002 mg L^{-1}, which is below the EPA standard [10]. In addition to atrazine, preozonation in combination with BAC filtration effectively removed pesticides such as diuron and bentazon [77].

9.5.5.2
Pharmaceuticals and Endocrine Disrupting Compounds (EDCs)

As discussed in Chapter 3, in order to determine the biodegradability of a compound, the biodegradation rate constant (k_{biol}) is commonly used, whereas for determination of adsorbability, the octanol–water coefficient, K_{ow}, is taken into consideration as an indicator of hydrophobicity. Also, in the evaluation of several pharmaceuticals and EDCs, these are the main criteria used.

The removal efficiencies of several pharmaceuticals were compared in sand biofiltration and BAC filtration based on the extent of biodegradability and adsorbability [78]. For almost all pharmaceuticals, the removal efficiency was higher for BAC treatment than for sand filtration, since in the former both adsorption and biodegradation took place. Paracetamol and salicylic acid were also efficiently eliminated in the nonadsorbing sand media, since these compounds have high biodegradation rates. Paracetamol is reported to have a biodegradation rate constant (k_{biol}) of 58–80 L/gSS.d. Compounds other than paracetamol and salicylic acid were hardly or even not at all removed by sand filtration. On the other hand, carbamazepine is essentially nonbiodegradable (k_{biol} < 0.01 L/gSS.d) but has a high adsorption potential (log K_{ow} = 2.3–2.5). In accordance with this, it was efficiently eliminated in BAC filtration, but not in sand filtration. Ibuprofen is both adsorbable (log K_{ow}=3.5–4.5), and biodegradable (k_{biol}: 9.0–35 L/gSS.d). Therefore, the removal of ibuprofen is expected to be higher in BAC treatment than in adsorption alone.

In another study, almost all pharmaceuticals could be removed below the detection limit by BAC treatment, except for erythromycin and gabapentin, which obviously are not very amenable to adsorption and biodegradation [78]. The effect of preozonation and EBCT was also examined. At EBCTs exceeding 83 min, BAC filtration was well capable of removing various types of pharmaceuticals. A high EBCT was needed for the removal of gabapentin, which is resistant to biodegradation. However, the additional removal of erythromycin achieved at high EBCTs was small. Preozonation did not have a significant impact on the removal of pharmaceutical compounds in BAC filters [78].

In another investigation, in a bench-scale system consisting of coagulation, flocculation, sedimentation, dual-media filtration, GAC treatment, and chlorination, the removal of the pharmaceuticals caffeine, trovafloxin mesylate, estradiol, and salicylic acid was studied. The addition of GAC filtration to the treatment train enhanced the removal of caffeine, trovafloxin mesylate, and estradiol, whereas it had limited effectiveness for salicylic acid, which was removed by biodegradation [79].

In recent years attention has been drawn to the detection and removal of EDCs from water sources. Nonylphenol (log K_{ow}: 5.76) and bisphenol A (log K_{ow}: 3.32) were effectively removed by adsorption onto virgin and used activated carbons. The extent of this removal depended on carbon type and time of operation. As expected, there was a relationship between the Freundlich adsorption constant K_F and the K_{ow} value of the EDCs. According to the study, mainly the pore volume of activated carbons influenced the adsorption capacity, but the surface charge of the carbon also seemed to be important. Among all the EDCs investigated, amitrol (log K_{ow}: −0.86) was poorly adsorbed onto activated carbon, but its degradation was evident in biological activated carbons, indicating that biodegradation was the removal mechanism [80].

9.5.5.3
Geosmin and MIB

Taste and odor problems in surface waters are frequently linked to the presence of geosmin (1,10-*trans*-dimethyl-*trans*-9-decalol) and MIB (2-methylisoborneol), which impart an earthy or musty odor to water. The raw water concentrations of these pollutants range from a few ng L^{-1} to more than 800 ng L^{-1} [81]. The odor threshold concentrations range from 4 to 15 ng L^{-1}.

Preozonation provides partial removal of geosmin and MIB. It is widely accepted nowadays that geosmin and MIB can be removed by biodegradation and adsorption onto activated carbon, though the extent of this is variable [82]. The treatment of a lake water by the O_3/BAC process led to 100% and 96.3% removal of geosmin and MIB, respectively [56]. Comparison of nonadsorptive crushed expanded clay with adsorptive GAC points to the significant role of adsorption in the removal of geosmin and MIB by the latter [83].

Geosmin removal has mostly been examined under conditions favoring the colonization of GAC media. A recent study additionally tested the ability of a biofilter to maintain a long-lasting biodegradation of geosmin, even under extended geosmin-free periods [81]. In the nonadsorbing anthracite–sand filter, geosmin removal took place due to biological activity only, whereas in the GAC filter both biodegradation and adsorption were observed. In the pilot-scale GAC filter, the actual bed life was much higher than predicted from batch tests and simulations due to biological degradation of geosmin. Another important observation was that microbial activity was restored after 40 days of geosmin-free feeding. Most probably, as outlined in Chapter 3, at low concentrations geosmin acted as a secondary substrate or a cometabolite, while the BOM in water served as the primary substrate. Presumably, bioregeneration of GAC took place by desorption and biodegradation of geosmin during extended geosmin-free periods. However, no clear evidence was provided under the conditions of the study.

MIB can also be removed in BAC filtration. Bench-scale experiments were conducted using two parallel filters containing fresh and exhausted GAC media and sand. Source water consisted of dechlorinated tap water to which geosmin and MIB were added as well as a cocktail of easily biodegradable organic matter in order to simulate water that had been subjected to ozonation prior to filtration. Using fresh GAC, total removal of geosmin and MIB ranged from 76% to 100% and 47% to 100%, respectively. The exhausted GAC initially removed less geosmin and MIB, but the removal increased over time. The potential of biofilters to respond to transient geosmin and MIB episodes was also demonstrated [82]. Other researchers pointed out that adsorption is the dominant mechanism in the removal of geosmin, whereas removal of MIB took place mainly by biodegradation [84].

9.5.5.4
Microcystins (Toxic Algal Metabolites)

Microcystins are cyanobacterial toxins which are resistant to removal in conventional water treatment units. Due to health effects, a guideline value of $1.0 \, \mu g \, L^{-1}$ has been issued for microcystins in drinking water by the World Health Organization (WHO).

In BAC filters, microcystins can be removed by both adsorption and biodegradation. Results obtained in slow sand and GAC filtration highlight the importance of biofilm formation in the removal of microcystins. Preexposure of biofilms to microcystins helps in eliminating the lag period in biodegradation [85].

The relative importance of adsorption and biodegradation in the removal of two mycrocystins was compared using sterile GAC, conventional GAC, and sand columns, respectively [86]. Biodegradation was an efficient removal mechanism whose rate was dependent on the temperature and initial bacterial concentration. However, formation of a biofilm layer on the GAC surface also had the effect of hindering adsorption. A high percentage of microcystins was still removed in the sterile GAC column after 6 months of operation, indicating that adsorption still played a vital role. In the GAC filter, the time required for the formation of a biofilm was much shorter (1 month) compared to a sand filter (7 months).

9.5.5.5
Other Organic Micropollutants

In studying the behavior of numerous organic micropollutants, the common approach is to use model compounds. For example, in one study, phenol and bromophenol were taken as representatives of organic micropollutants in BAC filtration [87]. Bromophenol could only be removed by adsorption, whereas phenol removal took place by both adsorption and biodegradation.

Ozonation can be assisted by the addition of GAC to the ozonator, a process referred to as the AC/O_3 process. The addition of activated carbon can initiate radical-type chain reactions that proceed in solution and accelerate the transformation of ozone into secondary oxidants, such as hydroxyl radical. The integration of this process with BAC filtration is designated as the AC/O_3-BAC process. It was shown to be more efficient than the integration of preozonation with BAC filtration (O_3 + BAC) alone. The AC/O_3-BAC process effectively removed more than 90% of phthalate esters (PAEs) and persistent organic pollutants (POPs) that were present at ng L^{-1} levels [88].

Methyl *tert* butyl ether (MTBE) is a gasoline additive that is accidentally released from underground storage tanks and pipelines into groundwater. MTBE was only degraded by 20% during ozonation of the water taken from Lake Zurich. Ozonation was followed by activated carbon filtration. Although the MTBE was well retained after 10 days of operation, the activated carbon soon become saturated with DOM, and after 150 days the adsorption capacity of carbon for MTBE was exhausted [5]. Another study demonstrated that the ozone/H_2O_2 concentrations required for a complete removal of MTBE from natural waters are much higher than the ozone levels applied nowadays in waterworks. MTBE was only poorly adsorbed on activated carbon. Therefore, GAC filtration was not regarded as efficient in eliminating MTBE [89]. However, a number of MTBE-degrading organisms have been reported in bioremediation of MTBE. However, the advantage of integration of activated carbon into biological processes is not obvious.

As outlined in Chapter 3, activated carbon also has the ability to retain Volatile Organic Compounds (VOCs). A recent study addresses the removal achieved in conventional parameters as well as in semi-volatile organic compounds (SVOCs) in BAC filters operated at HLRs varying from 1 to 8 m h^{-1}. Based on COD_{Mn} and AOC removal, the optimal HLR was found to be 3 m h^{-1}. The total concentration of SVOCs was 9012 ± 4837 ng L^{-1}. In total, 24 SVOCs were analyzed consisting of 12 polycyclic aromatic hydrocarbons, 6 phthalates, and 6 benzene derivatives or heterocyclic compounds. The removal of individual SVOCs depended strongly on the HLR. The removal of individual SVOCs was discussed taking into account their molecular structure. The maximum removal efficiencies of the main SVOCs, namely di-*n*-butyl phthalate, bis (2-ethylhexyl) phthalate and 2,6-dinitrotoluene, were 71.2%, 84.4%, and 65%, respectively [90].

9.5.6
Removal of Precursors and Disinfection By-Products (DBPs) in BAC Filtration

Trihalomethanes (THMs) and haloacetic acids (HAAs) are the dominant DBPs in waters that are disinfected with chlorine. Some of them have been identified as possible human carcinogens. In some countries, such as the USA, the sum of five haloacetic acids (HAA5), that is, monochloroacetic acid (MCAA), monobromoacetic acid (MBAA), dichloroacetic acid (DCAA), dibromoacetic acid (DBAA), and trichloroacetic acid (TCAA), is regulated, while in others, such as Korea, only the sum of DCAA and TCAA, that is, HAA2, is taken into consideration [91]. The possibility of DBP formation is usually indicated by the THM Formation Potential and the HAA Formation Potential, abbreviated as THMFP and HAAFP, respectively.

A strong relationship was shown to exist between the molecular weight distribution of DOM and the trihalomethane formation potential (THMFP) of water [92]. Further, the specific UV absorbance at 280 nm ($SUVA_{280}$) was shown to correlate with the total THM (TTHM) concentration. TTHM consisted of chloroform, bromodichloro- and dibromochloro-methanes, and bromoform [93].

In general, two different procedures can be followed to prevent the occurrence of DBPs in the water distribution network.

1. removal of precursors for prevention of DBP formation,
2. removal of DBPs after their formation.

As may be imagined, the first measure is the preferred one. However, the second alternative may also be chosen if ever DBP formation is unavoidable.

9.5.6.1
Removal of DBP Precurcors in BAC Filters
The removal of DBP precursors by adsorption and/or biodegradation in BAC filters leads to reduction in THMFP and HAAFP.

The combination of ozonation, coagulation, and biofiltration (through anthracite–sand and GAC–sand media after exhaustion of adsorption capacity) successfully reduced the THMFP and HAAFP up to 60–70% [94]. The positioning of ozonation, either as preozonation before rapid mixing or post-ozonation after sedimentation, did not affect this reduction.

Another study shows that the combination of pre- and post-ozonation with BAC filtration decreased the THMFP and HAAFP of water by about 70% and 50%, respectively, concurrently with DOC removal [46]. Preozonation and post-ozonation were more effective in the removal of THMFP and HAAFP, respectively.

Sand filtration and the combination of ozonation with BAC filtration were shown to be the most effective methods for reducing THMFP and HAAFP. Upon chlorination, the HMW organic fractions were inclined to form HAAs, whereas LMW fractions were mainly responsible for the formation of THMs [47].

Preozonation of water in the presence of bromide alone can lead to the formation of bromoform ($CHBr_3$) and thus affect THM speciation prior to post-chlorination [95]. On the other hand, in the absence of bromide, chloroform ($CHCl_3$) is formed. The long-term operation of a BAC filter indicated a good reduction in chloroform formation potential. In contrast to this, the reduction achieved in brominated THMFP was rather poor. The removal achieved in total THMFP was restricted to about 7–14% because of the presence of the bromide ion [96].

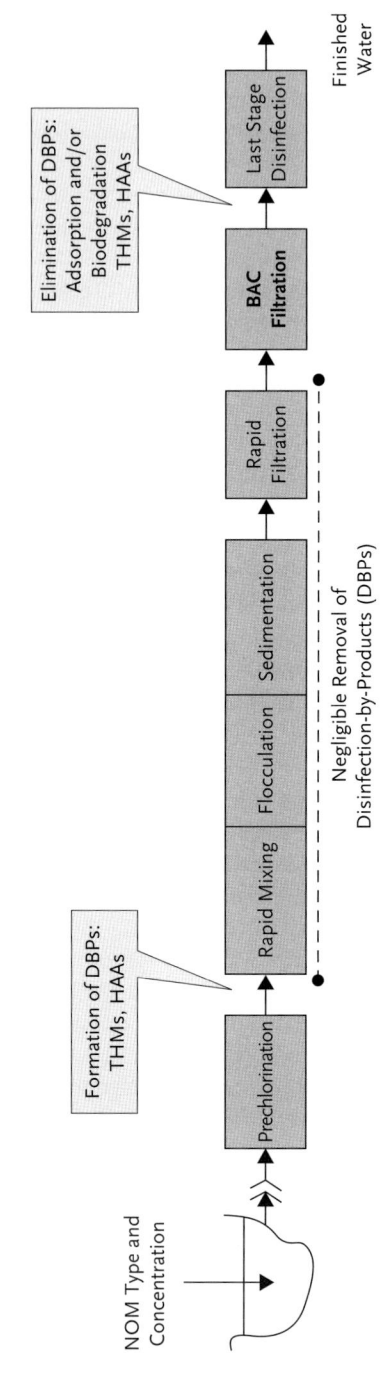

Figure 9.11 Removal of prechlorination products in the water treatment train.

9.5.6.2
Removal of Trihalomethanes (THMs) and Haloacetic Acids (HAAs) in BAC Filters

9.5.6.2.1 Metabolic Removal If prechlorination is applied after water intake, as shown in Figure 9.11, formation of chlorinated DBPs is very probable. However, compounds within this definition are possibly eliminated in subsequent BAC filtration.

Early studies point out that biological activity is improved upon preozonation, but is lowered by prechlorination due to the formation of chlorinated compounds [8]. The biodegradation potential of halogenated substances such as HAAs is primarily dependent on the chemical structure and extent of halogen substitution. In general, increased substitution of chlorine and bromine atoms into a molecule increases the biological stability.

The removal of five HAAs (HAA5) was investigated in BAC columns by spiking each of these compounds to the influent at 50 µg L^{-1} [97]. The BAC column effectively removed four of these compounds. In the initial periods of filter operation, removal took place due to adsorption, but later, biodegradation became the removal mechanism. Only the highly chlorinated trichloroacetic acid was incompletely removed and was detected at a level of 10 µg L^{-1} in the effluent.

Particularly, HAAs are shown to undergo biological degradation in BAC filtration, since bacteria that are responsible for their biodegradation commonly inhabit drinking water [98]. In BAC filtration, the breakthroughs of DOC, HAAs, and THMs were compared [99]. The capacity of the filter was much higher for DOC removal than for the removal of total HAAs. In turn, the HHA removal far exceeded the removal achieved in total THM. While THMs were removed by adsorption, some of the removal seen in HAAs and DOC was attributed to biodegradation. The study pointed mainly to desorption of THMs from activated carbon. To some extent, HAAs also desorbed. Desorption took place when the influent concentrations of these two groups went down. Since most HAAs are biodegradable, this desorption may eventually lead to bioregeneration of activated carbon.

In a treatment plant consisting of prechlorination, coagulation, sedimentation, filtration, and post-chlorination units, prechlorination at high doses raised concern about DBPs. Therefore, a conventional rapid sand filter was converted into a GAC filter to remove these substances. This configuration outperformed former operation in terms of DOC and DBP removal. Breakthrough of THMs was noticed after 3 months of GAC operation; their removal then became minimal (<10%). On the other hand, the removal efficiency of HAA5 exceeded that of THMs. At the early stages of GAC operation, HAA5 removal took place largely due to physical adsorption, but later on biodegradation seemed to dominate [91, 100]. In general, a higher reduction is achieved in BAC filtration in HAAs than THMs, as supported by other studies. In particular, brominated THMs are not removed in BAC filtration [56].

9.5.6.2.2 Cometabolic Removal of THMs in BAC Filters As outlined in Chapter 3, in biological systems cometabolic removal is often encountered in the case of various chlorinated aliphatics and aromatics [101].

Although biodegradation of HAAs is commonly observed in BAC filtration, no evidence is provided on the biodegradation of THMs via metabolic pathways. Nevertheless, in BAC filters some THMs were shown to be cometabolically removed in the presence of nitrifiers [102]. The BAC biofilters degraded four THMs (chloroform, bromodichloromethane, dibromochloromethane, and bromoform). In this specific case, ammonia in water acts as the primary or growth substrate for microorganisms; the ammonia monooxygenase enzyme (AMO) catalyzes the oxidation of THMs.

9.6
Removal of Ionic Pollutants in BAC Filtration

The biological activity in BAC filters also has the potential for the elimination of inorganic contaminants such as nitrate, bromate, and perchlorate. In biological reactions these ions serve as electron acceptors in the presence of an electron donor. However, it is essential to control the concentration of DO, the most preferable electron acceptor, in order to achieve successful reduction of these ions.

9.6.1
Nitrate Removal

In surface water BAC filtration, nitrate is mainly formed as a result of nitrification. For such cases, the denitrification of nitrate is discussed in Section 9.4. On the other hand, nitrate is a commonly encountered pollutant in groundwaters that are contaminated with fertilizers. Often, denitrifying biofilm reactors are used for nitrate removal from groundwater.

Since nitrate is not adsorbed onto GAC, the additional benefit of GAC over sand as a biofilm support has been documented in few cases. For example, in the treatment of reverse osmosis brine concentrate using a high-rate fluidized bed, denitrification in GAC media was superior to that of sand under transient-state conditions. However, both media were comparable under steady-state conditions [103]. The main advantage of GAC over nonadsorbing media was that it would potentially enhance denitrification by adsorption of inhibitory and toxic constituents.

9.6.2
Bromate Removal

The major anions of interest today that can be removed by BAC filtration are bromate and perchlorate. As shown in Figure 9.12, both ions are reduced in BAC filtration to give off bromide and chloride, respectively.

Natural processes and anthropogenic activities contribute to increased bromide (Br^-) levels in natural waters. Bromate (BrO_3^-) is an important DBP that forms as a result of ozonation of water containing the bromide ion. Bromate is formed by oxidation of bromide through a combination of ozone and hydroxyl radicals.

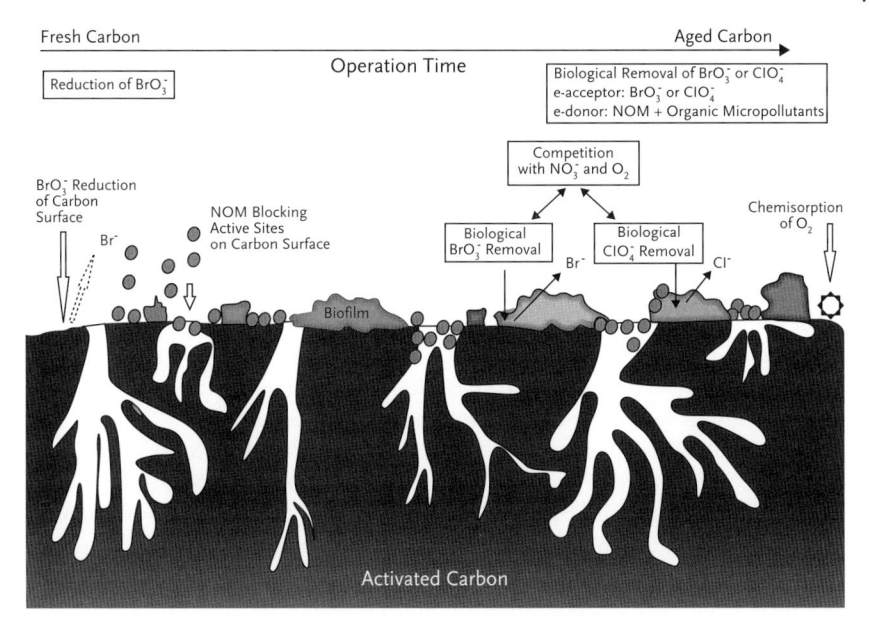

Figure 9.12 Removal mechanisms of bromate and perchlorate in BAC filtration.

During ozonation of NOM-containing water, brominated organic compounds will also form, but their concentrations are negligibly small.

Bromate is declared a potential human carcinogen and is therefore strictly regulated in drinking waters. The WHO standard is set at 25 µg L^{-1} for this ion while the European Union and the USEPA established a maximum contaminant level of 10 µg L^{-1} [104, 105].

In water treatment plants employing ozonation, bromide levels are reported to range from 2 to 658 µg L^{-1}. Depending on the level of bromide and the ozonation conditions, the bromate concentration may vary from 0.1 to 40 µg L^{-1}. Usually, for bromide levels in the range of 50–100 µg L^{-1}, excessive bromate formation may become a problem. However, at these levels the optimization and control of bromate formation may be possible to some extent. On the other hand, bromate formation can become a serious problem for bromide levels above 100 µg L^{-1}, although such high levels are rarely encountered. In European and US water works, only in about 6% of the more than 150 investigated water works, bromate concentrations were reported to be above 10 µg L^{-1} [105].

In BAC filters the removal of bromate is regarded as a controversial issue. Some studies provide evidence of only chemical reduction of this ion whereas others point to concurrent biological removal.

In biological reactions, bromate is consumed as an electron acceptor, as shown in Eq. (9.1) [106]. Under standard conditions and at pH 7, the Gibbs free energy gained from this reaction (-453 kJ mol^{-1}) is slightly lower compared to aerobic respiration (-501.6 kJ mol^{-1}) and denitrification (-476.5 kJ mol^{-1}).

$$6CH_2O + 4BrO_3^- \rightarrow 6H_2O + 6CO_2 + 4Br^- \qquad (9.1)$$

However, mostly oxic conditions are prevalent in BAC filters due to the presence of DO in the influent water. Since microorganisms prefer to use DO as an electron acceptor, bromate reduction is hindered, particularly in full-scale BAC operation where the influent contains high levels of DO.

It is also recognized that the formation of biofilm significantly affects bromate reduction. In one study it was shown that the GAC newly placed in a filter had the capacity to reduce the bromate ion to bromide [26]. However, the bromate removal rate apparently decreased in the long-term use of GAC following ozonation. In about 3 months, the GAC was colonized and converted into a BAC filter; the same ability to reduce bromate to bromide was then not observed. Obviously, the removal of bromate depended on a chemical reduction taking place on the GAC surface, and this was hindered by the formation of biofilm.

Bromate removal depends on the GAC type, the contact time, and the quality of the source water. Further, at high levels of background DOC and anions, competition for active sites on the GAC surface decreases bromate removal [107]. As shown in Figure 9.12, the presence of NOM hinders the reduction of bromate by blocking the reduction sites on the GAC surface. Therefore, pretreatment of NOM in a BAC filter could slightly improve bromate reduction in the resulting fresh GAC filter [108]. The presence of chloride, sulfate, bromide, and nitrate caused a decrease in the kinetics of bromate reduction since these anions occupied ion exchange sites on the carbon, reducing the rate at which bromate accessed the reduction sites. Besides chemical reduction, biological removal of bromate is also possible in BAC filters. At a 20 min EBCT, pH 7.5, and influent DO and nitrate concentrations of 2.1 and 5.1 mg L^{-1}, respectively, 40% of bromate was removed at an influent concentration of 20 µg L^{-1} [109].

The simultaneous removal of bromate and AOC was studied in the laboratory in rapid small-scale columns filled with five different GAC grades. The bromate concentration in the influent of GAC filters was relatively high (100 µg L^{-1}), while the EBCT ranged from 3.5 to 12 min. Then, the removal of bromate was also tested at pilot-scale. Bromate and AOC concentrations in ozonated water fed to the GAC filters were in the range of 18–163 µg L^{-1} and 91–226 µg CL^{-1}, respectively. In the pilot plant, new GAC was transformed into BAC over a period of 3–12 months, as determined from bacterial counts. Bromate removal was positively affected at high EBCT. The presence of DOC and competing anions had adverse effects. The bromate removal capacity of the BAC filter apparently decreased with respect to operation time. Probably, as shown in Figure 9.12, the formation of biofilm layer and/or adsorption of organic compounds were the factors hindering bromate reduction [110]. In contrast to bromate, AOC removal apparently increased with operation time.

However, in full-scale BAC filtration, data about bromate removal are limited [44, 59]. In a treatment plant, the bromate level was on average 6.69 µg L^{-1}, and decreased to 5.54 µg L^{-1} in the BAC effluent [44]. In the same plant, the AOC was reduced from 69 to 10 µg C L^{-1} as a result of BAC filtration.

Considering all the information about the removal of bromate in BAC filtration, chemical reduction of bromate seems to prevail. There is still little evidence of consistent and long-lasting biological removal of bromate in BAC filtration.

9.6.3
Perchlorate Removal

Perchlorate ion (ClO_4^-) is a pollutant that can interfere with iodide uptake into the thyroid gland. The ion is extremely soluble in water and polar organic solvents. In addition, perchlorate is highly mobile and persistent under typical environmental conditions [111]. In some polluted groundwaters, perchlorate levels may exceed 100 $\mu g\,L^{-1}$. Perchlorate contamination problems in groundwater result from unregulated discharges of ammonium perchlorate (NH_4ClO_4) from rocket fuel manufacturing plants or demilitarization of missiles. Another source of perchlorate contamination is potassium perchlorate, which is used as a solid oxidant for rocket fuel as well as for other purposes such as pyrotechnics, flares, and airbag igniters [112].

For perchlorate removal from polluted groundwaters, bioremediation techniques are preferred. As illustrated in Figure 9.12, biological reduction of perchlorate takes place in the presence of an electron donor and suitable redox conditions where perchlorate serves as an electron acceptor. A variety of electron donors can be utilized for this purpose such as ethanol, acetate, lactate, pyruvate, or H_2 gas. The reaction is carried out by perchlorate-reducing microorganisms, which are classified as facultative anaerobes or microaerophiles.

Various biofilm systems have been developed such as hollow-fiber membrane reactors, fluidized-bed reactors (FBR), and packed-bed reactors (PBR) for removing perchlorate. In biofilm reactors consisting of either GAC or sand, perchlorate was shown to decrease to very low levels. Laboratory-scale GAC-FBR-treated groundwater containing high concentrations of perchlorate, between 11 000 and 23 000 $\mu g\,L^{-1}$, had effluent concentrations below 350 $\mu g\,L^{-1}$; for the majority of experiments, they were below 5 $\mu g\,L^{-1}$. Perchlorate was reduced to chloride by an acclimated biomass under anoxic conditions in the presence of acetic acid and ethanol as electron donors [113]. In general, the aim is to reduce perchlorate concentrations to below 4 $\mu g\,L^{-1}$ in treated waters [112].

The use of dechlorinated tap water or BAC filter effluent was sufficient to develop biological activity; no specific seeding was required in laboratory-scale filters [111]. Further, excessive biomass growth diminished the perchlorate-reducing ability of BAC filters; however, this was restored upon filter cleaning.

As in the case of bromate, microorganisms gain more energy through respiration with oxygen or nitrate than with perchlorate. Therefore, aerobic or anoxic conditions inhibit perchlorate removal in BAC filters since DO and nitrate compete with perchlorate as an electron acceptor, as shown in Figure 9.12. In such cases, effective perchlorate removal was achieved by using acetate or ethanol as external electron donors, provided that the DO and nitrate in the influent were taken into consideration in calculation of electron donor addition. Similarly, sulfate may act as a competing ion for perchlorate and be reduced to sulfide, which has also to be removed from water due to odor problems. At 50 $\mu g\,L^{-1}$

concentrations and an EBCT of 5 min, 80% of perchlorate was removed, whereas complete removal was possible for EBCT $\geq$ 9 min. At an EBCT of 15 min the BAC filtration was robust even in the case of 300 µg L^{-1} perchlorate in the influent, and 4 µg L^{-1} was not exceeded [111].

Concurrent removal of bromate and perchlorate in BAC filtration is also addressed. In contrast to bromate removal, perchlorate removal was biological only. This ion was not removed by reduction on GAC surface or ion exchange [114].

Using bench-scale BAC filters following ozonation, complete bromate and perchlorate removal was observed when the influent concentrations were in the range of 10–50 µg L^{-1}, EBCTs were at 15 and 25 min, and influent DO was 0–2 mg L^{-1}. No external carbon addition was required to achieve this removal. In pilot-scale treatment, complete removal of 25 and 50 µg L^{-1} bromate and perchlorate was achieved, respectively, without the addition of an external electron donor [114].

The advantages of GAC over nonadsorbing media are also obvious in the case of perchlorate, particularly under dynamic loading conditions. Chemisorption of oxygen onto the GAC surface can enhance the stability of biological perchlorate reduction. It was shown that with nonadsorptive glass beads perchlorate reduction was negatively affected when DO was raised from 1 to 4 mg L^{-1}. In contrast to this, the BAC reactor was robust to short-term increases in influent DO up to 8 mg L^{-1} since the surface of GAC chemisorbed a substantial fraction of the oxygen, a competing electron acceptor, as shown in Figure 9.12. However, long-term exposure to influent DO concentrations of 8.5 mg L^{-1} led to slow increases in effluent perchlorate and DO concentrations. Subsequent exposure of the BAC reactor bed to low DO concentrations partially regenerated the capacity of GAC for oxygen chemisorption [18].

9.7
Integration of PAC and GAC into Biological Membrane Operations

9.7.1
Effect of PAC on Membrane Bioreactors

In general, microfiltration (MF) and ultrafiltration (UF) processes are the widely employed membrane processes for production of drinking water. Initially, they were introduced to provide a high level of particle, turbidity, and microorganism removal. Currently, they are being involved with other unit processes to provide removal of inorganic and organic materials [115].

The Membrane Bioreactor (MBR) for drinking water treatment is considered to be a new technology. These reactors can also be used in conjunction with PAC adsorption [116, 117]. The benefits of PAC addition to MBR can be summarized as follows: (i) removal of specific organic micropollutants, (ii) removal of DBP precursors, (iii) decrease in DOC and SUVA, and (iv) reduction in membrane fouling and flux decline.

As in the case of wastewater treatment (Chapter 3), compared to the stand-alone MBR process, PAC has the benefit of relieving the accumulation of organics in the mixed liquor, particularly the dissolved organics, and improving the removal efficiency. For example, in one study PAC addition was shown to decrease the SUVA value by 40%. This correspondingly lowered the overall DBP formation potential [116].

Another study reports the addition of PAC to bench- and small pilot-scale MBRs. The removal of BOM and THM precursors from preozonated water was studied. Biodegradation was the dominant mechanism in the removal of DOC and aldehydes. Addition of PAC did not appreciably enhance the removal of aldehydes and AOC. On the other hand, the THMFP was reduced by over 98% when 0.3% PAC was added. However, continuous PAC dosing was necessary for long-lasting THMFP removal. Addition of PAC enhanced DOC removal and prevented the decrease in permeate fluxes [117]. The underlying reasons are discussed in Chapter 3.

9.7.2
BAC Filtration Preceding Membrane Bioreactor Operation

BAC filtration can also be introduced as a pretreatment step before membrane filtration in drinking water. This hybrid process consisting of BAC filtration and a submerged membrane bioreactor (sMBR) was evaluated for the treatment of polluted raw water for drinking purposes, with the respective hydraulic retention time (HRT) of 0.5 h. The results confirmed the synergistic effects between BAC and sMBR. The removal of ammonium (54.5%) was moderate in the BAC, while the total removal efficiency was increased to 89.8% after further treatment by the sMBR. In the hybrid process, adsorption onto GAC, two-stage biodegradation (in BAC and sMBR), and separation by the membrane (in sMBR) contributed to the removal of organic matter. As a result, the hybrid process reduced the influent DOC, UV_{254}, COD_{Mn}, TOC, BDOC, and AOC by 26.3%, 29.9%, 22.8%, 27.8%, 57.2%, and 49.3%, respectively. BAC filtration also led to mitigation of membrane fouling in the downstream sMBR [118]. The underlying reasons are discussed in Chapter 3.

BAC and MBR were systematically compared under the same HRT of 0.5 h. MBR exhibited excellent turbidity removal capacity, while only 60% of influent turbidity was intercepted by BAC. Perfect nitrification was achieved by MBR, with 89% reduction in ammonia. By contrast, BAC only eliminated 54.5% of influent ammonia. BAC was able to remove more dissolved organic matter (DOM) from water than MBR, due to simultaneous adsorption and biodegradation. However, particulate organic matter (POM) was detected in BAC effluent. On the other hand, BAC and MBR displayed essentially the same capacity for BOM removal. BAC operation was recommended if the raw water had a high DOM consisting mainly of hydrophobic compounds. In comparison to this, MBR operation was recommended for the case of high ammonia and/or POM in water [119].

9.8
Integration of GAC into Groundwater Bioremediation

GAC filters are often employed for decontamination of groundwater. Surprisingly, the integration of activated carbon adsorption with biological processes has rarely been addressed in such cases. An example of such efforts is the laboratory- and on-site study conducted on an aquifer in Bitterfeld, Germany. The groundwater containing pollutants such as chlorobenzene (CB), 1,4-dichlorobenzene, 1,2-dichlorobenzene, benzene, vinyl chloride, *cis*-1,2-dichloroethene, and *trans*-1,2-dichloroethene was passed through GAC filters, which acted as barriers [120]. The aim of a complementary study, in which trichloroethylene (TCE), chlorobenzene (CB), and benzene were used as model pollutants, was to develop a long-lasting, sequential anaerobic/aerobic biological activated carbon barrier. In this biobarrier, adsorption of pollutants and biodegradation occurred simultaneously. TCE was dechlorinated under anaerobic conditions followed by aerobic conditions. During periods of low biological activity, elimination of TCE and CB took place by adsorption. The striking observation was that TCE and CB were desorbed from activated carbon and then removed biologically [11]. Although bioregeneration of activated carbon with regard to NOM is a debated issue (Section 9.1.3), these results point to bioregeneration of activated carbon by the release of such small compounds.

9.9
Biomass Characteristics in BAC Filtration

9.9.1
Microbial Ecology of BAC Filters

The presence of BOM in water leads to the dominance of heterotrophic bacteria in the biofilm that surrounds activated carbon. Additionally, nitrifiers too are embedded in the biofilm, usually in deeper parts than heterotrophs. GAC also harbors organisms other than bacteria, such as molds, yeast-like fungi, protozoa, rotifers, and nematodes [121–126]. The presence of protozoa and rotifers in filters is generally an indication of high bioactivity and appreciable organic carbon removal by bacteria [65]. Also important is the grazing activity of protozoa on heterotrophic bacteria. This grazing was in turn shown to favor the growth of nitrifiers which compete with heterotrophs [125].

Various figures for the number of bacteria in BAC filters are reported [12, 121, 127, 128]. The commonly shared opinion is that activated carbon filters are occupied by much more bacteria than nonadsorptive sand filters [124]. Extensive information about biofilm composition, activity and biofilm control in BAC filters can be found in a literature review [65].

The efficiency of biodegradation depends on the concentration and activity of biomass. In BAC filtration, a relationship is usually sought between removal efficiency and biomass characteristics. Early reviews demonstrate that measuring the biomass alone, for example by the phospholipid technique, particularly at the top of the filters, does not yield adequate information on BOM removal capacity. If the biomass amount is above a minimum level, no rate limitation is usually recorded. Factors such as the composition of BOM, formation of soluble microbial products (SMP), and stratification of microorganisms inside the filter may also affect the removal of BOM [12]. Operational parameters such as filtration rate, contact time, GAC properties, and temperature may determine the properties of the microbial community in a BAC filter.

9.9.2
Control of Biofilm Growth in BAC Filters

Excessive biofilm growth is unwanted in BAC filters since it leads to clogging, pressure drop, and the breakthrough of substrates and microorganisms including pathogens [27, 65]. It is essential to maintain an active biofilm inside a BAC filter while simultaneously avoiding filter clogging. For this purpose, backwashing is commonly applied to remove excessive biomass.

The most important parameters in achieving biofilm control are DO, pH, and a correct balance of nutrients [27]. In biofilters, organic carbon is the most important nutrient. In that regard, the AOC is the rate-limiting substrate. Also important is the concentration of microbially available phosphorus (MAP). Both AOC and MAP concentrations are increased by ozone treatment [129]. On the other hand, MAP is very efficiently removed by chemical treatment, which sometimes even results in shortages in subsequent BAC filtration if the amount of DOC is comparatively high. Therefore, phosphorus has to be supplemented in some cases to mineralize the organic carbon in BAC filtration [130].

9.9.3
Determination of Biomass and Microbial Activity in BAC Filters

The total amount of biomass in a BAC filter can be estimated from the following [131]:

- Physical parameters such as biofilm thickness, total dry weight, and biofilm density.
- Physicochemical determination of biomass in terms of TOC, COD, or both.

The concentration of attached biomass can also be estimated by the following microbial measurements [132]:

- Heterotrophic plate count (HPC) on various culture media.
- Total direct count using epifluorescence microscopy.

Further, at high biofilm concentrations, typical biomass components can be analyzed such as exopolysaccharides, proteins, peptidoglycan, lipopolysaccharides, and lipids.

Each method used in biomass characterization has its own limitations. HPC techniques only detect culturable microorganisms, whereas typical biomass components do not yield information on the viability of biomass [132].

The use of Scanning Electron Microscopy (SEM) is also common in the monitoring and characterization of bacteria in drinking water biofilters.

The presence of viable organisms in a BAC filter can be determined by the following biochemical procedures:

- Measurement of certain specific enzymes or specific products of bacterial metabolism [131];
- Measurement of adenosine triphosphate (ATP) in the biomass;
- Measurement of viable organisms by the phospholipid method;
- Utilization of assays such as the INT-dehydrogenase activity test;
- Measurement of the synthesis of bacterial nucleic acids;
- Monitoring the relative changes in the quantity of bacteria and protozoa.

However, viability and microbial activity are regarded as two different criteria [133]. For example, the phospholipid method is reported to measure viable biomass, but not to provide a measure of microbial activity. The following methods provide an indirect estimation of biological activity:

- Measurement of DO across and/or along the filter;
- Measurement of organic parameters such as DOC, BDOC, and AOC across, and/or along the filter.

The biomass respiration method (BRP) developed by Urfer and Huck shows good sensitivity in terms of BOM removal capacity [133]. However, measurements based on oxygen consumption alone may not always reflect the true biological activity, since abiotic processes, such as chemisorption of oxygen by activated carbon, may also play a role.

Among biological activity tests, the ATP test is one of the most widely used. Magic-Knezev and van der Kooij report that at least 6–125 kg of dry weight of microbial biomass carbon is present in a filter bed containing 100 m^3 of GAC [134].

Pilot plant measurements showed that development of maximum biomass activity in drinking water BAC filters took about 8 months. Biomass developed in the entire filter, but at the top of the filters biomass activity was higher than in the lower sections. In general, the total volume of the biomass occupying a BAC filter is rather small. In pilot-scale operation this volume was roughly estimated as 0.0032 m^3 per m^3 of filter material using ATP measurements [135].

Care should be taken that activated carbon does not interfere in biomass measurements. Therefore, removal of biomass from GAC is necessary before commencing with ATP measurement. In recent years a method has been suggested which is based on the direct measurement of total ATP content of a GAC sample and other support media. According to this method, flow-cytometric absolute cell counting and ATP analysis are combined to derive case-specific ATP/cell conversion values [136].

9.9.4
Determination of Microorganisms by Classical and Molecular Microbiology Methods

9.9.4.1
Heterotrophic Biomass and Activity in BAC Filters
Identification of microorganisms by cultivation-dependent methods usually leads to biased results on microbial communities. From many aspects, culture-independent molecular tools are considered superior. Molecular tools have the

property of revealing many unknown and uncultured bacteria. However, cultivation methods still have the advantage that organisms are obtained for further studies [137]. Therefore, many studies rely on a combined use of culture-dependent and culture-independent techniques.

Initial studies on the identification of microbial communities and activity center around sand filtration. In bench-scale sand biofilters, total viable biomass was quantified as extractable lipid phosphate, and phospholipid fatty acid profiles were determined. Low temperature not only decreased the rate of substrate metabolism, but also changed the microbial community [40]. In another case, tetrazolium reduction assays, phospholipid analysis, and 16S rRNA (rDNA) sequence analysis were applied to assess the distribution, composition, and activity of microbial communities in sand biofilters treating non-ozonated and ozonated drinking water [138]. The main goal was to assess microbiological activity and relate it to operating conditions, performance, and microbial community structure.

In comparison to sand biofilters, microbial communities in BAC filters have rarely been investigated in detail. The disadvantage of relying on culture-dependent techniques alone is also addressed in studies dealing with BAC filtration. Until now, as reported in a recent study, in many studies bacteria isolated from GAC filters have been identified as members of the genera *Pseudomonas, Acinetobacter, Caulobacter, Alcaligens, Flavobacterium,* and *Bacillus.* However, attention is drawn to the fact that all these bacteria were isolated from a substrate-rich environment. In the classic tests at high substrate concentrations, it is unlikely that aquatic isolates growing in an oligotrophic environment are identified [137].

A recent study was aimed at identification of the predominating cultivable bacteria in GAC filters for selecting representative strains to study the role of bacteria in the removal of dissolved organic matter [137]. Bacterial isolates were collected from 21 GAC filters in 9 water treatment plants treating either groundwater or surface water with or without oxidative pretreatment. Enrichment of samples in dilute liquid medium improved the culturability of bacteria. Genomic fingerprinting and 16S rDNA sequence analysis revealed that most (68%) of the isolates belonged to the *Betaproteobacteria*, and 25% were identified as *Alphaproteobacteria*. The number of different genera within the *Betaproteobacteria* was higher in the GAC filters treating ozonated water than in those treating nonozonated water. *Polaromonas* was observed in nearly all of the GAC filters (86%), and the genera *Hydrogenophaga, Sphingomonas,* and *Afipia* were observed in 43%, 33%, and 29% of the filter beds, respectively. Amplified Fragment Length Polymorphism (AFLP) analysis revealed that the predominating genus *Polaromonas* included a total of 23 different genotypes. The predominance of ultra-oligotrophic bacteria in GAC filters also suggests that bacteria were able to establish themselves at very low organic carbon levels.

In another study, the aim was to investigate the microbial colonization of BAC filters at a treatment plant receiving lake water. Bacteria were characterized by morphology, physiological tests, whole cell protein profiles, and phospholipid fatty acid composition, and identified by partial 16S rRNA gene sequencing [139]. Epifluorescence revealed prothecate bacteria to dominate in BAC filters. The majority of the isolates belonged to the order *Burkholderiales* of *β-Proteobacteria*, a few to *Comamonadaceae*, but the majority to an undescribed family, and the related sequences belonged mainly to uncultured bacteria. Among the less common α-*Proteobacteria* the genus *Sphingomonas* and the genera *Afipia*, *Bosea*, or *Bradyrhizobium* of the *Bradyrhizobiaceae* family were detected. The majority of cultured bacteria persisting in the BAC biofilter were *Burkholderiales*, which according to ecological information are efficient in the mineralization of dissolved organic matter in BAC.

Other researchers have shown by phylogenetic analysis that *Escherichia coli*, *Shigella* sp., *E. fergusonii* and *Firmicutes* bacteria predominated in the BAC reactors that were fed with the effluent of a pilot-scale sand filter [90].

The effect of backwashing on the biofilm community was studied by means of bacterial cell enumeration and terminal-restriction fragment length polymorphism (T-RFLP) fingerprinting analysis of bacterial and eukaryotic ribosomal RNA genes (rDNA) [140]. The BAC samples were taken from a plant employing coagulation sedimentation, rapid sand filtration, ozonation, BAC treatment, chlorination, and secondary rapid sand filtration. Additional chlorination is conducted before coagulation (prechlorination) or after the secondary sand filtration (postchlorination), depending on the situation. The enumeration of bacterial cells attached to the BAC revealed that 36% of the bacterial cells were lost after backwashing. The T-RFLP analysis of bacterial and eukaryotic communities associated with the BAC demonstrated that the relative abundances of some terminal-restriction fragments (T-RFs) changed significantly after backwashing.

Using Polymerase Chain Reaction (PCR), a gene involved in microcystin degradation, the *mlrA* gene, was detected in sand biofilters, providing supporting evidence that the primary removal mechanism was biodegradation [85].

9.9.4.2
Nitrifying Biomass and Activity in BAC Filters

Although in BAC filters much attention is paid to heterotrophs, there is still a lack of information on nitrifiers. Nitrifiers have mostly been investigated in biofilters with nonadsorbing media. Moreover, the amount of nitrifying biomass has rarely been estimated. A method developed by Kihn and co-workers aims at estimation of fixed nitrifying biomass by using potential nitrifying activity measurement [141].

In a recent study, the occurrence and diversity of ammonia oxidizing bacteria were investigated using 16S rDNA and amoA gene-based molecular techniques in BAC filters in which organic carbon removal and nitrification were taking place concurrently [6, 60]. Filters were fed with both raw and preozonated water. Details about these filters can be found in the Examples 9.1 and 9.2.

In all BAC filters, ammonia was almost completely removed by nitrification, as seen from Figure 9.8. Most of ammonia removal took place in the upper part of the filters. This finding coincided with the results of real-time PCR analysis, which showed that the amoA gene numbers were high at the top (28.7%) and decreased toward the bottom (2.1%).

Ammonia oxidizing bacteria (AOB) and nitrite oxidizing bacteria (NOB) in biofilters were most closely related to *Nitrosomonas* sp. and *Nitrospira* sp. as determined by cloning and slot blot analysis, respectively. Although the same inoculum was used in each of four BAC filters, different species dominated in the course of operation. Ammonia removal did not change with preozonation or the type of GAC used. However, BAC filters fed with preozonated water harbored different types of AOB than those fed with raw water. Thus, with regard to the dominance of nitrifier species, the type of feed (raw or preozonated) played a more important role in the establishment of the microbial community than the GAC grade [60].

9.9.5
Microbiological Safety of Finished Water

In the early periods of BAC filtration, the main concern raised was that some biofilm attached onto fine activated carbon particles would be released from filters. Activated carbon particles were thought to provide a suitable habitat for coliform or pathogenic microorganisms and would protect them from inactivation during post-disinfection. Therefore, careful monitoring of the breakthrough of GAC particles was emphasized. The role of preozonation in preventing significant colonization of BAC filters was discussed as well as the probability of multiplication of coliforms in the distribution system biofilm [142].

In a study undertaken to characterize the bacterial population in a GAC filter, the use of a GAC bed presented no observable adverse effect on the bacteriological quality of the water compared to sand filtration, which is commonly the last operation before disinfection [143]. In another study, a physicochemical desorption technique was used to assess the impact of colonized GAC fines released from BAC reactors on disinfection performance. It was concluded that, after disinfection, attached bacteria in biologically treated waters have little impact on public health, either due to a low number of released fines or a low number of organisms attached to these fines [144].

Additional filtration through sand is recommended to prevent escape of activated carbon fines in the finished water. This can be accomplished by having a layer of sand media (6 to 9 in.) as a support for the GAC media [145].

Experiences gathered to date from laboratory-, pilot-, and full-scale studies point to the ability of biological drinking water treatment to produce microbiologically safe water, especially if post-disinfection is practiced [146]. In general, bacteria exiting BAC filters are easily eliminated by post-disinfection.

References

1 Nishijima, W. and Speitel, G.E. Jr (2004) Fate of biodegradable dissolved organic carbon produced by ozonation on biological activated carbon. *Chemosphere*, **56**, 113–119.

2 Yapsakli, K., Çeçen, F., Aktaş, Ö., and Can, Z.S. (2009) Impact of surface properties of granular activated carbon and preozonation on adsorption and desorption of natural organic matter. *Environmental Engineering Science*, **26**, 489–500.

3 Carlson, G. and Silverstein, J. (1998) Effect of molecular size and charge on

biofilm sorption of organic matter. *Water Research*, **32** (5), 1580–1592.
4 Carlson, G. and Silverstein, J. (1997) Effect of ozonation on sorption of natural organic matter by biofilm. *Water Research*, **31** (10), 2467–2478.
5 EAWAG (Publ.) (2009) Wave21 final report – Drinking Water for the 21st Century. Schriftenreihe Nr. 20.
6 Yapsakli, K. (2008) Application of biological activated carbon (BAC) in drinking water treatment. Ph.D. Thesis, Bogazici University, Institute of Environmental Sciences, Istanbul, Turkey.
7 Yapsakli, K. and Çeçen, F. (2010) Effect of type of granular activated carbon on DOC biodegradation in biological activated carbon filters. *Process Biochemistry*, **45**, 355–362.
8 Sontheimer, H., Crittenden, J., and Summers, R.S. (1988) Activated carbon for water treatment, 2nd edn, Forschungstelle Engler – Bunte-Institute, Universität Karlsruhe, Karlsruhe, Germany.
9 Klimenko, N.A., Savchina, L.A., Kozyatnik, I.P., Goncharuk, V.V., and Samsoni-Todorov, A.O. (2009) The effect of preliminary ozonization on the bioregeneration of activated carbon during its long-term service. *Journal of Water Chemistry and Technology*, **32** (4), 220–226.
10 Herzberg, M., Dosoretz, C.G., Tarre, S., Michael, B., Dror, M., and Green, M. (2004) Simultaneous removal of atrazine and nitrate using a biological granulated activated carbon (BGAC) reactor. *Journal of Chemical Technology and Biotechnology*, **79**, 626–631.
11 Tiehm, A., Gozan, M., Müller, A., Schell, H., Lorbeer, H., and Werner, P. (2002) Sequential anaerobic/aerobic biodegradation of chlorinated hydrocarbons in activated carbon barriers. *Water Science and Technology: Water Supply*, **2** (2), 51–58.
12 Urfer, D., Huck, P.M., Booth, S.D.J., and Coffrey, B.M. (1997) Biological filtration for BOM removal: a critical review. *Journal of American Water Works Association*, **89** (12), 83–98.
13 Dussert B.W. and VanStone, G.R. (1994) The biological activated carbon process for water purification. *Water Engineering and Management*, **141** (12), 22–24.
14 Steele, M.E.J., Evans, H.L., Stephens, J., Rachwal, A.J., and Clarke, B.A. (2006) Dissolved oxygen issues with granular activated carbon sandwich™ slow sand filtration, in *Recent Progress in Slow Sand and Alternative Biofiltration Processes* (eds R. Gimbel, N. J.G. Graham, M.R. Collins), IWA Publishing, London, pp. 83–94.
15 LeChevallier, M.W., Becker, W.C., Schorr, P., and Lee, R.G. (1992) Evaluating the performance of biologically active rapid sand filters. *Journal of American Water Works Association*, **84** (4), 136–146.
16 Chien, C.C., Kao, C.M., Dong, C.D., Chen, T.Y., and Chen, J.Y. (2006) Effectiveness of AOC removal by advanced water treatment systems: a case study. *Desalination*, **202**, 318–325.
17 Krasner, S.W., Schlimenti, M.J., and Coffey, B.M. (1993) Testing biologically active filters for removing aldehydes formed during ozonation. *Journal of American Water Works Association*, **85** (5), 62–73.
18 Choi, Y.C., Li, X., Raskin, L., and Morgenroth, E. (2008) Chemisorption of oxygen onto activated carbon can enhance the stability of biological perchlorate reduction in fixed bed biofilm reactors. *Water Research*, **42**, 3425–3434.
19 Grens, B.K. and Werth, C.J. (2001) Durability of wood-based versus coal-based GAC. *Journal of American Water Works Association*, **93** (4), 175–181.
20 Cleasby, J.L. (1990) Filtration, in *Water Quality and Treatment* (ed. F.W. Pontius) Mc Graw Hill, NY.
21 Boere, J., van den Dikkenberg, J., and Joon, G. (2002) BAC and granular activated carbon types in potable water treatment; about microorganisms and nano-pores. Norit Technical paper. Biological Activated Carbon Filtration Workshop, International Water Association, Delft University of

Technology, The Netherlands, May 29–31, 2002.
22 Hendricks, D. (2006) *Water Treatment Unit Processes: Physical and Chemical*, CRC Press. Printed in the USA.
23 AWWA (American Water Works Association and American Society of Civil Engineers) (2005a) *Water Treatment Plant Design* (ed. E.E. Baruth), 4th edn, New York: McGraw-Hill Handbooks.
24 Nishijima, W., Kim, W.H., Shoto, E., and Okada, M. (1998) The performance of an ozonation-biological activated carbon process under long term operation. *Water Science and Technology*, 38 (6), 163–169.
25 Kameya, T., Hada, T., and Urano, K. (1997) Changes of adsorption capacity and pore distribution of biological activated carbon on advanced water treatment. *Water Science and Technology*, 35 (7), 155–162.
26 Asami, M., Aizawa, T., Morioka, T., Nishijima, W., Tabata, A., and Magara, Y. (1999) Bromate removal during transition from new granular activated carbon (GAC) to biological activated carbon (BAC). *Water Research*, 33 (2), 2797–2804.
27 Scholz, M. and Martin, R.J. (1997) Ecological equilibrium on biological activated carbon. *Water Research*, 31 (2), 2959–2968.
28 Yavich, A.A., Lee, K.H., Cehn, K.C., Pape, L., and Maten, S.J. (2004) Evaluation of biodegradability of NOM after ozonation. *Water Research*, 38, 2839–2846.
29 DeWaters, J.E. and Digiano, F.A. (1990) Influence of ozonated natural organic matter on the biodegradation of a micropollutant in a GAC bed. *Journal of American Water Works Association*, 82 (8), 69–80.
30 Huck, P.M., Zhang, S., and Price, M.L. (1994) BOM removal during biological treatment: a first order model. *Journal of American Water Works Association*, 86 (6), 61–74.
31 Zhang, M. and Huck, P.M. (1996) Biological water treatment: a kinetic modeling approach. *Water Research*, 30 (5), 1195–1205.
32 Hozalski, R.M., Goel, S., and Bouwer, E.J. (1995) TOC removal in biologically active sand filters: effect of NOM source and EBCT. *Journal of American Water Works Association*, 87 (12), 40–54.
33 Carlson, K.H. and Amy, G.L. (1998) BOM removal during biofiltration. *Journal of American Water Works Association*, 90 (12), 42–52.
34 van der Aa, L.T.J., Kolpa, R.J., Magic-Knezev, A., Rietveld, L.C., and van Dijk, J.C. (2003) Biological activated carbon filtration: pilot experiments in the Netherlands. Water Quality Technology Conference, Philadelphia, November 2–6, 2003.
35 Huck, P.M., Finch, G.R., Hrudey, S.E., Peppler, M.S., Amirtharajah, A., Bouwer, E.J., Albritton, W.L., and Gammie, L. (1997) Design of Biological Processes for Organics Control. Submitted to American Water Works Association Research Foundation, Report No. 90722. Denver, Colorado.
36 Ahmad, R. and Amirtharajah, A. (1998) Detachment of particles during biofilter backwashing. *Journal of American Water Works Association*, 90 (12), 74–85.
37 Ahmad, R., Amirtharajah, A., Al-Shawwa, A., and Huck, P.M. (1998) Effects of backwashing on biological filters. *Journal of American Water Works Association*, 90 (12), 62–73.
38 Miltner, R.J., Summers, R.S., and Wang, J.Z. (1995) Biofiltration performance: Part II, Effect of backwashing. *Journal of American Water Works Association*, 87 (12), 64–70.
39 Liu, X., Huck, P.M., and Slawson, R.M. (2001) Factors affecting drinking water biofiltration. *Journal of American Water Works Association*, 93 (12), 90–101.
40 Moll, D.M., Summers, R.S., Fonseca, A.C., and Matheis, W. (1999) Impact of temperature on drinking water biofilter performance and microbial community structure. *Environmental Science and Technology*, 33, 2377–2382.
41 Weinberg, H., Glaze, W.H., Krasner, S. W., and Schlimenti, M.C. (1993)

Formation and removal of aldehydes in plants that use ozonation. *Journal of American Water Works Association*, **85** (5), 72–85.

42. Boere, J.A. (1991) Reduction of oxidants by granular activated carbon filtration. Proceedings of the Tenth Ozone World Congress, Monaco, 19–21 March 1991.

43. Urfer, D., Courbat, R., Walther, J.L., and Pärli, A. (2002) Fluidized bed GAC filtration for the control of ozone residuals. *Ozone Science and Engineering*, **23**, 429–438.

44. Chien, C.C., Kao, C.M., Chen C.W., Dong, C.D., and Wu, C.Y. (2008) Application of biofiltration system on AOC removal: column and field studies. *Chemosphere*, **71**, 1786–1793.

45. Shi-hu, S., Min, Y., Nai-yun, G., and Wen-Jie, H. (2008) Molecular weight distribution variation of assimilable organic carbon during ozonation/BAC process. *Journal of Water Supply: Research and Technology—AQUA*, **57** (4), 253–258.

46. Xu, B., Gao, N.Y., Sun, X.F., Xia, S.J., Simonnot, M.O., Causserand, C., Rui, M., and Wu, H.H. (2007), Characteristics of organic material in Huangpu river and treatability with the O_3-BAC process. *Separation and Purification Technology*, **57** 348–355.

47. An, D., Song, J., Gao, W., Chen, G., and Gao, N. (2009) Molecular weight distribution for NOM in different drinking water treatment processes. *Desalination and Water Treatment*, **5**, 267–274.

48. Hozalski, R.M., Bouwer, E.J., and Goel, S. (1999) Removal of natural organic matter (NOM) from drinking water supplies by ozone-biofiltration. *Water Science and Technology*, **40** (9), 157–163.

49. van der Aa, L.T.J., Rietveld, L.C., and van Dijk, J.C. (2011) Effects of ozonation and temperature on the biodegradation of natural organic matter in biological granular activated carbon filters. *Drinking Water Engineering and Science*, **4**, 25–35.

50. Carlson, K.H. and Amy, G. (1997) The formation of filter removable biodegradable organic matter during ozonation. *Ozone Science and Engineering*, **19** (2), 179–199.

51. Kim, W.H., Nishijima, W., Shoto, E., Okada, M., and Okada, M. (1997) Pilot plant study on ozonation and biological activated carbon process for drinking water treatment. *Water Science and Technology*, **35** (8), 21–28.

52. Melin, E., Eikebrokk, B., Brugger, M., and Odegaard, H. (2002) Treatment of humic surface water at cold temperatures by ozonation and biofiltration. *Water Science and Technology: Water Supply*, **2** (5–6), 451–457.

53. Newcombe, G., Ho, L., Hoefel, D., Saint, C., Slyman, N., and Oemcke, D. (2005) The role of biofiltration in the removal of algal metabolites using GAC, in Occasional Paper 9: Biological Filtration, Ozone and Activated Carbon. Cooperative Research Center for Water Quality and Treatment. City West Campus, University of South Australia. 18–24.

54. Persson, F., Heinicke, G., Uhl, W., Hedberg, T., and Hermansson, M. (2006) Performance of direct biofiltration of surface water for reduction of biodegradable organic matter and biofilm formation potential. *Environmental Technology*, **27**, 1037–1045.

55. Laurent, P., Kihn, A., Andersson, A., and Servais, P. (2003), Impact of backwashing on nitrification in the biological activated carbon filters used in drinking water treatment. *Environmental Technology*, **24** (3), 277–287.

56. Yang, J.S., Yuan, D.X., and Weng, T.P. (2010) Pilot study of drinking water treatment with GAC, O_3/BAC and membrane processes in Kinmen Island, Taiwan. *Desalination*, **263**, 271–278.

57. Andersson, A., Laurent, P., Kihn, A., Prevost, M., and Servais, P. (2001) Impact of temperature on nitrification in biological activated carbon (BAC) filters used for drinking water

treatment. *Water Research*, **35** (12), 2923–2934.

58 Tränckner, J., Wricke, B., and Krebs, P. (2008) Estimating nitrifying biomass in drinking water filters for surface water treatment. *Water Research*, **42**, 2574–84.

59 Marda, S., Kim, D., Kim, M.J., Gordy, J., Pierpont, S., Gianatasio, J., Amirtharajah, A., and Kim, J.H. (2008) Plant conversion experience: ozone BAC process installation and disinfectant residual control. *Journal of the American Water Works Association*, **100** (4), 117–128.

60 Yapsakli, K., Mertoğlu, B., and Çeçen, F. (2010) Identification of nitrifiers and nitrification performance in drinking water biological activated carbon (BAC) filtration. *Process Biochemistry*, **45**, 1543–1549.

61 van Dijk, J.C. and van der Kooij, D. (2004) Water Quality 21 research programme for water supplies in the Netherlands. *Water Science and Technology: Water Supply*, **4** (5–6), 181–188.

62 Ding, L., Marinas, B.J., Schideman, L.C., and Snoeyink, V.L. (2006) Competitive effects of natural organic matter: parametrization and verification of the three-component adsorption model COMPSORB. *Environmental Science and Technology*, **40** (1), 350–356.

63 Weber, W. (2002) Preloading, competition, adsorption analysis, ideal adsorption solution theory. Biological Activated Carbon Filtration Workshop, International Water Association, Delft University of Technology, The Netherlands, May 29–31, 2002.

64 Simpson, D.R. (2008) Review: biofilm processes in biologically active carbon water purification. *Water Research*, **42**, 2839–2848.

65 Schreiber, B., Schmalz, V., Brinkmann, T., and Worch, E. (2007) The effect of water temperature on the adsorption equilibrium of dissolved organic matter and atrazine on granular activated carbon. *Environmental Science and Technology*, **41**, 6448–6453.

66 Schreiber, B., Schmalz, V., and Worch, E. (2006) *Exportorientierte F&E auf dem Gebiet der Wasserver- und -entsorgung Teil I: Trinkwasser Band 2: Leitfaden (R&D in the field of water supply and waste water treatment under regional conditions Part I: Drinking water, Vol. 2: Recommendations)*, ISBN 3-00-015478-7, p. 200.

67 Hess, F., Müller, U., and Worch, E. (1997) Untersuchungen zum Einfluss der Adsorbierbarkeit natürlicher organischer Substanzen auf die konkurrierende Adsorption mit Spurenstoffen. *Vom Wasser*, **88**, 71–76.

68 Worch, E. (2009) Adsorptionsverfahren in der Trinkwasseraufbereitung. *Wiener Mitteilungen, Band*, **212**, 207–230.

69 Cook, D., Newcombe, G., and Sztajnbok, P. (2001) The application of powdered activated carbon for MIB and geosmin removal: predicting PAC doses in four raw waters. *Water Research*, **35** (59), 1325–1333.

70 Yu, Z., Peldszus, S., and Huck, P.M. (2008) Adsorption characteristics of selected pharmaceuticals and an endocrine disrupting compound—naproxen, carbamazepine and nonylphenol—on activated carbon. *Water Research*, **42**, 2873–2882.

71 Yu, Z., Peldszus, S., and Huck, P.M. (2009) Adsorption of selected pharmaceuticals and an endocrine disrupting compound by granular activated carbon: Adsorption capacity and kinetics. *Environmental Science and Technology*, **43** (5), 1467–73.

72 Simazaki, D., Fujiwara, J., Manabe, S., Matsuda, M., Asami, M., and Kunikane, S. (2008) Removal of selected pharmaceuticals by chlorination, coagulation–sedimentation and powdered activated carbon treatment. *Water Science and Technology* **58** (5), 1129–1135.

73 Karanfil, T., Kitis, M., Kilduff, J.E., and Wigton, A. (1999) The role of GAC surface chemistry on the adsorption of organic compounds: Natural organic matter. *Environmental Science and Technology*, **33** (18), 3225–3233.

74 Karanfil, T. and Kilduff, J.E. (1999) Role of granular activated carbon surface chemistry on the adsorption of organic compounds. 1. Priority pollutants. *Environmental Science and Technology*, **33**, 3217–3224.

75 Quinlivan, P.A., Li, L., and Knappe, D.R.U. (2005) Effects of activated carbon characteristics on the simultaneous adsorption of aqueous organic micropollutants and natural organic matter. *Water Research*, **39**, 1663–1673.

76 van der Aa, L.T.J., Rietveld, L.C., and van Dijk, J.C. (2010) Simultaneous removal of natural organic matter and atrazine in biological granular activated carbon filters: model validation. IWA Specialist Conference 'Water and Wastewater Treatment Plants in Towns and Communities of the XXI Century: Technologies, Design and Operation', Moscow, Russia, 2–4 June 2010.

77 Bonné, P.A.C., Beerendonk, E.F., van der Hoek, J.P., and Hofman, J.A.M.H. (2000) Retention of herbicides and pesticidess in relation to aging of RO membranes. *Desalination*, **132**, 189–193.

78 Nugroho, W.A., Reungoat, J., and Keller, J. (2010) The performance of biological activated carbon in removing pharmaceuticals in drinking water treatment. *Journal of Applied Sciences in Environmental Sanitation*, **V**, **N**, 123–133.

79 Bundy, M.M., Doucette, W.J., McNeill, L., and Ericson, J.F. (2007) Removal of pharmaceuticals and related compounds by a bench-scale drinking water treatment system. *Journal of Water Supply: Research and Technology – AQUA*, **56** (2), 105–115.

80 Choi, K.J., Kim, S.G., Kim, C.W., and Kim, S.H. (2005) Effects of activated carbon types and service life on removal of endocrine disrupting chemicals: amitrol, nonylphenol, and bisphenol-A. *Chemosphere*, **58**, 1535–1545.

81 Scharf, R.G., Johnston, R.W., Semmens, M.J., and Hozalski, R.M. (2010) Comparison of batch sorption tests, pilot studies, and modeling for estimating GAC bed life. *Water Research*, **44**, 769–780.

82 Elhadi, S.L.N., Huck, P.M., and Slawson, R.M. (2004) Removal of geosmin and 2-methylisoborneol by biological filtration. *Water Science and Technology*, **49** (9), 273–280.

83 Uhl, W., Persson, F., Heinicke, G., Hermansson, M., and Hedberg, T. (2006) Removal of geosmin and MIB in biofilters-on the role of biodegradation and adsorption, in *Recent Progress in Slow Sand and Alternative Biofiltration Processes* (eds R. Gimbel, N.J.G. Graham and M.R. Collins), IWA Publishing, London, pp. 360–368.

84 Drikas, M., Dixon, M., and Morran, J. (2009) Removal of MIB and geosmin using granular activated carbon with and without MIEX pre-treatment. *Water Research*, **43**, 5151–5159.

85 Ho, L., Hoefel, D., Meyn, T., Saint, C.P., and Newcombe, G. (2006) Biofiltration of microcystin toxins: an Australian perspective, in *Recent Progress in Slow Sand and Alternative Biofiltration Processes* (eds R. Gimbel, N.J.G. Graham and M.R. Collins), IWA Publishing, London, pp. 162–170.

86 Wang, H., Hob, L., Lewis, D.L., Brookes, J.D., and Newcombe, G. (2007) Discriminating and assessing adsorption and biodegradation removal mechanisms during granular activated carbon filtration of microcystin toxins. *Water Research*, **41**, 4262–4270.

87 Kim, W.H., Nishijima, W., Baes, A.U., and Okada, M. (1997b) Micropollutant removal with saturated biological activated carbon (BAC) in ozonation-BAC process. *Water Science and Technology*, **36** (12), 283–298.

88 Li, L., Zhu, W., Zhang, P., Zhang, Q., and Zhang, Z. (2006) AC/O$_3$-BAC processes for removing refractory and hazardous pollutants in raw water. *Journal of Hazardous Materials*, **B135**, 129–133.

89 Baus, C., Hung, H., Sacher, F.F., Fleig, M., and Brauch, H.J. (2005)

MTBE in drinking water production-occurrence and efficiency of treatment technologies. *Acta hydrochimica et hydrobiologica*, 33 (2), 118–132.
90 Zhang, X.X., Zhang, Z.Y., Ma, L.P., Liu, N., Wu, B., Zhang, Y., Li, A.M., and Cheng, S.P. (2010) Influences of hydraulic loading rate on SVOC removal and microbial community structure in drinking water treatment biofilters. *Journal of Hazardous Materials*, 178 (1–3), 652–657.
91 Kim, J. and Kang, B. (2008) DBPs removal in GAC filter-adsorber. *Water Research*, 42 (1–2), 145–152.
92 El-Rehaili, A.M. and Weber, W.J. Jr (1987) Correlation of humic substance trihalomethanes formation potential and adsorption behavior to molecular weight distribution in raw and chemically treated waters. *Water Research*, 21, 573–582.
93 Kitiş, M., Karanfil, T., Kilduff, J.E., and Wigton, A. (2001) The reactivity of natural organic matter to disinfection byproducts formation and its relation to specific ultraviolet absorbance. *Water Science and Technology*, 43 (2), 9–16.
94 Chaiket, T., Singer, P.C., Miles, A., Moran, M., and Pallota, C. (2002) Effectiveness of coagulation, ozonation and biofiltration in controlling DBPs. *Journal of the American Water Works Association*, 94 (12), 81–95.
95 Amy, G.L., Tan, L., and Davis, M.K. (1991) The effects of ozonation and activated carbon adsorption on trihalomethane speciation. *Water Research*, 25 (2), 191–202.
96 Zhang, Z.H. and Shao, L. (2008) Study on control of micro-pollutants by BAC filtration. *Water Science and Technology*, 58(3), 677–682.
97 Xie, Y.F. and Zhou, H. (2002) Biologically active carbon for HAA removal: Part II, column study. *Journal of the American Water Works Association*, 94 (5), 126–134.
98 Xie, Y. (2004) *Disinfection Byproducts in Drinking Water: Formation, Analysis, and Control*, Lewis Publishers.
99 Babi, K.G., Koumenides, K.M., Nikolaou, A.D., Makri, C.A., Tzoumerkas, F.K., and Lekkas, T.D. (2007) Pilot study of the removal of THMs, HAAs and DOC from drinking water by GAC adsorption. *Desalination*, 210, 215–224.
100 Kim, J. (2009) Fate of THMs and HAAs in low TOC surface water. *Environmental Research*, 109, 158–165.
101 Çeçen, F., Kocamemi, B.A., and Aktaş, Ö. (2010) Metabolic and co-metabolic degradation of industrially important chlorinated organics under aerobic conditions, in *Xenobiotics in the Urban Water Cycle: Mass Flows, Environmental Processes, Mitigation and Treatment Strategies* (eds K. Despo-Fatto, B. Kai, and K. Klaus) Book Series: Environmental Pollution, Vol. 16, Springer Science + Business Media, pp. 161–194.
102 Wahman, D.G., Henry, A.E., Katz, L.E., and Speitel, G. E. Jr (2006) Cometabolism of trihalomethanes by mixed culture nitrifiers. *Water Research*, 40, 3349–3358.
103 Ersever, I., Ravindran, V., and Pirbazari, M. (2007) Biological denitrification of reverse osmosis brine concentrates: II. Fluidized bed adsorber reactor studies. *Journal of Environmental Engineering and Science*, 6, 519–532.
104 Gottschalk, C., Libra, J.A., and Saupe, A. (2000) *Ozonation of Water and Waste Water, A Practical Guide to Understanding Ozone and its Application*, WILEY-VCH.
105 von Gunten, U. (2003) Review: ozonation of drinking water: Part II. disinfection and by-product formation in presence of bromide, iodide or chlorine. *Water Research*, 37, 1469–1487.
106 Butler, R., Godley, A., Lytton, L., and Cartmell, E. (2005) Bromate environmental contamination: review of impact and possible treatment. *Critical Reviews in Environmental Science and Technology*, 35, 193–217.
107 Bao, M.L., Griffini, O., Santianni, D., Barbieri, K., Burrini, D., and Pantani, F. (1999) Removal of bromate ion from

108 Kirisits, M.J., Snoeyink, V.L., and Kruithof, J.C. (2000) The reduction of bromate by granular activated carbon. Water Research, 35 (17), 4250–4260.

109 Kirisits, M.J., Snoeyink, V.L., Inan, H., Chee-Sanford, J.C., Raskin, L., and Brown, J.C. (2001) Water quality factors affecting bromate reduction in biologically active carbon filters. Water Research, 35 (4), 891–900.

110 Huang, W.J., Peng, H.S., Peng, M.Y., and Chen, L.Y. (2004) Removal of bromate and assimilable organic carbon from drinking water using granular activated carbon. Water Science and Technology, 50 (8), 73–80.

111 Brown, J.C., Snoeyink, V.L., Raskin, L. M., and Lin, R. (2003) The sensitivity of fixed-bed biological perchlorate removal to changes in operating conditions and water quality characteristics. Water Research, 37, 206–214.

112 California Environmental Protection Agency, Office of Pollution Prevention and Technology Development Department of Toxic Substances Control (2004) Perchlorate contamination treatment alternatives (draft). Sacramento, CA

113 Polk, J., Murray, J., Onewokae, C., Tolbert, D.E., Togna, A.P., Guarini, W. J., Frisch, S., and Del Vecchio, M. (2001) Case study of ex-situ biological treatment of perchlorate-contaminated groundwater, Presented at the 4th Tri-Services Environmental Technology Symposium, June 18–20, San Diego, USA. http://www.cluin.org/contaminantfocus/default.focus/sec/perchlorate/cat/Treatment_Technologies.

114 Kirisits, M.J. (2001) Removal of Bromate and Perchlorate in Conventional ozone/GAC Systems, AWWA Research Foundation and American Water Works Association, USA.

115 AWWA (American Water Works Association and American Society of Civil Engineers) (2005b) Microfiltration and ultrafiltration membranes for drinking water, Manual of Water Supply Practices-M 53, 1st edition. Printed in the USA.

116 Sagbo, O., Sun, Y., Hao, A., and Gu, P. (2008) Effect of PAC addition on MBR process for drinking water treatment. Separation and Purification Technology, 58 (3), 320–327.

117 Williams, M.D. and Pirbazari, M. (2007) Membrane bioreactor process for removing biodegradable organic matter from water. Water Research, 41, 3880–3893.

118 Tian, J.Y., Chen, Z.L., Yang, Y., Liang, H., Nan, J., Wang, Z.Z., and Li, G.B. (2009b) Hybrid process of BAC and sMBR for treating polluted raw water. Bioresource Technology, 100 (24), 6243–6249.

119 Tian, J.Y., Chen, Z.L., Liang, H., Li, X., Wang, Z.Z., and Li, G.B. (2009) Comparison of biological activated carbon (BAC) and membrane bioreactor (MBR) for pollutants removal in drinking water treatment. IWA Publishing. Water Science and Technology, 60 (6), 1515–1523.

120 Lorbeer, H., Starke, S., Gozan, M., Tiehm, A., and Werner, P. (2002) Bioremediation of chlorobenzene-contaminated groundwater on granular activated carbon barriers, Water, Air and Soil Pollution :Focus, 2, 183–193.

121 Servais, P.G., Bilen, G., Ventresque, C., and Bablon, G.P. (1991) Microbial activity in GAC filters at the Choisy-le-Roi treatment plant. Journal of the American Water Works Association, 83 (2), 62–68.

122 Beaudet, J.F., Prevost, N., Arcouette, N., Niquette, P., and Coallier, J. (1996) Controlling annelids in biological activated carbon filters. Proceedings of the American Water Works Association WQTC, Boston, MA.

123 Schreiber, H., Schoenen, D., and Traunspurger, W. (1997) Invertebrate colonization of granular activated carbon filters. Water Research, 31 (4), 743–748.

124 Uhl, W. (2000) Simultane biotische und abiotische Prozesse in

Aktivkohlefiltern der Trinkwasseraufbereitung in: 'Sorptionsverfahren der Wasser- und Abwasserreinigung - Schriftenreihe des SFB 193 (Biologische Behandlung industrieller und gewerblicher Abwässer) der TU Berlin', 13, 151–179, TU-Berlin Universitätsbibliothek, Berlin, ISBN3-7983-1827-1.

125 Madoni, M., Davoli, D., Cavagnoli, G., Cucchi, A., Pedroni, M., and Rossi, F. (2000) Microfauna and filamentous microflora in biological filters for tap water production. *Water Research*, **34** (14), 3561–3572.

126 Grabinska-Loniewska, A., Perchue, M., and Kowalska, T.K. (2004) Biocenoesis of BACs used for groundwater treatment. *Water Research*, **38**, 1695–1706.

127 Findlay, R.H., King, G.M., and Watling, L. (1989) Efficacy of phospholipid analysis in determining microbial biomass in sediments. *Applied and Environmental Microbiology*, **55** (11), 2888–2893.

128 Lu, P. and Huck, P.M. (1993) Evaluation of methods for measuring biomass and biofilm thickness in biological drinking water treatment, Proceedings of the American Water Works Association WQTC, Miami, FL.

129 Lehtola, M.J., Miettinen, I.T., Vartiainen, T., Myllykangas, T., and Martikainen, P.J. (2001) Microbially available organic carbon, phosphorus, and microbial growth in ozonated drinking water. *Water Research*, **35** (7), 1635–1640.

130 Juhna, T. and Rubulis, J. (2004) Problem of DOC removal during biological treatment of surface water with a high amount of humic substances. *Water Science and Technology: Water Supply*, **4** (4), 183–187.

131 Lazarova, V. and Manem, J. (1995) Biofilm characterization and activity analysis in water and wastewater treatment. *Water Research*, **29** (10), 2227–2245.

132 van der Kooij, D., Vrouwenvelder, J.S., and Veenendaal, H.R. (2003) Elucidation and control of biofilm formation processes in water treatment and distribution using the unified biofilm. *Water Science and Technology*, **47** (5), 83–90.

133 Urfer, D. and Huck, P.M. (2001) Measurement of biomass activity in drinking water biofilters using a respirometric method. *Water Research*, **36** (6), 1469–1477.

134 Magic-Knezev, A. and van der Kooij, D. (2004) Optimisation and significance of ATP analysis for measuring active biomass in granular activated carbon filters used in water treatment. *Water Research*, **38**, 3971–3979.

135 van der Aa, L.T.J., K-Magic-Knezev, A., Rietveld, L.C., and van Dijk, J.C. (2006) Biomass development in biological activated carbon filters, in *Recent Progress in Slow Sand and Alternative Biofiltration Processes* (eds R. Gimbel, N. J.G. Graham, and M.R. Collins), IWA Publishing, London, pp. 293–301.

136 Velten, S., Hammes, F., Boller, M., and Egli, T. (2007) Rapid and direct estimation of active biomass on granular activated carbon through adenosine tri-phosphate (ATP) determination. *Water Research*, **41**, 1973–1983.

137 Magic-Knezev, A., Wullings, B., and van der Kooij, D. (2009) Polaromonas and hydrogenophaga species are the predominant bacteria cultured from granular activated carbon filters in water treatment. *Journal of Applied Microbiology*, **107**, 944–953.

138 Fonseca, A.C., Summers, R.S., and Hernandez, M.T. (2001) Comparative measurements of microbial activity in drinking water biofilters. *Water Research*, **35** (16), 3817–3824.

139 Niemi, R.M., Heiskanen, I.R., Heine, R., and Rapala, J. (2009) Previously uncultured b-Proteobacteria dominate in biologically active granular activated carbon (BAC) filters. *Water Research*, **43**, 5075–5086.

140 Kasuga, I., Shimazaki, D., and Kunikane, S. (2007) Influence of backwashing on the microbial community in a biofilm developed on

biological activated carbon used in a drinking water treatment plant. *Water Science and Technology*, **55** (8–9), 173–180.

141 Kihn, A., Laurent, P., and Servais, P. (2000) Measurement of potential activity of fixed nitrifying bacteria in biological filters used in drinking water production. *Journal of Industrial Microbiology and Biotechnology*, **24**, 161–166.

142 Morin, P., Camper, A., Jones, W., Gatel, D., and Goldman, J.C. (1996) Colonization and disinfection of biofilms hosting coliform-colonized carbon fines. *Applied and Environmental Microbiology*, **62**, 4428–4432.

143 Elzanfaly, H.T., Reasoner, D.J., and Geldreich, E.E. (1998) Bacteriological changes associated with granular activated carbon in a pilot water treatment plant. *Water, Air, and Soil Pollution*, **107**, 73–80.

144 Pernitsky, D.J., Finch, G.R., and Huck, P.M. (1997) Recovery of attached bacteria from GAC fines and implications for disinfection efficacy. *Water Research*, **31** (3), 385–390.

145 Egli, T. (2008) New methods for assessing the safety of drinking water, *EAWAG News 65/ Research Report*, 20–23.

146 Camper, A. (2005) Chapter 18 Biofilms in drinking water treatment and distribution, in *Biofilms: Recent Advances in their Study and Control* (ed. L.V. Evans), Harwood Academic Publishers, pp. 322–344.

10
BAC Filtration Examples in Full-Scale Drinking Water Treatment Plants

Ferhan Çeçen

BAC filtration aims primarily at establishing the biostability of water. Water is considered as biologically stable only if, in the distribution system, it does not support the growth of microorganisms to a significant extent. As BDOC and AOC can indicate the potential for re-growth in treated water, this chapter first discusses the limits set for these parameters. Some examples are then demonstrated from full-scale BAC operation for the removal of organics and micropollutants.

10.1
Limits for BDOC and AOC as Indicators of Re-growth Potential in Water Distribution

In several countries, the BDOC in treated water spans a wide range from 0.01 to about $1\,mg\,L^{-1}$. Similarly, the AOC parameter ranges from about 1 to more than $400\,\mu g\,C\,L^{-1}$ [1]. The concept of a safe BDOC and AOC limit is rather relative. A maximum BDOC concentration of about $0.15\,mg\,L^{-1}$ was proposed to ensure biologically stable water [1]. However, the maximum allowable BOM concentration entering the distribution system is highly site specific, in that it depends on the physicochemical parameters of the water and the distribution system. As a treatment goal for BOM reduction in biofilters, in the absence of better information, many engineers have specified that after post-ozonation the BDOC value in water should be reduced to the pretreatment level.

Undesirable bacteria have been shown to multiply in water, even at few $\mu g\,L^{-1}$ of AOC. Therefore, the AOC value should be kept at very low levels [2]. However, for the occurrence of coliform bacteria, not only is the absolute AOC concentration decisive, but so also is the level of the disinfectant residual [3]. In the case of relatively high AOC in treated water, a high level of disinfectant residual should be maintained. Today, taking all information into account, an AOC concentration of $10\,\mu g\,C\,L^{-1}$ has been derived as a reference value for biological stability [2]. Although in former years the common belief was that compounds within the

---

Activated Carbon for Water and Wastewater Treatment: Integration of Adsorption and Biological Treatment.
First Edition. Ferhan Çeçen and Özgür Aktaş
© 2011 WILEY-VCH Verlag GmbH & Co. KGaA, Weinheim.
Published 2011 by WILEY-VCH Verlag GmbH & Co. KGaA

domain of BDOC or AOC parameters only would contribute to biological re-growth, humic substances have also been shown to lead to biofilm formation and contribute to re-growth potential [4].

10.2
BAC Filtration Experiences in Full-Scale Surface Water Treatment

10.2.1
Mülheim Plants, Germany

As discussed in Chapter 8, the Mülheim process is an original biofiltration process developed in Germany, mainly for the reduction of DOC. In fact, the Mülheim process acts as a multibarrier in the treatment of polluted Ruhr water and leads to a very effective reduction in organic, inorganic, and microbiological parameters, namely DOC, turbidity, ammonia, nitrite, nitrate, HPC, coliforms, *E. coli*, parasites, and pesticides. Figure 10.1 illustrates the first variation of the Mülheim process, in which the water is first pretreated and then infiltrated into groundwater. After re-catchment from groundwater, the water is disinfected with chlorine and deacidified with NaOH [5].

The original Mülheim process at the Dohne works has been modified in other treatment works according to arising needs. Thus, in the second variation, as applied at the Styrum-East Water Works, the Ruhr river water first undergoes slow sand filtration and artificial groundwater recharge. This is then followed by ozonation, biological double layer and BAC filtration. The water is then disinfected with UV-light and deacidified with NaOH [5].

10.2.2
Leiden Plant, Amsterdam, the Netherlands

Removal of organics, namely DOC, taste and odor, pesticides, and the adsorbable halogenated organics (denoted by AOX), is possible by partial oxidation followed by BAC filtration. If ozonation is not followed by BAC filtration, re-growth problems are common. It is reported that in all Dutch plants ozonation is applied in combination with BAC filtration. Reduction of AOC by BAC filtration is one of the objectives to prevent the re-growth of bacteria [6].

The Leiden plant in Amsterdam provides a good example of the removal of specific organics in BAC filtration (Figure 10.2). In the full-scale plant, Rhine water is first pretreated, and the subsequent post-treatment comprises rapid sand filtration, ozonation, hardness removal, BAC filtration, and slow sand filtration. The carbon reactivation frequency was initially set at 18 months, based on removal efficiencies for AOX, DOC, pesticides, and micropollutants. Activated carbon filtration was applied as a two-stage process [7]. In parallel to the full-scale plant, a pilot plant receiving pretreated water was operated with the purpose of investigating the remaining removal capacity and the breakthrough profile of carbon

10.2 BAC Filtration Experiences in Full-Scale Surface Water Treatment

Process flow diagram:

Ruhr River → Premixing Chamber / Preozonation → Flocculation and Sedimentation → Main Ozonation → Biological Double Layer Filtration → BAC Filtration → Infiltration for Artificial Groundwater Recharge → (Re-catchment) → Safety Disinfection by Chlorine → Deacidification by NaOH → Pumping and Distribution

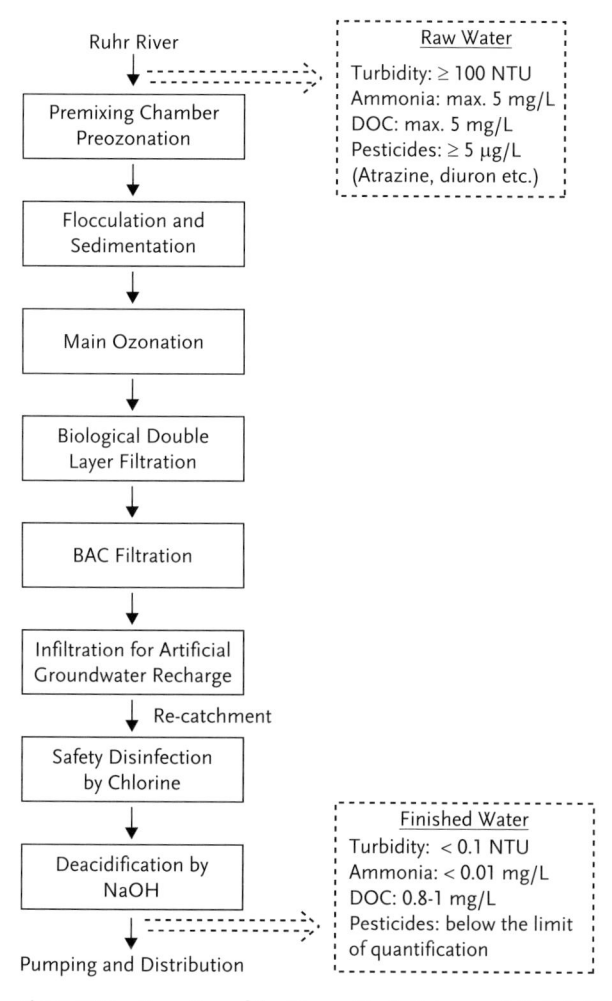

Raw Water:
- Turbidity: ≥ 100 NTU
- Ammonia: max. 5 mg/L
- DOC: max. 5 mg/L
- Pesticides: ≥ 5 µg/L (Atrazine, diuron etc.)

Finished Water:
- Turbidity: < 0.1 NTU
- Ammonia: < 0.01 mg/L
- DOC: 0.8-1 mg/L
- Pesticides: below the limit of quantification

Figure 10.1 Operation of the Styrum-West, Dohne, and Kettwig Water Works according to the Mülheim process (drawn based on Bundermann [5]).

filters. In contrast to the full-scale plant, no carbon reactivation was applied in the pilot plant during 4 years operation. Pesticides were spiked into the influent of BAC filtration where the concentrations varied between 2 and 10 µg L^{-1}. Even in the absence of carbon reactivation the filter effluent still complied with standards and guidelines: DOC was less than 2 mg L^{-1} and AOX remained below 5 µg L^{-1}. After 4 years of spiking with 2 µg L^{-1}, still no pesticide breakthrough was observed in the two-stage BAC filtration. For all the tested pesticides, the most consistent removal was seen in the case of atrazine. It was concluded that a running time of 3 years between two reactivations in the two-stage BAC filtration is achievable without negatively affecting the finished water quality. Average DOC concentrations would increase up to 1.2 mg L^{-1}, from 1 mg L^{-1} with running times of 2 years. After 4 years

322 | 10 BAC Filtration Examples in Full-Scale Drinking Water Treatment Plants

Full Scale Plant
Flow Rate: 70×10^6 m^3/y

Pilot Plant
Flow Rate: 10 m^3/h

River Rhine
↓
Pretreatment by Coagulation, Sedimentation, Rapid Filtration, Dune Filtration
↓ ↓
Rapid Sand Filtration | Rapid Sand Filtration
↓ ↓
Ozonation | Ozonation
↓ ↓
Softening | Softening
↓ ↓
BAC Filtration | BAC Filtration
↓ ↓
Slow Sand Filtration | Slow Sand Filtration
↓ ↓
Drinking Water | Drinking Water

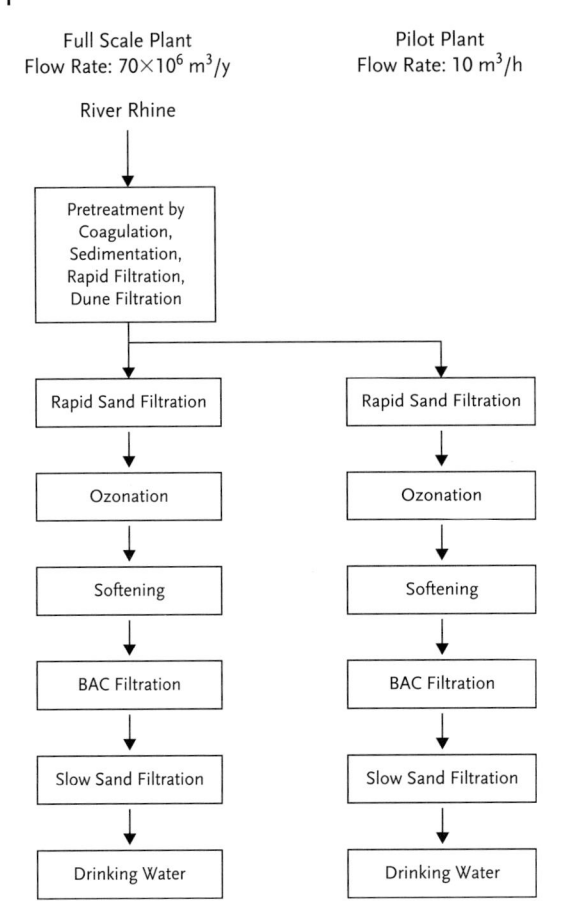

Figure 10.2 Treatment scheme of the Leiden full- and pilot-scale treatment plant (redrawn after [7]).

or 100 000 bed volumes the AOC content would be equal to or lower than 10 μg C L^{-1} after BAC filtration. With slow sand filtration as a polishing step AOC would be less than 10–5 μg C L^{-1} [7].

10.2.3
Plants in the Suburbs of Paris, France

The performance of BAC filters preceded by an ozonation step was evaluated in three treatment plants located in the suburbs of Paris (Méry-sur-Oise, Choisy-le-Roi, and Neuilly-sur-Marne). In these plants, having treatment capacities between 240 000 and 800 000 m^3 d^{-1}, river waters were treated by flocculation and settling, rapid sand filtration, ozonation, BAC filtration, and final disinfection using chlorine. In the ozonated water the DOC ranged from 1.74 to 2.29 mg L^{-1} in different treatment works. The corresponding BDOC in ozonated water was in the range of 0.40–0.48 mg L^{-1}. BDOC was composed of 4–13% of readily biodegradable substrate, 53–80% rapidly hydrolyzable substrate, and 16–33% of slowly hydrolyzable substrate [8]. In BAC filtration the EBCT varied in the range from 10 to 30 min. The carbon grade Picabiol GAC (Pica Company, France) was used in the operation. BAC filtration decreased the BDOC by about 0.19–0.27 mg L^{-1}. However, BAC filtration did not modify the refractory DOC that was present at about 1.34–1.81 mg L^{-1} in ozonated waters. This indicated that organic matter was no longer adsorbed in filters that had been functioning for several years without GAC regeneration [8].

10.2.4
Ste Rose Plant in Quebec, Canada

In BAC filtration in the Ste Rose treatment plant (Quebec, Canada), where river water (Milles Iles River) is treated by flocculation, settling, rapid sand filtration, ozonation, BAC filtration, and final disinfection with chlorine dioxide, the temperature was shown to affect the performance significantly. For temperatures between 1 and 10.0 °C, average BDOC was 0.32 mg L^{-1} in the inlet water before ozonation and 0.20 mg L^{-1} in the outlet water. BDOC levels showed seasonal variability. For temperatures between 10.0 °C and 25.5 °C, average BDOC was 0.54 mg L^{-1} in the inlet water and 0.21 mg L^{-1} in the outlet water [8].

10.2.5
Plant in Zürich-Lengg, Switzerland

The full-scale treatment plant at the water works Lengg comprises preozonation, rapid sand filtration, ozonation, GAC filtration, and slow sand filtration. Figure 10.3 shows the seasonal changes in the raw water fed to this plant. The major part of the organic carbon in the raw water stems from humic material. As Lake Zurich water is relatively clean, the highest DOC concentration during the whole pilot study was 1323 µg L^{-1}. In the same plant, after preozonation, AOC increased about threefold to an average concentration of 171 ± 57 µg C L^{-1} and decreased almost to the original level after sand filtration [9].

The pilot plant which simulated the full-plant operation includes prefiltration of raw water, ozonation, GAC filtration, and ultrafiltration steps. It was shown that prefiltration had no influence on NOM fractions. LC-OCD chromatograms of raw water and ozonated water showed that ozonation transformed higher-molecular-weight compounds (humics) to lower-molecular-weight compounds (building blocks, LMW humics). The SUVA of humics was about 40% lower in the ozonated sample than that in the raw water. Except for the biopolymers, there was a decrease in all NOM fractions from the top to the bottom of the GAC filter in the flow direction. However, the biopolymer fraction was not removed during startup or after 5 months of GAC filtration in spite of development of biological activity. Therefore, this fraction was regarded as nonbiodegradable and nonadsorbable [10]. However, this fraction was generally less than 10% in the raw water.

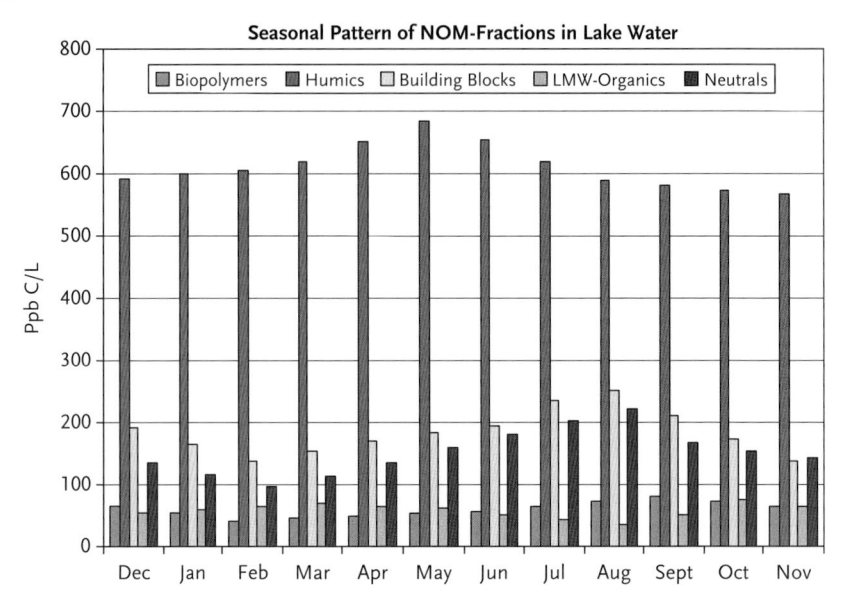

Seasonal Pattern of NOM-Fractions in Lake Water

[Bar chart showing Ppb C/L on y-axis (0-800) vs months (Dec-Nov) on x-axis, with bars for Biopolymers, Humics, Building Blocks, LMW-Organics, and Neutrals.]

Figure 10.3 Seasonal changes in NOM fractions in the Lengg raw water taken from Lake Zurich (at a depth of 30 m) [10].(permission received).

10.2.6
Weesperkarspel Plant, Amsterdam, the Netherlands

In some countries, such as the Netherlands and Switzerland, drinking water is produced without chlorine, and thus is distributed without disinfectant residual. In such cases, it is particularly important that no easily degradable organic matter should enter the water distribution system, as this could promote bacterial re-growth. The operation of the Weesperkarspel treatment plant of Waternet (the water cycle company for Amsterdam and surrounding areas) provides a good example of this (Figure 10.4).

The treatment plant at Weesperkarspel receives pretreated water, as shown in Figure 10.4 [11]. The pretreatment consists of an intake of seepage water from the Bethune polder, sometimes mixed with Amsterdam-Rhine Canal water, and a consecutive treatment of coagulation and sedimentation, self-purification in a lake-water reservoir, and rapid sand filtration. The pretreated water is transported to the Weesperkarspel treatment plant without chlorination. At this treatment plant, the first process is ozonation to provide disinfection and oxidation of organic matter, which results in an increase in the biodegradability of NOM.

After ozonation, pellet reactors are used for softening the water. This is followed by BAC filtration, which is applied to remove NOM and organic micropollutants. The last step in the treatment is slow sand filtration for nutrient removal and reduction of suspended solids. This process is regarded as the second important barrier in the treatment against pathogens, and is especially important for

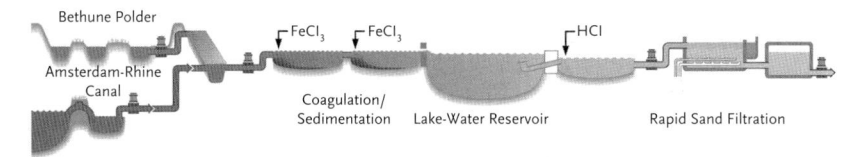

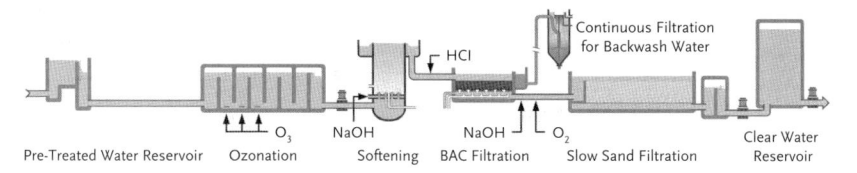

10.2 BAC Filtration Experiences in Full-Scale Surface Water Treatment

Figure 10.4 Weesperkarspel plant receiving pretreated water (by courtesy of Waternet, the Netherlands).

Table 10.1 DOC, UV_{254}, and $SUVA_{254}$ values in raw water and across the treatment train (adapted from [12]).

Sample	DOC (mg C L^{-1})	UV_{254} (cm^{-1})	SUVA (L/mgC.m)
Raw surface water	9.0	0.271	3.5
Pretreatment			
Coagulation effluent	7.1	0.191	3.0
Rapid sand filtration effluent	6.5	0.161	3.0
Surface reservoir effluent	6.0	0.151	2.8
Treatment at Weesperkarspel			
Ozonation effluent	5.7	0.090	1.8
Pellet softening effluent	5.4	0.087	1.7
Biological activated carbon filter effluent	3.0	0.039	1.5
Treated water	2.7	0.038	1.5

removing persistent pathogens with low susceptibility to ozone. The drinking water is transported and distributed without residual chlorine.

Table 10.1 shows the change in various parameters across treatment units. The bulk DOC is mainly removed by coagulation during pretreatment (about 22% of the influent DOC) and by BAC filtration during post-treatment, which removes about 45% of the BAC filter influent DOC. However, this type of characterization alone gives no hint about the biostability of water. Fluorescence excitation-emission matrix (EEM) showed that in raw and treated waters the major fraction of NOM was composed of humic-like material, whereas the contribution of protein-like material was small. Liquid chromatography with online organic carbon detection (LC-OCD) was able to quantify the changes in the DOC concentrations

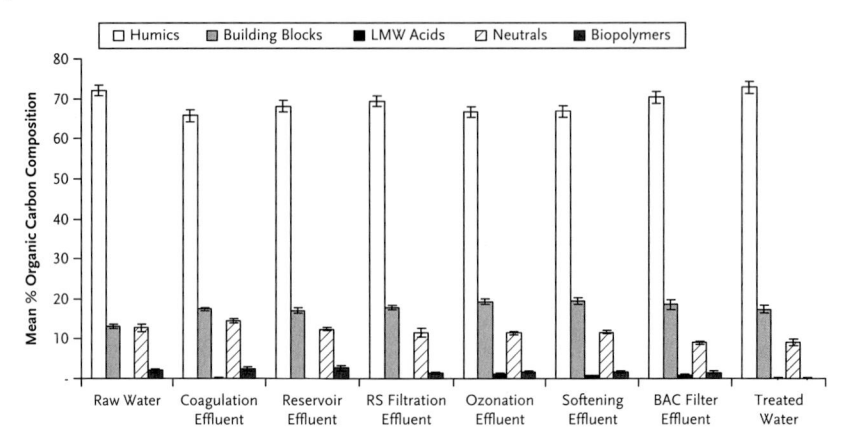

Figure 10.5 The mean percent contribution of each of the five LC-OCD NOM fractions to the bulk water across the treatment train [12](permission received).

of five NOM fractions: humic substances, building blocks (hydrolyzates of humics), biopolymers, low-molecular-weight acids, and neutrals, as shown in Figure 10.5 [12].

The main part of the NOM was composed of humics across the entire treatment. The humic fraction, contributing about 70% of the total DOC, was mainly removed by coagulation (30%) and by BAC filtration (42%), indicating that a significant part was biodegradable. No low-molecular-weight (LMW) acids were initially present in the raw water, but these formed as a result of ozonation and contributed very little to the total DOC. They were largely removed by softening, BAC filtration, and SS filtration.

10.2.7
Drinking Water Treatment Plants, Bendigo, Castlemaine, and Kyneton, Victoria, Australia

The Aqua project in Australia incorporates three water treatment plants using microfiltration, ozone, and BAC filtration technologies. The main plant in Bendigo has a capacity of 126 ML d^{-1} and features submerged microfiltration technology (CMF-S), while the Kyneton 8 ML d^{-1} plant and Castlemaine 18 ML d^{-1} plant use conventional CMF designs [13]. The raw water supplied from reservoirs is screened, then dosed with lime and carbon dioxide to stabilize the water to prevent corrosion. Coagulation is then the next step for removal of particulates, metals, and color. Microfiltration is introduced to remove particulates down to 0.2 μm and to guarantee the removal of *Cryptosporidium* and *Giardia* contamination from unprotected rural catchments. All plants use ozone and BAC filtration in order to remove taste and odor compounds and blue-green algae toxins. Ozonation breaks down complex organics. Subsequently, BAC filtration reduces organic carbon, eliminates taste and odor compounds, and reduces blue-green algae toxins, resulting in high-quality stable water. Then, lime is added to control the final pH and provide corrosion protection. Chlorine and ammonia are dosed to provide chloramination disinfection to ensure the quality of water within the distribution system.

10.3
New Approaches in the Evaluation of Ozonation and BAC Filtration

In recent years, the merits and disadvantages of processes used in water treatment are being evaluated not on the basis of reduction of contaminants alone. A modern approach is to exploit the life cycle assessment (LCA) and life cycle cost (LCC) methodologies to assess the environmental impact and costs, respectively. Adoption of this approach in drinking water treatment has led to the conclusion that optimization of the operation of drinking water treatment plants should focus on water quality and not on environmental impact or costs.

This approach was documented for ozonation in combination with BAC filtration at the drinking water treatment plant Weesperkarspel in Amsterdam [14]. The water quality parameters that were taken into account were AOC, DOC, and pathogens. The operational parameters considered were the ozone dosage and the regeneration frequency of BAC filters. Increasing the fraction of biodegradable DOC in the water by ozonation led to a later breakthrough of DOC. The current ozone dose and regeneration frequency were 2.5 mg O_3 L^{-1} and 18 months, respectively. The DOC standard for the drinking water produced is 2.5 mg L^{-1}, whereas the desired AOC concentration is 10 µg C L^{-1} in finished water. The effluent AOC concentration of the BAC filters should be a maximum of 15 µg C L^{-1}, since in the subsequent slow sand filters the additional AOC removal is 5 µg C L^{-1} on average. As shown in Figure 10.6, the effluent AOC concentration in BAC filters varied with regeneration frequency, while the dose of preozonation also played a major role. The maximum regeneration frequency for activated carbon in BAC filtration was found to be 8 months at an ozone dose of 0.95 mg L^{-1} to achieve these objectives [14].

10.4
BAC Filtration Experiences in Full-Scale Groundwater Treatment

Integrated activated carbon adsorption and biological removal is also applied in groundwater treatment. GAC-based fluidized-bed reactors (GAC-FBR), whose characteristics are discussed in Chapter 3, are frequently used in aerobic or anoxic treatment of groundwaters that are contaminated with petroleum hydrocarbons and pentachlorophenol, respectively [15].

As outlined in Chapter 9, perchlorate contamination gives rise to great concern in groundwater supplies. Based on the success of the laboratory test, a full-scale GAC-FBR was installed at the Longhorn Army Ammunition Plant (LHAAP) in Texas, with the capacity to treat 274 m^3 d^{-1} of groundwater contaminated with high concentrations of perchlorate (between 11 000 and 23 000 µg L^{-1}) from past

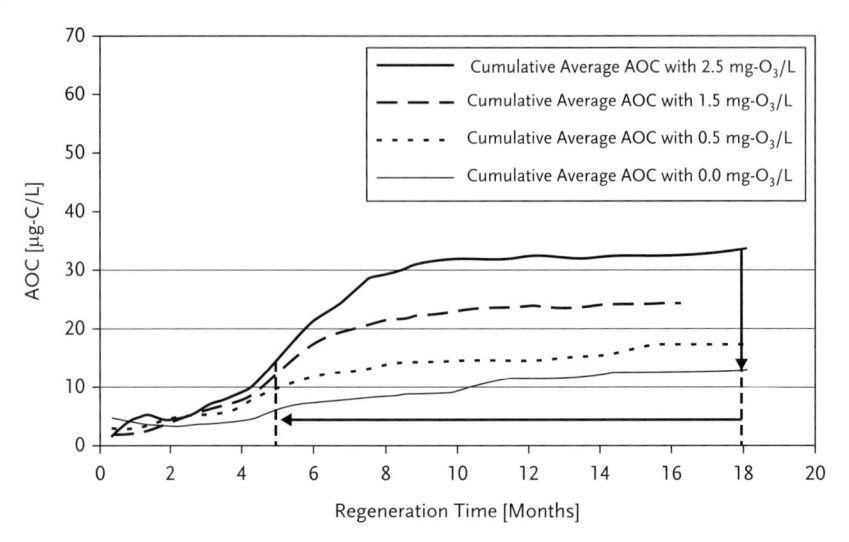

Figure 10.6 Variation of effluent AOC with regeneration frequency and preozonation dose [14](permission received).

operations [16]. The reactor vessel had a diameter and height of 1.5 and 6.4 m, respectively (Figure 10.7). The influent to the FBR was distributed through a proprietary distribution header at the bottom of the tank, which was designed to distribute flow evenly across the cross-sectional area of the reactor with a minimum amount of turbulence. When biological growth occurs on the fluidized-bed media, the diameter of the media increases and its effective density is reduced, resulting in an expansion of the media bed. Two biomass separation systems at the top of the bed and a third in the bed media, which could be positioned anywhere vertically along the bed, functioned to remove excess biomass from the surface of the carbon particles, preventing them from being carried out of the reactor. As outlined in Chapter 9, perchlorate was reduced to chloride by an acclimated biomass under anoxic conditions in the presence of acetic acid as electron donor. Full-scale GAC-FBR treated groundwater gave effluent perchlorate concentrations of less than 4 µg L^{-1} [17]. At another site, in Rancho Cordova, two full-scale GAC-FBR reactors operating at influent groundwater flow rates of 3240 m^3 d^{-1} decreased perchlorate concentrations from 6–7 mg L^{-1} to 4–40 µg L^{-1} with methanol as the electron acceptor [18].

In groundwater treatment, BAC reactors are also used for the elimination of various chlorinated organics. For example, at the first attempt, 240–360 m^3 d^{-1} groundwater contaminated with up to 20 mg L^{-1} 1,2-dichloroethane (1,2-DCA) was treated with two full-scale abiotic GAC reactors to obtain effluent levels below the required limit of 10 µg L^{-1}. Later, however, these reactors were converted to BAC reactors by inoculation with microorganisms able to mineralize 1,2-DCA. To provide appropriate conditions, in addition to nutrients, hydrogen peroxide was added to reactors as an oxygen source in order to avoid stripping of 1,2-DCA. In

Figure 10.7 Full-scale GAC-FBR system to treat groundwater at LHAAP in Texas, USA [17](permission received).

BAC reactors the required effluent limits were achieved, and the service life was extended more than 40-fold compared to abiotic GAC reactors [19].

References

1 Servais P., Billen G., Laurent P., Levi Y., and Randon, G. (1993) Bacterial regrowth in distribution systems. Proceedings of American Water Works Association WQTC Conference, Miami, FL, 7–10 November, 1993.

2 van der Kooij, D. (2003) Managing regrowth in drinking water distribution systems, in *Heterotrophic Plate Counts and Drinking-water Safety* (eds J. Bartram, J. Cotruvo, M. Exner, C. Fricker, and A. Glasmacher), World Health Organization (WHO), IWA Publishing, London, UK, pp. 199–232.

3 LeChevallier, M.W. (2003) Conditions favouring coliform and HPC bacterial growth in drinking-water and on water contact surfaces, in *Heterotrophic Plate Counts and Drinking-water Safety* (eds J. Bartram, J. Cotruvo, M. Exner, C. Frickesr, and A. Glasmacher), World Health Organization (WHO), IWA Publishing, London, UK, pp. 177–197.

4 Camper, A.K. (2004) Involvement of humic substances in regrowth. *International Journal of Food Microbiology*, **92** (3), 355–364.

5 Bundermann, G. (2006) 30 years of RWW's practical experience with an advanced micobiological water treatment system for Ruhr river water – 'The Mülheim treatment process 1976–2006', in *Recent Progress in Slow Sand and Alternative Biofiltration Processes* (eds R. Gimbel, N.J.G. Graham, and M.R. Collins), IWA Publishing, London, pp. 30–38.

6 Kruithof, J.C. and IJpelarr, G.F. (2002) State of the art of the application of

ozonation in dutch drinking water treatment. Proceedings of the IWA Conference on BAC filtration, International Water Association, Delft University of Technology, The Netherlands, May 29–31, 2002.

7 Bonné, P.A.C., Hofman, J.A.M.H., and van der Hoek, J.P. (2002) Long term capacity of biological activated carbon filtration for organics removal. *Water Science and Technology: Water Supply*, **2** (1), 139–146.

8 Servais, P., Laurent, P., and Gatel, D. (2002) Removal of biodegradable dissolved organic carbon in BAC filters: a modelling approach. Proceedings of the IWA Conference on BAC filtration. International Water Association, Delft University of Technology, The Netherlands, May 29–31, 2002.

9 Hammes, F., Salhi, E., Köster, O., Kaiser, H.P., Egli, T., and von Gunten, U. (2006) Mechanistic and kinetic evaluation of organic disinfection by-product and assimilable organic carbon (AOC) formation during the ozonation of drinking water. *Water Research*, **40**, 2275–2286.

10 EAWAG (Publ.) (2009) Wave21 final report – Drinking Water for the 21st Century. Schriftenreihe Nr. 20.

11 van der Helm, A.W.C., van der Aa, L.T.J., van Schagen, K.M., and Rietveld, L.C. (2009) Modeling of full-scale drinking water treatment plants with embedded plant control. *Water Science and Technology: Water Supply*, **9** (3), 253–261.

12 Baghoth, S.A., Dignum, M., Grefte, A., Kroesbergen J., and Amy, G.L. (2009) Characterization of NOM in a drinking water treatment process train with no disinfectant residual. *Water Science and Technology: Water Supply-WSTWS*, **9** (4) 139–146.

13 VEOLIA WATER AUSTRALIA PTY LIMITED, Municipal Drinking Water AQUA PROJECT, Bendigo, Castlemaine and Kyneton, Victoria, http://www.veoliawater.com.au/lib/vw-australia/8AC04YI8Vk8Xz1I027cR29qC.pdf. (access on July 2, 2010).

14 van der Helm, A.W.C., Rietveld, L.C., Bosklopper, T.G.J., Kappelhof, J.W.N.M., and van Dijk, J.C. (2008) Objectives for optimization and consequences for operation, design and concept of drinking water treatment plants. *Water Science and Technology: Water Supply*, **8** (3), 297–304.

15 Sutton, P.M. and Mishra, P.N. (1994) Activated carbon based biological fluidized beds for contaminated water and wastewater treatment: a state-of-the-art Review. *Water Science and Technology*, **29** (10–11), 309–317.

16 Polk, J., Murray, J., Onewokae, C., Tolbert, D.E., Togna, A.P., Guarini, W.J., Frisch, S., and Del Vecchio, M. (2001) Case study of ex-situ biological treatment of perchlorate-contaminated groundwater. Presented at the 4th Tri-Services Environmental Technology Symposium, June 18–20, San Diego, USA. http://www.cluin.org/contaminantfocus/default.focus/sec/perchlorate/cat/Treatment_Technologies.

17 Polk, J., Onewokae, C., Guarini, W.J., Murray, C., Tolbert, D.E., and Togna, A.P. (2002) Army success story: ex-situ biological treatment of perchlorate-contaminated groundwater, *Federal Facilities Environmental Journal/Summer 2002*, 85–94. Wiley Periodicals, Inc. Published online in Wiley InterScience (www.interscience.wiley.com) doi: 10.1002/ffej.10036.

18 Greene, M. and Pitre, M.R. (1999) Treatment of groundwater containing perchlorate using biological fluidized bed reactors with GAC or sand media. Perchlorate in the Environment: Microbial Remediation and Risk Management for Perchlorate- Preprints of Extended Abstracts, Vol.39 (2), pp. 105–107, August 22–26, 1999, New Orleans, LA, USA.

19 Stucki, G. and Thüer, M. (1994) Increased removal capacity for 1,2-dichloroethane by biological modification of the granular activated carbon process. *Applied Microbiology and Biotechnology*, **42**, 167–172.

11
Review of BAC Filtration Modeling in Drinking Water Treatment
Ferhan Çeçen

In drinking water treatment GAC filtration can remove natural organic matter (NOM) and synthetic organic compounds (SOCs) as well as some nutrients and inorganic substances. Traditionally, the primary aim of modeling in water treatment is to predict the removal of NOM; relatively few models consider the removal of nutrients and micropollutants.

The design of a GAC filter for water treatment is based on laboratory or pilot studies. The type of activated carbon has to be selected, and the phase and the optimum EBCT and filter bed height have to be determined. However, on the laboratory-or pilot-scale only a limited number of configurations can be tested. Therefore, the main advantage of developing GAC filtration models is the extrapolation of results to full-scale applications. Modeling can also be used for monitoring, development of operational strategies, and trend analysis [1].

In this chapter, some basic models are reviewed, mainly following the chronological order. The aim of this chapter is not to provide a detailed account of individual models, but to present the fundamental approaches adopted in drinking water biofilter modeling. Extensive reviews are available in the literature providing in-depth information on GAC/BAC filtration [2–5].

GAC/BAC modeling in water treatment bears many similarities to that in wastewater treatment. Therefore, for insight into the main issues the reader is advised to refer to the equations presented in Chapter 6.

11.1
Substrate Removal and Biofilm Formation

The major difference between GAC/BAC filters operated in wastewater and water treatment is that substrate concentrations are very low in the latter. In water treatment, the influent BDOC and AOC concentrations of a GAC/BAC filter range mostly from 0.1 to 1.5 mg L^{-1} and 10 to 150 µg L^{-1}, respectively. For the formation of a biofilm on the GAC surface, the AOC parameter is more crucial than the BDOC, since the former reflects compounds that can be readily assimilated, as discussed in Chapter 8. In the major part of GAC filtration studies, a

Activated Carbon for Water and Wastewater Treatment: Integration of Adsorption and Biological Treatment.
First Edition. Ferhan Çeçen and Özgür Aktaş
© 2011 WILEY-VCH Verlag GmbH & Co. KGaA, Weinheim.
Published 2011 by WILEY-VCH Verlag GmbH & Co. KGaA

period of 3 months is required to attain the maximum amount of biomass in a filter fed with surface water [6].

Generally, the low AOC concentration in water favors the establishment of a rather thin film over the surface of carbon. Therefore, there is often a debate about the appropriateness of the term 'biofilm' when used in drinking water BAC filtration. Some researchers even believe that this microorganism layer should not be considered as a biofilm since it is usually discontinuous or patchy.

Biological removal of NOM and other pollutants takes place mainly in the biofilm phase as a result of diffusion of substrate into the biofilm. Although the removal of organic matter does not directly relate to the amount of biomass kept in a GAC filter, a minimum biomass should be maintained in drinking water biofilters above which no rate limitation occurs in BOM removal. This minimum likely depends on the temperature and biodegradability of influent [2]. Therefore, the formation of a biofilm should be closely monitored in a drinking water biofilter.

In drinking water filtration, removal of organic matter takes place mainly aerobically, where the biodegradable part of NOM, namely biodegradable organic matter (BOM), serves as a carbon source and electron donor. Under normal operating conditions, the dissolved oxygen (DO) is the electron acceptor that diffuses from the bulk water into the biofilm and is consumed in parallel with BOM (Figure 11.1).

In drinking water biofilters, the diffusion of substrates into biofilm and the consumption in biofilm follows the same rules as those outlined in Chapter 6. However, in biofilm reactors used in wastewater treatment, the concentration of organic matter in the bulk liquid is usually much higher than that of DO.

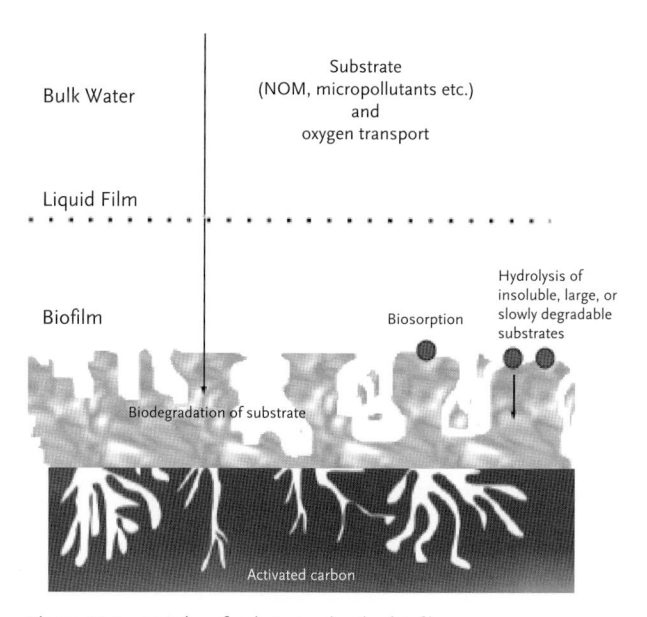

Figure 11.1 Uptake of substrates by the biofilm.

Therefore, oxygen is usually exhausted inside the biofilm and easily becomes the rate-limiting substrate [7]. In contrast to this, in water treatment, the influent and bulk water DO concentrations are high compared to the concentration of NOM. In surface water treatment, for example, the influent DO in BAC filters lies in the range of 7–8 mg L^{-1}. Preozonation of water is another factor increasing the influent DO and providing more suitable conditions for substrate removal. Thus, in drinking water biofilters, no rate limitations arise due to DO.

In BAC filters, microorganisms may also grow and function in the bulk liquid surrounding the biofilm. However, it is estimated that the main part of the biodegradation occurs in the biofilm rather than by suspended biomass found in the interstitial water.

11.2
Modeling of BAC Filtration

The models used in drinking water GAC filtration generally fall into two categories:

- Process-based models. These models are based on mass transfer theories and consist of the three components: the differential mass balance equation, the equilibrium adsorption relationship (isotherm), and a set of equations describing external and internal transport in the carbon particle [5].
- Empirical models. These models are derived from statistical analyses of experimental breakthrough data [8].

The emphasis of this chapter is on process-based models. These models describing BAC filtration in water treatment are based on the same mass transfer, adsorption, and biodegradation principles as those outlined in Section 6.1 of Chapter 6 for the case of wastewater treatment. The model equations involve terms related to dispersion, advection, biodegradation in the biofilm and by suspended biomass, and removal by adsorption. Depending on the stage of operation, either adsorption or biodegradation may be the predominant mechanism.

Adsorptive removal only If biological activity is absent, the GAC filter behaves as an adsorber only. The mass balance for the substrate in an infinitesimal slice of Δz in a GAC adsorber (see Figure 6.3 in Chapter 6), as shown by Eq. (6.37), simplifies to the following equation:

$$\frac{\partial S_b}{\partial t} = v_i \frac{\partial S_b}{\partial z} - \frac{(1-\varepsilon_B)}{\varepsilon_B} \rho_p \frac{\partial \bar{q}}{\partial t} \tag{11.1}$$

where

S_b = concentration of substrate in the bulk liquid (M_s/L^3),
v_i = interstitial velocity: $Q/A\varepsilon_B$ (L/T),
z = distance along the flow path (L),
ε_B = void ratio in the bed (unitless),
A = cross-sectional area of the reactor (L^2),

$\frac{\partial \bar{q}}{\partial t}$ = the mean substrate uptake rate onto carbon ($M_s/M_c \cdot T$),
ρ_p = apparent density of GAC granules (M_c/L^3).

In this equation the dispersion term is excluded from the mass balance because under the typical hydrodynamic conditions existing in water filtration dispersion effects can be neglected [5].

The models considering adsorption of organic matter only are valid in the initial stages of operation. Such models are also used to describe the removal of micropollutants in drinking water biofilters, because the common assumption is that they not amenable to biological removal.

Adsorptive and biological removal Most existing models of drinking water biofiltration account for both adsorptive and biological removal in a filter. The common aim in BAC filtration models is to calculate the effluent substrate concentration for a set of given and/or assumed conditions. Since the biofilm formation is much lower and uneven, the estimation of biological removal in these filters is a more challenging task than that in wastewater treatment.

Biological removal only There are also models that apply to later stages of operation where biological activity is dominant. These models consider the drinking water filters as simply biofilm reactors, disregarding the adsorption of substrate.

The models describing adsorptive removal are well established. Therefore, the emphasis in this chapter will be on the models involving biological activity with or without adsorption.

11.2.1
Models Emphasizing Biological Processes in Biofilters

11.2.1.1
Initial Models in Biofiltration

Initial models on biofiltration were developed and tested for nonadsorbing media. One of the first models, developed by Rittmann and McCarty, described steady-state conditions in completely mixed fixed-bed bioreactors [9, 10]. Although the model was not proposed for drinking water biofiltration originally, it has often been used in such applications. This model considers the transport of substrate into biofilm, simultaneous diffusion, and reaction of the substrate in the biofilm according to the Monod model, as shown in Chapter 6.

The model proposed by Rittmann and McCarty in the 1980s is commonly referred to as the S_{min} model. The definition of S_{min} can be found in Chapter 3. This model assumes a minimum concentration of a single substrate (S_{min}) in the bulk liquid below which no growth would occur and therefore no biomass could be kept in the system [11]. The persistence of many organic micropollutants in natural waters and wastewaters was explained, in part, by the inability of microorganisms to extract energy from very dilute environments.

The pseudo-analytical solution for the S_{min} model was further developed by Rittmann and co-workers [12, 13]. The S_{min} concept requires special attention in drinking water biofiltration since both the AOC and BDOC levels are low. Moreover, this concept is of importance with regard to the removal of many micropollutants that are found at concentrations of ng L^{-1} to low μg L^{-1}. However, nowadays it is recognized that substrates, even though at very low concentrations, may be removed as a secondary substrate (Chapter 3).

11.2.1.2
The CHABROL Model

Biosorption and biodegradation are the two mechanisms that can act either singly or simultaneously. They are effective in the removal of NOM as well as organic micropollutants. Both particulate (insoluble) and dissolved organic matter may be sorbed onto the biofilm. Biosorption is a mechanism paving the way for subsequent biodegradation. Particulate organic matter may undergo hydrolysis first and then diffuse into the biofilm. As a result of the long retention time on the surface, large, slowly biodegradable or even apparently nonbiodegradable soluble organics are also hydrolyzed by extracellular enzymes and diffuse into the biofilm (Figure 11.1).

The CHABROL model fractionates the BDOC in the influent water into three biodegradability classes: S: the readily biodegradable substrate, H1: the rapidly hydrolyzable substrate, and H2: the slowly hydrolyzable substrate [14]. The model considers both attached and suspended biomass in the interstitial water. It refers only to situations when the adsorption capacity of a GAC filter is exhausted. As such, it focuses mainly on extracellular hydrolysis of dissolved macromolecular organic substrate, growth of attached and suspended bacteria on the readily biodegradable substrate, adsorption and desorption of bacteria, and mortality caused mainly by protozoan predators. The model calculates the vertical distribution of attached bacteria and the effluent BDOC from the characteristics of influent water for given values of contact time and temperature.

Experimental results showed that modifications had to be made in the CHABROL model to account for the acclimation of biomass to very cold temperatures and to prevent underestimation of BDOC removal [15]. The applicability of the CHABROL model was shown by comparison of model calculations with data obtained from many waterworks in France such as the Neuilly-sur-Marne treatment plant [16].

11.2.1.3
Uhl's Model

This model developed by Uhl emphasizes the importance of attachment and detachment in drinking water biofilters [3, 17]. The model includes the processes biodegradation, attachment, and detachment of bacteria taking place in the filter bed as a function of bed depth. According to this model, NOM is removed by attached and suspended biomass.

Filters in drinking water treatment are commonly operated as fixed-bed downflow filters. Uhl's model treats the filter bed as a fixed-bed plug flow reactor. The following equation exemplifies the substrate mass balance for an infinitesimal reactor element of length Δz such as the one shown in Section 6.1 of Chapter 6:

$$\frac{\partial S_b}{\partial t} = r_{s,gr} + r_{s,maint.} + r_{s,cat} - \frac{Q}{A \varepsilon_B} \left(\frac{\partial S_b}{\partial z} \right) \tag{11.2}$$

where

S_b = concentration of substrate in the bulk liquid (M_s/L^3),
z = distance along the flow path (L),
$r_{s,gr}$ = rate of substrate consumption by bacterial growth ($M_s/L^3.T$),
$r_{s,maint}$ = rate of substrate consumption by maintenance requirements ($M_s/L^3.T$),
$r_{s,cat}$ = rate of biodegradable substrate production by surface catalytic processes ($M_s/L^3.T$),
Q = flow rate (L^3/T),
A = cross-sectional area of the reactor (L^2),
ε_B = void ratio in the bed (unitless).

In these terms, the subscript s stands for substrate.

Similarly, the changes in attached and suspended biomass are expressed by the following equations:

$$\frac{\partial X_{att}}{\partial t} = (\alpha J_{x,att} - \alpha J_{x,det}) + r_{x,gr} + r_{x,mort.} + r_{x,graz} \quad (11.3)$$

$$\frac{\partial X_{susp}}{\partial t} = -\frac{1}{\varepsilon_B}(\alpha J_{x,att} - \alpha J_{x,det}) - \frac{Q}{A\varepsilon_B}\left(\frac{\partial X_{susp}}{\partial z}\right) \quad (11.4)$$

where in these equations:

X_{att} = concentration of attached biomass per unit filter volume (M_x/L^3),
$r_{x,gr}$ = rate of bacterial proliferation ($M_x/L^3.T$),
$r_{x,mort}$ = rate of bacterial mortality ($M_x/L^3.T$),
$r_{x,graz}$ = rate for grazing by protozoa ($M_x/L^3.T$),
$J_{x,att}$ = flux of suspended bacteria to the media surface ($M_x/L^2.T$),
$J_{x,det}$ = flux of bacterial detachment from the media surface ($M_x/L^2.T$),
α = specific surface area (surface area of the media per reactor volume) (L^2/L^3),
X_{susp} = concentration of suspended biomass (M_x/L^3).

In these units, the subscript x stands for the biomass.

In Eq. (11.4), growth of suspended bacteria is neglected as their concentration was found to be several orders of magnitude lower than the concentration of attached bacteria.

The model proposed by Uhl does not incorporate adsorption of substrate onto the GAC surface. Thus, it applies mainly for nonadsorptive media such as sand or for the later stages of BAC filtration in which only biological mechanisms are relevant.

Uhl notes that bacteria are capable of degrading the substrate at low concentrations. He also states that, if the S_{min} model were applicable, the minimum AOC concentration should be about 340 µg C L^{-1}, and below this value no stable biofilm could be sustained in drinking water biofilters. Also, very low substrate concentrations of a few µg C L^{-1}, commonly observed in the effluent of BAC filters, would not be achieved [17]. However, attention is drawn to biodegradation occurring at concentrations below this value. The reduction of substrate below the theoretical S_{min} value is attributed to the deposition of biomass in the filter. This deposition of biomass is also regarded to be a crucial factor for the startup of deep-bed filters [3, 17].

11.2.1.4
The Model of Wang and Summers

Wang and Summers stated that a single-substrate approach is not representative of NOM and proposed a model which is suitable to describe the removal of rapidly and slowly biodegradable fractions in DOC along the depth of a biofilter [18].

As shown in Eq. (11.5), for each NOM fraction, a mass balance was constructed for steady-state conditions in a way similar to Eq. (6.38) in Chapter 6. The dispersion was neglected under the hydrodynamic conditions of drinking water filtration.

$$0 = -v_i \frac{\partial S_b(z)}{\partial z} - \frac{3(1-\varepsilon_B)}{R_p} \cdot k_{fc}(S_b(z) - S_R(z)) \tag{11.5}$$

where

v_i = interstitial velocity: $Q/A \, \varepsilon_B$ (L/T),
S_b = substrate concentration in the bulk liquid (M_s/L^3),
S_R = substrate concentration at the surface of filter media (M_s/L^3),
z = distance along the flow path (L),
k_{fc} = external mass transfer coefficient (L/T),
ε_B = void ratio in the bed (unitless),
R_p = radius of filter media (L).

The model further assumed that biofiltration is a two-step process in which the external mass transfer (liquid film diffusion), shown on the right-hand term in Eq. (11.5), is followed by a Monod type biodegradation on the surface. The model considered that a thin biofilm was present in the filter. The thickness of the biofilm was neglected in Eq. (11.5). Therefore, the diffusion of substrates into biofilm was neglected. The biomass concentration along the filter was assumed to be constant. Based on these assumptions, the substrate mass balance results in the following equation:

$$\frac{3(1-\varepsilon_B)}{R_p} \cdot k_{fc}(S_b(z) - S_R(z)) = \frac{k_{max} X \, S_b(z)}{K_s + S_b(z)} \tag{11.6}$$

where

k_{max} = maximum rate of specific substrate utilization ($M_s/M_x \cdot T$),
X = biomass concentration in biofilter (M_x/L^3),
K_s = half-velocity constant (M_s/L^3).

The external mass transfer coefficient k_{fc} appearing in Eq. (11.5) had little impact on DOC removal along the filter depth if it exceeded 5×10^{-6} m s^{-1}. In rapid filtration, k_{fc} is reported to be greater than this value [18]. Model results indicated that the biodegradation rate was the rate-limiting step in DOC removal rather than external mass transfer. The reader may refer to Chapter 6 for a full discussion of the relative importance of external mass transfer and biodegradation.

The model was validated over a wide range of hydraulic loading rates (HLRs) using an ozonated humic substance solution where the total biodegradable fraction was 45%. The removal of rapidly and slowly biodegradable fractions in NOM was examined. Model predictions were in good agreement with experimental data in the HLR range from 1.5 to 15 m h^{-1}.

11.2.1.5
The Dimensionless Empty-Bed Contact Time Concept

Zhang and Huck extended the model developed by Rittmann and McCarty [9] to deep-bed filters that show the characteristics of a plug flow reactor [19]. They regarded in their model the AOC as the sole carbon and energy source and the growth-limiting substrate. The successful fitting of the model to data demonstrates that AOC can be used as a surrogate for BOM.

In the modeling of drinking water biofilters, the Empty-Bed Contact Time (EBCT) is one of the most important design parameters. Additionally, Zhang and Huck have introduced the concept of the dimensionless EBCT, X^* [19].

$$X^* = \theta \, \alpha \, D_f^{1/2} (k_{max} \, X_f / K_s)^{1/2} \tag{11.7}$$

where

X^* = dimensionless EBCT (unitless),
θ = Empty-Bed Contact Time (EBCT) (T),
α = specific surface area (surface area of the media per reactor volume) (L^2/L^3),
D_f = molecular diffusivity of substrate in the biofilm (L^2/T),
k_{max} = maximum specific substrate utilization rate ($M_s/M_x \cdot T$),
X_f = biofilm density (M_x/L^3),
K_s = half-velocity constant (M_s/L^3).

Factors affecting BOM removal were identified as EBCT, surface area, media, temperature, nature and concentration of the influent BOM, and biofilm disruption due to backwashing [4]. As seen in Eq. (11.7), X^* unifies many of the relevant factors, the EBCT, the specific surface area of the medium, and parameters related to substrate biodegradability and diffusivity in the biofilm in a single parameter [4]. The researchers have shown that there was a correlation between X^* and substrate removal.

Zhang and Huck identified that in drinking water biofilters the liquid film mass transfer resistance (shown in Chapter 6) could be ignored and first-order removal kinetics could be used within the biofilm rather than Monod kinetics [19]. Under these conditions the X^* parameter in Eq. (11.7) was simplified to the following form:

$$X^* = \theta \, \alpha \, K \, (k_{max} \, X_f)^{1/2} \tag{11.8}$$

where K was introduced as a constant required to preserve the nondimensionality of X^*.

The importance of this simplified dimensionless EBCT for practice was also emphasized [4]. For a given substrate at a given temperature (i.e., given k_{max}), the performance of the biofilter is essentially directly proportional to EBCT and media surface area. As such, this expression allows comparison among different process conditions.

As indicated in Chapter 9, water bodies contain many micropollutants at trace concentrations. Conceptually, it was shown that the X^* concept and its simplified form could be extended to the removal of such contaminants that are removed by secondary utilization (e.g., pharmaceuticals, endocrine disrupting substances, or odorous compounds such as geosmin) [4].

11.2.1.6
The BIOFILT Model

Biofilm processes in drinking water biofilters involve an initial biomass development period followed by a dynamic steady-state period. During the initial period, attachment and growth of microorganisms plays an important role. However, some biomass is simultaneously removed from the surface of the GAC in this period.

Fluid shear and backwashing are identified as the main factors causing biomass loss in drinking water biofilters. The BIOFILT model proposed by Hozalski and Bouwer was the first to simulate the effect of filter backwashing on BOM removal [20]. Unlike other models, this model addresses non-steady-state conditions. However, the model does not incorporate the adsorption of substrate onto filter media such as GAC.

The model considers the phenomena dispersion, advection, removal by suspended biomass, and removal in the biofilm phase. The substrate concentration is predicted as a function of time and distance along the reactor length:

$$\frac{\partial S_b}{\partial t} = D'\frac{\partial^2 S_b}{\partial z^2} - v_i\frac{\partial S_b}{\partial z} - \frac{k_{max} S_b X_{susp}}{K_s + S_b} - \frac{\alpha}{\varepsilon_B}J_f \tag{11.9}$$

where

S_b = concentration of substrate in the bulk liquid (M_s/L^3),
D' = dispersion coefficient (L^2/T),
v_i = interstitial velocity: $Q/A\,\varepsilon_B$ (L/T),
z = distance along the flow path (L),
X_{susp} = concentration of suspended biomass (M_x/L^3),
K_s = half-velocity constant (M_s/L^3),
J_f = flux of substrate into the biofilm ($M_s/L^2.T$),
α = specific surface area (surface area of the media per reactor volume) (L^2/L^3),
ε_B = void ratio in the bed (unitless).

In this equation J_f shows the flux of substrate through a liquid film into the biofilm. Accordingly, it represents the mass that is removed by the biofilm.

The model also calculates the attached and suspended biomass as a function of time and filter depth. In the following equation, advective transport, biofilm shear, and washout of suspended biomass due to filtration are taken into consideration, whereas the effect of dispersion, growth, and decay of suspended biomass is neglected:

$$\frac{\partial X_{susp}}{\partial t} = -v_i \frac{\partial X_{susp}}{\partial z} + \frac{\alpha X_f \sigma}{\varepsilon_B} L_f - \frac{X_{susp} \beta}{\theta \varepsilon_B} \qquad (11.10)$$

where

X_{susp} = suspended biomass concentration (M_x/L^3),
X_f = density of biomass within biofilm (M_x/L^3),
L_f = biofilm thickness (L),
β = filtration efficiency (unitless),
θ = EBCT (T),
ε_B = void ratio in the bed (unitless),
σ = biofilm shear loss coefficient (1/T),
α = specific surface area of the media (surface area per reactor volume) (L^2/L^3).

The researchers also modeled the net accumulation of attached biomass at any point of the biofilter due to growth, deposition, decay, and shear loss. The time-dependent change in biofilm thickness was expressed as follows [20]:

$$\frac{\partial L_f}{\partial t} = \frac{Y J_f}{X_f} + \frac{X_{susp} \beta}{X_f \alpha \theta} - b_{tot} L_f \qquad (11.11)$$

where

b_{tot} = overall loss rate of bacteria due to decay and fluid shear (1/T),
Y = yield coefficient (M_x/M_s).

The first term on the right-hand side of Eq. (11.11) shows that biofilm grows as a result of substrate flux (J_f) into the biofilm. The term in the middle represents the increase of biofilm thickness due to deposition of suspended biomass on filter media. The last term is the loss of biofilm by decay and fluid shear.

The BIOFILT model was used in simulation of BOM removal in full-scale biofilters. The effect of BOM composition, water temperature, and biomass removal during backwashing was also examined. BOM removal was shown to depend strongly on influent BOM composition [21]. The temperature decrease resulted in the decline of pseudo-steady-state BOM removal and increase of biofilter startup time. One of the innovative features of this model is the ability to simulate the effects of a sudden loss in attached biomass due to filter backwashing. Biofilter performance was not affected unless biomass removal by backwashing was less than 60%. Therefore, for practically oriented modeling the steady-state approach was considered adequate [4].

11.2.1.7
Consideration of Multiple Species Inside Drinking Water Biofilters

Most models of drinking water biofiltration disregard the variation of bacterial species inside the biofilm. The transient-state, multiple-species biofilm model (TSMSBM), on the other hand, describes how biofilms change with time [22]. This model includes six features: (i) four biomass types: heterotrophs, ammonia oxidizers, nitrite oxidizers, and inert biomass; (ii) seven chemical species: input biodegradable organic matter (BOM), NH_4^+-N, NO_2^--N, NO_3^--N, the Soluble Microbial Products (SMP) consisting of the fractions Utilization-Associated Products (UAP) and Biomass-Associated Products (BAP), and dissolved oxygen; (iii) eight reactions that describe the rates of consumption or production of the different species, as well as the stoichiometric linkages among the rates; (iv) reaction with diffusion of all the soluble species in the biofilm; (v) growth, decay, detachment, and flux of each biomass type by location in the biofilm; and (vi) constant or periodic detachment of biofilm, both of which allow for protection of biomass deep inside the biofilm.

The following equation exemplifies how the accumulation rate of ammonia oxidizers is expressed in an infinitesimal slice of the filter having a length of Δz.

$$\frac{\partial X_{ns}}{\partial t} = -\frac{\partial (X_{ns} v_i)}{\partial z} + Y_{ns} M_{ns} X_{ns} - b_{ns} X_{ns} \tag{11.12}$$

where

X_{ns} = concentration of ammonia oxidizers (represented by *Nitrosomonas*) ($M_{OD,x}/L^3$),
v_i = interstitial velocity (L/T),
z = distance along the flow path (L),
Y_{ns} = true yield coefficient ($M_{OD,x}/M_{OD,s}$),
M_{ns} = specific rate of ammonia removal ($M_{OD,s}/M_{OD,x} \cdot T$),
b_{ns} = specific rate of biomass decay (1/T).

The units $M_{OD,s}$ and $M_{OD,x}$ denote the substrate and biomass, respectively, all expressed in Oxygen Demand (OD) units.

The first term on the right-hand side shows the net flux of ammonia oxidizers. The second term represents the synthesis rate of ammonia oxidizers due to NH_4^+-N utilization. The third term stands for decay rate of ammonia oxidizers due to endogenous respiration and BAP formation.

The TSMSBM could provide information about the transient development of multiple-species biofilm, the roles of SMP and detachment in controlling the distribution of biomass types and process performance, and how backwashing affects the biofilm in drinking water biofiltration. Backwashing did not remove more than 25% of the biofilm under any condition simulated [23]. The model also predicted the washout of bacterial species under presumed detachment rates. However, the TSMSBM does not include adsorption and desorption of substrate onto either GAC or biomass.

The TSMSMB was used to interpret the results of a pilot study which investigated the removal of color, DOC, and BOM from a groundwater by ozonation and

Table 11.1 Comparison of effluent DOC components in sand, anthracite, and GAC filters receiving ozonated groundwater (adapted from [23]).

Parameter	Sand biofilter	Anthracite biofilter	GAC biofilter
Experimental results			
DOC (mg L^{-1})	2.10	2.20	2.00
BOM (mg L^{-1} as C)	0.35	0.35	0.38
Model Simulation			
DOC (mg L^{-1})	2.04	2.07	2.02
Refractory DOC (mg L^{-1})	1.71	1.71	1.71
BOM (mg L^{-1} as C)	0.33	0.36	0.31

subsequent filtration using either sand, anthracite, or GAC biofilters with EBCT up to 9 min. Of all media, the GAC biofilter gave the best performance with respect to the removal of color and DOC, probably because of initial adsorption (Table 11.1). The effluent DOC consisted of refractory organic carbon and effluent BOM (sum of original BOM and SMP). Model simulations showed that SMP comprised a significant part of the effluent BOM.

11.2.2
Models Integrating Adsorption and Biological Processes

11.2.2.1
Initial Models

In the early 1980s, Sontheimer and co-workers studied the behavior of natural waters in GAC columns with and without preozonation [24]. Sontheimer points out that the initial models proposed for integrated adsorption and biodegradation by Ying and Weber [25], Andrews and Tien [26], as discussed in Chapter 6, and DiGiano and co-workers [27] consider single solutes only and can therefore be only partially adapted to a drinking water source which contains compounds displaying a range of adsorbability and biodegradability. On the other hand, as Sontheimer indicates, the Hubele model developed for fixed beds reflects the multisolute nature of most natural waters and considers the differences in adsorbability and biodegradability [24]. The Hubele model also included the variation in biofilm thickness with time and distance, the sequential uptake of substrates as a function of column depth, and the displacement of biomass in the GAC filter.

Using a humic acid solution which represented a multicomponent mixture similar to natural waters, Sontheimer and co-workers examined the breakthrough of DOC in GAC columns. The modeling was aimed at predicting the breakthrough profiles in GAC filters for the case of nonozonated and ozonated humic acid solution. Since no single-solute approach was possible, the compounds in the mixture were divided into five groups, the first of which consisted of nonadsorbable compounds and the other four groups had different degrees of adsorbabilities as expressed by Freundlich isotherm constants. This type of approach of fractionating the influent water into adsorptivity classes by the use of Freundlich constants is also adopted and applied by other researchers such as Worch and co-workers [5].

Sontheimer and co-workers have shown that ozonation of the humic acid solution changed the adsorbability and biodegradability of organics. If biodegradation was disregarded in GAC filter modeling, ozonation had the effect of leading to a faster increase in effluent concentration, since the adsorbability of the humic acid was lowered. However, experimental results pointed to the occurrence of biodegradation in GAC filters. Biodegradation was an important mechanism in lowering the concentrations in filter effluent for a long period of time. The researchers also examined the effect of EBCT on the adsorbability and biological removal of the various fractions in DOC [24, 28].

Also in the 1980s, the original biofilm model developed by Rittmann and McCarty [9, 10] was extended by Chang and Rittmann to include the adsorption of substrate onto activated carbon, as outlined in Chapter 6 [29]. These models are regarded as basic in drinking water treatment also.

11.2.2.2
Models Involving Adsorption, Biomass Development, and Biodegradation

The determination of biofilm formation in a filter is a challenging task, in particular in drinking water filtration where a thin biofilm exists. Figure 11.2 shows the determination of biofilm with respect to operation time [30]. As shown in this figure, and in other studies, attachment, growth, decay, and detachment of biomass results first in a steady increase in biofilm amount with respect to operation time [31]. When substrate removal levels out and the exhaustion capacity of the filter is approached, as shown in Figure 11.2, the biofilm amount reaches a dynamic steady-state.

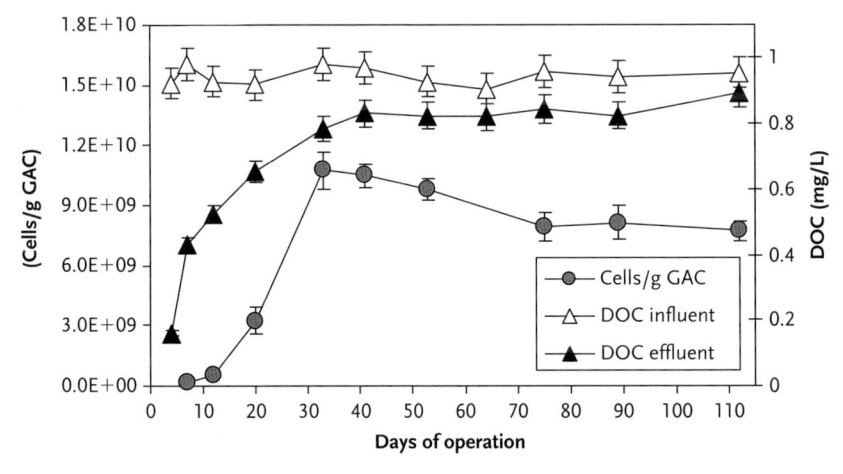

Figure 11.2 Biofilm formation in relation to DOC Removal in a pilot-scale GAC Filter [30] (permission received).

11.2.2.2.1 Biomass Development in Relation to BDOC Removal

Few models consider the time-dependent development of biomass on the surface of activated carbon. Pilot-scale studies carried out by van der Aa and co-workers aimed at gaining knowledge of biomass development in BAC filters along with BDOC removal [32]. A dynamic BAC model was developed for simultaneous simulation of DOC adsorption, biomass development, and biodegradation of DOC.

The biodegradation was modeled by assuming the formation of a uniform biofilm on the outer surface of the GAC. In the initial phases of BAC filtration, pilot plant data and model predictions for adsorption of DOC were in good agreement with each other. The model also included the development of biomass in dependence of NOM biodegradation. However, the model overestimated biomass growth and BDOC removal compared to pilot plant data. While in the model the biomass developed only in the upper part of the filters, in reality significant biomass concentrations were measured in the middle part. The model predicted that biomass concentration would reach the maximum value within two weeks, whereas in reality the development took much longer[32].

11.2.2.2.2 Modeling NOM Removal and the Effect of Preozonation

In the 2000s van der Aa and co-workers worked on the development of a dynamic simulation model for simultaneous NOM adsorption and biodegradation. One of the aims was to account for the changes in the breakthrough curves upon ozonation of water at various doses. Model parameters were calibrated from laboratory experiments, from pilot experiments with exhausted activated carbon, and from literature data. Four pilot BAC filters were operated with different preozonation doses to generate different NOM breakthrough curves with simultaneous adsorption and biodegradation for model validation. In the model the NOM was divided into the following fractions: nonadsorbable and nonbiodegradable (NOM_{NA-NB}), adsorbable and nonbiodegradable (NOM_{A-NB}), and adsorbable and biodegradable (NOM_{A-B}). The nonadsorbable fraction was assumed to be also nonbiodegradable. The model included convection and liquid film diffusion of dissolved substances. In the model, NOM_{A-B} is biodegraded according to Monod growth kinetics based on the locally available concentrations NOM_{A-B}, oxygen, and phosphate.

In the model, biomass growth rate increases with water temperature. Suspended biomass, attachment, and detachment of bacteria are neglected. Biomass maintenance is also neglected. NOM_{A-NB} and the remaining NOM_{A-B} diffuse into activated carbon via surface diffusion, as outlined in Chapter 6. Inside the pores of the activated carbon, nonreversible adsorption of NOM_{A-NB} and NOM_{A-B} takes place according to Freundlich adsorption isotherms combined with the ideal adsorption solution theory (IAST) [33]. The Freundlich isotherm constants were derived from NOM isotherms for ozonated and nonozonated waters. In the model, the filter was divided into a number of continuous-flow stirred tank reactors (CSTRs) in series. For each CSTR, mass balances were constructed in order to calculate all model variables dynamically.

The model was able to roughly describe the S-shaped NOM breakthrough curves in BAC filters. However, the model was not accurate enough to predict the effect of preozonation dosage on NOM removal in the adsorption phase. At filter run times longer than 200 days, the model predicted a reduction of adsorbed NOM at increased preozonation dosages. It was stated that this would favor the removal of micropollutants since preloading, pore blocking, and competitive effects due to the presence of NOM are reduced.

In another study, the main aim was to study the effect of ozonation and water temperature on the biodegradation of NOM in BAC filters. For this purpose the researchers used pilot-scale data to develop multivariate linear regressions (MLRs) for the target parameters DOC, AOC, oxygen, CO_2, and pH [34]. Oxygen consumption, CO_2 production, and DOC were higher than predicted from theoretical calculations. Bioregeneration of NOM was thought to be the reason for this excessive oxygen consumption and CO_2 production.

11.2.2.2.3 Modeling the Substrate Removal in Different Types of Water A non-steady-state model was developed by Liang and co-workers to differentiate between adsorption and biodegradation in a BAC column [35]. The mechanisms considered in this model are dispersion, advection, adsorption, and biodegradation in the biofilm phase as well as in the suspended biomass, as shown in Eq. (6.37) in Chapter 6. The influence of the major parameters, the packing media size, and the superficial velocity on the adsorption and biodegradation of DOC was evaluated by simulation.

Three different target waters were considered in the study. The first water contained the highly biodegradable ozonation by-product glyoxalic acid, which has a low adsorptivity. The second water contained p-hydroxybenzoic acid and its ozonation intermediates, representing compounds with different biodegradability and adsorptivity. In the third scenario, alachlor was used as the target compound with acetate to support the growth of microbes. Data were taken from another study [36]. The dimensionless Freundlich intensity constant (n), together with the maximum specific substrate utilization rate (k_{max}) and the diffusion coefficient (D_f), were the most sensitive variables affecting BAC performance.

The same researchers also showed that the biofilm that had developed around the GAC granules, as shown in Figure 6.1 in Chapter 6, can hinder the mass transfer of the substrate onto the carbon surface. Although increasing the specific biodegradation rate would be capable of increasing the performance of biodegradation, the adsorption efficiency would be decreased by lowering the boundary concentration in the biofilm–GAC interface. In contrast, increasing the diffusivity into the biofilm was able to increase both the adsorption and biodegradation efficiencies simultaneously, so that the overall removal efficiency could be promoted through the improvement of mass transfer [37].

Rhim and co-workers aimed at predicting the effluent quality in BAC filtration using the plug flow stationary solid phase column (PSSPC) model. The biodegradable fraction in the influent shown by the BDOC/DOC ratio was about 0.42. The model had a good accuracy for prediction of breakthrough profiles [38]. The effect of biomass density (X_f) in the range of 750–3750 mg L^{-1} on breakthrough profiles was also simulated in the model.

11.2.3
Models Describing the Removal of Micropollutants

11.2.3.1
Pesticides/Herbicides Removal

11.2.3.1.1 Alachlor Removal Badriyha and co-workers discussed the mathematical modeling and design protocol for bioactive expanded-bed GAC adsorbers employed in the treatment of drinking water contaminated by a chlorinated herbicide [36]. The model compound chosen was alachlor, while acetate represented the background NOM. As outlined in Chapter 9, NOM and micropollutants compete with each other for adsorption sites. However, as acetate is poorly adsorbed in contrast to alachlor, competition between the two could be ignored. The researchers proposed a thin biofilm model in which the biofilm thickness varied from 0.5 to 4 µm. The model assumed that alachlor was removed as a cometabolite while acetate served as the primary substrate. The model took into account the following processes: film transfer from the bulk liquid to the adsorbent particle, diffusion through the biofilm, and adsorption of the contaminant into the GAC particle. The temporal variation of biofilm thickness was also considered in the model. Eq. (6.38) in Chapter 6 was used in describing the change in the concentration of bulk substrate.

Experiments were conducted to verify the predictive capability of the model, evaluate the combined effects of adsorption and biodegradation, and compare the performances of nonbioactive and bioactive adsorbers [36]. Aerobic microbial activity was sustained in the biofilm, since in the surrounding liquid the DO was about 7–8 mg L^{-1}. No undesirable organic by-products or alachlor metabolites were detected in the treated effluent, confirming that they were effectively removed by adsorption and/or biodegradation. Comparison of bioactive adsorbers with nonbioactive ones showed that biofilm growth on GAC particles could substantially extend the useful service life of GAC filters.

Sensitivity analyses for bioactive adsorbers illustrated that the biological parameters, Monod parameters, μ_{max}, and K_s significantly influenced the process performance. However, the maximum biofilm thickness, $L_{b,max}$, moderately affected the process dynamics. Adsorption was the dominant mechanism for substrate removal in the initial steps of operation. Model results showed that progressive increase of maximum biofilm thickness from 4 to 6 µm did not influence the results in this phase because adsorption dominated over biodegradation. Moreover, the overall process dynamics was not affected by any diffusional resistance, the liquid film, the biofilm, or surface diffusion inside the carbon particle.

After development of a biofilm, biodegradation became the dominant mechanism in substrate removal, and steady-state removal was no longer dependent on adsorption mechanisms. In these cases, the effluent concentration declined rapidly.

11.2.3.1.2 Atrazine Removal In order to study the behavior of atrazine, two laboratory-scale fluidized-bed (FB) continuous-flow reactors were operated using *Pseudomonas* ADP, a fast atrazine degrading bacteria, with citrate as the main carbon and energy source [39]. In one FB reactor, GAC was used as a biofilm carrier while in the other, a nonadsorbing carbon carrier with the same surface area was employed. The researchers took into consideration the fact that biofilm grows only gradually and presented a conceptual model of adsorption and desorption with a patchy biofilm coverage. The model proposed included three interfaces: (i) Biofilm–bulk liquid, (ii) GAC–bulk liquid, and (iii) GAC–biofilm.

The model is based on non-steady-state material balance. It involves diffusive mass transfer through the biofilm and removal of substrate according to Monod kinetics. Adsorption and diffusion were described using the homogeneous surface diffusion model (HSDM) discussed in Chapter 6. Adsorption was assumed to occur only on areas not covered by the biofilm. The model describes transient conditions resulting from a sharp decrease in influent concentration. Under such conditions the biofilm was only partially penetrated by the substrate. In this case, the preadsorbed atrazine was desorbed from the GAC and biodegraded, pointing to bioregeneration of activated carbon, a phenomenon discussed in Chapter 7. Although bioregeneration is a debated issue in the case of NOM components, as discussed in Section 9.1.3 in Chapter 9, in some cases, such as this small-sized micropollutant, activated carbon is amenable to bioregeneration. The adsorption and desorption ability of GAC was a factor enhancing biofilm activity and biodegradation of atrazine. Thus, experimental and model results clearly showed the superiority of an adsorbing/desorbing medium. In the BAC reactor, the effluent concentration of atrazine was stabilized at $2\ \mu g\,L^{-1}$ whereas in the nonadsorbing carrier this was about $5–6\ \mu g\,L^{-1}$.

Pore blockage by NOM, as covered in Chapter 9, plays an important role in adsorption of trace organic compounds such as atrazine. A bisolute kinetic model was developed to predict atrazine adsorption in a continuous-flow powdered activated carbon (PAC)/membrane filtration system, incorporating the effect of pore blockage. The model was also experimentally verified. However, the model did not incorporate the effect of biological activity [40, 41].

Rietveld and co-workers presented a model for the removal of pesticides by GAC filtration in full-scale water treatment. The model describes GAC filtration in a pseudo-moving-bed configuration. The model assumed that only adsorption was taking place in the removal of the pesticide bentazon [42].

Recently, a model was presented that accounted for simultaneous removal of NOM and atrazine in BAC filters [43]. In the model ,NOM was removed by both biodegradation and adsorption. Atrazine was assumed to be removed by adsorption only. Due to competitive effects, the adsorption of atrazine was limited by the amount of adsorbed NOM. According to model results, better atrazine removal was observed in BAC filtration than in GAC filtration alone, since the negative effects of NOM were reduced in BAC filtration.

11.2.3.2
Perchlorate Removal

Fluidized-bed reactors (FBR) with biological activity are often employed in the removal of perchlorate from contaminated groundwater. In this specific case, GAC serves as a biofilm support only since no adsorption/desorption interactions occur between GAC and perchlorate. Consequently, the particle size and the total surface area of the GAC are of importance. A numerical model was developed for the sequential removal of oxygen, nitrate, and perchlorate in the biofilm surrounding GAC particles [44]. The model is based upon fundamental aspects of mass transfer from the fluid to the biofilm, mass transfer of electron acceptors and donor within the biofilm, reaction stoichiometry, and biofilm reaction kinetics for microorganism growth and substrate utilization. The model was further evaluated by the results of a full-scale treatment system. The model indicates that GAC particle size, reactor diameter, and perchlorate concentration affect the performance. With a 10-min retention time, the effluent goal of 4 $\mu g\, L^{-1}$ should be achievable even with influent perchlorate concentration as high as $10\,mg\, L^{-1}$.

11.2.3.3
Microcystin Removal

Wang and co-workers studied the adsorption and biodegradation of microcystin toxins in GAC filtration. Their study showed that adsorption was prevalent during the initial stages of operation. Adsorption of microcystins was modeled using the HSDM. The model provided evidence that an active biofilm on the surface of GAC would hinder the transfer of microcystin into the internal adsorption sites on GAC [45].

11.2.3.4
THM Removal

As outlined in Chapter 9, the biodegradation potential of trihalomethanes (THMs) as a primary substrate is rather low, whereas they can be removed by cometabolism. The removal of four THMs (chloroform, bromodichloromethane, dibromochloromethane, and bromoform) in BAC filtration was modeled. These substances would be removed cometabolically in the upper layers of GAC on which nitrifiers reside. The model developed for THM degradation was verified and used in simulation of full-scale operation. Total THM removals ranged from 16% to 54% in these simulations with influent total THM concentrations of 75–82 $\mu g\, L^{-1}$; this finding illustrated the potential of THM cometabolism [46].

References

1 Rietveld, L.C. (2005) Improving operation of drinking water treatment through modelling. Technical University Delft. Dissertation.

2 Urfer, D., Huck, P.M., Booth, S.D.J., and Coffrey, B.M. (1997) Biological filtration for BOM removal: a critical review. *Journal of the American Water Works Association*, **89** (12), 83–98.

3 Uhl, W. (2001) Modeling biological DOM-removal in drinking water treatment filters – a new approach. Proceedings AWWA Water Quality Technology Conference, Nashville, USA. CDROM, Wed7, Paper American Water Works Association, Denver, CO.

4 Huck, P.M. and Sozanski, M.M. (2008) Biological filtration for membrane pretreatment and other applications: towards the development of a practically-oriented performance parameter. *Journal of Water Supply: Research and Technology – AQUA*, **57** (4), 203–224.

5 Worch, E. (2008) Fixed-bed adsorption in drinking water treatment: a critical review on models and parameter estimation. *Journal of Water Supply: Research and Technology – AQUA*, **57** (3), 171–183.

6 Chaudhary, D.S., Vigneswaran, S., Ngo, H.H., Shim W.G., and Moon, H. (2003) Review: biofilter in water and wastewater treatment. *Korean Journal of Chemical Engineering*, **20** (6), 1054–1065.

7 Çeçen, F. and Gönenç, E. (1995) Criteria for nitrification and denitrification of high-strength wastes in two upflow submerged filters. *Water Environment Research*, **67** (2), 132–142.

8 Zachman, B.A., Rajagopalan, B., and Summers, R.S. (2007) Modeling NOM breakthrough in GAC adsorbers using nonparametric regression techniques. *Environmental Engineering Science*, **24** (9), 1280–1296.

9 Rittmann, B.E. and McCarty, P.L. (1980a) Evaluation of steady-state-biofilm kinetics. *Biotechnology and Bioengineering*, **22**, 2359–2373.

10 Rittmann, B.E. and McCarty, P.L. (1980b) Model of steady-state-biofilm kinetics. *Biotechnology and Bioengineering*, **22**, 2343–2357.

11 Rittmann, B.E. and McCarty, P.L. (2001) *Environmental Biotechnology: Principles and Applications*, McGraw-Hill.

12 Saez, P.B. and Rittmann, B.E. (1988) Improved pseudoanalytical solution for steady-state biofilm kinetics. *Biotechnology and Bioengineering*, **32**, 379–385.

13 Saez, P.B. and Rittmann, B.E. (1992) Accurate pseudoanalytical solution for steady-state biofilms. *Biotechnology and Bioengineering*, **39** (7), 790–793.

14 Billen, G., Servais, P., Bouillot, P., and Ventresque, C. (1992) Functioning of biological filters used in drinking-water treatment – the Chabrol model. *Journal of Water Supply: Research and. Technology – AQUA*, **41** (4), 231–241.

15 Laurent, P., Prevost, M., Cigana, J., Niquette, P., and Servais, P. (1999) Biodegradable organic matter removal in biological filters: evaluation of the CHABROL model. *Water Research*, **33** (6), 1387–1398.

16 Servais, P., Laurent, P., and Gatel, D. (2002) Removal of biodegradable dissolved organic carbon in BAC filters: a modelling approach. Biological Activated Carbon Filtration, IWA Workshop, 29–31 May 2002, Paper No: 9.

17 Uhl, W. (2002) The importance of bacterial attachment and detachment in GAC bioreactors: modeling and comparison with pilot investigations. Biological Activated Carbon Filtration, IWA Workshop, 29–31 May 2002, Paper No: 13.

18 Wang, J. and Summers, R.S. (1994) Modeling of biofiltration of natural organic matter in drinking water treatment. Critical issues in water and wastewater treatment. Proceedings of the 1994 National Conference on Environmental Engineering held in Boulder, Colorado, July 11–13, Eds. Joseph N. Ryan and Mare Edwards.

19 Zhang, S. and Huck, P.M. (1996) Removal of AOC in biological water treatment processes: a kinetic modeling approach. *Water Research*, **30** (5), 1195–1207.

20 Hozalski, R.M. and Bouwer, E.J. (2001) Non-steady state simulation of BOM removal in drinking water biofilters: model development. *Water Research*, **35** (1), 198–210.

21 Hozalski, R.M. and Bouwer, E.J. (2001) Non-steady state simulation of BOM removal in drinking water biofilters: applications and full-scale validation. *Water Research*, **35** (1), 211–223.

22 Rittmann, B.E., Stilwell, D., and Ohashi, A. (2002) The transient-state, multiple-species biofilm model for biofiltration processes. *Water Research*, **36**, 2342–2356.

23 Rittmann, B.E., Stilwell, D., Garside, J. C., Amy, G.L., Spangenberg, C., Kalinsky, A., and Akiyoshi, E. (2002) Treatment of a colored groundwater by ozone-biofiltration: pilot studies and modeling interpretation. *Water Research*, **36**, 3387–3397.

24 Sontheimer, H., Crittenden, J., and Summers, R.S. (1988) Activated carbon

for water treatmento, 2nd edn, Forschungstelle Engler-Bunte Institute, Universität Karlsruhe, Karlsruhe, Germany.

25 Ying, W.C. and Weber, W.J. Jr, (1979) Bio-physicochemical adsorption model systems for wastewater treatment. *Journal of the Water Pollution Control Federation*, **55** (11), 2661–2677.

26 Andrews, G.F. and Tien, C. (1981) Bacterial film growth in adsorbent surfaces. *American Institute of Chemical Engineers Journal*, **27** (3), 396–403.

27 DiGiano, F.A., Dovantzis, K., and Speitel, G.E. (1984) Influence of adsorption on biofilm development. Proceedings of the 1984 National Conference on Environmental Engineering of the American Society of Civil Engineers, held in Los Angeles, California, June 25–27, 1984, 382–391 (eds M. Pirbazari and J.S. Devinny).

28 Hubele, C. and Sontheimer, H. (1984) Adsorption and biodegradation in activated carbon filters treating pre-ozonated humic acid. Proceedings of the 1984 National Conference on Environmental Engineering of the American Society of Civil Engineers, held in Los Angeles, California, June 25–27, 1984, 376–381, Eds. M. Pirbazari and J.S. Devinny.

29 Chang, H.T. and Rittmann, B.E. (1987) Mathematical modeling of biofilm on activated carbon. *Environmental Science and Technology*, **21**, 273–280.

30 Velten, S., Hammes, F., Boller, M., and Egli, T. (2007) Rapid and direct estimation of active biomass on granular activated carbon through adenosine tri-phosphate (ATP) determination. *Water Research*, **41**, 1973–1983.

31 Hozalski, R.M. (1996) Removal of biodegradable organic matter in drinking water biofilters: experimental studies and model development, Johns Hopkins University, Baltimore, MD, USA, Dissertation.

32 van der Aa, L.T.J., Magic-Knezev, A., Rietveld, L.C., and van Dijk, J.C. (2006) Biomass development in biological activated carbon filters, in *Recent Progress in Slow Sand and Alternative Biofiltration Processes* (eds R. Gimbel, N.J.D. Graham and M.R. Collins), IWA Publishing, pp. 293–302.

33 van der Aa, L.T.J., Rietveld, L.C., and van Dijk, J.C. (2009) Adsorption and biodegradation of natural organic matter in biological granular activated carbon filters: model validation, in TECHNEAU: Safe Drinking Water from Source to Tap State-of-art & Perspectives (eds T. van den Hoven and C. Kazner), IWA Publishing, pp. 377–389.

34 van der Aa, L.T.J., Rietveld, L.C., and van Dijk, J.C. (2011) Effects of ozonation and temperature on the biodegradation of natural organic matter in biological granular activated carbon filters. *Drinking Water Engineering and Science*, **4**, 25–35.

35 Liang, C.H., Chiang, P.C., and Chang, E.E. (2007) Modeling the behaviors of adsorption and biodegradation in biological activated carbon filters. *Water Research*, **41**, 3241–3250.

36 Badriyha, B.N., Ravindran, V., Den, W., and Pirbazari, M. (2003) Bioadsorber efficiency, design, and performance forecasting for alachlor removal. *Water Research*, **37**, 4051–4072.

37 Liang, C.H. and Chiang, P.C. (2007) Mathematical model of the non-steady-state adsorption and biodegradation capacities of BAC filters. *Journal of Hazardous Materials*, **B139**, 316–322.

38 Rhim, J.A. and Yoon, J.H. (2007) Application of a plug-flow column model to a biologically activated pilot plant. *Korean Journal of Chemical Engineering*, **24** (4), 600–606.

39 Herzberg, M., Carlos, G., Dosoretz, C.G., and Green, M. (2005) Increased biofilm activity in BGAC reactors. *American Institute of Chemical Engineers Journal (AIChE J)*, **51**(3), 1042–1047.

40 Li, Q., Marinas, B.J., Snoeyink, V.L., and Campos, C. (2004) Pore blockage effects on atrazine adsorption in a powdered activated carbon/membrane system. I: model development. *Journal of Environmental Engineering*, **130** (11), 1242–1252.

41 Li, Q., Marinas, B.J., Snoeyink, V.L., and Campos, C. (2004) Pore blockage effects

on atrazine adsorption in a powdered activated carbon/membrane system. II: model verification and application. *Journal of Environmental Engineering*, **130** (11), 1253–1262.

42 Rietveld, L.C. and de Wet, W.W.J.M (2009) Dynamic modeling of bentazon removal by pseudo-moving-bed granular activated carbon filtration applied to full-scale water treatment. *Journal of Environmental Engineering ASCE*, **135** (4), 243–249.

43 van der Aa, L.T.J., Rietveld, L.C., and van Dijk, J.C. (2010) Simultaneous removal of natural organic matter and atrazine in biological granular activated carbon filters: model validation. IWA Specialist Conference 'Water and Wastewater Treatment Plants in Towns and Communities of the XXI Century: Technologies, Design and Operation', Moscow, Russia, 2–4 June 2010.

44 McCarthy, P.L. and Meyer, T. (2005) Numerical model for biological fluidized-bed reactor treatment of perchlorate-contaminated groundwater. *Journal of Environmental Science and Technology*, **39**, 850–858.

45 Wang, H., Hob, L., Lewis, D.L., Brookes, J.D., and Newcombe, G. (2007) Discriminating and assessing adsorption and biodegradation removal mechanisms during granular activated carbon filtration of microcystin toxins. *Water Research*, **41**, 4262–4270.

46 Wahman, D.G., Katz, L.E., and Speitel, G.E. Jr (2007) Modeling of trihalomethane cometabolism in nitrifying biofilters. *Water Research*, **41**, 444–457.

12
Concluding Remarks and Future Outlook
Ferhan Çeçen and Özgür Aktaş

In this chapter we attempt to put the topics discussed in the preceding chapters in perspective, at the same time identifying specific areas for future research. We first delineate the current state of the art in integrated adsorption and biological treatment of domestic wastewaters, industrial wastewaters, landfill leachates, and drinking waters. We also include discussions on more specific issues such as removal mechanisms, micropollutant elimination, importance of activated carbon type, biological regeneration (bioregeneration), and physicochemical regeneration of activated carbon in integrated systems.

12.1
Overview of Applications in Wastewater and Water Treatment: PACT and BAC Systems

The PACT and BAC processes find widespread application in the field of wastewater treatment, more for industrial wastewaters and leachates than for domestic wastewaters. This happens because the former two wastewaters often contain specific pollutants that necessitate the implementation of integrated processes. In the past, most studies conducted on industrial wastewater treatment focused primarily on reduction of total organic parameters such as COD, BOD, and TOC, though attention was also paid to the removal of specific organic groups that may reach high concentrations in industrial effluents, such as phenolics. Relatively less attention has been paid, however, to the removal of more specific pollutants that are found at trace levels.

An important issue in integrated treatment is the decision whether to choose PACT or BAC processes. This decision depends on the characteristics of the wastewater and the aim of the treatment. Up to now, in full-scale industrial wastewater treatment and in laboratory- and pilot-scale studies, PACT has found wider application than BAC.

In industrial wastewaters, both the flow rates and the concentrations fluctuate drastically according to the production campaigns, which results in a highly unstable wastewater. Similarly, the characteristics of landfill leachates vary

Activated Carbon for Water and Wastewater Treatment: Integration of Adsorption and Biological Treatment.
First Edition. Ferhan Çeçen and Özgür Aktaş
© 2011 WILEY-VCH Verlag GmbH & Co. KGaA, Weinheim.
Published 2011 by WILEY-VCH Verlag GmbH & Co. KGaA

strongly depending on solid waste composition and landfill age. Hence, in the treatment of such wastewaters, the PAC dose to an activated sludge system often needs to be adjusted in order to meet discharge standards, and the flexibility of the PACT process is one of the main reasons why it is so often preferred for industrial wastewater treatment. Besides this, many temporary problems encountered in industrial wastewater or leachate treatment plants are easily relieved by upgrading existing activated sludge systems with PAC dosing, without the necessity for great process changes or expenditure.

On the other hand, in domestic wastewater treatment there is generally no need for the introduction of activated carbon, since domestic wastewater often only contains pollutants that can be eliminated by relatively conventional techniques. However, some large-scale WWTPs treating municipal wastewaters receive both domestic and industrial wastewaters or landfill leachates. The input of the latter two can often exert adverse effects on biological treatment and affect sludge quality.

Moreover, in the last decades, the observation is often made that domestic wastewaters themselves contain numerous micropollutants, even if there is no other input. The use of anthropogenic chemicals in our daily life and the excretion of pharmaceuticals and hormones are the underlying reasons. The fate of organic micropollutants has gained special importance in recent years, particularly if these possess carcinogenic, bioaccumulating, or endocrine disrupting properties in the environmental media to which they are discharged. Therefore, their identification and removal from wastewater are among 'cutting edge' topics. Often the term 'micropollutants' is used interchangeably with 'emerging pollutants' or 'xenobiotics.'

Surprisingly, integrated adsorption and biological treatment has received very little attention so far in domestic or municipal wastewater treatment despite the fact that much research is being carried out on the fate and removal of micropollutants. Nowadays, the most emphasized micropollutants are the pharmaceuticals, hormones, and endocrine disrupting compounds. The amenability of such compounds to biodegradation and adsorption varies widely, although fortunately some of them are readily biodegradable and/or adsorbable, so that activated carbon adsorption and biological processes can be integrated into the treatment system. This integration can be realized at two different treatment steps: secondary or tertiary treatment.

For secondary treatment of domestic or municipal wastewaters, the potential of the PACT process for pollutant or micropollutant removal has to be assessed. As discussed in preceding chapters, besides the elimination of pollutants from the liquid phase, other impacts, such as improvement of sludge settling, are also expected in the simultaneous presence of PAC and biomass. Even so, the PACT process has not been widely researched as an alternative for the upgrading of secondary treatment.

At the present time, integrated processes mainly receive attention for the treatment of domestic or municipal wastewater only if the goal is reuse of the treated water, and such processes are introduced as a tertiary rather than a secondary treatment step. A tertiary treatment process with considerable potential is BAC filtration. BAC filtration of secondary effluents may lead to further reduction in organic sum parameters and inorganic parameters such as ammonia. However, other factors have also to be considered when deciding about the potential reuse of treated wastewater. Although effluents may satisfy the criteria based on conventional parameters, much work has still to be done on the elimination of specific micropollutants from secondary effluents. Fortunately, sophisticated analytical methods are now available that can detect micropollutants at $\mu g\,L^{-1}$ or $ng\,L^{-1}$ levels. At very low concentrations of micropollutants, the BAC process, which essentially combines the merits of biofilm reactors with those of GAC adsorbers, is particularly likely to lead to efficient concurrent adsorption and biodegradation.

However, other options are also available as integrated configurations as well as the PACT and BAC processes. Although in biological wastewater treatment the use of PAC was initially confined to activated sludge operation, interest has also recently grown in assisting membrane bioreactor operation (MBR) with PAC, a configuration designated as the PAC-MBR system, a hybrid process giving improved removal efficiency and sludge characteristics compared to that achieved by conventional MBR, especially as far as the removal of micropollutants is concerned. Another alternative is to assist membrane operation with GAC, although this configuration (designated as the GAC-MBR system) has received attention only recently.

If micropollutants are not effectively eliminated in treatment plants, they will leave with the effluent and enter receiving waters, leading to pollution of the water resources from which drinking water is abstracted, and, therefore, the subject of micropollutants in drinking water is increasingly receiving attention today. In drinking water treatment, integrated adsorption and biological removal mainly takes the form of BAC filtration, in contrast to the variety of configurations exploited in wastewater treatment, although in recent years, other configurations, such as PAC-assisted MBR operation, have also been studied, though to a limited extent. Moreover, relatively little attention has been paid to integration of activated carbon adsorption with biological removal in groundwater bioremediation, although it may yield satisfactory results in the elimination of groundwater pollutants.

Further research on integrated adsorption and biological removal should not ignore the practical demands of water and wastewater treatment, and the following sections are intended to give an idea of the kind of further research most likely to be of practical value. To profit from the vast capability of integrated systems in removing inorganic and organic pollutants, close collaboration seems to be needed between people from the various disciplines, in particular, between engineers and microbial ecologists.

12.2
Further Research on Removal Mechanisms and Micropollutant Elimination

12.2.1
Wastewater Treatment

In biological wastewater treatment systems incorporating activated carbon, generally three elimination processes are of importance for pollutants as well as micropollutants: adsorption to sludge (biosorption), biodegradation, and adsorption onto activated carbon. However, if the wastewater contains volatile or semivolatile compounds, another important mechanism, volatilization, would be of interest. Although in domestic or municipal wastewater treatment systems most of the micropollutants that are of concern today are nonvolatile, volatilization may be an important elimination mechanism in industrial wastewaters or the like.

The synergism created in concurrent adsorption and biodegradation of micropollutants is not yet being studied. The main difficulties in biodegradation of organic micropollutants arise due to their very low concentrations. Below a minimum concentration, biodegradable organics can only be eliminated as a secondary substrate. As shown in preceding chapters, in integrated systems, the primary role of activated carbon would be the enrichment of micropollutants, a major mechanism that enables biodegradation to take place even though the bulk concentrations are low. It would therefore be of interest to study how enrichment affects the biological removal of micropollutants. For micropollutants, enrichment on a carbon surface would probably lead to enhanced biological removal. The conversion of micropollutants into transformation or end products by suspended or attached biomass also needs to be investigated, and more detailed information is needed on the time-variable uptake of micropollutants by activated carbon under non-steady operating conditions.

In many places in this book it has been demonstrated that inherent substrate properties such as molecular structure put important constraints on biological removal. Nevertheless, in drinking water or wastewater treatment, even under such unsuitable conditions, some biological pathways can still accomplish the removal of pollutants.

Cometabolic removal is likely to occur in degradation of pollutants that cannot be used as energy or carbon sources but have a structural similarity to growth (primary) substrate. In wastewater treatment the biological sludge consists of a number of heterotrophs and nitrifiers that have the ability to produce oxygenase enzymes which can initiate the cometabolic degradation of various micropollutants. Although cometabolic degradation of industrially important compounds, particularly the chlorinated ones, has often been addressed, information about some emerging micropollutants, such as PPCPs and hormones, is still very limited. These could possibly be removed by cometabolism, although this has not yet been established.

Surprisingly, cometabolism has not been investigated in biological systems that incorporate activated carbon adsorption. Also in the case of cometabolic removal, adsorption of a pollutant onto a carbon surface is likely to provoke its removal, and therefore efforts should be made to maximize the cometabolic removal of micropollutants in the presence of proper growth substrates. Among the numerous organic compounds found in domestic, municipal, or other types of wastewaters, many would likely behave as growth substrates. In particular, adsorption characteristics of growth and cometabolic substrates and the competitive adsorption between them need to be elaborated. In fact, competitive adsorption is not only important in cometabolic removal, it also largely determines when a compound is utilized as a growth substrate. Often there is competition between growth and cometabolic substrates regarding biological removal. Moreover, drinking water or wastewaters do not contain one type of micropollutant, but many micropollutants at variable concentrations. Therefore, competition for adsorption and biodegradation would take place among several micropollutants.

In a water or wastewater matrix, an accurate determination of biodegradation rates of the numerous micropollutants is extremely difficult. Most studies conducted so far have focused on the removal of micropollutants in secondary treatment of domestic or municipal wastewaters and center around 'biodegradation potential' of single micropollutants only. A survey of the literature indicates that great discrepancies exist among biological treatment studies with regard to micropollutant removal efficiencies, and special emphasis is put on the biodegradation potential of pharmaceuticals. On the other hand, the capability of activated carbon for adsorption of pharmaceuticals or the like should also be fully exploited by integrating adsorption into biological processes. For example, PAC dosing to activated sludge was shown to reduce the persistent compound carbamazepine very effectively. Yet, for this compound and many others, the exact mechanisms of removal have been not sufficiently elucidated. An insight into mechanisms would help to optimize the removal and widen the potential of modeling.

Nowadays, antibiotics too are among important micropollutants. Their presence in biological treatment systems raises concerns mainly associated with the proliferation of antibiotic-resistant microorganisms. While for some antibiotics tested so far, biodegradation is the main elimination mechanism in activated sludge, literature results indicate that for others sorption seems to predominate. As in the case of pharmaceuticals, the removal of antibiotics has not been adequately studied in integrated systems. For pharmaceuticals, antibiotics, and other micropollutants, laboratory tests could be conducted to observe the probable enhancement in biodegradability upon adsorption.

Moreover, for many micropollutants, such as flame retardants, complexing agents, and perfluorinated tensides, information on the impact of PAC or GAC is still very limited. Neither is any information available on how the metabolites or daughter products of micropollutant degradation are adsorbed, desorbed, and biodegraded in biological systems that are assisted by activated carbon, in particular under variable loadings. Temporary adsorption and desorption of micropollutants would not only affect effluent quality, but would also have implications for sludge management.

In systems that integrate activated carbon with biological treatment, another potential sorbent for micropollutants would be the biomass. Although activated carbon has probably a much higher adsorption potential for micropollutants than biomass, it is worth examining the relative importance of carbon adsorption and sludge biosorption. In earlier studies, both the sludge partitioning and the activated carbon adsorption coefficients have been determined in the presence of single solutes, but the influence of a multicomponent mixture and the coexistence of activated carbon and biological sludge have yet to be elucidated.

MBRs are regarded as possibly better technologies to eliminate many micropollutants since they operate at very high sludge ages. Most of the research done on MBRs that are dosed with PAC has focused so far on the adsorption of EPS and SMP to improve effluent quality, sludge settling, and dewaterability. In addition, the adsorption, desorption, and biodegradation behavior of micropollutants is worthy of study in these reactors. An interesting point for future research would be the bioregeneration of activated carbon in MBRs; one would expect bioregeneration to be favored in such reactors for the reasons discussed in preceding chapters.

It may be noted that most of the research on PACT and BAC processes primarily aims to characterize the enhanced removal in aerobic systems. In future, it would be beneficial to conduct more research on anoxic and anaerobic processes. Almost all of the opinions that are put forward concerning aerobic conditions can be extended to anoxic or anaerobic systems.

Moreover, it is important to examine the microbial diversity and activity in integrated systems. Acclimation of biomass is of particular importance in the treatment of industrial wastewaters that contain resistant substances. In PACT and BAC applications, a proper acclimation procedure may increase the removal efficiency of slowly biodegradable organic compounds. For example, for complex phenolic wastewaters, acclimation of biomass can be achieved with a relatively more biodegradable compound such as phenol. Following a successful period of acclimation with phenol, acclimation may be continued by the step-wise addition of slowly biodegradable and inhibitory phenolics in addition to phenol. It is important to monitor the response of the biomass to the increase in inhibitory compounds, for example, by using respirometric techniques such as the measurement of oxygen uptake rate. The step-wise increases in concentration should be continued only after the biomass can tolerate the previous concentration and should be stopped otherwise. To mimic the real conditions, in a BAC system, for example, it would also be worth studying how the presence of activated carbon affects the acclimation process.

12.2.2
Drinking Water Treatment

The issues discussed for integrated adsorption and biological removal of wastewaters apply to a large extent to drinking water treatment also. Compared to wastewater treatment, many issues need still to be clarified in drinking water treatment systems that incorporate BAC filtration. In drinking water treatment, the biodegradation potential of micropollutants is often inferred from their biodegradation rates in wastewater treatment systems. For example, in biological wastewater treatment, the pharmaceutical carbamazepine is regarded as essentially nonbiodegradable but adsorbable, while ibuprofen is considered to be biodegradable and adsorbable. It is recognized, however, that the decision on biodegradability is commonly based on data taken from activated sludge systems where the microbial composition is entirely different from that in a drinking water biofilter. In addition, the background organic matter differs significantly with respect to both concentration and composition. In any case, studies on the biodegradation potential of pollutants in drinking water biofilters should be carried out under conditions pertaining to drinking water.

In many cases, studies on drinking water BAC filtration disregard biodegradation as an effective micropollutant elimination mechanism since the concentrations of micropollutants are too low to support the growth of bacteria. For instance, studies dealing with atrazine often assume that biodegradation is not occurring. However, as mentioned at several places in this book, such micropollutants could still be removed as secondary or cometabolic substrates.

In BAC filtration of drinking water, the cometabolic removal of several micropollutants such as PPCPs and hormones by the enzymes produced by nitrifiers or heterotrophs or both would be an interesting point to study. A very interesting issue may be the cometabolic removal of micropollutants in the presence of nitrifiers which use ammonia as a growth substrate. As shown in this book, the AMO enzyme of nitrifiers catalyzes the oxidation of THMs that are widely regarded as nonbiodegradable. Similarly, the same enzyme might catalyze the oxidation of a number of other chlorinated or nonchlorinated organic micropollutants found in water resources today. In the realm of wastewater treatment, there are some very limited studies dealing with the cometabolic removal of some micropollutants by nitrifier activity, such as hormones, but no information is available on drinking water treatment. The cometabolic degradation rate of micropollutants is often linked to the nitrification rate in BAC filtration. At low concentrations of micropollutants and ammonia, cometabolic degradation would take place at very slow rates. If experiments were to be conducted with pure or enriched nitrifying cultures in the presence of activated carbon, they would certainly lead to interesting results. The enrichment ability of activated carbon may accelerate cometabolic degradation.

As in the case of aerobic wastewater treatment, in drinking water BAC filtration the enzymes produced by heterotrophs, either mono- or dioxygenase enzymes, may bring about the cometabolic degradation of various micropollutants. However, in drinking water the biodegradable fraction in background organic matter is very low. This fraction serves as a growth substrate and is characterized by the BDOC, or more precisely, by the AOC parameter. The constant or intermittent addition of growth substrate is one of the key factors in cometabolic removal. If a micropollutant is eliminated by cometabolism only in the presence of sufficient AOC and not by other means, even small decreases in AOC would seriously slow down its removal. It would be worth studying how micropollutants that are commonly found in drinking water today are temporarily adsorbed, desorbed, and biodegraded in the presence of background organic matter whose concentration is varying. The information gained from such studies would also expand the capacity for modeling.

In drinking water BAC filtration, the mechanisms leading to the removal of inorganic pollutants also need to be further clarified. For example, no consistent information about the removal of bromate in BAC systems is available. Moreover, very limited data exists on bromate removal in full-scale plants.

12.3
Further Research on Regeneration of Activated Carbon

This section addresses some specific questions that are related to the carbon grade, bioregeneration, and physicochemical regeneration of activated carbon. As such, the ideas presented here apply mostly to wastewater as well as drinking water treatment.

12.3.1
Importance of Activated Carbon Grade

The characteristics of activated carbon are usually ignored in water and wastewater treatment works. However, as shown in many places in this book, carbon characteristics, like the method used for carbon activation, strongly influence adsorption, desorption, and bioregeneration. In future, the type of activated carbon should be given more consideration in laboratory- and pilot-scale studies. In the design of studies, researchers can follow the guidelines on activated carbon characteristics provided by carbon producers. This, however, is an insufficient approach. In all cases, comparative adsorption isotherms should be constructed to determine the optimum carbon grade. Depending on the characteristics of wastewater, the selection of the most efficient activated carbon grade in full-scale applications will significantly decrease the costs related to the replacement and regeneration of activated carbon.

Activated carbon users usually investigate the adsorption characteristics of a carbon grade for a specific wastewater or pollutant. However, they often overlook the desorption from a carbon grade, although this can be as important as adsorption in many cases. One activated carbon grade may be quite effective in terms of adsorptive removal of a specific wastewater or a specific compound, but if adsorption is highly reversible (highly desorbable) and the loaded carbon is disposed to landfills without any further treatment, environmental problems are likely to arise. In fact, this is a common problem encountered in activated carbon applications that do not integrate biological treatment. In contrast to this, reversibility of adsorption (desorption) may be advantageous if activated carbon applications are integrated into biological processes. This happens because in such cases activated carbon may be bioregenerated upon desorption. For optimum bioregeneration to take place, those activated carbons should be preferred that adsorb pollutants more reversibly. Particularly, chemically activated carbons and carbons with higher macro- and mesopore volumes seem to be more prone to desorption. On the other hand, thermally activated carbons generally exhibit higher adsorption and lower desorption than chemically activated carbon grades. Depending on wastewater characteristics, the most bioregenerable activated carbon has to be determined in each case.

Bioregenerated as well as physicochemically regenerated carbons will eventually contribute to less pollution when they are disposed of since they would have adsorbed mostly nondesorbable substances. In any case, desorption isotherms should be taken into consideration when determining the optimum carbon grade.

For selection of the more environmentally friendly activated carbon grades, it is very important to assess both the adsorption and desorption characteristics.

Modeling can be an effective part of the selection of the right carbon grade. Mathematical models developed for activated carbon systems with biological activity should also take into account the desorption properties. Therefore, for prediction of removal efficiency and bioregeneration, both adsorption and desorption data have to be incorporated into the mathematical models that have been previously described in this book, or into newly developed ones.

12.3.2
Bioregeneration of Activated Carbon

Bioregeneration is a specific topic in integrated systems and is handled in this book from various viewpoints. Bioregeneration is dependent on several factors including biodegradability, adsorbability, and desorbability of sorbate, characteristics of activated carbon, and process configuration. Literature studies have shown that bioregeneration of activated carbon can increase service life; it can further be optimized by taking flexible operation measures. Yet, future research is still required to determine the optimum conditions leading to enhanced bioregeneration. In particular, the type of activated carbon, the nature of the microbial community, the optimum process configuration, and the operational conditions need to be further investigated.

12.3.2.1
Discussion of the Term 'Bioregeneration' and Hypotheses

Some researchers consider bioregeneration as a process that inevitably could be encountered in integrated activated carbon biodegradation processes. Other researchers highlight the important role of bioregeneration in increasing the service life of activated carbon. Disputes about the appropriate use of this term are still common. Some authors prefer to use the term 'bioregeneration' as long as a direct reaction has been demonstrated between microorganisms and adsorbed compound. However, most authors employ the term when the adsorptive capacity of activated carbon is renewed through the indirect action of microorganisms. According to their view, as long as the adsorbate/substrate is desorbed and is removed from the bulk liquid by microorganisms, the process should be called 'bioregeneration.'

There are also literature studies showing that bioregeneration may be caused by the activity of exoenzymes. Exoenzymatic reactions inside the carbon pores result in higher bioregeneration rates than would be expected from desorption due to the development of a concentration gradient. However, the effect of exoenzymatic reactions on bioregeneration is still a hypothesis because it was not evidenced by a direct measurement of enzyme reactions inside the pores. The characteristics of the microbial community and/or the nature of the microbial ecosystem seem to be important factors for exoenzymatic reactions.

Research is still needed to verify the validity of exoenzymatic reactions inside carbon pores. For this purpose, activated carbons can be initially loaded with a model compound such as phenol. Then, the reversibly adsorbed fraction can be removed through successive desorption steps. Finally, this activated carbon should be contacted with exoenxymes only and not with the total biomass. At the end of the process, phenol remaining on activated carbon can be measured or calculated following a further adsorption experiment. If exoenzymatic bioregeneration is valid, the amount of irreversibly adsorbed phenol should decrease. Enzyme activity assays or protein measurements can also be used in such studies to determine the presence of exoenzymes inside or on carbon surfaces. If the validity of exoenzymatic bioregeneration can be proved, intensive research will still be required to determine the optimum conditions.

The efficiency of bioregeneration depends on the desorbability of substrate. If not enough substrate has desorbed from the carbon, as seen in many cases, the efficiency of bioregeneration will be low. Particularly in offline bioregeneration, this handicap may be overcome by the engineered use of enzymes, which, however, could be a challenging task. If exoenzymatic reactions could be provoked in this way, more effective bioregeneration would be achieved.

12.3.2.2
Bioregeneration of Activated Carbon in the Case of Micropollutants

The competitive adsorption and biodegradation of micropollutants is worthy of study, as well the possibility of bioregeneration of activated carbon in the presence of multiple micropollutants. For example, it is likely that nonbiodegradable but adsorbable micropollutants such as carbamazepine are preferentially adsorbed. However, such nonbiodegradable micropollutants would not lead to bioregeneration of activated carbon. On the other hand, slowly biodegradable micropollutants such as nitrophenols would have the opportunity to be biodegraded upon adsorption. There are also few studies in the literature regarding competitive adsorption and cometabolic removal during bioregeneration.

For bioregeneration of activated carbon that has adsorbed micropollutants, adsorption, desorption, and biodegradation properties of the micropollutants would be of interest. The adsorption and desorption characteristics of a single pollutant are relatively easy to determine. In contrast to this, in a multicomponent system, the adsorption and desorption behavior of each pollutant may vary strongly. For many micropollutants, such as PPCPs, hormones, and endocrine disrupting compounds, the sorption hysteresis, or the reversibility of adsorption needs to be clarified. In particular, desorption of nonbiodegradable micropollutants is very important, since this would affect effluent quality. Moreover, as mentioned earlier in this chapter, the biodegradation potential of emerging micropollutants has so far been studied to a limited extent. Since biodegradability of a pollutant is the key factor for bioregeneration, more attention should be paid to the biodegradation potential of each pollutant in a multicomponent medium.

12.3.2.3
Bioregeneration Conditions

A very important determinant for effective bioregeneration of activated carbon is the microbial ecology. For effective bioregeneration, many studies in the literature have used either specific microorganisms that target the pollutant or a mixture of microorganisms that are acclimated to wastewater. In addition, in future studies molecular tools can be used to determine which type of microorganisms induce bioregeneration more effectively.

Several studies showed that the process configuration is particularly important for bioregeneration. Bioregeneration seems to be more easily realizable in offline bioregeneration systems, since the control of the process is easier. On the other hand, bioregeneration in simultaneous (concurrent) biodegradation and adsorption systems can also be optimized by changing the process configuration, for example, by applying variable organic loadings or changing hydraulic and/or sludge retention times.

12.3.2.4
Bioregeneration of Activated Carbon in Drinking Water Treatment

Although bioregeneration of activated carbon is a widely accepted concept in biological wastewater treatment, a review of the literature reveals that in drinking water BAC filtration bioregeneration of activated carbon is still debated. At the present time, drinking water BAC filtration relies merely on physicochemical regeneration techniques.

For bioregeneration to take place, NOM components should desorb from activated carbon and be biodegraded. However, literature studies to date indicate that NOM components that are adsorbed onto activated carbon are often nondesorbable and nonbiodegradable. As shown in preceding chapters, bioregeneration of activated carbon is most probable under conditions when activated carbon has reached saturation level, after which it starts to desorb. However, in drinking water BAC filters such optimum conditions leading to desorption are often not created. Compared to wastewater treatment, adsorption onto GAC occurs very slowly, and saturation is often not reached. Obviously, the composition of NOM and the type of activated carbon also play a role in bioregeneration. Most studies conducted so far have aimed to characterize NOM components, but have ignored the characteristics of activated carbon itself.

12.3.3
Physicochemical Regeneration of Biological Activated Carbon

An important problem encountered in integrated activated carbon adsorption–biological processes is the generation of PACT sludge or waste BAC. Although bioregeneration is a way to increase the service life of activated carbon, it is highly conditional. As a consequence, total regeneration of carbon often cannot be achieved by biological means only. On the other hand, most of the conventional physicochemical regeneration techniques used for nonbiological GACs can be applied under dry conditions only. Therefore, they are usually not suitable for sludges emerging from PACT or BAC operations, since these contain a significant amount of water.

Particularly, in the past, regeneration via wet air oxidation, which is referred to as Wet Air Regeneration (WAR), was used in full-scale PACT applications. Wet air oxidation is also advantageous in terms of sludge management, since it decreases the amount of waste sludge. Hence, whenever a wet air oxidation facility is constructed at full scale, it may serve for both the regeneration of activated carbon and sludge treatment. Therefore, use of regeneration facilities such as wet air oxidation may be cost-effective in PACT facilities where waste sludge is an important problem. However, in most full-scale PACT or BAC applications, regeneration processes such as WAR are not applied. It is also recognized that laboratory- or pilot-scale studies consider the liquid phase only, without paying enough attention to the regeneration of sludges from PACT or BAC applications.

The optimization of the WAR process as well as the development of new regeneration techniques should be given more attention in future studies. For example, future research could examine ultrasonic treatment of waste PACT sludges or biological GACs as a new regeneration technique. Ultrasonic treatment may provide regeneration of activated carbon through destruction of adsorbates. It may also be used for disintegration of PACT sludge through destruction of microbial cells. Hence, as in the case of WAR, it may also serve as a means of sludge management in a PACT facility.

Index

A
Abiotic degradation, 53–54
Abiotic desorption, 191, 211
ABS. *See* Acrylonitrile butadiene styrene (ABS)
AC. *See* Activated carbon (AC)
Acclimation, 49, 56, 57–58, 63, 68, 80, 108, 193–194, 199, 207, 221, 335, 358
Acetic acid, 112, 301, 328
Acetone, 96
Acid Orange II, 111
Acid Orange 7, 111
Acinetobacter, 307
Acrylonitrile butadiene styrene (ABS), 99
Acrylonitrile manufacturing wastewater, 103
Activated carbon (AC), 13
– adsorption, 7–8, 18–19
– in anaerobic treatment, 81
– ash content, 15
– biodegradable and nonbiodegradable organics on, 56
– biological regeneration of, 59
– biological sludge properties with, 81–82
– bioregeneration of, 217–229, 361–363
–– in drinking water treatment, 363
–– in micropollutants, 362
– carbon activation, 29, 197
–– chemically activated carbon, 21, 29, 193, 202, 257, 259, 269, 360

–– thermally activated carbon, 21, 29, 193, 197, 256, 257, 274, 360
– characteristics of, 14–15
– density, 153
– in drinking water treatment, 4–5
– enhancement of denitrification by, 80
– enhancement of nitrification by, 78
– enhancement of organic carbon removal by, 78
– in environmental media, 7
– in environmental pollution, 4
– grade, 202, 360–361
– heterogeneity of, 21
– historical appraisal of, 1–3
– in industrial wastewater treatment, 6
– for inorganic species, 80
– for membrane filtration, 85–86
– for microbial products, 83–85
– microcrystallites, 14–15
– microorganisms
–– attachment and growth of, 59
– natural organic matter
–– adsorbability of, 255–258
–– desorbability of, 258–259
– particle size, 15, 16, 85
– physicochemical regeneration of, 363
– pollutants, control of, 8–10
– pore volume, 13, 292
– porosity, 1, 203

– preparation of, 13–14
– raw materials
–– charcoal, 14
–– coal, 14
–– coconut shell, 2, 13, 99
–– coke, 13, 14
–– lignite, 13, 14
–– wood, 13, 14
– reactivation, 37–39
– regeneration, 37–39
– saturation, 204, 213
– in secondary treatment, 46
– sorption onto, 53
– substrates concentration, 58
– surface area, 1
– surface functional groups, 15, 21
– in tertiary treatment, 46–47
– for toxic and inhibitory substances, 56–58
– types, 15
–– granular activated carbon (GAC), 16
–– powdered activated carbon (PAC), 15
– use of, 1, 3–4
– volatile organic compounds (VOCs) on, 58
– in wastewater treatment, 5–6
– in water treatment, 246–247
–– granular activated carbon (GAC) filtration, 247
–– powdered activated carbon (PAC) addition, 246

Activated Carbon for Water and Wastewater Treatment: Integration of Adsorption and Biological Treatment.
First Edition. Ferhan Çeçen and Özgür Aktaş
© 2011 WILEY-VCH Verlag GmbH & Co. KGaA, Weinheim.
Published 2011 by WILEY-VCH Verlag GmbH & Co. KGaA

Activated carbon adsorption, 6, 18, 43, 127, 177
– with biological processes, 7
– – in wastewater treatment, 7
– – in water treatment, 8
Activated carbon manufacturing, 13
– Bayer, 3
– Norit, 2–3
– Purit, 2
Activated carbon reactors, 30
– GAC adsorbers, 30
– – breakthrough curves, 34–37
– – operation of, 31–34
– – purpose of use, 30
– – types of, 30–31
– PAC adsorbers, 30
Activated sludge process, 70–71
– conventional, 68, 72, 73, 74, 102, 111, 134
– Powdered Activated Carbon Treatment (PACT), 7–8, 69, 98, 99, 100–101, 102, 110, 127–128
Adenosine triphosphate (ATP), 242, 306
Adsorbability, 255
– characteristics of, 54
– of Natural Organic Matter (NOM), 254, 255–258
– of organic matter
– – in raw and ozonated waters, 255–259
Adsorbable Organic Xenobiotics (AOX), 10, 62, 112
Adsorbate, 16
– aromaticity, 255
– hydrophobicity, 21
– physical and chemical characteristics of, 18–19
– polarity, 19
– solubility, 17, 18–19
Adsorbed organics, 59, 87, 190, 215
Adsorbents, 16
– abrasion number, 33
– apparent density, 33
– iodine number, 33
– molasses number, 33
– particle density, 33
– porosity, 20–21
– shape, 20, 33
– size, 20, 33
– surface area, 33
– type, 33
Adsorbers
– bed area, 33
– bed volume (BV), 34
– breakthrough curve, 34–37
– breakthrough time, 36
– carbon depth, 31
– carbon usage rate (CUR), 34
– contact time, 32–33
– effective contact time, 33
– empty bed contact time (EBCT), 32–33
– filter operation time, 34
– filter velocities, 33–34
– hydraulic loading rate (HLR), 31, 33
– interstitial velocity, 34
– mass transfer zone (MTZ), 34, 37
– reactivation, 37–38
– superficial linear velocity, 33
– throughput volume, 34
– void ratio, 160
– wave front, 35
Adsorbing media, 153
Adsorption, 16, 53
– adsorption energy, 17, 211
– adsorption enthalpy, 18
– adsorption equilibrium and isotherms, 24
– – BET (Brunauer, Emmett, Teller), 27
– – Freundlich, 25–27
– – Fritz–Schluender, 228
– – Langmuir, 26–27
– adsorption kinetics, 18, 22, 36
– adsorption reversibility. See Hysteresis
– and biological removal, 268–269
– chemical adsorption, 17, 18
– chemical bonding, 15, 17
– competitive adsorption, 28–29
– coverage, 18
– equilibrium and isotherms, 24–27
– factors influencing, 18
– – chemical surface characteristics, 21–22
– – pH value, 19–20
– – physical and chemical characteristics of adsorbate, 18–19
– – porosity of adsorbent, 20–21
– – surface area of adsorbent, 18
– – temperature, 20
– irreversible adsorption, 29, 196, 197, 198, 200, 202, 204, 205, 208, 211, 222, 259
– kinetics of, 22
– – adsorption bond, 24
– – bulk solution transport (advection), 22–23
– – external diffusion, 23
– – intraparticle (internal) diffusion, 23
– mono- and multilayer adsorption, 20
– multisolute adsorption, 28–30
– of nonphenolic compound, 197
– of phenolic compounds, 197
– physical adsorption, 17, 18, 24
– reversibility of, 195–198
– single solute adsorption, 27–28
– types of, 16–17
Adsorption and biodegradation characteristics of water, 250–260
– determination of, 255–259
– NOM characteristics, ozonation impact on, 250–251
– – adsorbability, change in, 255
– – biodegradability, increase in, 251–254
– raw water NOM, 250
Adsorption capacities
– direct measurement by, 213

Index | 367

Adsorption-induced acclimation, 193–194
Adsorption isotherm, 24, 154, 211–212, 256
Adsorption models
– Ideal Adsorbed Solution (IAS) model, 28
– Ideal Adsorbed Solution Theory (IAST), 28, 177, 227, 334
– Improved Simplified Ideal Adsorbed Solution (ISIAS) model, 28
– Simplified Ideal Adsorbed Solution (SIAS) model, 28, 29
Advanced Oxidation Processes (AOPs), 47, 97, 118
Advective transport, 170
Aerobic PACT, 46
Aerobic systems, respirometry in, 215
Afipia, 307
AFLP. *See* Amplified Fragment Length Polymorphism (AFLP)
Alcaligens, 306
Aldehydes, 252, 273, 302
Algae, 239, 326–327
Aliphatics, 6
– aliphatic acids, 6
– and aromatic hydrocarbons, 5, 288
– chlorinated aliphatics, 14, 79, 185
– – *cis*-1,2-dichloroethene, 304
– – *trans*-1,2-dichloroethene, 304
– – trichloroethylene (TCE), 54, 105, 215, 290
– – vinyl chloride, 133
17 α-Ethinylestradiol (EE2), 52, 53, 55
α-Proteobacteria, 307
Amines, 288
Ammonia in water treatment systems, 278–282
Ammonia monooxygenase enzyme (AMO), 298
Ammonia oxidizing bacteria (AOB), 78, 308

Ammonia removal, 285, 308
– importance of, 285–286
Ammonium, 78, 79, 238, 301
AMO. *See* Ammonia monooxygenase enzyme (AMO)
amoA (ammonia monooxygenase gene), 308
Amoco refinery, 98
Amplified Fragment Length Polymorphism (AFLP), 307
Anaerobic–aerobic PACT, 46
Anaerobic PACT®, 46, 74
Anaerobic systems, bioregeneration in, 216–217
Anaerobic treatment, activated carbon in, 81
Aniline, 54, 198, 288
Anoxic systems, bioregeneration in, 216–217
Anthracene, 209
Anthracite, 66, 68, 99, 157, 273
AOB. *See* Ammonia Oxidizing Bacteria (AOB)
AOC. *See* Assimilable Organic Carbon (AOC)
AOCl (Adsorbable Organic Chlorine), 10
AOC limits, in water distribution, 319–320
AOPs. *See* Advanced Oxidation Processes (AOPs)
AOX. *See* Adsorbable Organic Xenobiotics (AOX)
Aromatics
– aromatic acids
– – benzoic acids, 14, 118
– benzene, 14, 22, 87, 104, 105, 106, 110, 138, 182, 183
– chlorinated aromatics, 5, 14, 130
– – benzenes
– – – 1,2-dichlorobenzene, 106, 110, 134, 182, 183, 304

– – – 1,4-dichlorobenzene, 304
– – – trichlorobenzene (TCB), 106, 134, 182, 183, 184
– – phenolics
– – – chlorophenol, 29, 50, 87, 107, 108, 134, 194, 199, 205, 207, 209
– – – pentachlorophenol (PCP), 108, 110, 206, 217, 327
– phenol
– – phenolics
– – – ethylphenol, 108
– – – isopropylphenol, 200
– – – methylphenol, 108
– – – nitrophenol, 87, 107, 110
– polynuclear aromatics, 5, 14
– toluene, 14, 50, 96, 105, 106, 118, 134, 138, 182, 183, 208, 224, 225
Arthrobacter viscosus, 108
Assimilable organic carbon (AOC), 240–242, 252
– modifications in, 242
ATP. *See* Adenosine triphosphate (ATP)
Atrazine, 206, 289–290, 291, 321, 346–347
Azo dyes, 110–111, 210

B
BAC-FBR, 68
BAC filters. *See* Biological activated carbon (BAC) filters
BAC filtration. *See* Biological activated carbon (BAC) filtration
BAC reactors. *See* Biological activated carbon (BAC) reactors
Bacillus, 307
Backwashing, 276
– biomass loss, 276
– chlorinated water, 276
– oxidant residuals, 277
BAC-Sequencing Batch Reactors (BAC-SBR), 112, 227
BAC sludge, 86–87

Bacteria
- ammonia oxidizing bacteria (AOB), 78, 308
- autotrophic bacteria, 78, 103
- cocci-shaped, 216
- filamentous, 216
- heterotrophic bacteria, 78, 80, 304
- nitrite oxidizing bacteria (NOB), 78, 308

Bacterial re-growth/aftergrowth, definition of, 238

BAF. See Biological Aerated Filter (BAF)

BAR. See Biofilm annular reactor (BAR)

BASM. See Biodegradation/adsorption–screening model (BASM)

BDOC. See Biodegradable Dissolved Organic Carbon (BDOC)

Bed configurations
- expanded bed, 31, 63, 69, 346
- fixed bed, 31, 175
- fluidized bed, 31, 68
- moving bed, 31, 347
- packed bed, 23, 166

Bed Volume (BV), 34
Bentazon, 291, 347
Bentonite, 111
Benzene, 87, 105, 110, 134, 182, 183, 210, 289, 304
Benzoic acids, removal of, 118
BET (Brunauer, Emmett, and Teller) isotherm, 25, 27
17 β-Estradiol (E2), 55, 109
β-Proteobacteria, 307
Bezafibrate, 52
BFAC model. See Biofilm on activated carbon (BFAC) model
BFR. See Biofilm Formation Rate (BFR)
BGAC. See Biological granular activated carbon (BGAC)
Bioavailability, 48
Biocarbone, 68
Biocides, 101, 288

Biodegradability, 199–200, 252–254
Biodegradable Dissolved Organic Carbon (BDOC), 240, 251, 265, 284
- modifications in, 242
- in water distribution, 319–320
Biodegradable fraction, 48, 78, 96, 114
- measurement of, in NOM, 239–243
Biodegradable organic compounds, 225, 358
Biodegradable organic matter (BOM), 48, 78, 199, 238, 265, 274, 332
- assimilable organic carbon (AOC), 240–242
- biodegradable dissolved organic carbon (BDOC), 240, 251, 265, 284
Biodegradation, 48, 293
- abiotic degradation/removal, 53–54
- biological removal
-- of micropollutants, 50–52
-- of organic substrates, 49
- characteristics of, 54
- cometabolism, 50, 208
- and concurrent adsorption
-- bioregeneration in, 218–219
- fractionation of organic matter in wastewater, 48
- models
-- first-order, 168, 228–229
-- Haldane, 228
-- Monod, 151, 183, 334
- organic substrates, biomass, and activation carbon, 54
- primary substrate, 49, 50
- reductive dechlorination, 105
- secondary substrate utilization, 49–50
- sorption onto activated carbon, 53
- sorption onto sludge, 52–53
Biodegradation/adsorption–screening model (BASM), 225, 227

Biodegradation products, measurement of, 214–215
Biofilm
- backwashing, 276, 308
- biofilm-carbon boundary, 67, 154, 172
- biofilm density, 155
- biofilm kinetics, 146, 151
- biofilm loss, 164
- biofilm thickness, 154–155, 161, 171, 174, 204, 220, 346
- detachment, 157
- dispersion coefficient, 33, 160, 163, 170, 171
- growth, 65, 154–155, 175, 223, 225, 304, 346
- mass transfer coefficient, 148, 166, 171, 224, 337
- models, 221, 222
-- BFAC model, 174, 175, 219, 223
-- MDBA model, 227
- molecular diffusivity, 148, 149, 151, 338
- substrate utilization, 151, 181
Biofilm annular reactor (BAR), 243
Biofilm–carbon system, dimensionless numbers in, 169–173
Biofilm formation, 57, 59
Biofilm Formation Rate (BFR), 243
Biofilm growth and loss, 154–155
Biofilm growth control, in BAC filters, 305
Biofilm on activated carbon (BFAC) model, 174, 175, 219, 223
Biofilters
- anthracite-sand filters, 276
- GAC-sand filters, 273, 276
- sand biofilter, 307, 308
BIOFILT model, 339–340
Biofouling, 8, 75–76, 195
Biological activated carbon (BAC), conversion of GAC into, 64–65
- advantages of BAC process, 66

– BAC versus GAC
  operation, 66
– removal mechanisms and
  biofilm formation in BAC
  operation, 65–66
Biological activated carbon
  (BAC) filters, 100, 108,
  141, 142, 146, 269,
  283–285, 291, 293, 297,
  299, 303, 304, 305
– adsorption and
  biodegradation, 269–270
– biofilm growth control, 305
– biomass and microbial
  activity determination in,
  305
– bioregeneration of, 272–273
– DBP precursors removal
  in, 295
– haloacetic acids (HAAs)
  removal in, 297
– heterotrophic biomass and
  activity in, 306–308
– microbial ecology of, 303
– microbiological safety, of
  finished water, 309
– nitrifying biomass and
  activity in, 308–309
– nutrient removal in, 285
–– ammonia removal,
  285–286
–– denitrification, 286–287
–– nitrification, factors
  affecting, 285–286
– organics removal in, 277
–– preozonation, relevance
  of, 284–285
– trihalomethanes (THMs)
  removal in, 297
Biological activated carbon
  (BAC) filtration, 8, 109,
  127, 140, 237, 238, 244,
  247, 273, 297, 331,
  354–355, 359
– adsorption and biological
  processes, models
  integrating, 342–345
– adsorptive and biological
  removal, 334
– adsorptive removal only,
  333–334
– BDOC and AOC limits,
  in water distribution,
  319–320

– BIOFILT model, 339–340
– biological processes in,
  334–342
– biological removal in, 334
– bromate removal in,
  298–301
– CHABROL model, 335
– current use of, in water
  treatment, 249–250
– domestic and industrial
  wastewaters reclamation
  for nonpotable uses,
  141–142
– domestic wastewater
  reclamation for drinking
  purposes, 140–141
– empirical models, 333
– empty-bed contact time
  (EBCT), 274–276, 338–339
– filter backwashing, 276
– in full-scale groundwater
  treatment, 327–329
– in full-scale surface water
  treatment, 320–326
– Granular Activated Carbon
  (GAC), 273
–– activation, 274
–– biomass attachment, 273
–– characteristics effects, on
  adsorption, 290
–– coal- and wood-based
  GAC, 274
–– removal performance,
  273
– haloacetic acids (HAAs)
  removal, in BAC filters,
  297
–– cometabolic removal,
  297–298
–– metabolic removal,
  297
– history of, in water
  treatment, 247–249
– hydraulic loading rate
  (HLR), 274–276
– initial models in, 334–335
– ionic pollutants removal,
  298
– micropollutants removal,
  models describing,
  346–348
– multiple species inside
  drinking water biofilters,
  341–342

– nitrate removal in,
  298
– organic micropollutant
  groups in, 291
–– Disinfection By-Products
  (DBP) precursors
  removal, 294, 295
–– geosmin and MIB, 292
–– microcystins, 293
–– pesticides, 291
–– pharmaceuticals and
  endocrine disrupting
  compounds (EDCs), 291
– organic micropollutants in
  water, 288
–– NOM and, 288–290
–– adsorption onto
  preloaded GAC, 290
– oxidant residuals, effect of,
  277
– ozonation and, 249, 327
– perchlorate removal in,
  301
– physicochemical
  regeneration of, 363–364
– process-based models, 333
– sewage treatment for reuse
  in agriculture, 140
– temperature, effect of, 276
– trihalomethanes (THMs)
  removal, 297
–– cometabolic removal,
  297
–– metabolic removal,
  297
– Uhl's model, 335–337
– Wang and Summers,
  model of, 337–338
Biological Activated Carbon
  (BAC) process, 8, 44
Biological activated carbon
  (BAC) reactors, 44, 99,
  100, 105, 111
– carbon properties, in
  modeling, 177–178
– characterization, by
  dimensionless numbers,
  164–173
– biofilm–carbon system,
  169–173
– dimensionless parameters,
  165
– first-order kinetics inside
  biofilm, 168–169

Biological activated carbon (BAC) reactors" (*continued*)
- mass transfer equations, 165–166
- reactor mass balance, 166–167
- zero-order kinetics inside biofilm, 167–168
- high substrate concentrations, 175–176
- initial steps in modeling, 173
- integration of adsorption into models, 174–175
- landfill leachate treatment in, 118–120
- substrate removal and biofilm formation, 173–174
- three-phase fluidized bed reactor, 177
- very low substrate conditions, 176–177

Biological activated carbon (BAC) systems, concurrent bioregeneration in, 195

Biological aerated filter (BAF), 68, 100

Biological granular activated carbon (BGAC), 237

Biologically treated water, 256, 258, 259

Biological Membrane Assisted Carbon Filtration (BioMAC), 76

Biological nitrogen removal, 250

Biological processes, in water treatment, 238

Biological rapid sand filtration (BSF), 249

Biological regeneration of activated carbon, 59

Biological removal of micropollutants, 50–52

Biological sludge, properties of, 81
- dewaterability of sludge, 82
- sludge settling and thickening, 81

Biological stability/instability, 238

- AOC, 10, 240–242, 251
- BDOC, 240, 251, 265, 284

BioMAC. *See* Biological Membrane Assisted Carbon Filtration (BioMAC)

Biomass
- acclimated, 51, 193–194, 199, 214, 301, 328
- activity
-- ATP measurement, 306
-- heterotrophic biomass, 306–308
-- nitrifying biomass, 308–309
- attachment, 273
- concentration, 68, 205–206, 242, 344
- configuration
-- fixed-growth, attached-growth (biofilm), 44, 66
-- hybrid, 81
-- suspended-growth, 44
- and microbial activity determination, in BAC filters, 305
- nonacclimated, 108, 199

Biomass respiration method (BRP), 306

Biomembrane operation assisted by PAC and GAC, 74
- membrane-assisted biological GAC filtration, 76
- membrane bioreactors (MBRs), 74
- PAC-MBR process, 75

Biopolymers, 244, 268, 323

Bioregenerated carbon, 107, 195, 208, 360

Bioregeneration, 59, 189, 211, 219, 361
- anaerobic/anoxic systems, 216–217
- bioregeneration conditions, 362–363
- concurrent bioregeneration, 195
- definition of, 189
- determination of, 209–216
-- adsorption capacities, direct measurement by, 213

-- adsorption isotherms, use of, 211–212
-- biodegradation products, measurement of, 214–215
-- quantification, 213–214
-- respirometry, in aerobic systems, 215
-- reversible adsorption, extent of, 211
-- scanning electron microscopy (SEM), 216
-- solvent extraction, direct measurement by, 213
- in drinking water treatment, 363
- *ex-situ*, 138
- factors affecting, 198–209
-- biodegradability, 199–200
-- biomass concentration, 205–206
-- carbon activation type, 202
-- carbon particle size, 200–201
-- carbon porosity, 201–202
-- concentration gradient and carbon saturation, 204–205
-- desorption kinetics, 203
-- dissolved oxygen concentration, 206
-- microorganism type, 206–207
-- multiple substrates, presence of, 208–209
-- physical surface properties, of carbon, 202–203
-- substrate, chemical properties of, 200
-- substrate and biomass associated products, 207–208
-- substrate–carbon contact time, 203–204
- groundwater, 7, 50, 250, 273, 300, 303–304, 355
- *in-situ*, 218
- mechanisms of, 189–194
-- acclimation of biomass, 193–194
-- concentration gradient, 189–191

-- exoenzymatic reactions, 191–193
- in micropollutants, 362
- modeling/models, 217–229
-- biodegradation/adsorption-screening model (BASM), 225, 227
-- biofilm model, 221, 222
-- biofilm on activated carbon (BFAC) model, 174, 175, 219, 223
-- concurrent adsorption and biodegradation, 218–219
-- kinetics, 228–229
-- multicomponent systems, 225–227
-- multi-solute MDBA model, 218
-- offline bioregeneration, 228
-- one-liquid film model, 221, 222
-- single solute systems, 220–225
-- two-liquid film model, 221–222
- offline bioregeneration, 194–195
- reversibility of adsorption, 195–198
- schematic representation of, 190
Bioremediation, groundwater, 50, 303–304
Bioremediation, soil, 7, 50
Biosorption, 100
- characterization of, 54
- NOM, 335
- potential, 52
- sorption onto sludge, 47
Biot number, 166, 172, 173
Biotransformation, 47, 48, 71, 156
Bis (2-ethylhexyl) phthalate, 111, 294
Bisphenol A (BPA), 53, 55, 109, 292
BKME. *See* Bleached kraft pulp mill effluent (BKME)
Bleached kraft pulp mill effluent (BKME), 98
BOD (biochemical oxygen demand), 8, 48, 113, 215, 240, 353

BOD_5/COD ratio, in landfill leachate, 81, 99, 100, 113
BOM. *See* Biodegradable Organic Matter (BOM)
Bosea, 308
Bound EPS, 83
Bradyrhizobiaceae, 308
Bradyrhizobium, 308
Breakthrough curve, 36, 162, 265
- initial stage of operation, 265
- intermediate and later stages of operation, 266
Breakthrough point, 36, 223
Bromate, 301
- bromide in water, 295
- brominated organic compounds, 298
- ozonation effects, 249
- removal of, 298–301
Bromide ion, 295, 298
Brominated flame retardants, 43
4-Bromo-3-chloroaniline, 111
Bromoform, 295, 298
Bromophenol, 107, 294
BRP. *See* Biomass respiration method (BRP)
BSF. *See* Biological rapid sand filtration (BSF)
BTEX (Benzene, Toluene, Ethylbenzene, Xylene), 99, 104, 138
BTX (Benzene, Toluene, p-Xylene), 104, 105
Bulk organic matter, 47, 48, 56
Burkholderiales, 308
Bussy, 2
Butyl cellosolve, 102
BV. *See* Bed Volume (BV)

C
CAA. *See* Chloroacetaldehyde (CAA)
Caffeine, 292
CAgran, 202, 257, 258, 259, 269
Candida, 206
Carbamazapine, 75
Carbamazepine (CBZ), 52, 55, 109, 290, 291, 358, 362

Carbon, physical surface properties of, 202–203
Carbon activation type, 29, 202
Carbonization process, 13
Carbon particle size, 200–201
Carbon porosity, 201–202
Carbon properties' consideration in modeling, 177–178
Carbon saturation
- concentration gradient and, 204–205
Carbon tetrachloride, 14, 105, 134, 209, 210
Carbon usage rate (CUR), 33, 34
Carbonyl oxygen, 15
Carboxylic acids, 241, 253
Carcinogen, 244, 295, 299, 354
Carman–Kozeny equation, 85
Castlemaine plant, 326
Cationic polyelectrolyte, 71
Caulobacter, 307
Caustic hydrolysate wastewater, 101
CB. *See* Chlorobenzene (CB)
CBZ. *See* Carbamazepine (CBZ)
CDOC. *See* Chromatographic DOC (CDOC)
CHABROL model, 335
Chambers Works PACT® system, 134
Characteristics of activated carbon, 14–15
Charcoal, 1, 2, 14
Chemical adsorption (chemisorption). *See* Adsorption
Chemical bonding, 15, 17
Chemically activated PAC, 201, 202
Chemical oxygen demand (COD)
- soluble chemical oxygen demand, (SCOD), 115
- total chemical oxygen demand (TCOD), 115

Index

Chemical surface characteristics, 21–22
'Chemische Werke', 2
Chemisorption, 17, 196
– oxygen, 15, 272, 302, 306
– perchlorate reduction, 301
Chloramine, 276
Chlorinated compounds, 50, 105, 215, 297
Chlorinated hydrocarbons, 129, 182
Chlorinated organic compounds (AOCl), 10, 215
Chlorinated phenols, 50, 108
Chlorine, 4, 112, 248, 276, 277, 285, 294, 297, 327
Chlorine dioxide, 277, 323
Chloroacetaldehyde (CAA), 110, 217
Chlorobenzene (CB), 14, 87, 110, 134, 183, 304
3-Chlorobenzoic acid, 197, 200, 202, 210
Chlorobenzoic acids, removal of, 118, 200, 202
Chloroform, 96, 134, 138, 185, 295, 298, 348
Chlorophenols, 4, 87, 108
2-Chlorophenol (2-CP), 29, 50, 87, 108, 134, 194, 199, 200, 207, 208, 210
4-Chlorophenol (4-CP), 107–108, 205, 207, 209, 210, 215
Chromatographic DOC (CDOC), 244
Chromium compounds, 113
Clofibric acid, 290
Clogging, 31, 47, 63, 68, 69, 305
Closed-loop offline bioregeneration system, 194
COD. See Chemical oxygen demand (COD)
Coke plant wastewaters, 102, 103
Coliform bacteria, 309, 319
Color, 62, 70, 78, 98, 100, 102, 114, 128, 130, 133, 239, 251, 289, 326
Comamonadaceae, 308
Cometabolic removal of substrate, 50

Cometabolism, 50, 208, 356
– ammonia, 78, 103, 133, 135, 248, 285
– AMO enzyme, 359
– chlorinated benzenes, 199
– chlorinated phenols, 50, 95, 108
– cometabolic regeneration, 362
– cometabolic substrate, 50, 356
Complexing agents, 109, 357
Concentration gradient
– bioregeneration due to, 189–191
– and carbon saturation, 204–205
Concurrent adsorption and biodegradation
– bioregeneration in, 218–219
Contaminated groundwater, remediation of, 7
Contaminated groundwaters, PACT for, 138–139
Contaminated surface runoff waters, PACT for, 139–140
Continuous-flow (CF) activated sludges, 115
Continuous-flow stirred tank reactor (CSTR), 70, 103, 177
Conventional water treatment, NOM removal in, 244–246
– extent of, 245–246
– rationale for, 244–245
Co-treatment, 82, 96, 136
Cresols, 108
Cryptosporidium, 326
CSTR. See Continuous-flow stirred tank reactor (CSTR)
CUR. See Carbon usage rate (CUR)
Cyanide
– biodegradation of, 112
– biological removal of, 102, 103
Cyanobacterial toxins
1,3-Cyclohexanediamine, 111

Cyctostatics, 55
Cytarabine (CytR), 53, 55

D
DAF. See Dissolved air flotation (DAF)
Damköhler number of type I, 170, 172
Damköhler number of type II, 170, 172
DBAA. See Dibromoacetic acid (DBAA)
DBPs. See Disinfection By-Products (DBPs)
DCAA. See Dichloroacetic acid (DCAA)
DCF. See Diclofenac (DCF)
DCM. See Dichloromethane (DCM)
de Cavaillon, Joseph, 2
Dechlorination, 105, 215
– of chlorinated water, 4
Denitrification, 49, 286
– enhancement of, by activated carbon, 80
– GAC-FBR configuration, 80
– groundwater, 238, 297
– nitrogen removal, 286–288
– surface water, 288, 298
– wastewater, 135, 141
Desorbability, of NOM
– from activated carbon, 258–260
– in raw and ozonated waters, 254–260
Desorption, 53, 196
– abiotic desorption, 191, 211
– of biodegradable organics, 53, 65, 255
– desorption isotherm, 211, 258, 260, 360
– kinetics, 203
– of organic matter, 191
Detachment rate, 155, 157
Dewatering, 73, 82, 137
Diazepam (DZP), 55, 109
Dibromoacetic acid (DBAA), 295
Dibutyl phthalate, 111
Dichloroacetic acid (DCAA), 295
2,4-Dichloro-benzenamine, 111

Dichlorobenzene, 134
1,2-Dichlorobenzene, 110, 183
Dichloroethane (1,2-DCE), 105, 110, 328
Dichloromethane (DCM), 105
Dichlorophenol, 84, 107, 197, 200, 227, 229
1,4-Dichlorophenol, 134
2,4-Dichlorophenol, 107, 110, 197, 200, 208, 210, 227, 229
3,5-Dichlorophenol, 84, 106, 107
2,4-Dichlorophenoxyacetate (2,4-D), 60
Diclofenac (DCF), 52, 55, 109, 290
Diethylene triamine pentaacetic acid (DTPA), 109
Diffusion
– biofilm diffusivity, 171
– intraparticle diffusivity, 171, 201
– of substrate in water, 171
Dimensionless Empty Bed Contact Time, 338–339
Dimensionless numbers
– Biot number, 166, 172, 173
– Damköhler number Type I, 170, 172
– Damköhler number Type II, 170, 172
– modified Stanton number, 169, 170
– Peclet number, 167, 170
– Sherwood number, 165, 170
– solute distribution parameter, 167, 170
– Stanton number, 169, 170
Dimensionless parameters, definition of, 165
Dimensionless separation factor, 26
3,5-Dimethoxyacetophenone, 111
N,N-dimethyl acetamide, 96
N,N-dimethyl formamide, 96
Di-n-butyl phthalate, 294

Dinitrophenol, 107, 110, 134
Dinitrotoluene, 110, 134, 294
1,4-Dioxane, 130
Disinfection
– chlorine, 4, 5, 294, 320, 322, 323
– UV/H_2O_2, 249
Disinfection By-Products (DBPs), 5, 10, 243, 244, 270, 294–295
– bromate, 298, 299
– brominated organic compounds, 298
– haloacetic acids (HAAs), 10, 244, 294–295, 297
– total trihalomethanes (TTHM), 295
– trihalomethane formation potential (THMFP), 10, 295, 303
– trihalomethanes (THMs), 10, 244, 294, 297, 348, 359
Dissociation constant, 19
Dissolved air floatation (DAF), 4, 6, 73, 138, 141
Dissolved Organic Carbon (DOC), 239, 265
– Biodegradable Dissolved Organic Carbon (BDOC), 240, 270
– mineralization of, 253–254
– Nonbiodegradable Dissolved Organic Carbon (NBDOC), 243
Dissolved Organic Matter (DOM), 85, 239, 303
Dissolved oxygen (DO), 70, 194, 266
Dissolved oxygen concentration, 206
Diuron, 291
DO. See Dissolved oxygen (DO)
DOC. See Dissolved Organic Carbon (DOC)
Dohne water works, 248
DOM. See Dissolved Organic Matter (DOM)
Domestic and industrial wastewaters, PACT for, 136
Domestic wastewaters reuse, PACT for, 139

Domestic wastewater treatment, 83, 354
Draft tube gas–liquid–solid fluidized bed bioreactor (DTFB), 177
Drinking water BAC filtration, 273, 275, 332, 359, 363
Drinking water treatment, 237, 358–359
– activated carbon in, 4–5, 246–247
– – granular activated carbon (GAC) filtration, 247
– – powdered activated carbon (PAC) addition, 246
– adsorption and biodegradation characteristics of water, 250–259
– – determination of, 254–259
– – NOM characteristics, ozonation impact on, 250–254
– – raw water NOM, 250
– biological activated carbon (BAC) filtration, 247
– – adsorption and biological processes, 342–345
– – biological processes in biofilters, 334–342
– – current use of, 249–250
– – history of, 247–249
– – micropollutants removal, 346–348
– – and ozonation, 249
– – substrate removal and biofilm formation, 331–333
– biological processes in, 238
– bioregeneration of activated carbon in, 363
– NOM removal, in conventional water treatment, 244–246
– – extent of, 245–246
– – rationale for, 244–245
– organic matter in, 238–244
– – biodegradable fraction, in NOM, 239–243
– – expression, in terms of organic carbon, 239

Drinking water treatment (*continued*)
-- fractionation, of NOM, 243–244
-- nonbiodegradable dissolved organic carbon (NBDOC), 243
DTFB. *See* Draft tube gas–liquid–solid fluidized bed bioreactor (DTFB)
DTPA. *See* Diethylene triamine pentaacetic acid (DTPA)
DuPont, 70, 101
DuPont's Chambers Works, 128, 134
Dyes
- acid dye, 100, 214
- Acid Orange, 111
- anthraquinone dye, 100, 214
- azo dye, 100, 110, 111, 214
- indigo carmine, 14
- methylene blue, 14, 111
- polymeric dye, 111
- Reactive Black, 110
- removal of, 110–111
- soluble organic dyes, 6, 14
DZP. *See* Diazepam (DZP)

E
E2 (17 β-Estradiol), 55
E3 (Estriol), 55
EBCT. *See* Empty-bed contact time (EBCT)
EC. *See* Expanded clay (EC)
EOCl. *See* Extractable organic chlorine (EOCl)
EDCs. *See* Endocrine disrupting compounds (EDCs)
EDTA (Ethylene diamine tetraacetic acid), 109
EE2 (17α-Ethinylestradiol), 52, 53, 55
EEM (Excitation-Emission Matrix), 325
Effective contact time, 33
Electrochemical oxidation, 101
Emerging substances, 5
Empty-bed contact time (EBCT), 32–33, 204, 274–276, 284, 338–339
Endocrine disrupters, 55

Endocrine disrupting compounds (EDCs), 5, 109, 288, 291–292
- bisphenol A, 53, 55, 109
- nonylphenol, 109, 290, 292
Endogenous respiration, 49
Engineered/mechanical treatment systems, 248–249
Envirex®, 68
Environmental media, activated carbon in, 7
Environmental pollution, activated carbon in, 4–10
- activated carbon adsorption, with biological processes, 7
-- in wastewater treatment, 7–8
-- in water treatment, 8
- drinking water treatment, 4–5
- environmental media, 7
-- flue gases, treatment of, 7
-- groundwater contamination, remediation of, 7
-- soil contamination, remediation of, 7
-- water preparation, for industrial purposes, 7
- pollutants control, 8–10
- wastewater treatment, 5–6
Environmental Scanning Electron Microscopy (ESEM), 216
Enzymes, 50
- ammonia monooxygenase. *See* Ammonia monooxygenase enzyme (AMO)
- extracellular enzymes (Exoenzymes), 191, 193
Eponit®, 2
EPS. *See* Extracellular polymeric substances (EPS)
Erythromycin, 292
Escherichia coli, 308
Escherichia fergusonii, 308
ESEM. *See* Environmental Scanning Electron Microscopy (ESEM)
Estradiol, 292

Estriol (E3), 55
Estrone, 52
Ethanol, 96, 108, 195, 301
Ethyl acetate, 96
Ethylbenzene, 134, 182
Ethylene diamine tetraacetic acid. *See* EDTA (Ethylene diamine tetraacetic acid)
Excitation-Emission Matrix. *See* EEM (Excitation-Emission Matrix)
Exhausted GAC, 205, 293
Exoenzymatic reactions
- bioregeneration due to, 191–193
- validity of, 361–362
Exopolymers. *See* Extracellular polymeric substances (EPS)
Expanded and fluidized-bed BAC reactors, 68–69
Expanded clay (EC), 105
Extracellular enzymes, 191
Extracellular polymeric substances (EPS), 77
- characteristics/definition, 83
- composition, 82–83
- effect on dewaterability, 83, 358
- retention by PAC, 84
Extractable organic chlorine (EOCl), 103

F
FA. *See* Free ammonia (FA)
FB. *See* Fluidized bed (FB)
FBR. *See* Fluidized bed reactor (FBR)
Fenoprofen, 290
Fick's law, 148, 149
Filter backwashing, 276
Filter operation time, 34
Filter velocities, 33–34
Finished water
- microbiological safety, 309
- re-growth potential, 242–243
Firmicutes, 308
First-order kinetics, 228, 229
- inside biofilm, 168–169
Five haloacetic acids (HAA5), 295

Fixed-bed BAC reactors, 67–68, 110–111, 112
Fixed-bed carbon adsorbers, 31
Flame retardants, 43, 109, 357
Flavobacterium, 207, 307
Flue gases
– treatment of, 7
Fluidized bed (FB)
Fluidized bed reactor (FBR), 31, 68, 300
Fluorescence spectroscopy, 243
Fluorobenzene, 210, 215
2-Fluorobenzoate (2-FB), 105
5-Fluorouacil (5-Fu), 53, 55
FNA. *See* Free nitrous acid (FNA)
Food-to-microorganism (F/M) ratio, 70, 71
Formaldehyde, 274
Fouling, 75
– *See also* Biofouling; Membrane fouling; Microbial fouling
Free ammonia (FA), 79
Free nitrous acid (FNA), 79
Freundlich adsorption, 227
Freundlich isotherm, 25, 26, 27, 154, 220, 225, 256, 258
Freundlich parameters, 29
Full-scale drinking water treatment plants
– BAC filtration in, 319–329
Full-scale PACT systems, 127
– for contaminated groundwaters, 138–139
– for contaminated surface runoff waters, 139–140
– for domestic and industrial wastewaters co-treatment, 136
– for domestic wastewaters reuse, 139
– for industrial effluents
– – organic chemicals production industry, 128–130
– – pharmaceutical wastewaters, treatment of, 135–136

– – priority pollutants, treatment of, 134
– – propylene oxide/styrene monomer (PO/SM) production wastewater, 130–131
– – refinery and petrochemical wastewaters, 131–134
– – synthetic fiber manufacturing industry, 130
– for landfill leachates, 136–138
Full-scale surface water treatment, BAC filtration in, 320–326
– Bendigo, Castlemaine & Kyneton, 326–327
– Leiden plant, 320–322
– Mülheim plants, 320
– Ste Rose plant in Quebec, 323
– suburbs of Paris, plants in, 322–323
– Weesperkarspel plant, 324–326, 327
– Zürich-Lengg, plant in, 323–324

G
Gabapentin, 292
GAC. *See* Granular activated carbon (GAC)
GAC-SBR, 112
GAC-Sequencing Batch Biofilm Reactor (GAC-SBBR), 68
Gas masks, 3
Gemfibrozil, 290
Geosmin, 292–293
Giardia contamination, 326
Gibbs free energy, 190, 198, 200
Glyoxal, 273, 274
Granular activated carbon (GAC), 3, 4, 6, 13, 16, 39, 97, 105
– BAC reactors, main processes in, 66
– biomass attachment, 273
– coal-based, 274
– conversion of, into BAC, 64–65

– – BAC process, advantages of, 66
– – removal mechanisms and biofilm formation in BAC operation, 65–66
– – versus BAC operation, 66
– expanded and fluidized-bed BAC reactors, 68
– filtration, 145, 247, 265, 331
– fixed-bed BAC reactors, 67–68
– GAC upflow anaerobic sludge blanket reactor (GAC-UASB), 81
– integration into groundwater bioremediation, 303–304
– positioning, in wastewater treatment, 59–63
– recognition of biological activity, 63–64
– removal performance, 273
– secondary treatment of wastewaters, 45
– tertiary treatment of wastewaters, 45
– wood-based, 274
Granular activated carbon (GAC) adsorbers, 30, 145
– breakthrough curves, 34–37
– integrated adsorption and biological removal, benefits of, 155–158
– modeling approaches in GAC/BAC reactors, 158
– – BAC reactors' characterization, by dimensionless numbers, 164–173
– – biomass balance in reactor, 164
– – substrate mass balance in liquid phase of reactor, 159–164
– models in BAC reactors involving adsorption and biodegradation, 173
– – carbon properties, in modeling, 177–178
– – high substrate concentrations, 175–176

Granular activated carbon (GAC) adsorbers (*continued*)
-- initial steps in modeling, 173
-- integration of adsorption into models, 174–175
-- substrate removal and biofilm formation, 173–174
-- three-phase fluidized bed reactor, 177
-- very low substrate (fasting) conditions, 176–177
- operation of, 31–34
-- bed volume (BV), 34
-- carbon usage rate (CUR), 34
-- effective contact time, 33
-- empty-bed contact time (EBCT), 32
-- filter operation time, 34
-- filter velocities, 33–34
-- throughput volume, 34
- processes around carbon particle surrounded by biofilm, 146
-- adsorption isotherms, 154
-- biofilm growth and loss, 154–155
-- diffusion and removal of substrate within biofilm, 149–152
-- diffusion into activated carbon pores and adsorption, 152–154
-- related to biomass and activated carbon, 146
-- related to substrate transport and removal, 146
-- transport of substrate to surface of biofilm, 148
- purpose of use, 30
- types of, 30–31
Granular activated carbon (GAC) reactors
- GAC/BAC, 158–173, 177, 267–268, 331
- GAC-FBR, 68, 99, 104, 118, 119, 120, 138, 208, 327
- GAC-MBR, 209

- GAC-SBBR, 68
- GAC-UASB, 81
- GAC-UFBR, 105
Granular Activated Carbon-Fluidized Bed Reactor (GAC-FBR) system, 68, 99, 104, 118, 119, 120, 138, 208, 327
Groundwater bioremediation, 303–304
Groundwater recharge, 248
Gum arabic powder, 111

H
HAA. *See* Haloacetic Acids (HAA)
HAA5. *See* Five haloacetic acids (HAA5)
HAAFP, 10, 295
Haloacetic Acids (HAA), 10, 244, 294–295, 297
- HAA5, 295, 297
- HAAFP (HAA Formation Potential), 295
Halogenated hydrocarbons, 288
HD 4000, 256, 258
Heavy metals, 80
- cadmium (Cd), 58, 80
- chromium (Cr), 58, 113
- copper (Cu), 58, 80
- nickel (Ni), 58, 80
- zinc (Zn), 58, 81
- *See also* Organic pollutants
Henry constant, 54, 184
Heterogeneity of activated carbon, 21
Heterotrophic plate count (HPC), 238, 305
Hexavalent chromium (Cr (VI)), 80
High substrate concentrations, modeling in, 175–176
Historical appraisal of activated carbon, 1–3
History of activated carbon
- Bussy, 2
- Hunter, 2
- Joseph de Cavaillon, 2
- Kayser, 2
- Lipscombe, 2
- Lowitz, 2
- Schatten, 2

- Scheele, 2
- Stenhouse, 2
- von Ostreijko, 2
HLR. *See* Hydraulic loading rate (HLR)
HMX (High Melting eXplosive), 195
Homogeneous surface diffusion model, 152
HPC. *See* Heterotrophic Plate Count (HPC)
Hubele model, 342
Hueco Bolson aquifer, 139
Humic acids, 64, 114, 254, 342, 343
Humic substances, 196, 240, 244, 338
- humic acid, 64, 111, 114, 254, 342, 343
Hunter, 2
Hydraulic loading rate (HLR), 33, 35, 72, 274–276
Hydraulic regime
- completely mixed, 146, 173
- dispersed plug flow, 146, 173
- plug flow, 173
Hydraulic residence time, 246
Hydrocarbons
- aliphatic and aromatic, 5, 288
- chlorinated, 129, 182
- halogenated, 288
- high molecular weight hydrocarbons, 14
- polycyclic aromatic hydrocarbon (PAH), 99, 138, 209
- total volatile hydrocarbon (TVH), 99
Hydrogenophaga, 307
Hydrogen peroxide, 196, 277, 328
Hydrogen sulfide (H_2S)
- biodegradation of, 112
Hydrolysis, 48, 83, 149, 335
Hydrophilicity, 15, 290
Hydrophobicity, 21, 239, 291
Hydrophobic nonylphenol, 290

4-Hydroxyacetophenone, 109
Hysteresis, 196, 201, 208, 211, 362

I
IAS model. *See* Ideal Adsorbed Solution (IAS) model
IAST. *See* Ideal Adsorbed Solution Theory (IAST)
IBP. *See* Ibuprofen (IBP)
IBPCT. *See* Integrated Biological-Physicochemical Treatment (IBPCT)
Ibuprofen (IBP), 52, 55, 290, 292
IC_{50}, 106
Ideal Adsorbed Solution (IAS) model, 28
Ideal Adsorbed Solution Theory (IAST), 28, 177, 227
iMBRs. *See* Immersed MBRs (iMBRs)
Immersed MBRs (iMBRs), 74–75
Improved Simplified Ideal Adsorbed Solution (ISIAS) model, 28, 29
Indomethacin, 290
Industrial wastewater treatment, 6, 95, 100–104
– paper and pulp wastewaters, 97–98
– petroleum refinery and petrochemical wastewaters, 98–99
– pharmaceutical wastewaters, 95–97
– textile wastewaters, 99–100
Infrared spectroscopy, 243
Inhibition
– competitive, 56, 61
– noncompetitive, 56, 61
– substrate, 56, 60
Inhibitory compound, 108
Inorganic pollutants, 10, 79, 145, 359
Inorganic species, activated carbon in, 80
Instrumental analysis
– Excitation-Emission Matrix (EEM), 325
– Gas Chromatography-Mass Spectrometry (GC-MS), 243
– Nuclear Magnetic Resonance (NMR), 243, 244
– Organic Carbon Detection (OCD), 243
– Organic Nitrogen Detection (OND), 243
– Size Exclusion Chromatography (SEC), 243, 244
– spectral measurements, 239
Integrated adsorption and biological removal, 47
– benefits of, 155–158
– biodegradation/biotransformation, 48
– – abiotic degradation/removal, 53–54
– – micropollutants, 50–52
– – organic matter fractionation, in wastewater, 48
– – organic substrates, 49
– – sorption onto activated carbon, 53
– – sorption onto sludge, 52–53
– biological regeneration of activation carbon, 59
– organic substrates, biomass, and activation carbon
– – interactions between, 54
– organic substrates, main removal mechanisms for, 47
– substrates, behavior and removal of, 59
– surface of activation carbon
– – biodegradable and nonbiodegradable organics retention on, 56
– – microorganisms, attachment and growth of, 59
– – substrates concentration on, 58
– – toxic and inhibitory substances retention on, 56–58
– – volatile organic compounds (VOCs) retention on, 58
Integrated Biological-Physicochemical Treatment (IBPCT), 173
Integrated systems, observed benefits of, 76
– anaerobic treatment, 81
– biological sludge in presence of activated carbon, 81
– – dewaterability of sludge, 82
– – sludge settling and thickening, 81
– denitrification, enhancement of, 80
– inorganic species, activated carbon addition on, 80–81
– membrane bioreactors (MBRs), effect of PAC on, 87
– – membrane filtration, effect of activated carbon on, 85
– – microbial products, effect of activated carbon on, 83
– – microbial products, importance of, 82
– nitrification, enhancement of, 78–80
– organic carbon removal, enhancement of, 78
Ion exchange, 300, 302
Iopromide, 52
Irreversible adsorption, 29, 196, 197, 198, 200, 202, 204, 205, 208, 211, 222, 259
Irreversible fouling resistance, 85, 86
ISIAS model. *See* Improved Simplified Ideal Adsorbed Solution (ISIAS) model
Isopropyl alcohol, 96
Isopropyl ether, 96
Isothiazolinones, 101

K
Kaolin, 111
KATOX, 63
Kayser, 2
Ketoprofen, 290

Kinetics
– adsorption, 22, 34, 36, 162, 169
– of bioregeneration, 228–229
– desorption, 203
Kyneton plant, 326

L
Laboratory-Scale PACT reactors, 114–118
Lake Zurich, 294, 323, 324
Landfill leachate treatment
– BAC filtration, 118–120
– PAC-MBR process, 118
– PACT, 114–118, 136–138
Langmuir isotherm, 25, 26, 154
LC_{50} tests, 101
LCC. See Life cycle cost (LCC)
LDF model. See Linear Driving Force (LDF) model
Leiden plant, 320–322
LHAAP. See Longhorn Army Ammunition Plant (LHAAP)
Life cycle assessment (LCA), 327
Life cycle cost (LCC), 327
Lignin, 13
Lignin sulfonate, 111
Lindane, 110, 182, 183
Linear Driving Force (LDF) model, 153
Lipscombe, 2
Liquid chromatography with online organic carbon detection (LC-OCD), 325
Liquid film-biofilm boundary
LMW organics. See Low-molecular-weight (LMW) organics
Longhorn Army Ammunition Plant (LHAAP), 327, 328
Lowitz, 2
Low-molecular-weight (LMW) organics, 244
Lyophobic character of solute, 17

M
Macropores, 20, 21, 23, 152, 193, 203, 216, 257
MAP. See Microbially available phosphorus (MAP)
Mass balance
– biofilm, 149
– liquid phase, 159–164
– in PACT process, 178–180
– reactor, 164
– solid phase, 158
Mass transfer coefficient, 148
Mass transfer equations, 165–166
Mass transfer zone (MTZ), 34, 37
Mass transport, 36, 145, 165, 171–172, 222
Maturation pond, 140
MBAA. See Monobromoacetic acid (MBAA)
MBRs. See Membrane bioreactors (MBRs)
MCAA. See Monochloroacetic acid (MCAA)
MDBA. See Multiple-Component Biofilm Diffusion Biodegradation and Adsorption model (MDBA)
Membrane-assisted biological GAC filtration (BioMAC process), 76
Membrane bioreactors (MBRs), 8, 74, 302–303
– biofouling, 75
– biological membrane assisted carbon filtration (BIOMAC), 76
– effect of PAC on, 82–86
– extracellular polymeric substances (EPS), 82–83, 84
– fouling, 75, 84
– GAC-MBR (GAC added membrane bioreactor), 209, 355
– immersed membrane bioreactor (iMBR), 74
– membrane flux, 84, 86
– microfiltration, 75
– PAC-MBR, 75–76
– permeability, 75, 76
– soluble microbial products (SMP), 358
– submerged membrane bioreactor (sMBR), 74, 303
– transmembrane pressure (TMP), 75, 85
– ultrafiltration, 75
Membrane filtration, activated carbon in, 85–86
Membrane fouling, 75, 245
MEP. See Metabolic end products (MEP)
Mesopores, 20, 21, 152, 192, 193, 201
Metabolic end products (MEP), 83, 84
Methanol, 96
2-Methyl-1-dioxolane, 130, 131
Methyl chloride, 134
Methylene chloride, 96, 213
Methyl ethyl ketone, 102
Methylphenols, 108
Methyl-tert-butylether (MTBE), 288, 294
MF. See Microfiltration (MF)
MIB (2-Methylisoborneol), 245, 292–293
Michigan Adsorption Design and Applications Model (MADAM) program, 173
Microbial ecology, 362–363
– of BAC filters, 304–305
Microbial fouling, 228
Microbially available phosphorus (MAP), 305
Microbial products, activated carbon on, 83–85
Microcrystallites, 14–15
Microcystins, 288, 293, 348
Microfiltration (MF), 74, 302
– continuous microfiltration (CMF), 326
– continuous microfiltration-submerged (CMF-S), 326
Microorganisms
– attachment and growth of, on surface of activated carbon, 59

- and bioregeneration, 206–207
- determination, 306
-- heterotrophic biomass and activity, 306–308
-- nitrifying biomass and activity, 308–309
Micropollutants
- adsorption and biodegradation of, 362
- removal from drinking water in BAC systems, 288–298
- in wastewaters, 55
Micropores, 14, 20, 21, 23, 27, 152, 192, 193, 201, 202, 203, 216
Microtox® test, 101
Mixed liquor (volatile) suspended solids (MLSS/MLVSS), 70, 133, 180
M-nitrophenol (MNP), 107
MNP. See M-nitrophenol (MNP)
Modeling, 145
- adsorption, 145, 152, 158, 173, 174–175, 177
- biofilm detachment, 155
- biological removal
-- attached-growth, 66
-- suspended-growth, 145, 157
- bioregeneration, 217–229
- diffusion, 152
- drinking water biofiltration, 331, 341
- GAC adsorbers, 145
-- GAC/BAC reactors, modeling approaches in, 158–173
-- integrated adsorption and biological removal, benefits of, 155–158
-- prevalent models in BAC reactors involving adsorption and biodegradation, 173–178
-- processes around carbon particle surrounded by biofilm, 146–155
- PACT process, modeling of, 178
-- mass balance for PAC in, 178–180

-- mass balances in, 178
-- models describing substrate removal in, 181–185
Models
- biodegradation/adsorption –screening model (BASM), 225
- biofilm on activated carbon (BFAC) model, 174, 175, 223
- BIOFILT model, 339–340
- homogeneous surface diffusion model (HSDM), 174, 347
- ideal adsorbed solution (IAS) model, 28
- improved simplified ideal adsorbed solution (ISIAS) model, 28
- michigan adsorption design and applications model (MADAM), 173
- model of Wang and Summers, 337–338
- models involving bioregeneration, 217–229
- Multiple-Component Biofilm Diffusion Biodegradation and Adsorption model (MDBA), 227
- plug flow stationary solid phase column (PSSPC) model, 345
- Pore Diffusion Model (PDM), 152
- S_{min} model, 334, 336
- Simplified Ideal Adsorbed Solution (SIAS) model, 28, 29,
- Uhl's Model, 335–337
Modified Biot number, 170, 173
Modified Stanton number, 169, 170
Molecular tools
- Amplified Fragment Length Polymorphism (AFLP), 307
- Polymerase Chain Reaction (PCR), 308
- terminal-restriction fragment length polymorphism (T-RFLP), 308
Molinate, 202, 203, 210
Monobromoacetic acid (MBAA), 295
Monochloramine, 277
Monochloroacetic acid (MCAA), 295
Monod model, 151, 183
MTBE. See Methyl-tert-butylether (MTBE)
MTZ. See Mass transfer zone (MTZ)
Mülheim process, 247–248, 320, 321
Multicomponent systems, bioregeneration modeling in, 225–227
Multiple-Component Biofilm Diffusion Biodegradation and Adsorption model (MDBA), 227
Multisolute Fritz–Schluender isotherm, 228
Multi-solute MDBA model, 218
Municipal wastewater treatment, 6

N
Naphthalene, 21, 118
Naproxen (NPX), 52, 55, 290
National Pollutant Discharge Elimination System (NPDES), 132
Natural Organic Matter (NOM), 5, 30, 200, 237, 363
- acidity, 255, 267
- adsorbability of, 255, 255–258
- aromaticity, 251, 255
- BAC filtration, 288–289
- biodegradability, 251, 253–254
- biodegradable fraction in, 239–243
- biological removal of, 331, 332, 344–345
- desorbability of, 258–260
- fractionation of, 243–244, 255
- fractions removal, 267

Natural Organic Matter (NOM) (continued)
– molecular size distribution of, 243
– ozonation effects, 248, 250–254
– polarity, 250, 255
– raw water NOM, 250
– removal in conventional water treatment, 244–246
– structure and composition, 251
– see also Organic matter
Natural treatment systems, 248
N-butanol, 102
NDMA. See N-nitrosodimethylamine (NDMA)
Neutrals, definition of, 244
Nitrate, 238
– removal, 298
– role in bromate reduction, 300
– role in perchlorate reduction, 301
Nitrification, 78–79, 285
– affecting factors, 285–286
– BAC filtration, 285–286
– drinking water, 359
– enhancement, by activated carbon, 78
– inhibition of, 79
– nitrifiers in BAC filtration, 285, 286
– PAC addition, 79–80
– wastewater treatment, 78–80
Nitrifiers, 285
– ammonia oxidizing bacteria (AOB), 78, 308–309
– molecular microbiology, 308–309
– nitrite oxidizing bacteria (NOB), 78, 308
Nitrifying biomass, 308
Nitrite Oxidizing Bacteria (NOB). See Nitrifiers
Nitrobenzene, 110, 134
Nitrophenol, 87, 107, 197, 200, 362
2-Nitrophenol (2-NP), 58, 108, 200, 207, 210

4-Nitrophenol (4-NP), 107, 110, 134, 198
N-nitrosodimethylamine (NDMA), 288
Nitrosomonas sp., 308
Nitrospira sp., 308
NMR spectroscopy. See Nuclear Magnetic Resonance (NMR) spectroscopy
NOM. See Natural Organic Matter (NOM)
Nonadsorbing media, 168, 334
Nonbiodegradable dissolved organic carbon (NBDOC), 243, 265
Nonbiodegradable organics, 56, 97, 111, 255, 256, 258, 259
Nongrowth substrate. See Cometabolism: cometabolic substrate
Nonsingularity, 196
Nonylphenol, 290, 292
Nonylphenol ethoxylates (NPEs), 101
Norit 1240, 258, 259, 260, 268, 270
Norit®, 2, 3
NPDES. See National Pollutant Discharge Elimination System (NPDES)
NPEs. See Nonylphenol ethoxylates (NPEs)
NPX. See Naproxen (NPX)
Nuclear Magnetic Resonance (NMR) spectroscopy, 243, 244

O

OCPSF. See Organic chemicals, plastics and synthetic fiber (OCPSF)
o-cresol, 108, 177, 197, 200, 202, 210
9-Octadecenamide, 111
Octanol-water partition coefficient, 52, 54, 110
– relation to adsorption, 53
– sorption to biological sludge, 52

Offline bioregeneration, 194–195, 204, 207, 228
Oil and grease, 98, 99, 140
Oilfield wastewater, treatment of, 99
'One-liquid film' model, 221, 222
Orange II, 110, 111
Organic carbon
– assimilable organic carbon (AOC), 240–242, 253
– biodegradable organic carbon (BDOC), 242
– dissolved organic carbon (DOC), 239
– nonbiodegradable organic carbon (non-BDOC or NBDOC), 243
– particulate organic carbon (POC), 239, 251
– total organic carbon (TOC), 239
Organic carbon removal, enhancement of
– by activated carbon, 78
Organic chemicals, plastics and synthetic fiber (OCPSF), 127
Organic chemicals production industry, full-scale PACT for, 128–130
Organic matter
– biodegradable organic matter (BOM), 48, 78, 199, 205, 238, 274, 293
– dissolved organic matter (DOM), 85, 239, 303, 308
– fractionation, in wastewater, 48
– natural organic matter (NOM), 5, 30, 200, 237
– particulate organic matter (POM), 48, 67, 303, 335
– in water treatment, 238–244
–– biodegradable fraction, 239–243
–– expression of, 239
–– fractionation of, 243–244
–– nonbiodegradable dissolved organic carbon (NBDOC), 243
Organic matter, removal of, 265

Index | 381

- BAC filters, bioregeneration of, 272–273
- breakthrough curves, 265
-- initial stage of operation, 265
-- intermediate and later stages of operation, 266–268
- main mechanisms, 265
Organic micropollutants, 43, 48
- adsorption of, onto preloaded GAC, 290
- in BAC filtration, 291
- biodegradation of, 356
- NOM and, 288–290
- in water, 288
Organic Nitrogen Detection (OND). See Instrumental analysis
Organic pollutants, 79
- in secondary effluents, 111
OUR. See Oxygen uptake rate (OUR)
Oxidant residuals, effect of, 277
Oxidative coupling, 196, 197, 200
- of phenolic compounds, 197
Oxidative polymerization, 197
Oxygenase enzyme
- dioxygenase, 359
- monooxygenase, 298
Oxygen uptake rate (OUR), 80, 96, 185, 358
Ozonated and biologically treated water, 256, 258
Ozonation, 62–63, 97, 118, 140–141, 294
- and BAC filtration, 249, 327
- intermediate ozonation, 267
- NOM characteristics, impact on, 250–255
- postozonation, 295
- preozonation, 63, 141, 251, 253, 283–285, 287, 292, 295, 297, 309
Ozone dose, 249, 250–251, 254

P
PAC. See Powdered activated carbon (PAC)
PAC added membrane bioreactor (PAC-MBR) process, 44, 75–76, 101, 104, 111, 118, 157, 355
Packed Bed Reactor (PBR), 300
Packing media, 32, 345
- GAC, 66, 345
- plastics, 66, 129
- sand, 214, 242, 273, 293, 300
PACT process. See Powdered activated carbon treatment (PACT) process
PAE. See Phthalate ester (PAE)
PAH. See Polycyclic aromatic hydrocarbon (PAH)
Paint wastewater, 102
Paper and pulp wastewaters, treatment of, 97–98
Paracetamol, 291
Partially exhausted GAC, 205
Partial penetration, 151
Particle size, 15, 16, 85
Particulate Organic Carbon (POC), 239, 251
Particulate organic matter (POM), 48, 67, 303, 335
Partition ratio, 24
PCBs. See Polychlorinated biphenyls (PCBs)
PCE. See Perchloroethylene (tetrachloroethylene) (PCE)
PCPs. See Personal care products (PCPs)
PCR. See Polymerase Chain Reaction (PCR)
p-cresol, 228
PDM. See Pore diffusion model (PDM)
Peclet number (Pe), 167, 170
Pentachlorophenol, 108, 110, 206, 217, 327
Perchlorate, 8, 274, 298, 299, 301, 347–348
- contamination, 327–328
- removal, 301–302

Perchloroethylene (tetrachloroethylene) (PCE), 105, 110, 134, 217
Persistent Organic Pollutant (POP), 294
Personal care products (PCPs), 43, 288
Pesticides, 288, 291
- BAC filtration, 247, 249
- chlorinated pesticides, 109
- drinking water, 5, 6, 107, 112, 140
- herbicides, 346
-- alachlor, 346
-- atrazine, 346–347
-- bentazon, 347
- insecticides, 6, 182
-- lindane, 110, 182, 183, 184
- organic pesticides, 100–101
- organophosphate pesticides, 109
- organosulfur pesticides, 109
Pesticides and polychlorinated biphenyls (PCBs), removal of, 109
p-ethylphenol, 200, 210
Petrochemical wastewaters, 98–99
- PACT treatment of, 131–134
PFR. See Plug flow reactor (PFR)
pH values, 19–20, 70, 78, 79, 100, 115, 119, 135, 299, 305
Pharmaceutical and Personal Care Products (PPCPs), 109, 288, 356, 362
Pharmaceuticals, 96
- 5-fluorouracil, 53, 55
- antibiotics, 76, 357
- BAC filtration, 291, 292
- biological rate constant, 291
- caffeine, 292
- carbamazepine, 52, 55, 109, 290, 357, 358, 362
- clofibric acid, 290
- diclofenac, 52, 55, 109, 290

Pharmaceuticals (continued)
- and endocrine disrupting compounds (EDCs), 109, 291
- estradiol, 292
- fenoprofen, 290
- gemfibrozil, 290
- ibuprofen, 52, 109, 290, 292, 358
- indomethacin, 290
- ketoprofen, 290
- naproxen, 52, 55, 109, 290
- octanol-water partitioning, 52, 54, 110, 291
- propyphenazone, 290
- salicylic acid, 291, 292
- sorption to sludge, 55
- trovafloxin mesylate, 292
- wastewater treatment, 95–97, 135
- water treatment, 44, 52

Pharmaceutical wastewaters, treatment of, 95–97, 135–136

Phenol, 5, 20, 21, 29, 50, 59, 87, 111, 134, 200, 208, 210, 217, 220, 224, 227, 228, 229, 288, 294
- removal of, 106–108, 110, 118

Phenolic compounds
- adsorption of, 197
- chlorophenol, 4, 87, 107
- nitrophenol, 87, 107, 110, 134, 197, 200, 362
- oxidative coupling of, 197
- oxidative polymerization of, 196
- wastewaters, 101

Phenol molecules (PhOH), 196
Phenoxy radicals (PhO*), 196, 197
Phosphorus, 70, 140, 198, 254, 304
Phragmitis communis, 207
Phthalate ester (PAE), 294
Phthalates, 294
Physical adsorption (Physisorption). *See* Adsorption
Physicochemically regenerated carbon, 360

Physicochemical regeneration of activated carbon, 363
Physisorption, 17, 193, 196
Pilot reactors, 102
Pilot-scale GAC-FBR, leachate treatment in, 119–120
p-isopropylphenol, 200, 210
Plane of zero gradient (PZG), 190
Plants in suburbs, of Paris, 322–323
Plug flow reactor (PFR), 70, 170, 177, 242, 335, 338
Plug flow stationary solid phase column (PSSPC) model. *See* Models
p-methylphenol, 195, 200, 210
p-nitrophenol (PNP), 107, 200, 210, 220, 221, 224, 229
PNP. *See p*-Nitrophenol (PNP)
POC. *See* Particulate Organic Carbon (POC)
Polarity, 21, 65, 250
- of adsorbate, 19
Polaromonas, 307
Pollutants
- biodegradation of, 49
- control of, 8–10
Pollutants and wastewaters, biological treatment of, 95
- industrial wastewaters, treatment of, 95, 100–104
-- paper and pulp wastewaters, 97–98
-- petroleum refinery and petrochemical wastewaters, 98–99
-- pharmaceutical wastewaters, 95–97
-- textile wastewaters, 99–100
- landfill leachate treatment, 113
-- in biological activated carbon (BAC) media, 118–120
-- in PAC-added activated sludge systems, 114–118

-- using PAC-MBR process, 118
- removal of specific chemicals, 104, 112–113
-- dyes, 110–111
-- organic pollutants, in secondary effluents, 111
-- pesticides and polychlorinated biphenyls (PCBs), 109
-- pharmaceuticals and endocrine-disrupting compounds, 109
-- phenols, 106–108
-- priority pollutants, 109–110
-- volatile organic compounds (VOCs), 104–106
Polychlorinated biphenyls (PCBs), 109
Polycyclic aromatic hydrocarbon (PAH), 99, 138, 209
Polyelectrolyte, 71
Polymerase Chain Reaction (PCR), 308
Polyoxyethylen, 210
Poly S119, 111
Polysaccharide, 84
POM. *See* Particulate organic matter (POM)
POP. *See* Persistent Organic Pollutant (POP)
Pore blockage, 347
Pore diffusion model (PDM), 152
'Pore diffusion', 23
Pore-filling, 21
Pores, 14
- macropores, 20, 21, 152, 193, 203, 216, 257
- mesopores, 20, 21, 152, 192, 193, 201
- micropores, 14, 20, 21, 27, 152, 192, 193, 201, 202, 203, 216
- pore blockage, 347
Pore volume, 13, 21, 193, 202, 257, 292
Porosity, 1, 203
Postchlorination, 308
Potable water, 139, 194

POTW. *See* Publicly Owned Treatment Works (POTW)
Powdered activated carbon (PAC), 2, 4, 13, 15, 96, 97, 99, 106, 109, 111, 129, 302, 347, 354, 355, 358
– addition, 246
– adsorbers, 30
– anaerobic PACT process, 74
– integration into biological wastewater treatment, 70–73
– sequencing batch PACT reactors, 73
– single-stage continuous-flow aerobic PACTs process, 70
– – activated sludge process, basic features of, 70
– – characteristics of, 72
– – development of, 70
– – process parameters in, 72
Powdered Activated Carbon Treatment (PACT) process, 7, 44, 69, 100, 101, 102, 109, 114, 127, 145, 157, 195, 199, 207, 209
– aerobic PACT, 46
– anaerobic PACT, 46, 74, 81, 86, 114, 157
– and BAC systems, 353–355
– concurrent bioregeneration in, 195
– for contaminated groundwaters, 138–139
– for contaminated surface runoff waters, 139–140
– for domestic and industrial wastewaters co-treatment, 136
– full-scale PACT systems, 127
– – for contaminated groundwaters, 138–139
– – for contaminated surface runoff waters, 139–140
– – for co-treatment of domestic and industrial wastewaters, 136
– – for domestic wastewaters reuse, 139

– – for industrial effluents, 128–136
– – for landfill leachates, 136–138
– general process diagram, 71
– for landfill leachates, 136–138
– mass balance for PAC in, 178, 180–181
– models describing substrate removal in, 181–185
– operation process parameters, 72
– PAC-MBR, 44, 75–76, 101, 104, 111, 118, 157, 355
– PACT®, 70–74, 87, 128, 131–132, 134
– PACT sludge, 70, 73, 81, 85, 87, 106, 139, 180, 363–364
– PACT/WAR, 87, 132, 136, 139
– for reuse of domestic wastewaters, 139
– SBR-PACT, 74, 135, 137
– typical conditions in, 72
PPCPs. *See* Pharmaceutical and Personal Care Products (PPCPs)
Prechlorination, 308
Precurcors, 295
– DBP formation, 5, 244, 270, 285, 295, 303
– humic substances, 196, 240, 244, 245
Preozonation, 141, 249, 251, 273, 283, 288, 291, 294, 295, 309, 323, 327, 333, 344–345
Preparation of activated carbon, 13–14
Pressure swing adsorption, 38
Primary substrate, removal as, 49, 108, 293, 356
Priority pollutants, 70, 76, 87
– removal of, 109–110
– treatment of, 134
Propylene oxide/styrene monomer (PO/SM) production wastewater, 130–131

Propyphenazone, 290
Pseudomonas, 206, 306
Pseudomonas chrysosporium, 206
Pseudomonas fluorescens, 112
Pseudomonas fluorescens P17, 241
Pseudomonas putida ATCC 70047, 107
Pseudomonas strains, 207
Publicly Owned Treatment Works (POTW), 43, 56, 101, 127, 135, 137
Purit®, 2–3
Pyrolysis/GC-MS, 243
PZG. *See* Plane of zero gradient (PZG)

R
Radiolabeled carbon, 214–215
Radiolabeled phenol, 215
Rapid sand filtration, 248, 249, 307, 320, 322, 323
Rate-limiting step, 16, 24, 199, 201, 337–338
Raw and ozonated waters, 250, 253, 267, 286–287
– adsorbability and desorbability of organic matter in, 255–259
Raw water, 5, 245, 255, 258, 285, 291, 302, 303, 323, 325, 326
– NOM, 250
RBF. *See* River bank filtration (RBF)
RDX. *See* Royal Demolition Explosive (RDX)
Reactivation, of activated carbon, 37–39
Reactive Black 5, 110
Reactor mass balance, 220
– dimensionless form of, 166–167
– total dimensionless expression of, 168–169
Reactors
– BAC, 37, 44, 56, 59, 64, 66, 67, 99, 110, 142, 155, 157, 164–173, 307, 328
– BAC-SBR, 227
– biomass balance in, 164

Reactors (*continued*)
- FBR, 68, 69
- GAC-FBR (Fluidized Bed Reactor packed with Granular Activated Carbon), 119, 327
- GAC-MBR (GAC added Membrane Bioreactor), 209, 355
- GAC-SBBR (GAC Reactor operated as a Sequencing Batch Biofilm Reactor), 68
- GAC-UASB (Upflow Anaerobic Sludge Blanket packed with Granular Activated Carbon), 81, 100
- GAC-UFBR (Upflow Fixed Bed Reactors packed with GAC), 105
- MBR (Membrane Bioreactor), 74–75, 86
- PAC-MBR, 111, 302
- PACT, 73–74, 101, 114, 129, 132, 137
- substrate mass balance in liquid phase of, 159–164
Reclamation
- BAC filtration, 141
- domestic wastewater, 140–141
- industrial wastewater, 141–142
- nonpotable use, 141–142
- ozonation, 140
Recycle fluidized bed (RFB), 175
Reductive dechlorination, 105
Refinery wastewaters, PACT treatment of, 98, 131–134
Regeneration
- of activated carbon, 360
-- activated carbon grade, importance of, 360–361
-- biological activated carbon, 363–364
-- bioregeneration of activated carbon, 361–363
- bioregeneration
- of PACT and BAC sludges 86
- frequency, 327, 328
- thermal, 39

- wet air oxidation (WAO), 38, 86, 103, 130, 131, 139, 363–364
- wet air regeneration (WAR), 71, 87, 128, 131, 363
Reichs Ford Road landfill, 137
Remediation, 7, 8
Removal mechanisms and micropollutant elimination, 355
- drinking water treatment, 358–359
- wastewater treatment, 355–358
Repsol Tarragona wastewater treatment plant, 131
Respirometry, 215
- OUR, 80, 96, 103, 106, 117, 185
- SOUR, 58
Reversible adsorption, 196
- bioregeneration on, 195–198
- extent of, 211
RFB. *See* Recycle fluidized bed (RFB)
Rheinisch–Westfälische Wasserwerksgesellschaft (RWW), 247–248
Rhodococcus rhodochrous, 206
Ribosomal RNA genes, 308
River bank filtration (RBF), 248
Row Supra, 256, 257, 258, 259
Royal Demolition Explosive (RDX), 195
RWW. *See* Rheinisch–Westfälische Wasserwerksgesellschaft (RWW)

S
Salicylic acid, 291, 292
Sand biofiltration, 277, 291
Sand filtration/filters, 5, 46, 137, 237, 238, 242, 266, 273, 275, 276, 286, 287, 291, 295, 297, 304, 307, 308, 322, 323, 327
Sandwich™ filter, 273

SAT. *See* Soil aquifer treatment (SAT)
SBR. *See* Sequencing batch reactor (SBR)
SBR-PACT, 74, 135, 137
Scale
- full-scale, 10, 74, 76, 97, 101, 127, 132–134, 141, 248, 255, 274, 286, 300, 319, 320, 321, 323, 327, 331, 340, 348, 360, 363, 364
- laboratory-scale, 80, 97, 101, 103, 112, 114, 119, 141, 269, 275, 300
- pilot-scale, 98, 102, 104, 109, 119, 133, 152, 268, 276, 277, 283, 284, 286, 293, 300, 308, 343, 344, 345
Scanning Electron Micrographs, 207, 217
Scanning Electron Microscopy (SEM), 64, 207, 216, 305
Scattered surface growth, meaning of, 222
SCFB. *See* Semi-Continuously Fed Batch (Reactor) (SCFB)
Schatten, 2
Scheele, 2
SCR. *See* Specific Cake Resistance (SCR)
SEC. *See* Instrumental analysis
Secondary substrate, 49–50, 356
Secondary treatment, activated carbon in, 46
- GAC, 46
- PAC, 46
SEM. *See* Scanning Electron Microscopy (SEM)
Semi-Continuously Fed Batch (Reactor) (SCFB), 115–116
Semi-volatile organic compounds (SVOCs), 294
Sequencing batch reactor (SBR), 73, 81, 103, 106, 195
- BAC-SBR, 112, 227
- GAC-SBBR, 68

- SBR-PACT system, 74, 135, 137
- sequencing batch biofilm reactor (SBBR), 68, 194, 195
Sequential adsorption–biodegradation approach, 215
Settling, 81–82, 103, 129, 322, 323, 354, 358
Sewage treatment plants (STPs), 43–44, 62, 114, 140
Sherwood number. See Dimensionless numbers
Shigella sp., 308
SIAS. See Simplified Ideal Adsorbed Solution (SIAS)
Siemens
Simplified Ideal Adsorbed Solution (SIAS), 28, 29
Single solute adsorption, 27–28, 37
Single solute systems, bioregeneration in, 220–225
Single-stage continuous-flow aerobic PACT® process, 70
- activated sludge process, basic features of, 70–71
- characteristics of, 71
- development of, 70
- process parameters in, 72–73
Size Exclusion Liquid Chromatography coupled to Organic Carbon Detection (SEC-OCD), 243
Slow sand filtration (SSF), 249, 320, 323
Sludge, biological
- dewaterability, 78, 82, 83, 87, 358
- mean floc size, 85
- settleability, 82, 84, 97, 181
- sludge volume index (SVI), 82, 97, 129
- specific cake resistance (SCR), 85–86
- specific resistance to filtration (SRF), 82, 83
Sludge age. See Sludge retention time (SRT)

Sludge retention time (SRT), 47, 52, 66, 68, 70, 72, 75, 85, 157, 179
Sludge-water partition coefficient, 54
sMBRs. See Submerged membrane bioreactors (sMBRs)
SMP. See Soluble microbial products (SMP)
SOC. See Synthetic Organic Compounds (SOC)
Sodium-2-(diisopropylamino) ethylthiolate, 101
Sodium ethylmethyl phophonate, 101
Soil aquifer treatment (SAT), 248–249
Soil contamination, remediation of, 7
Solid–water partition coefficient, 52
Solubility, 18–19, 48, 113
Soluble EPS, 83
Soluble microbial products (SMP), 77, 83, 208, 304, 341
- biomass associated products (BAP), 341
- drinking water biofiltration, 341
- membrane bioreactors, 302
- utilization associated products (UAP), 341
Solute, 16, 17, 18, 158
- bisolute, 29, 37, 208, 209
- multisolute, 27, 28–30, 37
- single solute, 27–28, 37, 220–225, 342
Solute distribution parameter. See Dimensionless numbers
Solvent, 96, 102, 103, 112, 138, 182, 213, 300
- solvent extraction, 213
Sontheimer, 248, 342, 343
Sorption, 21, 47, 52, 53, 69, 104, 184, 185, 203, 357, 362
- biosorption, 47, 52, 54, 100, 183, 335
- onto activated carbon, 53
- adsorption, 3, 8, 10, 16–39, 43, 96, 97, 112, 145, 189,

237, 265, 342, 347, 348, 353, 354, 357, 358, 360–363
- ion exchange, 66, 300, 302
SOUR. See Specific Oxygen Uptake Rate (SOUR)
South Caboolture Water Reclamation Plant, 141, 142
Speciation, 79, 113, 295
Specific Cake Resistance (SCR), 85–86
Specific Oxygen Uptake Rate (SOUR), 58
Specific ozone dose, 251, 253
Specific Resistance to Filtration (SRF), 82, 83
Specific surface area, 18, 20–21, 27, 273, 338
Specific Ultraviolet Absorbance (SUVA), 239, 245, 250, 252, 253, 272, 302, 323
Spectral measurements, organic matter expression by, 239
Speitel, 205, 220, 221, 227
Sphingomonas, 109, 307
Spirillum sp. strain NOX, 241
SRF. See Specific Resistance to Filtration (SRF)
SRT. See Sludge retention time (SRT)
SSF. See Slow sand filtration (SSF)
Stanton number. See Dimensionless numbers
Stenhouse, 2
Ste Rose treatment plant, in Quebec, 323
Stewartby Landfill Site, 138
Stripping, 7, 54, 96, 102, 134, 184, 185
STPs. See Sewage treatment plants (STPs)
Styrum-East Water Works, 320
Submerged membrane bioreactors (sMBRs), 74, 303
Substituent groups, 19, 200
Substrate and biomass associated products, of biodegradation, 207–208

Substrate–carbon contact time, 203–204
Substrate concentration, 65, 67, 149, 151, 160, 161, 174, 185, 190, 201, 215, 218, 223, 307, 339
Substrate removal and biofilm formation, 173–174, 331–333
Substrates, 149, 159, 173–174, 177–178, 193, 208, 223, 225, 305, 307, 323, 332, 339, 340, 345, 356, 359, 361, 362
– behavior and removal of, 59, 60–61
– chemical properties of, 200
Substrates' concentration on the surface of activated carbon, 58
Sulfate, 15, 135, 300, 301
Sulfonol, 210
Surface acidity, 21, 203, 290
Surface diffusion, 23
Surface diffusion coefficient, 153, 166, 178, 221
Surface functional groups, 15, 20, 21, 196
Surface loading rate, 33
Surface runoff waters, 139–140, 285
Surfactants, 5–6, 128, 195, 200
Surfactants mixture, 210
SUVA. See Specific Ultraviolet Absorbance (SUVA)
SVOCs. See Semi-volatile organic compounds (SVOCs)
Synthetic carbonaceous adsorbent, 66
Synthetic fiber manufacturing industry, wastewater of, 130
Synthetic organic compounds (SOC), 5, 30, 175, 209, 237, 331

T
Tannic acid, 111
Taste and odor, 3, 4, 5, 238, 245, 246, 247, 288, 292, 320, 326–327

TCA. See Trichloroethane (TCA)
TCAA. See Trichloroacetic acid (TCAA)
TCB. See Trichlorobenzene (TCB)
TCE. See Trichloroethylene (TCE)
TDS. See Total dissolved solids (TDS)
Temperature, 15, 17, 20, 22, 38, 70, 79, 87, 119, 138, 197, 215, 248, 274, 276–277, 284, 286, 305, 323, 335, 340, 345
Tensides, 109, 288, 357
Terminal-restriction fragment length polymorphism (T-RFLP), 308
Terminal-restriction fragments (T-RFs), 308
Tert-butyldimethylsilanol, 111
Tertiary treatment, activated carbon in, 46
– GAC, 47
– PAC, 46–47
Tetrachloroethylene (PCE), 105, 110, 134, 138, 210, 217
Tetrahydrofurane, 96
Textile wastewaters, treatment of, 99–100, 111
Thermally activated PACs, 202
Thermal volatilization, 38
Thiocyanate, biological removal of , 102, 103
THMs. See Trihalomethanes (THMs)
THMFP, 10, 295, 303
Three-phase fluidized bed reactor, modeling step input of substrate in, 177
Throughput volume, 34
$TiO_2/UV/O_3$, 111
TKN. See Total Kjeldahl Nitrogen (TKN)
TMP. See Transmembrane pressure (TMP)
TN. See Total Nitrogen (TN)
TOC. See Total Organic Carbon (TOC)

Toluene, 50, 96, 102, 104, 105, 118, 134, 138, 182, 183, 202, 210, 224, 225, 289
Total dissolved solids (TDS), 135
Total Kjeldahl Nitrogen (TKN), 96, 139
Total Nitrogen (TN), 100, 141, 287, 288
Total Organic Carbon (TOC), 239, 243
Total THM (TTHM), 295, 297, 348
Total volatile hydrocarbons (TVH), 99
Toxic algal metabolites, 288, 293
Trace organics, 47, 347
Transient-state, multiple-species biofilm model (TSMSBM), 341–342
Transmembrane pressure (TMP), 75, 85
Transport mechanisms
– adsorption, 24
– boundary layer, 22–23
– bulk solution transport (Advection), 22–23
– external diffusion, 23
– external film, 23
– intraparticle (internal) diffusion, 23
– liquid film, 23
– molecular diffusion, 23
– pore diffusion, 23
– surface diffusion, 23
T-RFLP. See Terminal-restriction fragment length polymorphism (T-RFLP)
T-RFs. See Terminal-restriction fragments (T-RFs)
Trichloroacetic acid (TCAA), 295, 297
Trichlorobenzene (TCB), 134, 182, 290
Trichloroethane (TCA), 112
Trichloroethylene (TCE), 54, 105, 134, 138, 208, 210, 215, 290, 304
Trihalomethanes (THMs), 10, 244, 294–295, 297, 348

Index | 387

- bromodichloromethane, 298, 348
- cometabolism
-- nitrifiers, 298
- dibromochloromethane, 298, 348
- trihalomethane formation potential (THMFP), 10, 295, 303
Trovafloxin mesylate, 292
TSMSBM. *See* Transient-state, multiple-species biofilm model (TSMSBM)
TTHM. *See* Total THM (TTHM)
TVH. *See* Total volatile hydrocarbons (TVH)
'Two-liquid film' model, 221–222

U
UASB. *See* Upflow anaerobic sludge blanket (UASB)
UF. *See* Ultrafiltration (UF)
Uhl's model, 335–337
Ultrafiltration (UF), 74, 75, 85, 118, 302, 323
Upflow anaerobic sludge blanket (UASB), 81, 100
Upflow fixed-bed reactors packed with GAC (GAC-UFBR), 105
Use of activated carbon, 1, 3–4
UV absorbance, 8, 239, 251, 295
- BAC filtration, 277
- specific ultraviolet absorbance (SUVA), 239, 245, 250, 252, 253, 272, 283, 295, 302, 323, 325
- UV_{254}, 96, 141, 239, 253, 255, 277, 303, 325
- UV_{280}, 96, 111

V
Valeric acid, 173, 218, 219
van der Kooij, 241, 306
van der Waals forces, 15, 17, 21, 27, 196
Very low substrate (fasting) conditions, modeling the case of, 176–177, 336
Vinyl chloride-containing wastewater, 133

Virgin GAC, 177, 205, 208
VOCs. *See* Volatile organic compounds (VOCs)
Void ratio, 160, 164
Volatile organic compounds (VOCs), 7, 38, 58, 96, 102, 104–106, 128, 135, 185, 294
Volatility, 54, 59, 104
- Henry constant, 54, 184
Volatilization, 38, 54, 183
von Ostreijko, 2

W
Wang and Summers, model of, 337–338
WAO. *See* Wet air oxidation (WAO)
WAR. *See* Wet Air Regeneration (WAR)
Warburg's apparatus, 215
Wastewaters
- hazardous landfill leachates, 6, 8, 43, 50, 69, 127
- industrial
-- acrylonitrile manufacturing, 103
-- alcohol distillery wastewater, 104
-- bactericide wastewater, 101
-- caustic hydrolysate wastewater, 101
-- coke oven plant wastewater, 103
-- dyes and pigments processing wastewater, 102
-- metal finishing industry wastewater, 112
-- mining industry wastewater, 112
-- organic chemicals production industry wastewater, 5, 8, 19, 127, 128, 129, 130, 134, 136, 182, 242
-- paint and ink industry wastewater, 112
-- paper and pulp wastewater, 97–98
-- pesticide manufacturing wastewater, 107, 128

-- petrochemical wastewater, 131–134
-- petroleum refinery wastewater, 98–99
-- pharmaceutical wastewater, 95–97, 135–136
-- phenolic wastewater, 101, 358
-- propylene oxide/styrene monomer (PO/SM) production wastewater, 130–131
-- steel mill coke plant wastewater, 102
-- synthetic fiber manufacturing industry wastewater, 130
-- tannery wastewater, 101
-- textile industry wastewater, 63, 214
- municipal, domestic, 50, 63
-- reuse, 6, 50, 76, 127, 139
- sanitary landfill leachates, 80, 82, 113
- treatment, 355–358
-- activated carbon adsorption in, 7
-- activated carbon in, 5–6
-- advanced, 141
-- biological, 59, 60, 69, 127, 355, 358, 363
-- industrial, 6, 31, 43, 46, 47, 50, 56, 62, 63, 68, 74, 76, 95, 99
-- municipal, 6, 43, 63, 79, 354, 356, 357
-- physicochemical, 6, 30, 62, 110, 238
-- primary, 44, 45, 46, 76
-- secondary, 43–45, 46
-- tertiary, 43–45, 46
Wastewater treatment plants (WWTPs), 6, 52, 75, 76, 101, 185, 354
Water preparation, for industrial purposes, 7
Water treatment
- activated carbon adsorption in, 8
- activated carbon in, 246–247

– BAC filtration, 8, 76, 109, 118, 127, 140, 195, 197, 237, 247, 249, 250, 251, 260, 265, 274, 284, 286, 288, 291, 294, 295, 297, 300, 303, 307, 309, 319, 321, 326, 327, 331, 333, 358, 359, 363
– biofiltration, 247, 250, 265, 283, 291, 295, 320, 321, 334, 337, 341
– coagulation, 6, 30, 101, 118, 141, 237, 245, 249, 277, 292, 295, 308, 324, 326
– disinfection, 4, 140, 141, 237, 244, 248, 285, 309, 322, 324
– double layer filtration, 248, 321
– filtration, 8, 85, 86, 100, 104, 108, 138, 140, 195, 237, 245, 247–249, 254, 284, 291, 297, 300, 303, 306, 319, 323, 326, 327, 333, 344, 354, 355, 359, 363
– flocculation, 63, 237, 245, 246, 248, 277, 292, 322, 323
– infiltration for groundwater recharge, 248, 321
– Mülheim process, 5, 247–248, 320
– sand filtration, 5, 46, 138, 140, 237, 248, 249, 266, 286, 291, 307, 308, 309, 320, 323

– sedimentation, 63, 139, 140, 245, 248, 277, 292, 308, 324
– water distribution, 238, 242–243, 245, 248, 284, 285, 295, 319, 324
– plants
–– Bendigo, Castlemaine & Kyneton (Victoria, Australia), 326–327
–– Leiden Plant (the Netherlands), 320–322
–– Mülheim (Germany), 320
–– Ste Rose Plant (Quebec, Canada), 323
–– suburbs of Paris, 322–323
–– Weesperkarspel (the Netherlands), 324–326, 327
–– Zürich-Lengg (Switzerland), 323-324
Weber, 63, 64, 69, 173, 174, 182, 221, 222
Weesperkarspel Plant, 276, 283, 324–326, 327
Wet air oxidation (WAO), 38, 86, 103, 130, 363–364
Wet Air Regeneration (WAR), 71, 87, 128, 131, 363, 364
WWTPs. See Wastewater treatment plants (WWTPs)

X
Xenobiotic organic compounds (XOCs), 43, 208, 288

– brominated flame retardants, 43
– endocrine disrupting compounds (EDCs), 5, 53, 76, 109, 140, 288, 290, 291, 354
– hormones, 43, 62, 354, 356, 359, 362
– personal care products (PCPs), 43, 109, 110, 206, 217, 288
– pesticides, 5, 43, 100, 103, 107, 109, 112, 128, 140, 288, 291, 320, 321, 346, 347
– pharmaceuticals, 43, 52, 75, 76, 95–97, 109, 135, 288, 290, 354, 357, 358
– pharmaceuticals and personal care products (PPCPs), 109, 288, 356, 359, 362
Xenobiotics, 43, 49, 207, 208, 288, 354
XOCs. See Xenobiotic organic compounds (XOCs)
Xylenols, 108

Z
Zernel Road Municipal Solid Waste Landfill, 137
Zero-order kinetics inside biofilm, 167–168
Zimpro® WAR, 70, 87
Zürich-Lengg, plant in, 323–324